Discovering Biology in the Lab

An Introductory Laboratory Manual

SECOND EDITION

Discovering Biology in the Lab

An Introductory Laboratory Manual

SECOND EDITION

Tara A. Scully
The George Washington University

Ecology labs and additional contributions by

Ryan Fisher
Salem State College

W. W. NORTON & COMPANY
NEW YORK • LONDON

W. W. Norton & Company has been independent since its founding in 1923, when William Warder Norton and Mary D. Herter Norton first published lectures delivered at the People's Institute, the adult education division of New York City's Cooper Union. The firm soon expanded its program beyond the Institute, publishing books by celebrated academics from America and abroad. By mid-century, the two major pillars of Norton's publishing program—trade books and college texts—were firmly established. In the 1950s, the Norton family transferred control of the company to its employees, and today—with a staff of four hundred and a comparable number of trade, college, and professional titles published each year—W. W. Norton & Company stands as the largest and oldest publishing house owned wholly by its employees.

Printed in the United States of America

Composition by Preparé

Manufacturing by Courier, Westford

Illustrations by Penumbra Design, Inc., and John McAusland

Editor: Betsy Twitchell

Developmental Editor: Andrew Sobel

Project Editor: Christine D'Antonio

Production Manager: Eric Pier-Hocking

Marketing Manager: John Kresse

Copy Editor: Stephanie Hiebert

Book Design: Lisa Buckley

Cover Design: Leah Clark

Custom Editor: Katie Hannah

ISBN 978-0-393-91817-5

W. W. Norton & Company, Inc., 500 Fifth Avenue, New York, NY 10110
www.wwnorton.com

W. W. Norton & Company, Ltd., Castle House, 75/76 Wells Street, London W1T 3QT

1 2 3 4 5 6 7 8 9 0

Contents

Preface

Discovering Biology in the Lab is designed for introductory biology courses for nonscience majors. It was written exclusively for students in this audience and caters to their particular needs. Although the labs are sequenced to be coordinated with the order of topics as presented in *Discover Biology*, they are also compatible with other introductory biology textbooks.

In developing *Discovering Biology in the Lab*, my goal was to convey the excitement, breadth, and relevance of biology to nonmajors without burying them in a level of detail that does not serve their interests. Rather, I wanted this manual to cultivate critical thinking skills so that, in addition to learning about biology as a scientific discipline, students learn an invaluable way of thinking about the world that will serve them for the rest of their lives. The lab activities cultivate critical thinking by constantly asking students to consider the following questions:

- Why are you doing what you're doing?
- How do your observations and results connect to the overall objectives of the lab activities and, more broadly, to the biological concepts you're learning about in lecture and in your textbook?
- How does what you're learning apply to the "real world"?

Many lab manuals command students to "do this," "observe that," "write some notes," and "move on to the next command." By constantly asking students to consider why they're doing what they're doing, *Discovering Biology in the Lab* strives to keep students engaged, develop critical thinking, and immediately demonstrate how working in the lab enriches and deepens their understanding of biology.

NEW Labs

The second edition of this manual includes three brand new labs on topics I hope instructors and students alike will find engaging and instructive. They are:

- Lab 23: Forensic Science
- Lab 24: Animal Behavior
- Lab 25: Food Science

Activity Features

In addition to being informed by the above goals throughout, each lab activity has the following features:

Clear Objectives

A list of objectives is presented at the beginning of each lab to help students recognize the most important concepts covered by the activities in that lab. This feature serves as a road map for students, highlighting the main points that will be taught and emphasized in each activity. Objectives are further reinforced through Concept Checks and Review Questions.

Concept Checks

Concept Checks after each activity teach students to connect what they're doing in the lab to key biological concepts and the lab objectives. This feature teaches students to step back from the details and consider the big picture. In doing so, students recognize that the key biological concepts emphasized in the lab connect to the key concepts they're learning about in lecture and in their textbook.

Review Questions

Review Questions at the end of each lab help students integrate the preceding activities and Concept Checks into a conceptual whole. They reinforce the objectives presented at the beginning of each lab by asking students to demonstrate an understanding of these key biological concepts. Review Questions can also serve as a diagnostic tool—an assignment that instructors can collect at the end of a lab.

Clear Art

Discovering Biology in the Lab features many of the clear and direct figures found in *Discover Biology*. Particular attention has been paid to focusing on key concepts and avoiding overly complex figures and captions.

Connections to the Real World

Lab activities in *Discovering Biology in the Lab* demonstrate to students how concepts apply to the real world. Applications include antibiotic resistance (Lab 13: "Evolution of Resistance"), birth control (Lab 18: "Human Urinary and Reproductive Systems"), nutrition (Lab 5: "Chemical Building Blocks and Nutrition"), and global change (Lab 22: "The Impacts of Global Change on Ecosystems"). By making connections between the labs and the real world, students are able to understand the ramifications of scientific inquiry and enhance their scientific literacy. As a result, students will gain a greater appreciation for how science directly affects their everyday lives.

An Emphasis on Fundamentals

The activities in *Discovering Biology in the Lab* focus on developing an understanding of fundamentals such as evolution and the scientific method.

Evolution

Evolutionary thinking is introduced early and incorporated throughout *Discovering Biology in the Lab*. Students are taught principles and mechanisms of evolution and shown concrete connections between this fundamental biological concept and their surrounding environment. For example, evolution is used to explain the development of biological diversity, the structure of eukaryotic cells, antibiotic resistance, and other phenomena.

The Scientific Method

The scientific method is introduced in Lab 1 as a foundational component of modern scientific inquiry. Teaching students the steps of the scientific method enables them to start thinking like scientists. Labs offer students several opportunities to think critically and scientifically, formulating hypotheses, running experiments, gathering and analyzing data, and drawing conclusions.

Customization and Flexibility

The following features of *Discovering Biology in the Lab* are designed to make it easy to tailor the lab activities to your course:

Self-contained Content

In its introduction and activities, each lab contains all the information necessary for students to learn and complete the lab without referring to additional labs or references. The stand-alone nature of each lab allows instructors to coordinate the labs with the order of topics taught in their course.

Activities That Fit Your Schedule

Discovering Biology in the Lab is designed for use in lab sessions of varying length. Each lab consists of a series of discrete activities that can be mixed and matched so that instructors can run labs that fit in their lab sessions without sacrificing educational objectives and quality.

Customization and Pricing

To keep prices down, you may customize *Discovering Biology in the Lab* by giving students only the labs required for your course. Custom manuals are offered for only $1 net per lab (e.g., $10 net for ten labs) when packaged with *Discover Biology*, Fifth Edition, by Singh-Cundy and Cain. Lab topics are offered for courses that run either one-semester or two-semester biology sequences. Instructors can also add their own material to further customize the manual. To build a custom manual, please visit wwnorton.com/books/978-0-393-91817-5.

Instructor Support

An *Instructor's Manual* is available for download at **wwnorton.com/instructors**. This manual includes

- Material lists needed to run each lab
- Setup instructions for each lab
- Safety information
- Helpful hints and ideas on how to coordinate each lab with your lecture
- For professors with shorter lab periods, suggestions on which activities to use in each lab

Acknowledgments

Preparing the second edition of this lab manual was not a solitary undertaking. I would like to thank my dedicated and detail-oriented developmental editor, Andrew Sobel, for his excellent suggestions for revising the three new labs in this second edition. Thanks also to the reviewers who provided in-depth and thoughtful comments on these new labs based on their own teaching. Thanks to the team at Norton, including Betsy Twitchell, Cait Callahan, and Christine D'Antonio for marshalling resources to make sure the second edition of this lab manual is even better than the first. Finally, thank you to my department at The George Washington University, especially Kamali, Alexis, and Dr. Ken Brown, as well as Dr. Jane Ferguson and Paul Spiegler, without whose guidences I would not have gotten here.

Reviewers

We would like to thank the following people for providing valuable feedback in the development of this lab manual:

Second Edition Reviewers

Paul Gier, Huntington College
Paige A. Mettler-Cherry, Lindenwood University
Lori Ann Rose, Sam Houston State University

First Edition Reviewers

Neil R. Baker, The Ohio State University
Eric A. Brown, Rollins College
Evelyn K. Bruce, University of North Alabama
Kelly S. Cartwright, College of Lake County
Judy M. Dacus, Cedar Valley College
Marirose T. Ethington, Genesee Community College
Richard Farrar, Idaho State University
Shirley A. Gangwere, Tarrant County College
Beverly Glover, Western Oklahoma State College
Maureen A. Leupold, Genesee Community College
Lee Likins, University of Missouri/Kansas City
Caroline Mackintosh, University of Saint Mary
Kerry C. McKenna, Lord Fairfax Community College
Alexie McNerthney, Portland Community College/Sylvania
Susan L. Meacham, University of Nevada, Las Vegas
Dorothy Quinn, Framingham State College
Lori Rose, Sam Houston State University
James L. Smith, Montgomery College/(Takoma Park/Silver Spring)
Doreen Soldato, Dean College
James Stegge, Rochester Community and Technical College
Cindy L. White, University of Northern Colorado
Daniel B. Williams, Winston-Salem State University
Silvia Wozniak, Winthrop University

Laboratory Safety

Familiarizing yourself with the rules and guidelines below will ensure that you remain safe while in the lab. Always follow the direction of your instructor and pay attention to any warning labels or special handling instructions on the chemicals and equipment that you use. If you are unsure of how to handle any chemicals or equipment, ask for clarification from your instructor—not a fellow student. Safety should be your first priority.

Always notify your instructor of any accident, no matter how minor.

KNOW the location of the following safety information and equipment:

1. **The first aid kit** Each lab is equipped with a first aid kit containing bandages, ointments, and other emergency medical care. Use these items to alleviate any minor medical issue. If there is an emergency situation you should notify your instructor immediately and contact 911.
2. **The Material Safety Data Sheets (MSDS) binder** Every chemical in the lab has a safety sheet in the MSDS binder. These sheets give instruction on how to handle, store, and dispose of a chemical. Additionally, these safety sheets indicate how flammable, corrosive, or toxic a chemical is, and its incompatibility with other chemicals. In the event of an emergency, you may be asked by a firefighter or an emergency medical technician (EMT) where the MSDS is located. Make sure to know the answer.
3. **The fire extinguisher** Notify your instructor immediately if there is a fire. Retrieve the fire extinguisher and hold it upright. Stay at least 8 feet away from any fire. Follow the **PASS** technique to operate your fire extinguisher: pull the ring pin to break the tamper seal; aim the nozzle low at the base of the fire; squeeze the handle to release the extinguishing agent; sweep the extinguisher spray from side to side at the base of the fire until it is out. Monitor the area where the fire occurred to ensure that it is completely extinguished. If there are any flare ups, repeat the PASS technique until the fire is finally out.
4. **The fume hood** When working with flammable chemicals or chemicals that generate a toxic gas during a reaction, you will work in a fume hood. The glass partition should not be opened beyond the specified level. DO NOT stick your head in the fume hood. If you ignore these safety regulations, you run the risk of inhaling harmful fumes, some of which may be odorless and colorless.
5. **The eye wash fountain** If your eyes come in contact with any chemicals, wash them for 15 minutes. Even if it is difficult, use your hands to force your eyelids to remain open.
6. **The safety shower** Be careful when working with chemicals, especially corrosives. If you spill a large quantity on you (any amount of concentrated acids or bases, or 50 ml or more of any chemical identified as dangerous), the room will be evacuated and you'll need to remove your clothing and stand under the shower.

"DO"s:

1. Do wear safety glasses or goggles at all times when handling chemicals. These safety glasses must provide protection against impacts and splashes. Your own glasses are NOT a satisfactory substitute. Contact lenses should not be worn, even if you are wearing safety glasses or goggles. Chemical fumes can get behind the lenses and damage the eyes.

2. Do confine long hair when in the laboratory.
3. Do wear protective clothing, such as a lab apron, to protect your clothes from spills.
4. Do wear gloves when instructed to do so. Do not touch any objects other than those indicated to you by your instructor. Remove and properly dispose of gloves any time you have to leave the laboratory.
5. Do wear adequate footwear to protect your feet from spilled chemicals and broken glass. Bare feet are prohibited in the lab. Sandals and open-toed shoes do not afford adequate protection.
6. Do label your beakers and flasks with the name of the chemical or substance, the date, and your name before you use them.
7. Do follow your instructor's directions when disposing of waste. Waste containers in the fume hood are labeled with the chemical waste name. Never pour anything down the drain, unless specifically told to do so.
8. Do immediately wash with running water whenever a chemical comes into contact with exposed skin or eyes. Ask your instructor for instruction if you spill any chemical on your clothing. Know the location of the eye wash fountains.
9. Do clean up all spills. Follow your instructor's directions for how to dispose of all papers, broken glass, etc. Place waste into the proper receptacles.
10. Do wash your hands at the end of each lab period.

"DO NOT"s:

1. Do not taste or ingest anything in the laboratory. This applies to food or chemicals. Don't bring food into the laboratory or eat or drink from laboratory glassware.
2. Do not smoke in the laboratory because some chemicals are flammable.
3. Do not directly breathe fumes of any kind. In some labs you may be asked to note the odor of a substance. To do so, exercise great care and follow your instructor's instructions on how to safely note the odor of the substance you are handling.
4. Do not use mouth suction in filling pipettes with chemical reagents. Use a suction bulb available from your instructor or laboratory staff.
5. Do not weigh chemicals directly on balance pans and do not take reagent bottles or balance pans to your bench.
6. Do not pour excess chemicals back into reagent bottles. Check with your instructor for proper disposal procedures.
7. Do not work in the laboratory alone.
8. Do not perform unauthorized experiments.
9. Do not bring visitors into the laboratory.

I have read and understand the safety rules and guidelines presented. By signing below I agree to abide by them and other instructions given by the lab instructor at all times while I am in the lab.

Name: ____ Date: ________________

Course/Lab Number: ____________________ Section: _________

Discovering Biology in the Lab

An Introductory Laboratory Manual

SECOND EDITION

LAB 1 The Scientific Method

Objectives

- Use each step of the scientific method to understand the world around you.
- Make observations of the world around you.
- Gather information, think critically about your observations, and formulate hypotheses.
- Interpret and apply data.
- Create graphical representations of data to visualize trends.
- Peform an experiment to test predictions.
- Draw conclusions based on the results of experiments.

Introduction

Science begins with curiosity. Scientists want to understand the world around them, so they ask questions about what they see and design experiments to help answer those questions. The scientific method describes how scientists learn more about the natural world. This process can be exciting and fulfilling, influencing all aspects of the modern world. Learning the steps of the scientific method will help you to think like a scientist.

Step 1: Make observations. The first step in the scientific method is **observation** (Figure 1.1). Scientists observe something curious or interesting in nature and want to understand what they see. Sometimes they observe with their unaided senses, and sometimes they use technology like microscopes to aid them in observing. EXAMPLE: Scientists observe that the populations of fish in North Carolina rivers have decreased in number. An infectious protist has also been detected in the fish populations. Observations then lead scientists to ask questions like "What is causing the decline in fish populations?" "Does the presence of protists have an effect on the survival of fish in North Carolina rivers?"

Step 2: Devise a hypothesis to explain observations. A **hypothesis** is an educated guess that attempts to answer a question about the natural world. It is a tentative explanation of what the scientist observes. A hypothesis must adequately explain an observation and must be testable. Scientists form hypotheses based on the questions that they ask in response to observations. EXAMPLE: The protist *Pfiesteria* kills fish in the rivers of North Carolina.

Step 3: Generate predictions from that hypothesis. From the hypothesis, the scientist specifies the possible outcomes—that is, makes **predictions**. EXAMPLE: *Pfiesteria* would be capable of killing healthy fish if introduced into the aquarium housing the fish in the laboratory.

Step 4: Test predictions. The scientist then tests these predictions by performing an experiment and/or making new observations. An **experiment** is a controlled and repeatable manipulation of nature that tests the scientist's hypothesis and predictions. Scientists design experiments to isolate the factor they want to test. Experiments typically consist of a **control group** and **experimental group(s)**. In the experimental group(s) the scientist deliberately changes the factor being tested to observe its effect, and in the control group that factor is not deliberately changed, so that the control scenario mimics a natural environment or condition and provides a point of comparison for the experimental groups. Experiments are performed repeatedly to determine if the results are valid. In addition, any scientist must be able to pick up the directions for an experiment, perform it, and obtain the same results as those published by another scientist. EXAMPLE: In an experiment designed to test the effect of *Pfiesteria* on fish populations in the rivers of North Carolina, scientists gather species of fish from these rivers. Some are kept in aquaria mimicking their environment without *Pfiesteria* (serving as a control group), and some are kept in aquaria that mimic their environment but contain *Pfiesteria*. Experimental groups would include varying concentrations of *Pfiesteria* to determine the effective amount of *Pfiesteria*. The effect is measured by the number of fish killed by the concentration of *Pfiesteria*.

Step 5: Draw conclusions. The scientist determines whether the results from the experiment support or reject the hypothesis—that is, draws **conclusions** from the results—and evaluates the hypothesis on that basis, revising it if necessary and starting the process of the scientific method again. EXAMPLE: The results show that all species of fish die shortly after exposure to *Pfiesteria*. In this case the result supports the hypothesis and the experiment may be repeated to further validify results.

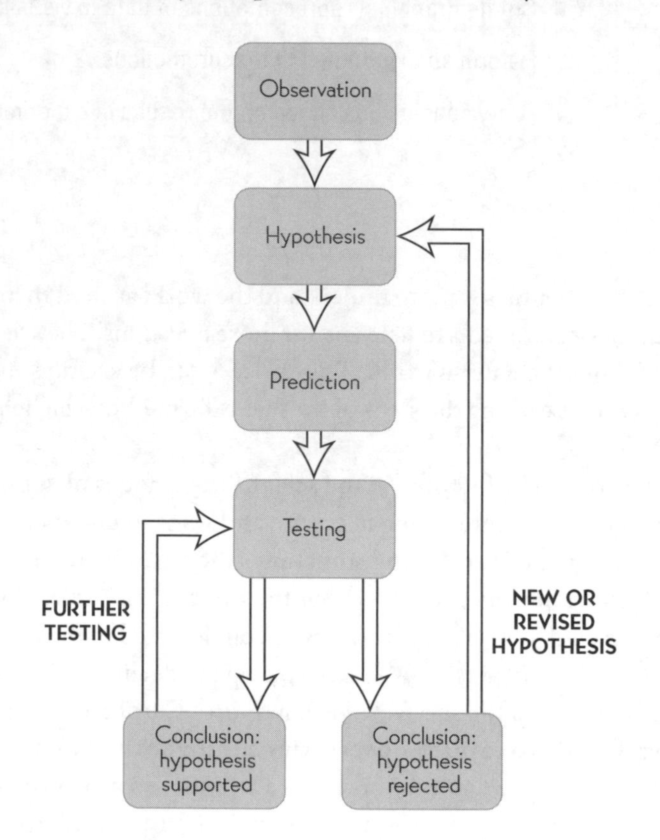

Figure 1.1
The Scientific Method

When scientists gather evidence from many experiments supporting a hypothesis (plural "hypotheses") and the scientific community collectively agrees that this evidence supports that hypothesis, the hypothesis is accepted as a **theory.** This is not the same term that we use in everyday conversation. Typically, we use the word "theory" to mean an "educated guess or hunch" (much like the definition of "hypothesis"), but this is not what "theory" means in the world of science. A theory is a well-supported explanation, generally accepted by the scientific community. The activities in this lab will help guide you through the process that a scientist engages in when doing research and will show you the importance of the scientific method, experimental design, and data analysis.

ACTIVITY 1 Observing and Hypothesizing

Scientific investigation begins when something an individual observes piques that person's curiosity. This interest leads to application of the scientific method to understand why, what, or how something is occurring in the natural world. In this activity you will examine an insect called a Bess beetle. These beetles are detritivores, meaning they consume dead or decaying organic matter called detritus. Examine this insect to familiarize yourself with its structure and normal behavior.

1. Remove a Bess beetle from the jar on your lab bench. Examine the head, body, appendages, eyes, and any other prominent structures. In the box below, sketch the beetle.

2. Place the beetle on your benchtop and examine its behavior, such as the way it moves its appendages or explores the benchtop. Take notes on your observations.

3. Now that you've observed the beetle, what features or behaviors do you find particularly interesting? What are some unique features of Bess beetles? Which characteristics of the beetle might make it a good detritivore?

4. What questions could you ask about the beetle that you might be able to answer through experimentation?

5. From your questions, develop two hypotheses explaining the questions you asked. Write them here.

   Hypothesis 1:

   Hypothesis 2:

Concept Check

1. How do scientists turn their questions into hypotheses?

2. Although you may not realize it, you use the scientific method on a daily basis. What are some explanations (hypotheses) you have come up with in the past 24 hours?

3. Why is observation the crucial first step in the scientific method?

4. What are some important characteristics of a hypothesis?

ACTIVITY 2 Understanding Data

Experiments allow scientists to test their hypotheses. Experimentation involves at least two different groups: one control group and one or more experimental groups. In the control group, a baseline of normal behavior or conditions is tested; in each experimental group, the scientists purposely change one condition.

The one condition that the scientists change in the experimental group is called the **independent variable**. The experimental group is compared to the control group to determine whether the independent variable has an effect. A characteristic that changes as a result of the independent variable is called the **dependent variable**. In the fish experiment described in the introduction to this lab, the independent variable is *Pfiesteria* and the dependent variable is the survival of fish species in North Carolina rivers. Whenever possible, scientists graph their results to determine whether the relationship between the independent and dependent variables reveals any trends. In the fish experiment, the rapid decline in the survivorship of all fish species indicates a negative impact of *Pfiesteria* and supports the hypothesis.

In this activity you will examine the data from an experiment performed with the aquatic organism *Daphnia*. You will identify the independent variable, dependent variable, and control group, and then perform calculations and properly graph the data that has been collected. Finally, you will decide whether the hypothesis was supported or rejected and draw a conclusion based on the trends revealed by the data.

1. *Observation:* Alcohol is a known physical depressant.

   Hypothesis: Alcohol reduces the heart rate of an individual.

   Experimental design:

   Each group contains two beakers with the same treatment (Table 1.1). All groups contain 20 *Daphnia* per beaker. The heart rate is tested 5 minutes after treatment.

TABLE 1.1 DESIGN OF THE *DAPHNIA* EXPERIMENT

Group	Treatment
1	100 ml of spring water
2	100 ml of 1% alcohol/spring water solution
3	100 ml of 2% alcohol/spring water solution
4	100 ml of 3% alcohol/spring water solution
5	100 ml of 4% alcohol/spring water solution
6	100 ml of 5% alcohol/spring water solution

   a. Which group is the control?

   b. The independent variable is under the control of the scientist. Which variable is the independent variable in this experiment?

   c. The dependent variable is the characteristic changed because of the independent variable. What is the dependent variable in this experiment?

2. *Results:*

   To ensure that the results recorded are valid, scientists often perform duplicate treatments. If two identical treatments yield the same results, those results have more strength. We calculate the **average** (mean) by adding together the results from each treatment and dividing by the number of treatments. In this experiment there are two results for each condition, so in each case add these two numbers and divide by 2. What is the average number of heartbeats for each treatment in this experiment? Record these averages in the third column of Table 1.2. If results are expressed in heartbeats per 20 seconds, what is the average heart rate for each group per minute? Record these averages in Table 1.2 as well.

TABLE 1.2 RESULTS OF THE *DAPHNIA* EXPERIMENT

Group	Heartbeats per 20 s	Heartbeats per 20 s	Average number of heartbeats per 20 s	Average heart rate (beats/min)
Group 1 (0%)	59	63		
Group 2 (1%)	51	48		
Group 3 (2%)	43	39		
Group 4 (3%)	33	32		
Group 5 (4%)	27	25		
Group 6 (5%)	16	14		

3. To determine whether the data exhibit a trend displaying the effect of alcohol on the heart rate of *Daphnia*, it is useful to graph the results. Graphs should provide enough information about your results that they can stand alone.

   There are many ways to display data. The two most common ways are bar graphs and line graphs. Sometimes either method can be used to display data, but when you're comparing a qualitative variable, such as type or kind, a bar graph represents the data in a clearer format than a line graph (Figure 1.2). On the other hand, when comparing a more quantitative variable, such as varying concentrations of one item, it is best to use a line graph so that any trend is easy to see (Figure 1.3).

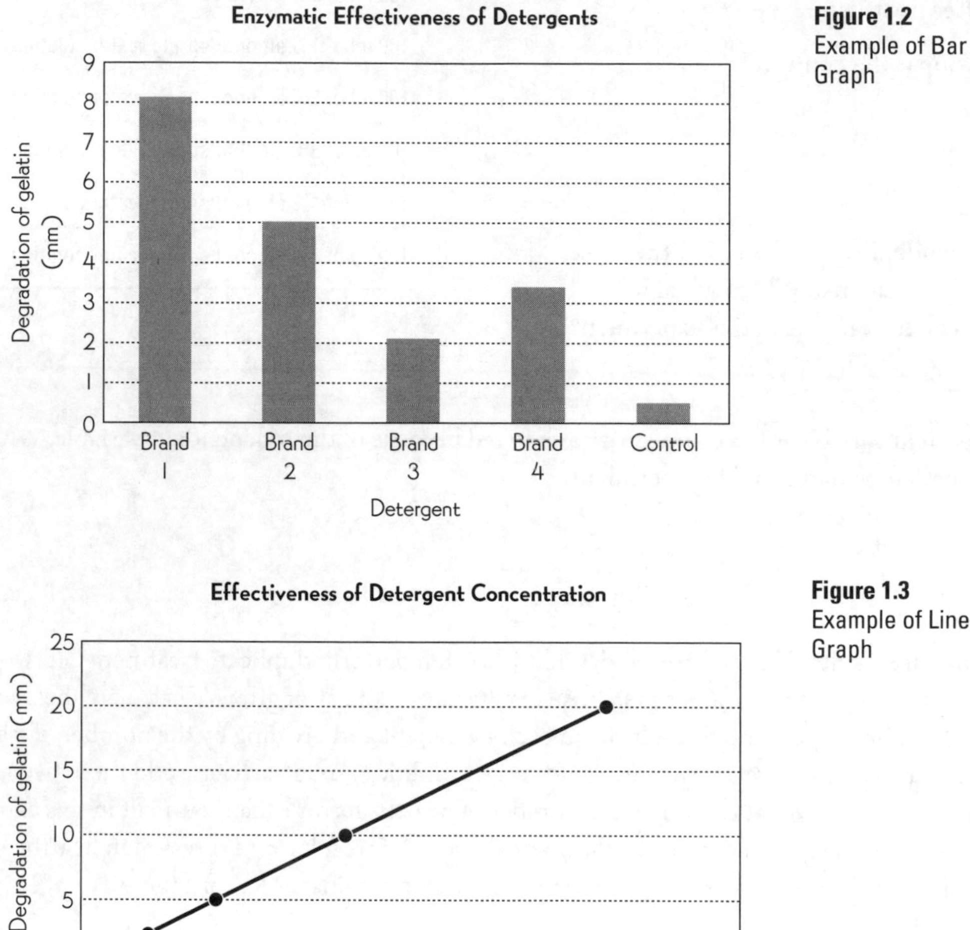

Figure 1.2
Example of Bar Graph

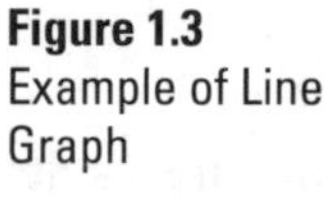

Figure 1.3
Example of Line Graph

4. Now let's graph the data from the *Daphnia* experiment. Use the following steps to guide you in plotting the data in Figure 1.4. Remember, anyone in the class should be able to look at your graph and determine if there is a relationship between alcohol and heart rate.

   a. The *x*-axis, or horizontal line, represents the independent variable; the *y*-axis, or vertical line, represents the dependent variable. In Figure 1.4 label your two axes; include after the label the units of measurement—for example, "Concentration of alcohol (%)."

b. To identify the scale appropriate for the *y*-axis, look at your data and determine the upper and lower limits of your results. You don't want your scale to end on either of these limits, so you need to create a buffer. For example, an appropriate scale for data with a lower limit of 10 and an upper limit of 87 would be 0 to 100.
c. Indicate the upper and lower limits of your scale along the *y*-axis. For proper interpretation of the results, equal intervals are also necessary. What interval is appropriate for each scale? On a 0–100 scale, possible intervals might be 5, 10, or 25. The interval that best displays your data is the one you should choose. Record the intervals of your choice along both axes.
d. Choose the upper and lower limits of the *x*-axis in the same way, according to the scope of the independent variable.
e. The title of your graph will provide the viewer with a quick understanding of your data. Choose a title that summarizes your results, the purpose of your experiment, and/or the variables of the experiment.
f. Record the average beats/min for each group on the graph. Do you see any trend in your results?

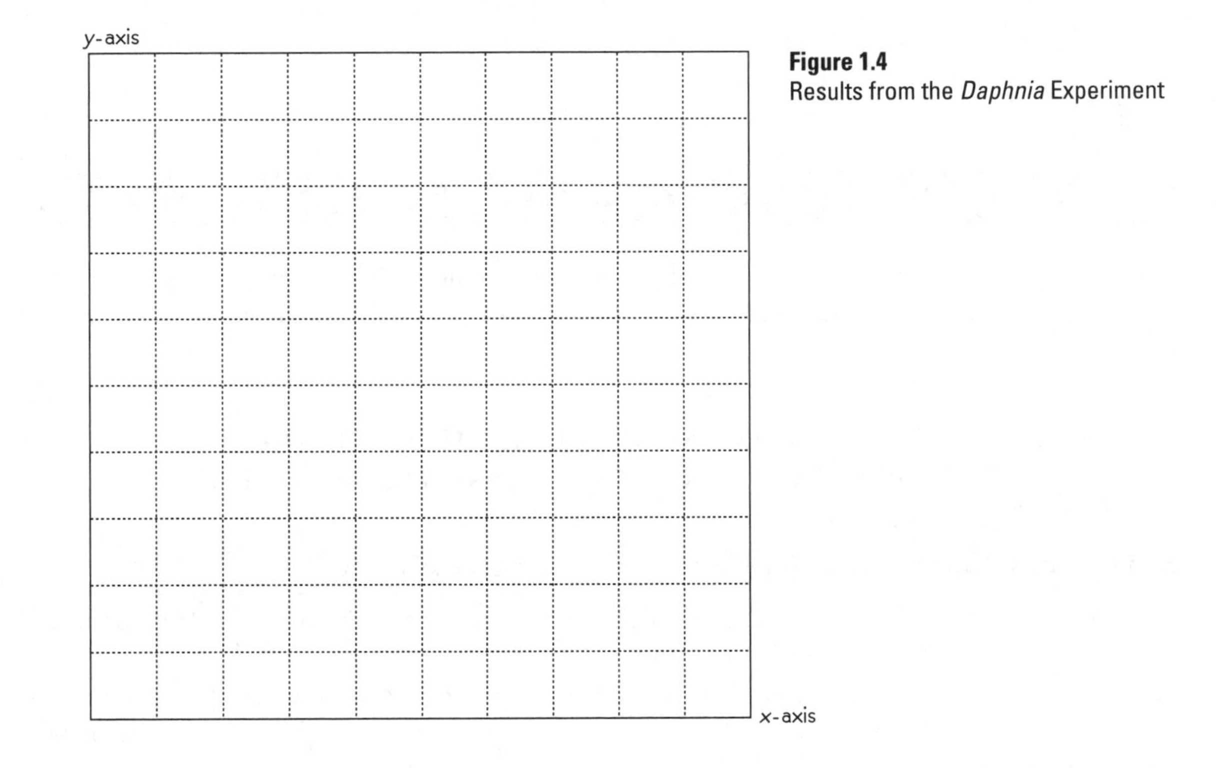

Figure 1.4
Results from the *Daphnia* Experiment

5. Conclude whether your data supports or rejects the hypothesis.

6. What additional information would be useful in determining the effects of alcohol on heart rate?

Concept Check

1. What is the relationship between the independent and dependent variables in an experiment?

2. What information is important to include on a graph?

3. What is the purpose of the control group?

4. How does displaying your data in the form of a graph assist your interpretation of experimental results?

ACTIVITY 3 Experimentation

Understanding experimental design is a crucial part of the scientific method. In this activity you will test how weight carried affects the Bess beetle's ability to travel. Bess beetles are known to be able to carry much more than their body weight.

NOTE: THIS EXERCISE IS COMPLETELY HUMANE, AND THE BEETLES WILL NOT BE HARMED DURING THE ACTIVITY.

Remember that scientists approach experimentation by having the following:

- *Control group.* This is the baseline, or normal, condition, where the factor to be tested in experimental groups is not manipulated by the scientist.
- *Experimental group(s).* This is typically a group or several groups in which the scientist has changed one factor/variable. Because the variable is controlled by the scientist, this factor is called the independent variable.
- *Means by which to determine change.* This can be a microscope, a fancy mass spectrophotometer, or a ruler. Any change observed and recorded corresponds to the dependent variable.

In addition, remember that performing duplicate treatments helps strengthen experimental results.

Start by developing a hypothesis and predicting the outcome. Remember that we're interested in learning how the amount of weight the beetle carries affects its ability to travel. Think of how the weight of your backpack affects you as you are walking to class, or how much extra gas the car in which your parents drove you to college used because it contained the weight of all the clothes and furniture you brought with you.

Hypothesis

1. With this in mind, what hypothesis would be an appropriate explanation for the question, "What effect does weight have on the distance a Bess beetle travels?"

Prediction

2. Predict how the Bess beetle will travel with less weight and more weight.

Testing

3. Obtain strings with weights (metal washers), a ruler, and two Bess beetles. To test the effect of weight on the Bess beetle, you will measure the distance the Bess beetle can travel while carrying different amounts of weight.

4. Put the two beetles in two separate jars. Place a ruler on the benchtop. Place the first beetle at the beginning of the ruler. Rotate the ruler as necessary if the beetle crawls in another direction. Observe the distance the beetle crawls over a 1-minute time span. Perform this test twice, and record in Table 1.3 the distances traveled by beetle 1 (listing "0" in the first row, under "Treatment"). Does the beetle travel the same distance each time? Now calculate the average distance traveled and record this in Table 1.3.

5. The instructor has provided strings that are the same length, with varying amounts of washers attached at one end and a loop on the other end. To effectively determine how the weight will affect the distance traveled by the beetle, how many weights should you test?

6. Loop the weight on one of the same beetle's horns. Now perform the test for each of the weights you chose. Complete the tests twice and calculate the average distance traveled. Record these results for beetle 1 in Table 1.3 on the next page.

7. To ensure that you're getting valid results, test the second beetle using the same conditions. What is the average distance for all of your tests on beetle 2? Record your results in Table 1.3.

TABLE 1.3 RESULTS OF THE BESS BEETLE EXPERIMENT

Treatment (weight carried)	Distance traveled in test 1 (mm)	Distance traveled in test 2 (mm)	Average distance traveled (mm)
BEETLE 1			
BEETLE 2			

Results

8. Graph your results in Figure 1.5. The independent variable is recorded on the *x*-axis. Which is the independent variable? Label the axis appropriately. The dependent variable is recorded on the *y*-axis. Which is the dependent variable? Label the *y*-axis accordingly. For each axis label, remember to include the unit of the variable in parentheses.

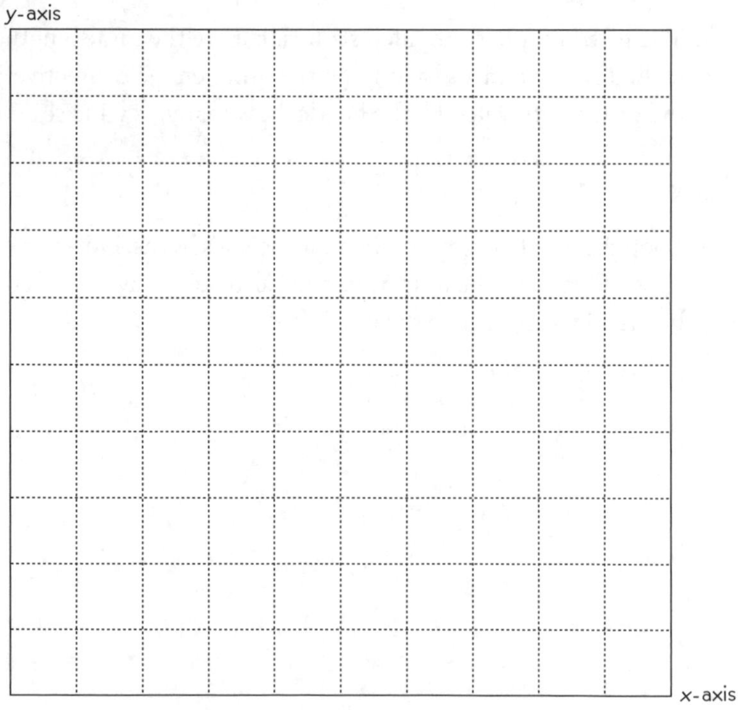

y-axis

x-axis

Figure 1.5
Results from the Bess Beetle Experiment

9. Should you use a bar graph or a line graph? Which one would best display your results? Using the average distance for each beetle, record the results on the graph. Do you see any trends in the data?

Conclusions

10. Draw conclusions from your recorded results. Do the results support or reject your hypothesis?

11. Did your duplicate experiments have the same results? If not, why not?

Concept Check

1. Which trials served as a control group in this experiment? How did your control group help you interpret the results of your experiment?

2. What is the value of testing duplicate conditions?

3. What features should be in place in the design of a reliable, useful, experiment?

4. What experiment might you design to test the hypothesis you formulated in Activity 1?

Key Terms

average (p. 1-5)
conclusion (p. 1-2)
control group (p. 1-2)
dependent variable (p. 1-4)
experiment (p. 1-2)
experimental group (p. 1-2)
hypothesis (p. 1-1)
independent variable (p. 1-4)
observation (p. 1-1)
prediction (p. 1-1)
theory (p. 1-2)

Review Questions

1. Why must a hypothesis be testable?

2. When should you use a bar graph versus a line graph?

3. Name the steps of the scientific method.

4. What is the difference between a hypothesis and a theory?

5. Define "independent variable" and "dependent variable."

6. Why is a control group necessary?

7. Why should experiments be repeatable?

8. Explain how a graph on its own completely depicts the data from an experiment.

9. ______________ is an explanation that identifies natural causes for an observation made by a scientist.

10. In an experiment, the ______________ variable changes as a result of the ______________ variable.

11. The ______________ is a flexible series of logical steps that helps scientists study the natural world.

12. In an experiment, the ______________ variable is controlled by scientists.

13. On a graph, the dependent variable is plotted on the ______________-axis, and the independent variable is plotted on the ______________-axis.

LAB **2**

Exploring Life under a Microscope

Objectives

- Explore the structure and function of simple dissecting and compound microscopes.
- Recognize the variety of applications of microscopes and manipulate focus, field of view, depth of field, and magnification of a microscope to achieve the desired image.
- Compare the magnification of dissecting and compound microscopes and choose the appropriate microscope for a specimen.
- Prepare a wet mount.
- Use stains to reveal cellular structures.
- Use dissecting and compound microscopes to observe living organisms.

Introduction

The development of the **microscope** dates from its beginnings in the sixteenth century. In the mid-1600s, Robert Hooke built his own microscope and was the first to describe cells. During the same time, Antoni van Leeuwenhoek also designed and built his own microscope, and he was the first to describe the microbes in the human mouth. The microscope is an optical instrument enabling detailed observation of microscopic organisms while allowing larger objects to be visualized. Without this instrument, humankind would not have discovered bacteria or cells, nor would we understand the complex composition of the human body. Today, there are many types of microscopes, using different approaches to examine the three-dimensional structure, surface detail, or sectional composition of a specimen. Advanced forms of microscopy reveal remarkable detail, including showing how all the parts of the cell fit together. In this lab you will examine the two basic microscopes similar to the one that Leeuwenhoek used to first reveal the tiny organisms living on his teeth.

Common Notes on Using a Dissecting or Compound Microscope

- Always carry a microscope with two hands—one hand on the arm, the other under the base.
- Begin your examination at the lowest magnification.
- Clean lenses only with microscope lens paper.
- Never unscrew anything on the microscope.
- You do not need to wear glasses if they correct for near- or farsightedness. But if your glasses correct for astigmatism, you must leave them on because microscope lenses cannot correct for this problem.
- If you have any problems with your microscope, err on the safe side and consult your instructor.
- Always remove the slide you have examined before storing your microscope.
- REMEMBER TO REPLACE THE COVER AND RETURN YOUR MICROSCOPE TO ITS ASSIGNED SPACE AT THE END OF THE LAB PERIOD.

ACTIVITY 1 Structures of the Dissecting Microscope

To examine greater details of small animals such as insects, scientists use a dissecting microscope. This instrument increases the size of a specimen in three dimensions. The structures of a dissecting microscope are simple but effective.

Figure 2.1 (see p. 2-4) identifies the three main structures of the dissecting microscope: the **base**, the **arm**, and the **head**. In this activity you will learn about the more specific structures of the dissecting microscope.

1. Remove your dissecting microscope from the cabinet, grasping it with one hand on the arm, the other under the base. Carry it to your lab bench and put it down gently, with its arm away from you.
2. Plug the microscope into the outlet on your bench. Locate the power switch on the base and turn it on. Which structure is emitting light? This is the light source, or **illuminator**, so named because the light is projected from above the specimen.
3. Specimens are placed on the glass **stage** located on the base. Light from the surface of the specimen reflects up into the head, where lenses magnify the image.
4. This magnified image is projected into the binocular eyepieces (**ocular lenses**). Because your microscope has two such lenses, it is referred to as a **binocular microscope**. Each eyepiece contains an additional magnification lens. The magnification for this lens is printed on the side of the eyepiece. What are the magnifications of your eyepieces?
5. Look around the room and note that people's eyes are different distances apart. To account for this variation, microscopes allow you to change the distance between the eyepieces. Bring the eyepieces together and then slowly pull them apart until you see one circle of light. If you wear eyeglasses, should you have them on?
6. The eyepieces on dissecting microscopes may be either stationary or adjustable. Adjustable eyepieces allow observers to correct for visual differences between their eyes by focusing one eyepiece. Place an M&M's® candy, with the "m"-stamped side facing up, on the glass stage and look into the

eyepieces. If you have stationary eyepieces, move on to step 7; if one of your eyepieces is adjustable, move on to step 8.

7. Is your candy in focus? Use the focus knobs at the bottom on either side of the arm to see the candy more clearly. Move on to step 9.

8. Close the eye above the adjustable eyepiece. Is your candy in focus? If not, use the **focus knobs** at the bottom on either side of the arm. Once your candy is in focus, swap eyes, opening the one above the adjustable eyepiece and closing the one above the stationary eyepiece. Turn the ring on the adjustable eyepiece until your candy is in focus. Note the size of the candy.

9. On the head of the microscope is another knob, called the **magnification knob**. This knob lowers or raises the head of the microscope to increase or decrease the magnification of the specimen. Adjust the magnification knob so that the head is the maximum distance from the candy. In the first circle below, sketch your view through the microscope. What amount of space does the candy take up and how much space is empty?

10. The area of the stage that you can see through the optical lenses is the **field of view**. The field of view changes as you adjust the distance of the head from the stage. The closer you are to the stage, the higher the magnification but the smaller the field of view. Adjust the magnification knob so that the head of the microscope is the minimum distance from the candy. Has the size of your candy changed? What happened to your field of view? In the second circle below, sketch what you see. Note the difference in the size of the candy and the surrounding space. For each drawing, indicate the magnification used in the line provided under the circle.

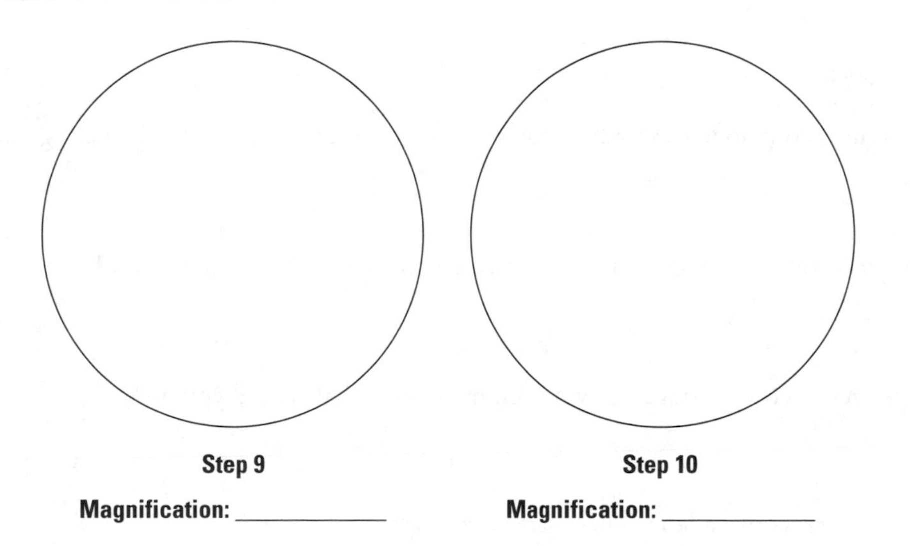

11. Label the dissecting microscope pictured in Figure 2.1.

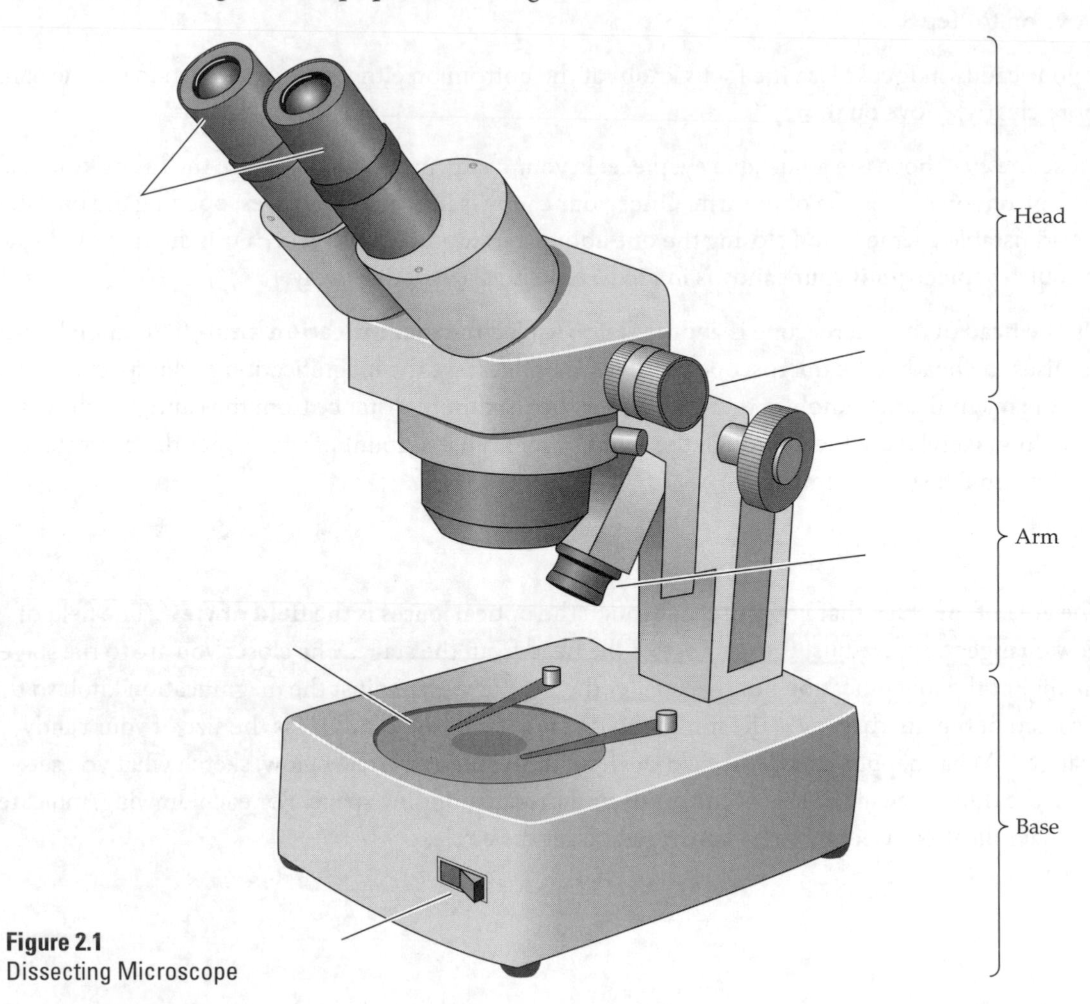

Figure 2.1
Dissecting Microscope

Concept Check

1. What structures contain lenses that magnify a specimen located inside the dissecting microscope?

2. How does adjusting the magnification knobs affect the head of the microscope?

3. If your specimen is on the stage but you cannot view it, what should you do?

4. Why is this microscope called a binocular microscope?

5. What is the purpose of the adjustable eyepiece?

6. Describe the field of view and how it changes as the magnification changes.

ACTIVITY 2 Using the Dissecting Microscope

The aquatic organism *Daphnia* is transparent, so at higher magnifications a dissecting microscope reveals its physiological processes. Lower magnifications show *Daphnia*'s interesting body structure. In this activity you will use a dissecting microscope to examine *Daphnia* at both high and low magnifications.

1. Using a pipette, remove one daphnia from the culture and place it on a concave/depression slide. Gently place a coverslip over it.
2. Take a few minutes to examine the daphnia in focus at the lowest magnification of the dissecting microscope.
3. In the first circle below, sketch the daphnia. Record your magnification in the space provided. What structures can you identify? Note these and any other observations.
4. Increase to the maximum magnification and examine the daphnia (in focus) again. In the second circle below, sketch the daphnia and record this new magnification.

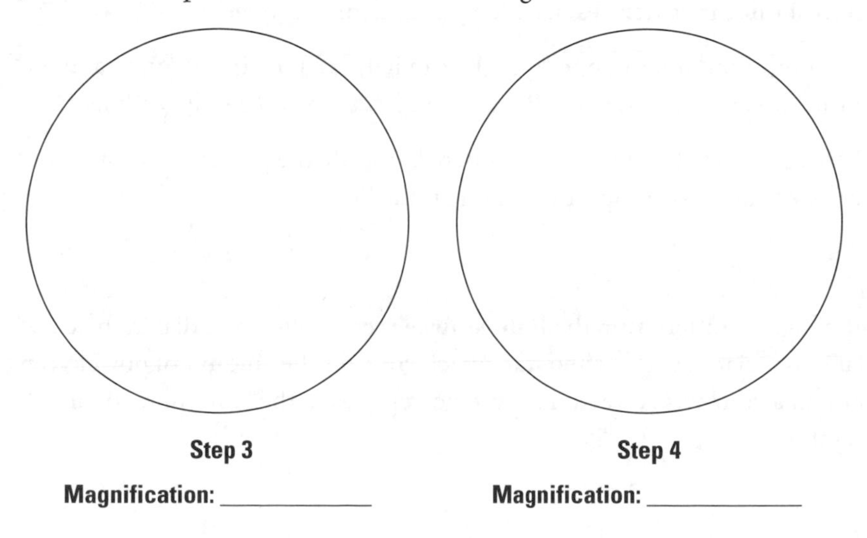

Step 3

Magnification: ___________

Step 4

Magnification: ___________

5. On the basis of your observations, what is unique about this organism? How might *Daphnia* acquire food? What appear to be the functions of the external structures that you've drawn?

Concept Check

1. Do you see external features more clearly at the lower or higher magnification?

2. Are specific internal structures more noticeable at the higher or lower magnification?

3. Can you see any physiological activity at either magnification?

ACTIVITY 3 Structures of the Compound Microscope

The compound microscope (Figure 2.2) is a delicate and expensive instrument. This type of microscope projects light from below a specimen. The light is then concentrated to illuminate the specimen. In order for an image to be visualized with a compound microscope, the specimen must be transparent in certain areas, so that the light can pass through to be magnified by the lenses. The light is magnified by lenses within an objective above the specimen and then again within the ocular lenses. The ocular lenses are contained within the eyepieces you use to examine the specimen. Listen carefully to the instructions of your lab instructor. Learning to use your microscope correctly will take both practice and patience. In this activity you will learn about the structures of the compound microscope.

1. Remove your compound microscope from the cabinet, grasping it with one hand on the arm, the other under the base. Carry it to your lab bench and put it down gently, with its arm toward you.

2. Plug the microscope into the outlet on your bench. Locate the power switch and turn it on. Compared to the dissecting microscope, how strong is the light?

3. The amount of light emitted from the **light source** is controlled by a dial in the base. Locate this dial. The dial is associated with a **rheostat**, which controls the amount of power going to the lamp. Turn the knob in a clockwise direction. More power goes to the lamp, and you'll notice that the light becomes brighter.

4. The light is directed through a lens into a structure called the **condenser**, which is located above the light source and hangs from the stage. The condenser focuses light onto the specimen.
5. A lever jutting out from the condenser is called the **iris diaphragm**. Like the iris of your eye, it opens and closes to allow more or less light past. Move the lever from right to left. What happens to the light?
6. The main part of the stage is a platform on which to place the slide containing your specimen. The slide is held in place by a **stage clip** located on the right-hand side of the platform. Open and slowly release the clip. The slide should be positioned all the way back against the bracket of the slide holder.
7. The stage is movable rather than stationary, so it is called a **mechanical stage**. This allows you to position the specimen directly under the objective, using two stage knobs located just below the stage. If you imagine the platform of the stage as a piece of graph paper, one knob moves the stage in the *x* plane (right or left) while the other moves it in the *y* plane (away from or toward you). Rotate these knobs to get a feel for how they function. How will they help you examine your specimen?
8. Above the stage and located on the head of the microscope is a revolving **nosepiece**. The **objective lenses (objectives)** project out and down from the nosepiece. These are the first structures that magnify the specimen according to the magnifying power printed on the outside of each lens. Your microscope may contain three or four objectives. Grasp the knurled ring surrounding the nosepiece and rotate it. Do you feel each objective click into position?
9. The name and function of each objective are listed in Table 2.1. The shortest objective is the scanning objective; next is the low-power objective and then the high-power objective. Note that the high-power lens is longer and might crash into a specimen if the stage is too close to the head of the microscope. If your microscope has four objectives, then the longest objective is the oil immersion objective. Find the magnification for each objective on your microscope and enter it in the second column of the table.
10. On top of the microscope head are the binocular eyepieces, or ocular lenses. Because your microscope has two such lenses, it is referred to as a binocular microscope. You can adjust these ocular lenses by means of a slider to position each one so that you can see more comfortably. Make this adjustment by grasping the grooved plates on either side of the oculars with thumb and forefinger. Push the oculars as closer together as possible while looking through the eyepieces. Slowly move the oculars away from each other until you see one circle of light.

11. The ocular lenses also provide another means of magnification. Ocular magnification is printed on the outside of the ocular lens. Record this number in Table 2.1. **Total magnification** is the magnification of the objective lens in place, multiplied by the magnification provided by the ocular lenses. Calculate the total magnification power of each objective and record the numbers in Table 2.1.

TABLE 2.1 OBJECTIVE LENSES OF THE COMPOUND MICROSCOPE

Name of objective	Magnification	Function	Ocular magnification	Total magnification
Scanning		To locate the specimen		
Low-power		To examine the specimen in some detail		
High-power		To examine the specimen in more detail		
Oil immersion		Highest magnification for a compound microscope		

12. You can focus the microscope by using the **coarse-** and **fine-adjustment knobs** located on either side of the arm near the base. The larger of the two knobs provides coarse adjustment, used when the scanning or low-power objective is in place. The smaller knob provides fine adjustment, used when the high-power objective is in place. Always begin focusing with the scanning objective in place, and then adjust your focus as you progress to the low- and high-power objectives. Turn the coarse-adjustment knob. What structure of the microscope is moving?

NEVER USE THE COARSE ADJUSTMENT WHEN THE HIGH-POWER OBJECTIVE IS IN PLACE!

13. Label the compound microscope pictured in Figure 2.2.

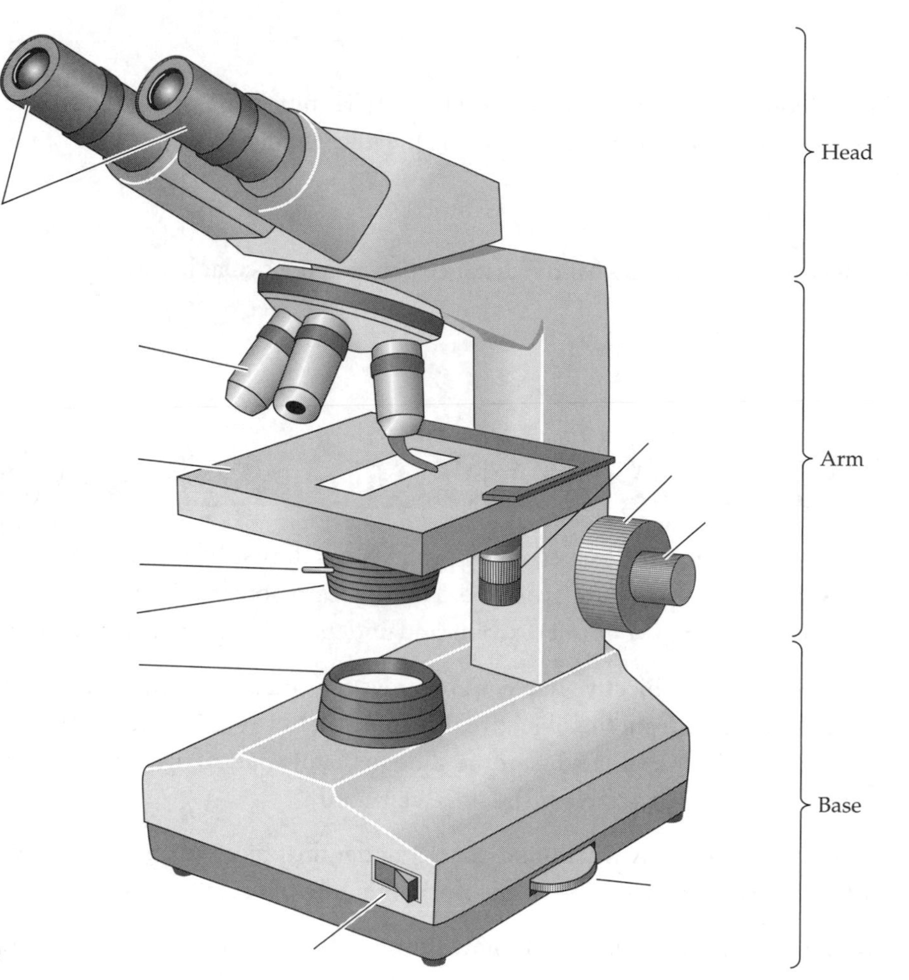

Figure 2.2
Compound Microscope

Concept Check

1. What structures contain lenses that magnify a specimen in the compound microscope?

2. Two structures control the amount of light emitted by the compound microscope. Name these two structures and describe their function.

3. Name the structure responsible for focusing light onto the specimen.

4. Write the formula for calculating total magnification.

5. Which objective should be in place when you're initially observing a specimen? Why is it important to remember this when you're focusing the microscope?

6. Why do you need to adjust the distance between the ocular lenses?

ACTIVITY 4 Using the Compound Microscope

Like most other human tasks, becoming proficient in the use of a microscope takes practice. In this activity you will use a compound microscope. You will learn how to adjust focus when necessary, examine the magnification power of your microscope, and understand the concept of magnification through lenses.

1. Put the scanning objective in place and move the coarse-adjustment knob so that the stage is the greatest distance from the objective. Locate the letter "e" slide on your lab bench and position it on the platform of your stage. Insert the slide so that it appears right side up—how you would read it. Remember to open and close the stage clip gently.
2. The letter most likely will not be directly under the objective for you to view. Use the *x*- and *y*-axis stage knobs to move the letter into position under the scanning objective.
3. Now look into the oculars and rotate the coarse-adjustment knob until the letter comes into focus. You may use the fine focus to clarify your view of the letter.
4. Note that one ocular is stationary and the other is adjustable, allowing you to correct for any visual difference between your eyes. Close the eye above the adjustable eyepiece. Use the fine focus to bring the letter "e" into perfect focus. Swap eyes, opening the one above the adjustable eyepiece, and closing the one above the stationary eyepiece. If the letter "e" is not in focus, turn the ring on the adjustable eyepiece until it is.

5. How does the image appear in the microscope? Now compare this to how it appears when not viewed through the microscope—that is, simply how it looks on the stage. In the end, the letter "e" is **inverted** and upside down. Figure 2.3 shows how the image becomes flipped and inverted during magnification.

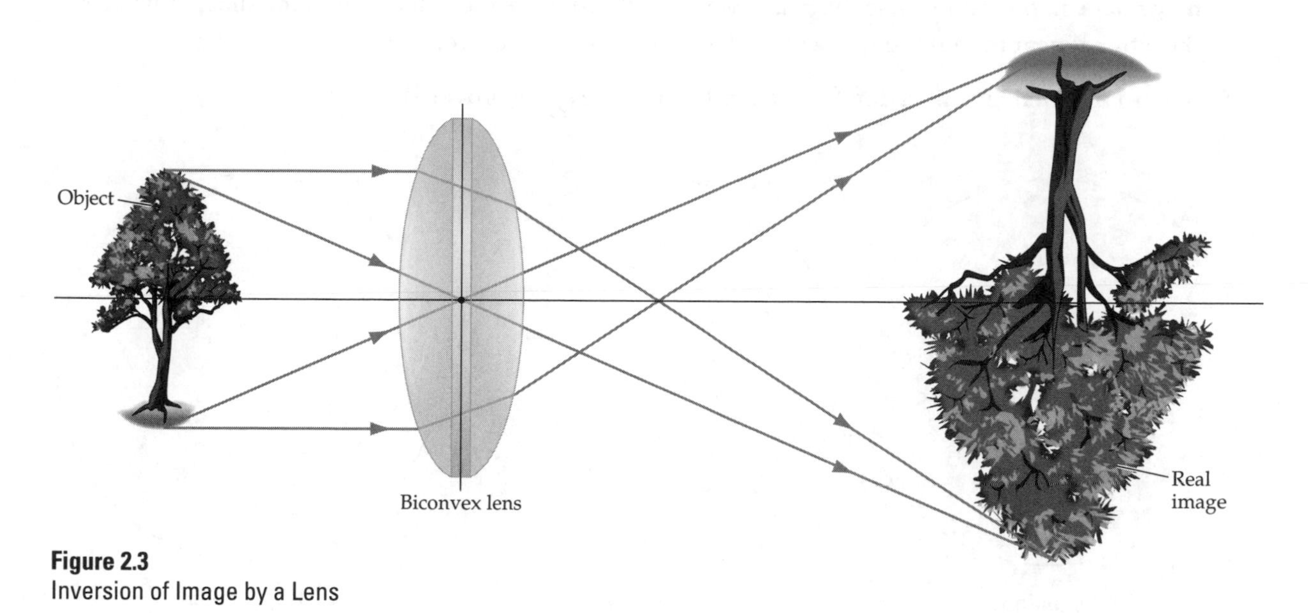

Figure 2.3
Inversion of Image by a Lens

6. Increase the magnification by moving the low-power objective into place. To adjust the amount of light at each magnification, open and close the iris diaphragm. What happens to the illumination of the letter?

7. In the three circles provided below, sketch the letter "e" as it appears under the scanning, low-power, and high-power objectives. As you switch to a higher objective, use only the fine focus to see the image clearly and adjust the iris diaphragm to increase the amount of light needed with each objective. Notice how close the higher-power objectives are to the slide. Also notice that the letter "e" may not be in the center. Use the *x*- and *y*-axis knobs to move the letter around the slide. How does the letter appear to move in the oculars? How does the stage move?
8. Record the total magnification for each instance in the space provided.

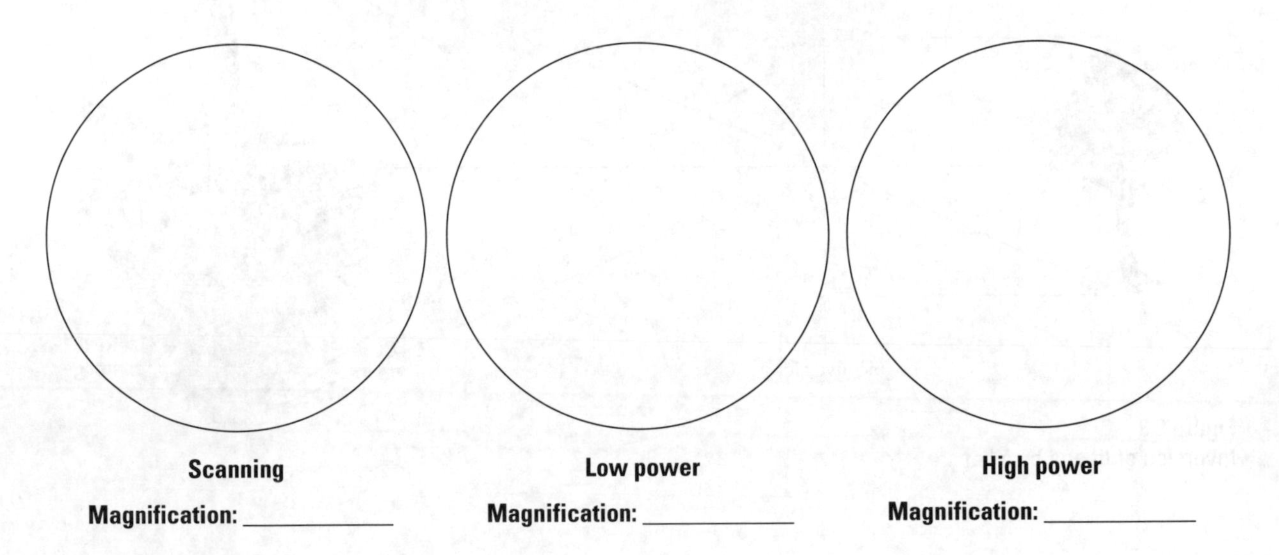

Concept Check

1. What differences do you notice about the images of the letter that you drew at the three different magnifications?

2. Does the field of view increase or decrease as you increase magnification?

3. Which objective has the greatest magnification power?

4. With each objective, the iris diaphragm needs to be adjusted to allow more or less light onto the specimen. How would you describe the relationship between each objective and the amount of light needed?

ACTIVITY 5 Depth of Field

The focus knob lowers and raises the stage to change the **depth of field**, the region of three-dimensional space that appears in reasonable focus. The thickness of the region in focus depends on the objective in place. Under the lower-power objective, more of the specimen will be in focus, but in turn, magnification is limited. Conversely, you can see greater detail using the high-power objective, but it has a thinner region of focus and therefore less of the specimen will be in focus at any given time. In this activity you will examine three threads laid on top of one another. You will get a sense of the limitations of magnification with respect to the depth of field. Use Figure 2.4 for help in identifying some of the common challenges related to depth of field.

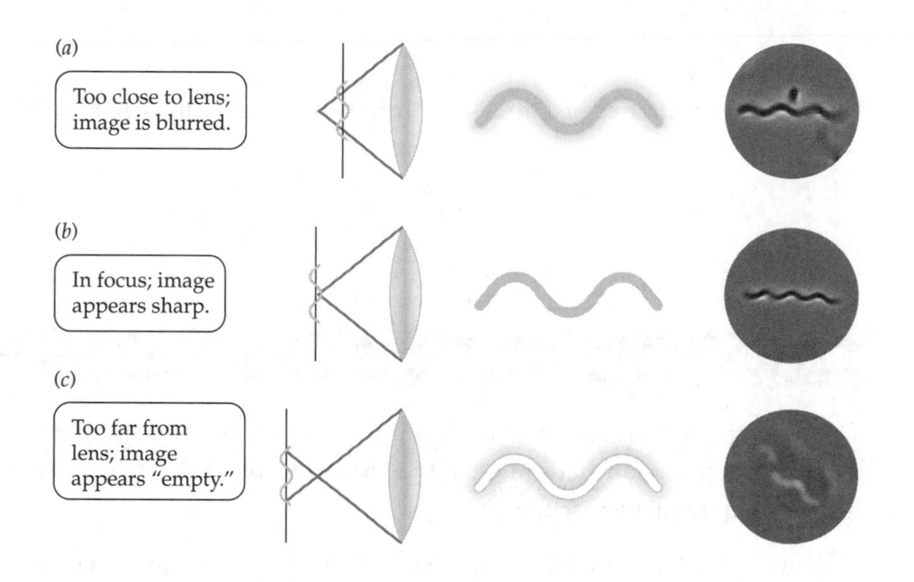

Figure 2.4
Depth of Field

1. Place the thread slide on the stage of a compound microscope. Each thread is a different color. Move the slide so that the point where the threads cross is in the middle of the field of view.
2. With the scanning objective in place, focus through the plane up and down. Notice how the threads change in clarity as you focus.
3. Now increase the magnification. For each magnification, note the clarity of the threads. Does it help to open or close the iris diaphragm as you increase magnification?

Scanning:

Low-power:

High-power:

4. To determine which thread is on top, move the slide/stage down or away using the coarse focus knobs until no thread is in focus. Now bring the slide/stage up by adjusting the fine focus. The first thread that comes into focus is the one that is closest to you (the one on top of the other threads). Continue to bring the stage up by adjusting the focus, and determine the order of all three threads. What is the order of the threads, from top to bottom?

Concept Check

1. Does the depth of field increase or decrease with increased magnification?

2. Explain how one area of a three-dimensional specimen can be in focus while another area is not.

ACTIVITY 6 Dissecting versus Compound Microscopes

As in Activity 2, in this activity you will examine the common aquatic organism *Daphnia*. As you proceed, refer to the sketches you made in Activity 2 and compare the detail of the structures that you're able to see with the compound microscope.

1. Using a pipette, remove one daphnia from the culture and place it on a concave/depression slide. Place a coverslip on top.

2. Take a few minutes to examine the daphnia using the scanning objective of the compound microscope. With the scanning objective in place, focus through the plane up and down, using first the coarse and then the fine focus knobs.

3. In the first circle below, sketch the daphnia. What structures can you identify? Note these and any other observations.

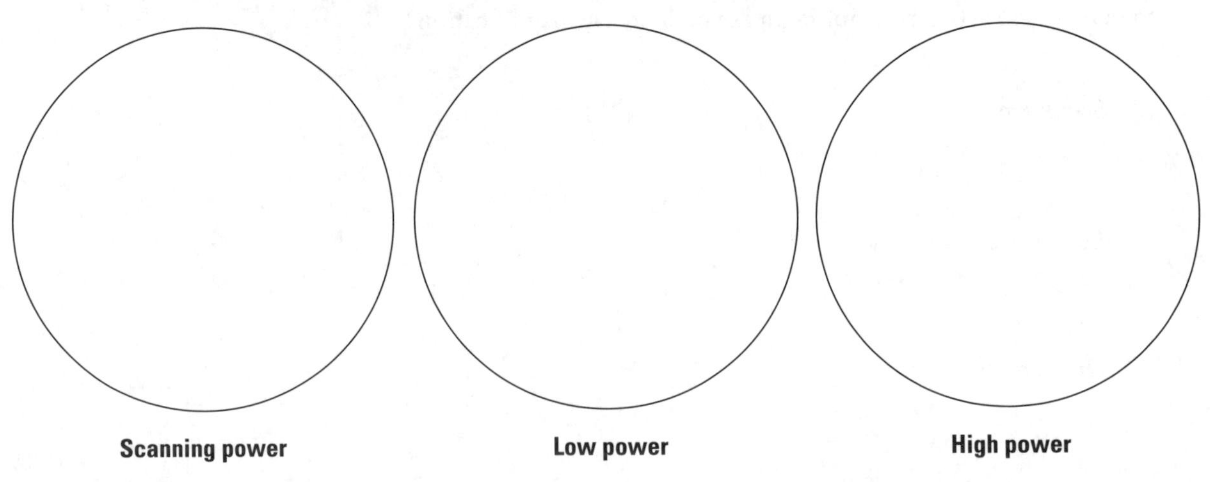

4. Increase to the low-power objective, then to the high-power objective, and examine the daphnia again. Remember to adjust the fine focus and iris diaphragm as you change objectives. In the remaining two circles, sketch the daphnia at these new magnifications.

Concept Check

1. Which microscope—dissecting or compound—enables you to see external features more clearly?

2. Are specific internal structures more noticeable with the compound microscope than with the dissecting microscope?

3. On the basis of your observations in Activities 2 and 6, what types of studies should use a dissecting microscope? A compound microscope?

ACTIVITY 7 Preparing a Wet Mount

Some specimens need a liquid or stain to help clarify structures. A slide that is prepared with any liquid applied to a specimen is called a **wet mount**. The sample is placed on the slide, and then a drop of liquid is added on top of the sample. A coverslip is gently laid down over the wet area. In this activity you will prepare a wet mount of human epithelial cells. You will prepare one of your wet mounts with water only, and the other with a stain that adheres to specific structures within a cell because of the chemical composition of those structures.

1. Gently scrape the inside of your cheek with the flattened end of a toothpick.
2. Wipe the whitish material from the toothpick onto the middle of a microscope slide. This material actually consists of some of the cells that line the inside of your mouth, called epithelial cells.
3. Using a pipette, place one drop of water on the cells.
4. Add a coverslip and examine the slide under your compound microscope. If it is difficult to see the cells clearly, move your iris diaphragm back and forth to get a sharper image by adjusting the illumination to increase contrast. In the first circle below, sketch what you see.

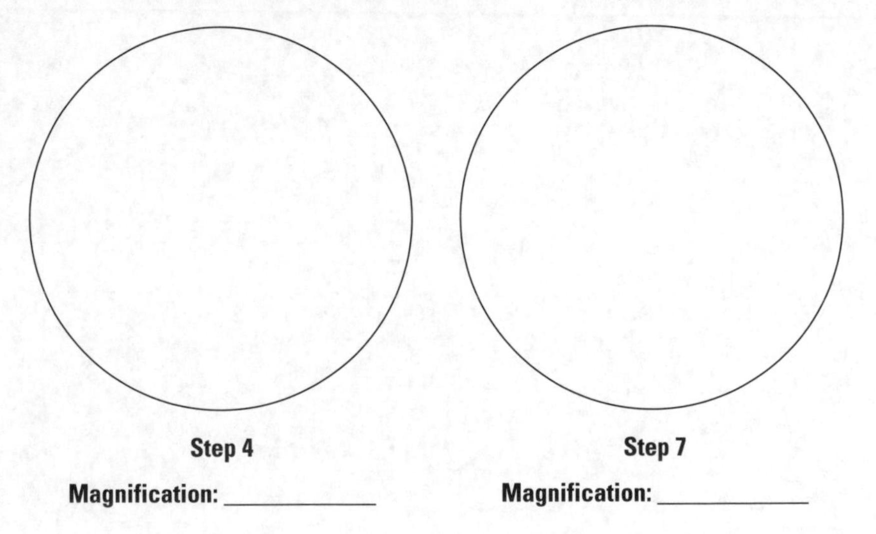

Step 4

Magnification: ___________

Step 7

Magnification: ___________

5. On another microscope slide, prepare another wet mount of your cheek cells, but this time place a drop of methylene blue stain on top of the cells. Use Figure 2.5 to help you.
6. Add a coverslip and let the stain bind to the cellular structures for 1 minute.

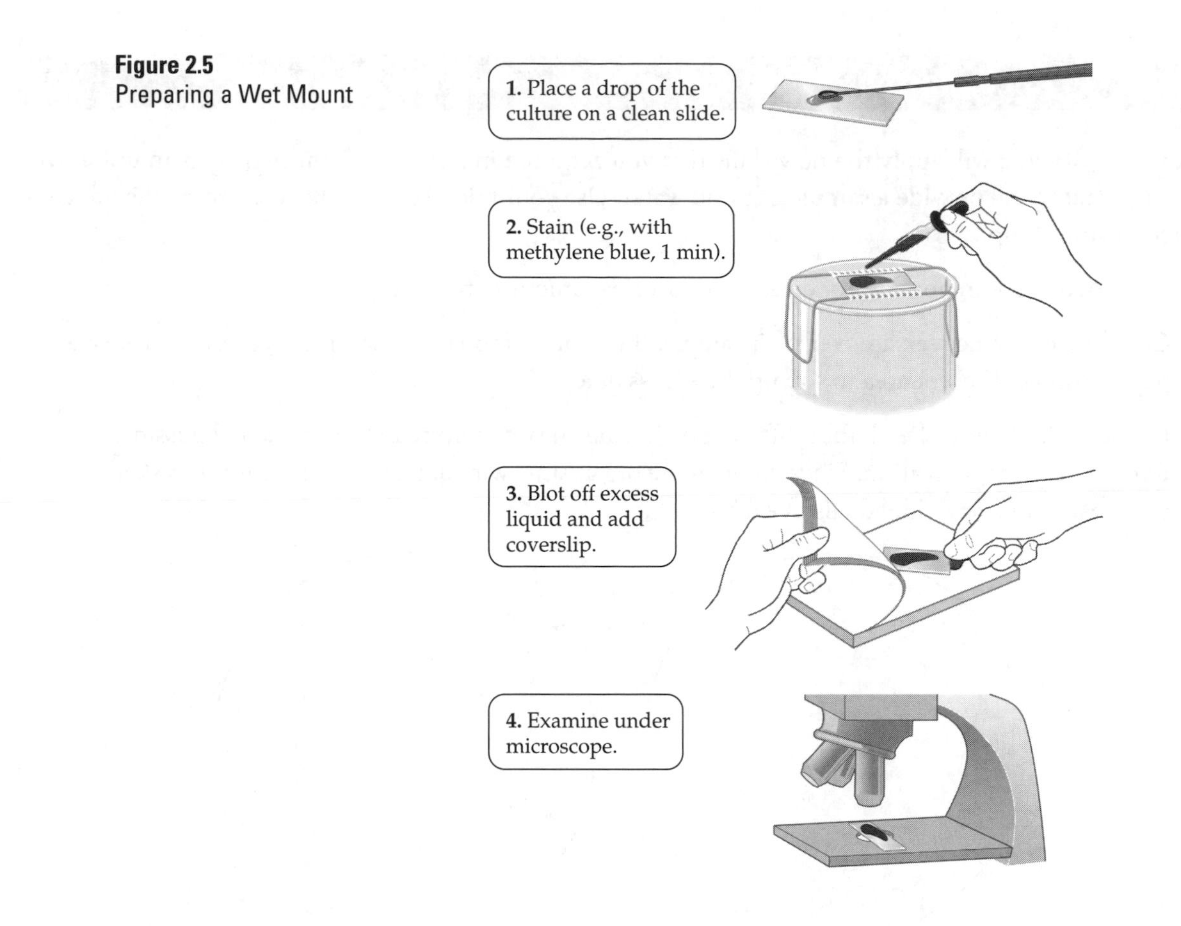

Figure 2.5
Preparing a Wet Mount

7. Examine the slide under your compound microscope. You should be able to locate the darkly stained nucleus in some of the cells. In the second circle below step 4 (see previous page), sketch what you see.

Concept Check

1. How are wet mounts useful?

2. Given that microbes are even smaller than the cells you've examined in this activity, what technique could you use to determine whether you have microbes on your slide?

3. Did the cells in the specimen stained with methylene blue look different from those that weren't stained? Describe any structures that were more distinct in one or the other of the two wet mounts.

ACTIVITY 8 Applying Your Skills

In this activity you will apply the new skills that you acquired in Activities 1 through 7 to an unknown. Your instructor will provide a sample of pond water. Use your microscopy skills to examine the life contained in your sample.

1. Place two drops of pond water directly onto a clean microscope slide.
2. Carefully lower a coverslip over your sample. If you placed too much sample on the slide, touch a paper towel to the wet area to sop up the excess fluid.
3. In the circles below, sketch four different organisms in your culture in the best detail possible. Record your magnification. Since some of the organisms swim quickly, you may want to sketch composites of a few individuals of each type.

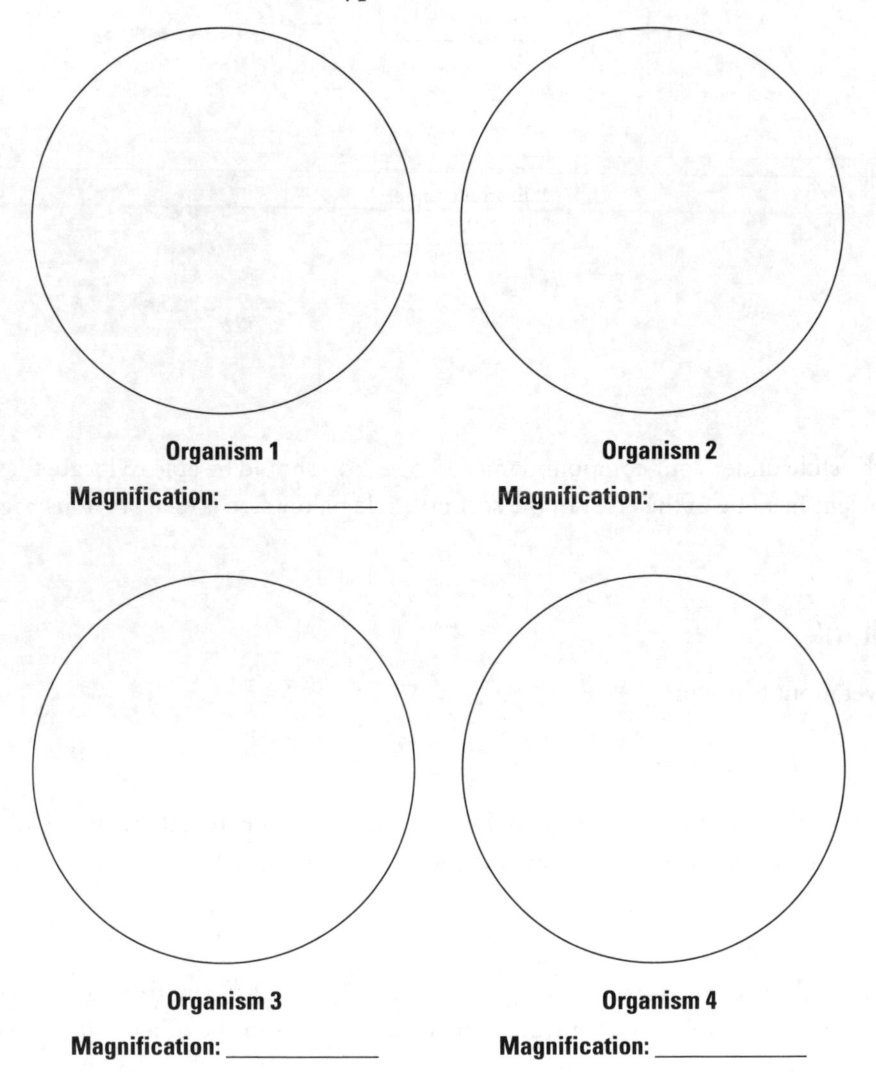

4. Finally, examine the poster of common microscopic aquatic organisms in your classroom. Identify the organisms that you've drawn and compare your findings with those of your benchmates.

Concept Check

1. Were you able to find more than four organisms?

2. What magnification was the best for sketching your organisms? Why?

3. How long did it take you to find your first organism?

4. Did it take a different amount of time to locate the next organism? Why or why not?

Key Terms

arm (p. 2-2)
base (p. 2-2)
binocular microscope (p. 2-2)
coarse-adjustment knob (p. 2-8)
condenser (p. 2-7)
depth of field (p. 2-13)
field of view (p. 2-3)
fine-adjustment knob (p. 2-8)
focus knob (p. 2-3)
head (p. 2-2)
inversion (p. 2-11)
illuminator (p. 2-2)
iris diaphragm (p. 2-7)
light source (p. 2-6)
magnification knob (p. 2-3)
mechanical stage (p. 2-7)
microscope (p. 2-1)
nosepiece (p. 2-7)
objective (p. 2-7)
objective lens (p. 2-7)
ocular lens (p. 2-2)
rheostat (p. 2-6)
stage (p. 2-2)
stage clip (p. 2-7)
total magnification (p. 2-8)
wet mount (p. 2-16)

Review Questions

1. Name the three types of objectives for a compound microscope and describe their functions.

2. The __________________ focuses light onto the specimen.

3. __________________ microscopes are microscopes that have two eyepieces, or ocular lenses.

4. The region of three-dimensional space that appears in reasonable focus is called the __________________.

5. We calculate total magnification of a compound microscope by __________________.

6. Why must you adjust the space between the ocular lenses?

7. What structure moves when you're focusing a dissecting microscope? A compound microscope?

8. If you cannot see enough of your specimen, what should you adjust?

9. How do stains help clarify the structure of cells?

10. Describe the steps of preparing a wet mount.

11. What structures does a compound microscope have that a dissecting scope does not? What structures are similar between the two?

12. If you cannot see enough detail of your specimen, what should you adjust?

13. How does the iris diaphragm control the amount of light hitting the specimen?

14. What type of microscope—dissecting or compound—would you use to magnify the following specimens?

- The cells from a hair follicle: ________________
- Your finger: ________________
- A throat culture of *Streptococcus* bacteria: ________________
- A dollar bill: ________________

LAB 3 Biodiversity and Evolution

Objectives

- Define the general characteristics of life.
- Describe common ancestry according to the evolutionary tree of life and the shared characteristics of all life-forms.
- Develop an evolutionary tree, mapping the relatedness of different groups of organisms.
- Observe the variety and interdependence of life-forms using living and preserved specimens.
- Describe the importance of biodiversity, its relationship to evolution, and its impact on human life.
- Assess the biodiversity of a local plot of land.

Introduction

Life is not defined by one word, phrase, or sentence, but there is unity in life. One unifying characteristic of all life is **evolution**: all life-forms evolve, or change through variations in the genetic characteristics of populations over time. Such changes can lead to the evolution of completely new groups of organisms. The identity of the first life-form is not known, but the basic characteristics common to all life-forms indicate that all life shares a **single common ancestor**. From this common ancestor, life has evolved over billions of years, establishing many lines of descendants, or **lineages**. There is no written record of how evolution shaped populations of organisms throughout time, so scientists use a number of forms of evidence to create evolutionary trees. Fossils provide direct evidence of ancestral life-forms that can be compared to living organisms, revealing shared features that indicate the common ancestry of different groups of organisms. Living organisms may also have **vestigial structures**, like the human tailbone, that provide evidence of descent from life-forms that still actively use those structures. Evidence of an evolutionary link to ancestral organisms is also provided by studies of the stages in embryonic development of animals, where ancestral links not visible in the adult organisms become evident. This technique allowed biologists to correctly place barnacles with crustaceans, for example. The barnacle adult form is quite different from that of shrimp and lobster adults, but the early embryonic stages are quite similar. The addition of new technologies, including the comparison of DNA and genes, also enables scientists to evaluate the relatedness of lineages.

As the variation of life-forms, or **biodiversity**, expanded over time, the physical environment and existence of other life-forms shaped the evolutionary success of surviving lineages. A **species** is a group of organisms that can breed, and a **population** is a group of interacting individuals of a single species located within a particular area. Species that exist today display inherited features that proved useful for the survival of their ancestors—these features were passed on through time precisely because older generations survived to pass them to their descendants.

Biodiversity is important to the success of life because greater diversity allows for more and different ways of coping with external challenges. In populations, sexual reproduction yields more diverse offspring, increasing the likelihood of offspring with new combinations of characteristics that may prove useful to continued survival.

Survival of a population depends not only on the population's diversity, but also on the diversity of the surrounding environment. Biodiversity is also important to the survival of whole ecosystems. Ecosystems with greater biodiversity are more stable and resilient than are less diverse ecosystems. It is easier for ecosystems to recover from droughts, disease, or other environmental challenges if they comprise several different species, because some of those species may be capable of coping with the environmental stress even if others are not.

In this lab you will study the bonds that all life-forms share through common characteristics and ancestry, and you will explore how the stability and success of life on Earth relies on biodiversity.

ACTIVITY 1 Common Characteristics of Life

All life-forms share certain features (Table 3.1). But while all organisms share common features that characterize life, organisms that existed further back in time have simpler embodiments of each feature. For example, the first organisms existed billions of years ago and were single-celled. A **cell** is a tiny, self-contained unit enclosed by a membrane. Singled-celled organisms still exist, but evolution led to more complex life-forms, including a vast array of multicellular organisms such as humans. The single common ancestor of all life-forms was the first cell.

TABLE 3.1 THE SHARED CHARACTERISTICS OF LIFE

Living organisms

- Are composed of cells
- Reproduce using DNA
- Grow and develop
- Actively take in energy from their environment
- Sense their environment and respond to it
- Maintain constant internal conditions
- Can evolve as groups

All living organisms have the ability to produce another organism through **reproduction**. Reproduction uses deoxyribonucleic acid (**DNA**) to pass information from parent to offspring. DNA houses all of the information needed to produce new cells or to run cellular processes. Every type of organism uses DNA, the macromolecule of inheritance. In the oldest, simplest organisms, reproduction started as an asexual process in which a cell copied its DNA and other cellular components to produce a new genetically identical offspring. In time, sexual reproduction evolved, in which different individuals combine DNA to form a new, genetically different offspring.

Growth and development is another characteristic shared by all life. Growth and development are the processes by which an organism increases in size and function and through which multicellular organisms undergo cell division and differentiation. Organisms have evolved many approaches to growth and development: they can be fast, such as during the short life of a fruit fly; or as complicated as the development of an elephant, which involves trillions of cell divisions, along with coordinated cell movement and even cell death to produce the complex body of an elephant.

Growth, development, and regular cellular functions require **energy** (chemical energy to fuel reactions). The energy fueling cellular processes comes from the environment; this energy may be solar or chemical, or originate from another source. The processes that organisms use to harness energy have become increasingly complex over time, from the simple process of breaking down small molecules to creating very large and complex molecules during a process like photosynthesis, which creates sugars for plants.

In addition, organisms detect and respond to changes to their external environment while managing their internal environment. The means of sensing the external environment has evolved from simple receptors in an organism's cell membrane to the complex network of sensory neurons that humans use in their five senses. An organism's management of constant internal conditions through many regulatory channels is **homeostasis**. Homeostasis is simple in a single-celled organism, as compared to the challenge of managing a trillion cells within multicellular organisms.

Finally, populations of organisms evolve. Evolution is the change over time in a lineage of organisms. Over time, populations of organisms change, ultimately through changes in the genetics of the population. All lifeforms either evolve to survive in changing environments, or go extinct.

In this activity you will describe in your own words the function of each shared characteristic of life and define some other important terms.

1. Describe the function of each characteristic (a–h) and define the important terms (i–k).
   a. *Cells:*

   b. *DNA:*

   c. *Reproduction:*

   d. *Growth and development:*

e. *Energy:*

f. *Sense:*

g. *Homeostasis:*

h. *Evolution:*

i. *Single common ancestor:*

j. *Species:*

k. *Population:*

Concept Check

1. How does DNA function in reproduction and cellular processes?

2. Give an example of how your body detects and responds to its environment. How would your pet or a friend's pet respond to the same environmental condition? How would a plant respond to that same condition? Is either organism better than you at detecting certain conditions of the environment?

3. In what ways do the characteristics of all life exemplify how all life-forms evolved from a single common ancestor?

ACTIVITY 2 Common Ancestry

Scientists called systematists study the relationships among groups of different organisms, depicting them in **evolutionary trees**. Like a family tree, an evolutionary tree traces lineage, or lines of descent (Figure 3.1). An evolutionary tree is a hypothesis of evolutionary relationships based on scientific studies, including fossils, embryonic development, vestigial traits, and genetic studies.

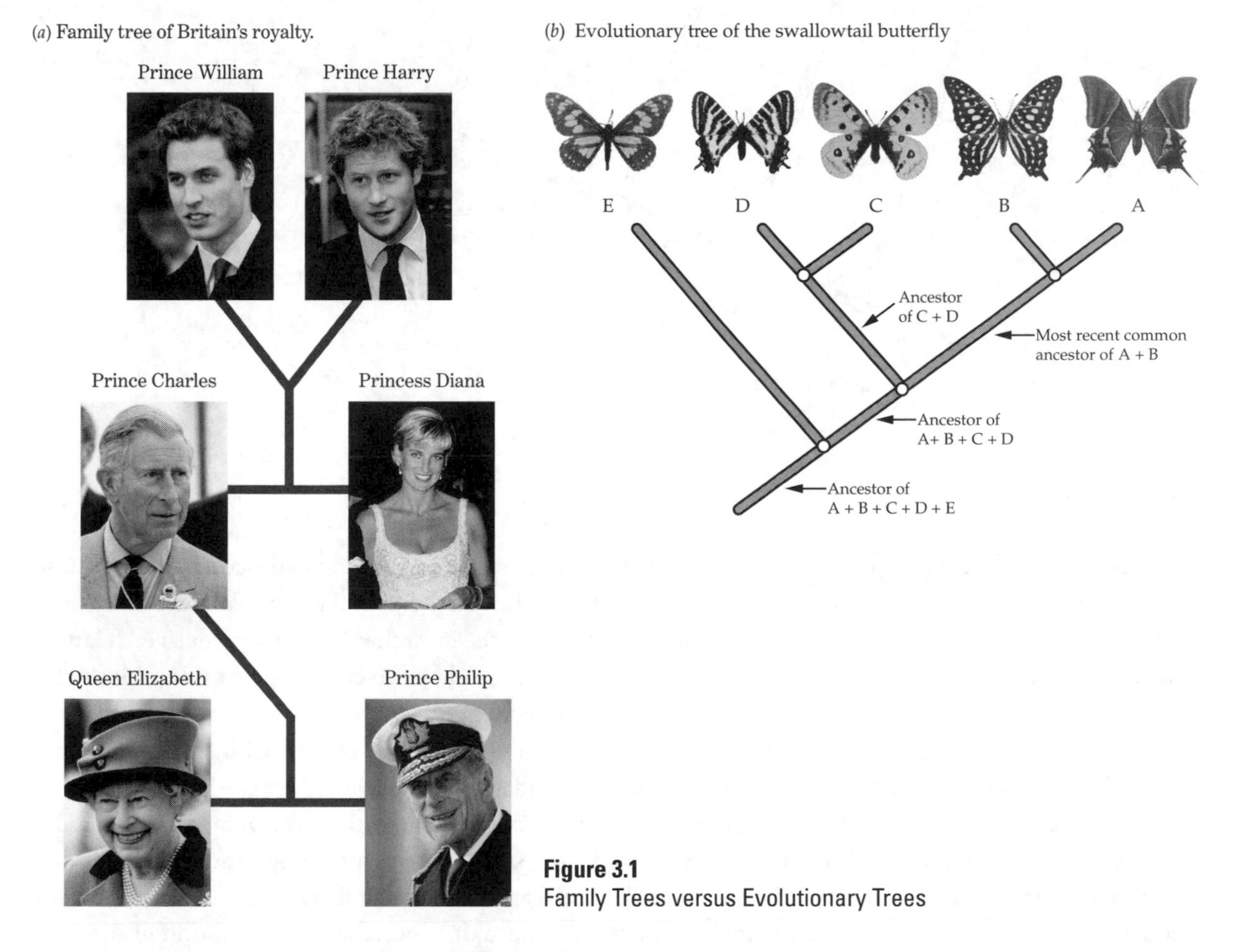

Figure 3.1
Family Trees versus Evolutionary Trees

Evolutionary relationships are defined by the presence of similar features in a group of organisms, which have evolved in lineages from a common ancestor. A feature that a common ancestor passes down to all of its descendants is a **shared derived feature** and clearly defines these descendants as a group of closely related organisms. Biologists use the term "shared derived feature" to refer to features that originate with the most recent common ancestor of a group. Shared derived features occur between the branches of an evolutionary tree because they evolve in one group and are then shared by all of the descendants of that group. For example, Figure 3.2 shows the evolutionary tree of all life.

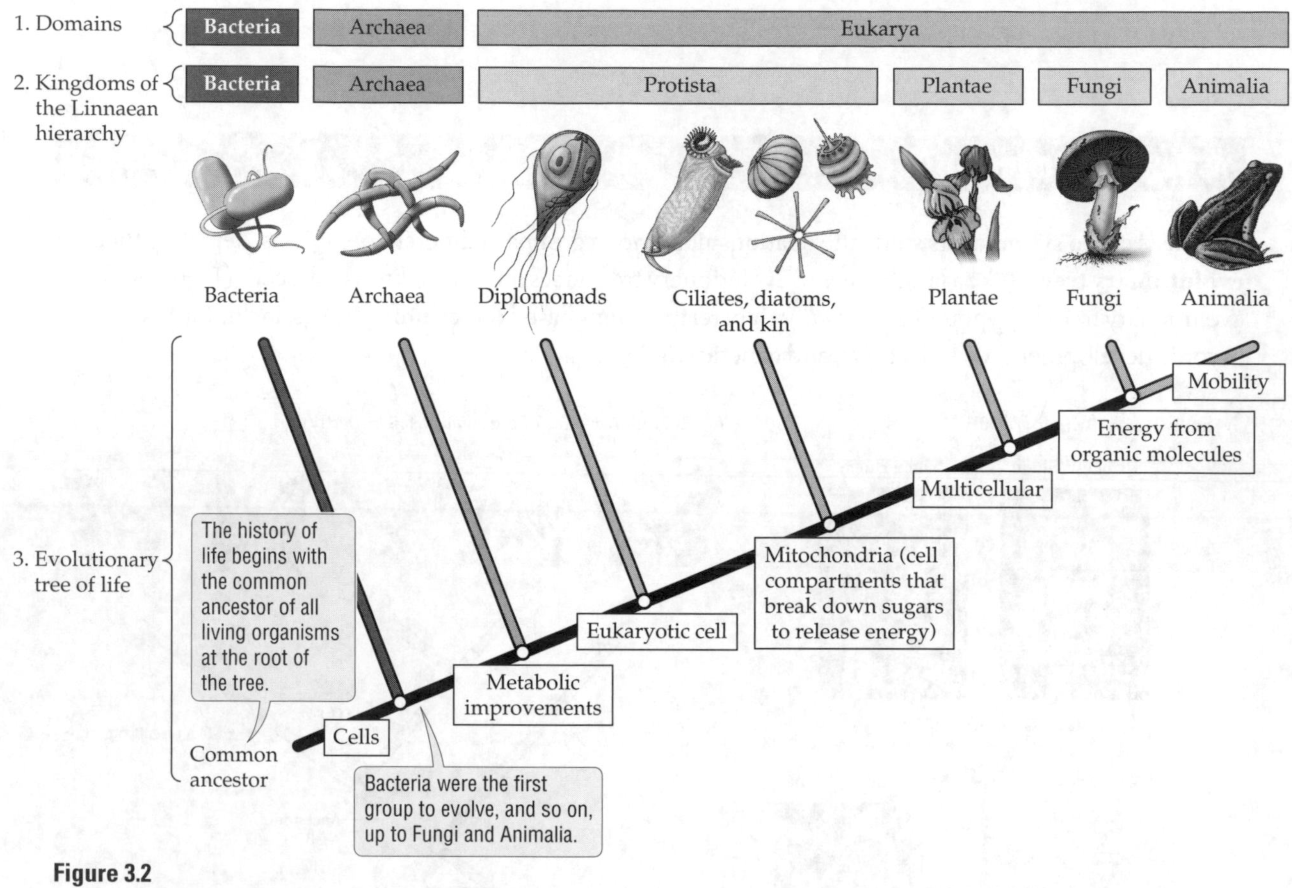

Figure 3.2
Evolutionary Tree of Life

The lineage of life has many branches, and each branch represents a new line of descendants. The base of the tree represents the common ancestor of all of the branches on the tree. Bacteria were the first group to evolve from early cells. The evolutionary innovations that we see in Archaea (its shared derived features that appeared after bacteria on the evolutionary time line) are all related to cellular processes such as metabolism. Instead of having simple chemical reactions to produce energy, Archaea evolved many more complex reactions to harness energy from their environment. This variety is shared with the groups that evolved later: the protists, plants, fungi, and animals—descendants of Archaea. Above the branching of Archaea, a more visible shared derived feature is the evolution of the eukaryotic cell. Again, this shared derived feature is retained within the groups that evolved later, such as the plants, fungi, and animals.

A major concept in this activity is that a shared derived feature, likely a more complex or more efficient characteristic, is passed on to descendants because it is a feature that increases the likelihood of survival. For example, all organisms in the domain Eukarya have eukaryotic cells containing a nucleus and various cell compartments. One far more efficient characteristic that evolved in these eukaryote cells was energy acquisition through specific cell compartments.

In this activity you will practice building an evolutionary tree of five organisms. These organisms are all quite different; from what you learn about them, you will choose a feature as a shared derived feature defining each branching along the evolutionary tree.

1. Look up the characteristics of each of the following five organisms using your textbook or the Internet: bacterium, yeast fungus, sponge, frog, human. Focus on topics like cell type and number, type of nutritional strategy, mode of reproduction, circulatory system, nervous system, and gas exchange features. (You do not have to limit yourself to these topics; they are provided only as a guide.)
2. Determine an evolutionary order for the organisms. Which organism evolved most recently? Place this one on the top right of the tree in Figure 3.3. Then work down the tree in order of most recently evolved to the organism that is "oldest" from an evolutionary perspective.
3. Choose a shared derived feature for each branch, using the example in Figure 3.2 to guide you. Remember that each shared derived feature is, by definition, passed to all of the descendants of the first group on the tree to inherit that feature. Write the shared derived features that you identify on the appropriate lines in Figure 3.3.
4. Explain why the two organisms on the left of the tree will have fewer shared derived features than the two organisms on the right.

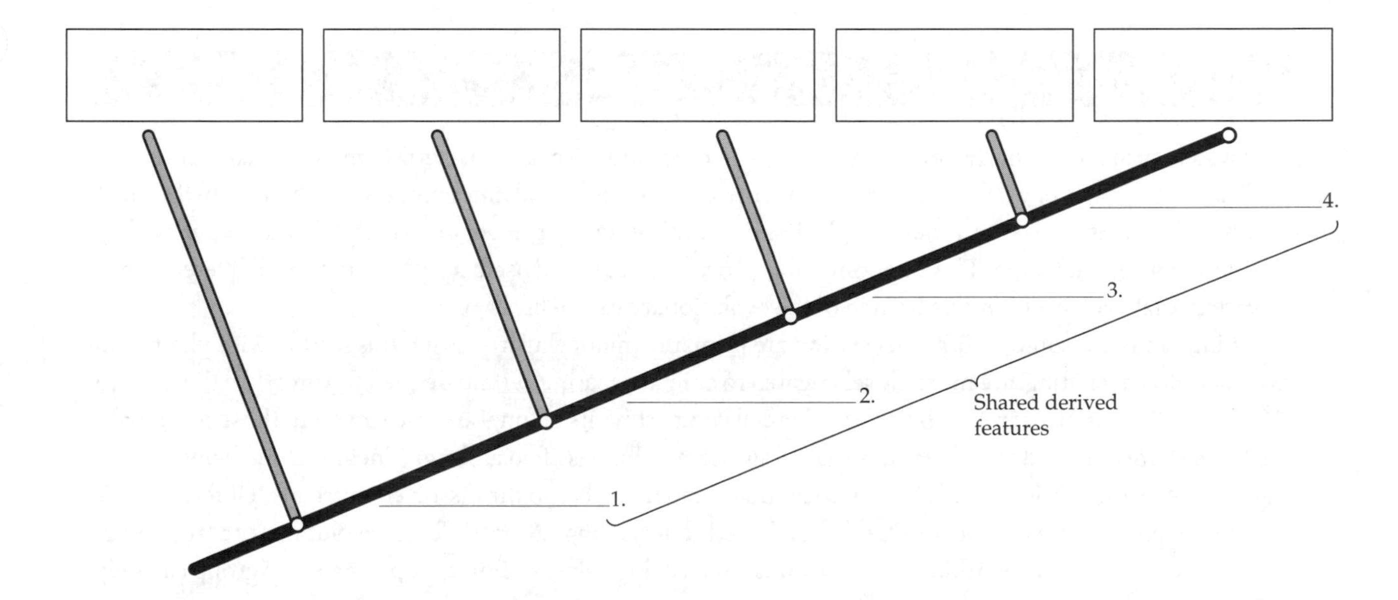

Figure 3.3
Shared Derived Features on an Evolutionary Tree

Concept Check

1. What is a shared derived feature?

2. How do evolutionary trees help display evolutionary relationships?

3. How does our ability to draw evolutionary trees support the theory of common ancestry?

4. Which groups on your tree are more complex—the most recently evolved (upper right) or the older groups (lower left)?

ACTIVITY 3 Variety of Life

The variety of life on Earth is astounding, from single-celled organisms like bacteria to multicellular mammoths such as the blue whale. Variety may be manifest in clear morphological differences, as can be seen in the great diversity of insects. However, variety may be less outwardly visible—present in a variety of strategies, including reproduction and nutrition. The kingdom Fungi provides an easy and useful way to understand different kinds of variety that evolve within one branch of the evolutionary tree of life.

Fungi number some 70,000 species and are primarily multicellular, sharing this feature with plants and animals. However, fungi are more closely related to animals, sharing certain derived features, like the need to find food, whereas plants make their own. The cell compartments of fungi are also similar to those of animals. All fungi share a set of derived features that allow them to be classified as fungi. These include being able to digest very resistant compounds in plants externally and then absorb the dissolved nutrients. Different environmental pressures and opportunities have allowed variety along several different evolutionary paths. Fungi have evolved numerous reproductive, nutritional, and ecological roles. For example, only one group can fully break down the toughest compounds in wood.

Many fungal species have evolved the vital ecological role of **decomposer**. Decomposers feed on **detritus**, any dead or dying organic matter. Fungi don't eat (consume) the detritus; that is the role of **detritivores**. Instead, a fungus secretes enzymes, breaking down the detritus into molecules that it can absorb into its cells.

Fungi include the molds and mushroom-producing fungi. Both of these fungal types are multicellular, composed of thin strings of cells incompletely separated by a partial divider called a **septum**. The strings of thread-like projections, **hyphae**, penetrate the soil, fruit, or dead animals, forming large mats called **mycelia** (singular "mycelium") (Figure 3.4). Although the mushroom is thought to be the body of the fungus, it is the large mat of hyphae, the mycelium, that is the actual body of a mold or mushroom-producing fungus. These organisms can penetrate just about any material and therefore are able to spread deep beneath the ground and vast

Figure 3.4
Typical Mushroom

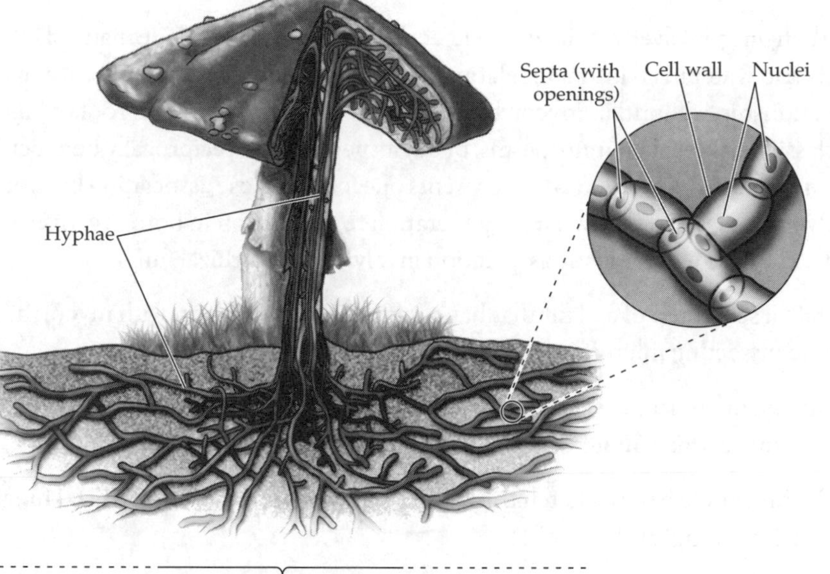

distances, and they have been identified as the largest organisms existing on Earth today, albeit mostly underground.

Molds and mushroom-producing fungi reproduce both asexually and sexually. Mushrooms produce **spores,** thickly coated reproductive cells. Puffball mushrooms eject spores; button mushrooms drop their spores to the ground through their gills. Sometimes spores wait years for favorable conditions to start growing into a new mycelium. This ability to lie dormant for so long makes it challenging to eliminate fungi such as the mold in your shower or black mold that crops up after flooding, because the spores lie in wait, blooming when ready.

Let's examine a typical mushroom.

1. Obtain a specimen of a fungus, including the whole mycelium and the reproductive structure (a mushroom), as well as a razor blade. Can you identify the hyphae of the mycelium?
2. Using the razor blade, cut a mushroom off of the growing fungus. Note how the hyphae are packed together within the mushroom structure but then spread out as the organism penetrates the dirt.
3. Cut open the mushroom. Does it drop spores or eject spores from its structure?
4. In the box below, sketch the hyphae penetrating the soil and the mushroom structure. Indicate where the spores are produced.

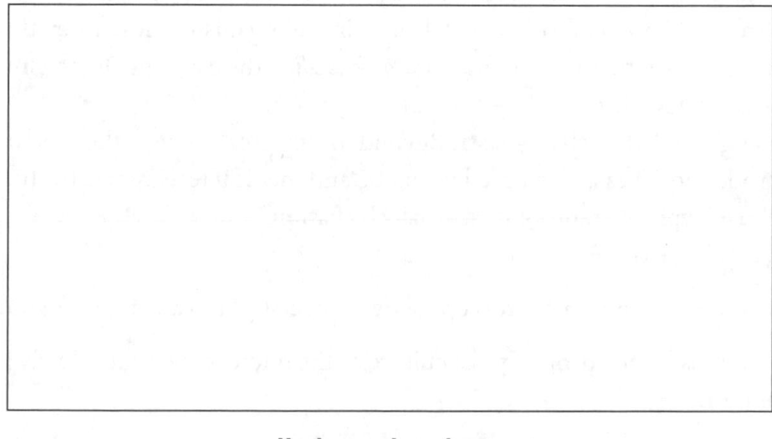

Hyphae and mushroom

Certain fungi evolved a quite different ecological role: they provide many plants and other organisms with nutrients and water in an intimate relationship that also supplies the fungi with the nutrients it requires. One such relationship is found in mycorrhizal fungi that associate with the roots of more than 95 percent of the plants classified as ferns, gymnosperms, or angiosperms. This reciprocally beneficial association with another organism is called **mutualism** and represents one form of close association between different organisms that has evolved and is passed down from generation to generation in both organisms.

Let's examine a mutualistic association involving mycorrhizal fungi.

5. Obtain a specimen of a plant that has a mutualistic association with mycorrhizal fungi, along with a simple dissecting microscope.
6. Using the microscope, examine the roots of the plant. Do you see any bumps along the roots? These are the mycorrhizal fungi.
7. In the first circle below, sketch the root structures where the mycorrhizal fungi are located. Label the mycorrhizal fungi and the roots.
8. In the second circle, sketch an area of the root where there is no mycorrhizal structure.
9. Why do you think the fungi attach to the plant's roots, versus other parts of the plant?

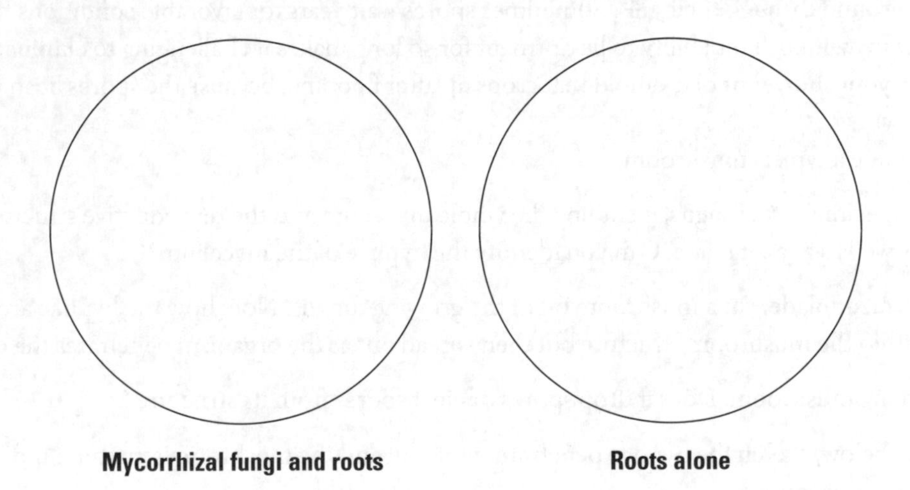

Mycorrhizal fungi and roots **Roots alone**

Another ecological grouping that has evolved in the fungi consists of **parasitic** associations, such as athlete's foot, in which the fungi obtain food directly from a living organism. More often than not, the cause of a major plant disease or devastation is a fungus. *Ceratocystis ulmi*, the cause of Dutch elm disease, destroyed many of the elm trees in the Midwest.

Some fungi are single-celled, such as **yeasts**. Beyond having great value in the food industry for alcohol, cheese, and bread production, yeasts are used by genetic and medical researchers to study disease, therapy, and development. Yeasts typically reproduce asexually by fission.

Let's examine a typical yeast.

10. Obtain a yeast culture, a pipette, a microscope slide, a coverslip, and a compound microscope.
11. Using the pipette, transfer one drop of yeast culture to the microscope slide. Slowly place the coverslip on top of the solution.

12. Using the microscope, examine the slide, starting on the lowest magnification and working your way up to the highest magnification. Remember to adjust the iris diaphragm.
13. In the circle below, sketch a few of the yeast cells you see.
14. The fungi that you examined earlier in this activity had structures with high surface area for the absorption of nutrients. Why do you think yeasts evolved a single-celled strategy?

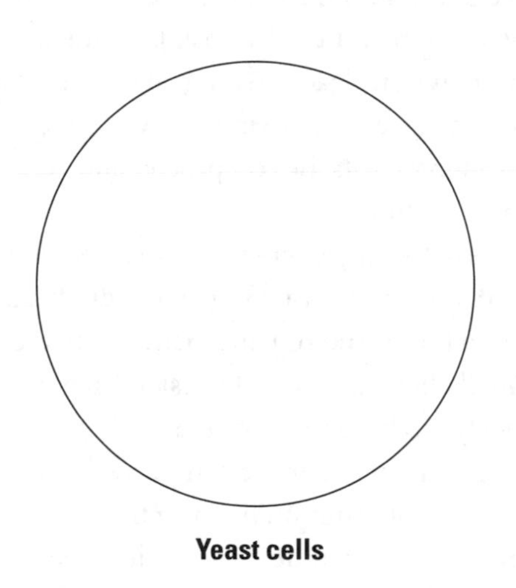

Yeast cells

Concept Check

1. Why may certain fungi be important to the diversity of land plants in any given area?

2. Describe each type of ecological role that fungi fill.

3. How do fungi affect your life?

4. From your observations and the information provided to you, what type of fungi evolved first?

ACTIVITY 4 Interconnectedness of Life

Mutualistic relationships are crucial to the existence of complex multicellular organisms. Humans have mutualistic relationships with bacteria in their digestive tract, which provide nutrients to the humans and a protective home to the bacteria. Plants establish several different types of mutualistic relationships—some with bacteria and others with fungi, both of which provide the plants with the nutrients needed for life. Hard corals evolved close mutualistic relationships with algae that allow them to lay down coral reefs. As organisms interacted with one another through evolutionary time, more and more of these close mutualistic relationships developed. Another good example of mutualistic relationships is the intimate relationship called a **lichen**. Lichens are mutualistic associations between algae and fungi in which the fungal hyphae wrap around the algal cells. This relationship protects the algae from drying out while providing them with essential nutrients and a supply of water. Meanwhile, the fungus gets the complex organic compounds it requires to live. The photosynthetic algae provide this food to the fungus.

This relationship permits lichens to live anywhere there are minerals (rocks) and sunlight. They inhabit the full range of terrestrial ecosystems and play a vital role in the establishment of the living components in an ecosystem. When no life exists and there is no organic matter in an area, such as after a volcanic eruption or a receding glacier, prokaryotes and lichens are the first to establish populations, arriving in numerous ways, including by air and water. This newly established ecosystem will develop soil and allow small, hardy plants to establish themselves. Through time, larger and more complex plants will move in as conditions, like soils, improve. This gradual, serial replacement of communities over time in an ecosystem is known as **succession.**

Lichens are able to thrive in extreme environments, including the dry, cold tundra, where the minerals in the rocks provide nutrients. The fungal cells retain enough moisture to enable lichens to survive even in hot, dry desert ecosystems. Scientists also use lichens as an indicator of air quality, referred to as the "lichen count." Lichens are sensitive to pollutants, so in most cities lichens have a hard time establishing a population, because of the poor air quality. In this activity you will examine three different types of lichens.

1. Obtain microscope slides of three different lichens, along with a dissecting microscope.
2. Using the microscope, examine the structures of each lichen type. What are the structural differences among the three types? Is there a difference in color?
3. In the three circles below, sketch each type. Label each circle with the name of the specimen you drew. Consider the name of each specimen, which is printed on the slide. How do you think lichen species are named?

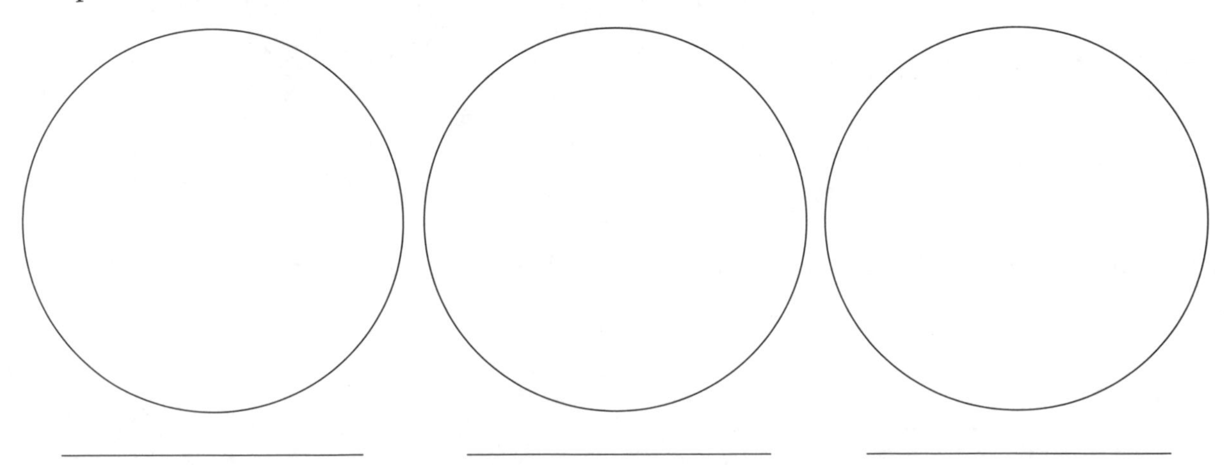

Concept Check

1. What is the importance of mutualistic relationships? Can you describe a type of mutualistic relationship not mentioned in this lab?

2. Why are lichens a mutualistic relationship and not a parasitic relationship?

3. Describe why lichens can live in areas where no other life exists.

4. Describe the evolutionary advantages of lichens.

ACTIVITY 5 Biodiversity

There is a lot of life out there! But throughout the history of life, much has been lost to extinction. Some lifeforms have been lost through mass extinctions triggered by a variety of causes, including asteroids. In Figure 3.5, the known number of species in each group is reflected by the size of its representative organism. Invertebrates encompass about half of the identified species. In reality, scientists have identified only a very small fraction, 1.5 million species, of the existing species on Earth today—never mind those that have gone extinct!

Which groups are most likely to survive the tests of time? Some organisms are **specialists**, groups that can survive in only a very specific environment with specific needs. Other organisms are **generalists**, those that can survive under a variety of conditions. Generalists are more likely to survive in changing conditions. An important factor in the race for survival is the amount of biodiversity in the area. Biodiversity measures three factors in a given area:

- *Species diversity,* the number of species within an area
- *Population diversity,* the relative amount of each population within an area
- *Genetic diversity,* the variation within the gene pool of each population

Research has proved repeatedly that the more diversity an ecosystem has, the healthier it appears to be. The greater the diversity is, the more likely it is that the ecosystem will bounce back from negative impacts—meaning that the ecosystem will be more **resilient.** Consider this point with respect to **species diversity.** What would happen if a plant species in an area went extinct in response to an environmental strain (less frequent rainfall perhaps) and the extinct plant had been the food source for a consumer? If there were no other species of plant that the consumer could feed on, the consumer might also go extinct. Now consider **population diversity.** If a community is composed mainly of consumers and not producers, consumers can drive the producers to localized extinction and at the same time eliminate a food source. High **genetic diversity** lowers

Figure 3.5
Diversity of Life

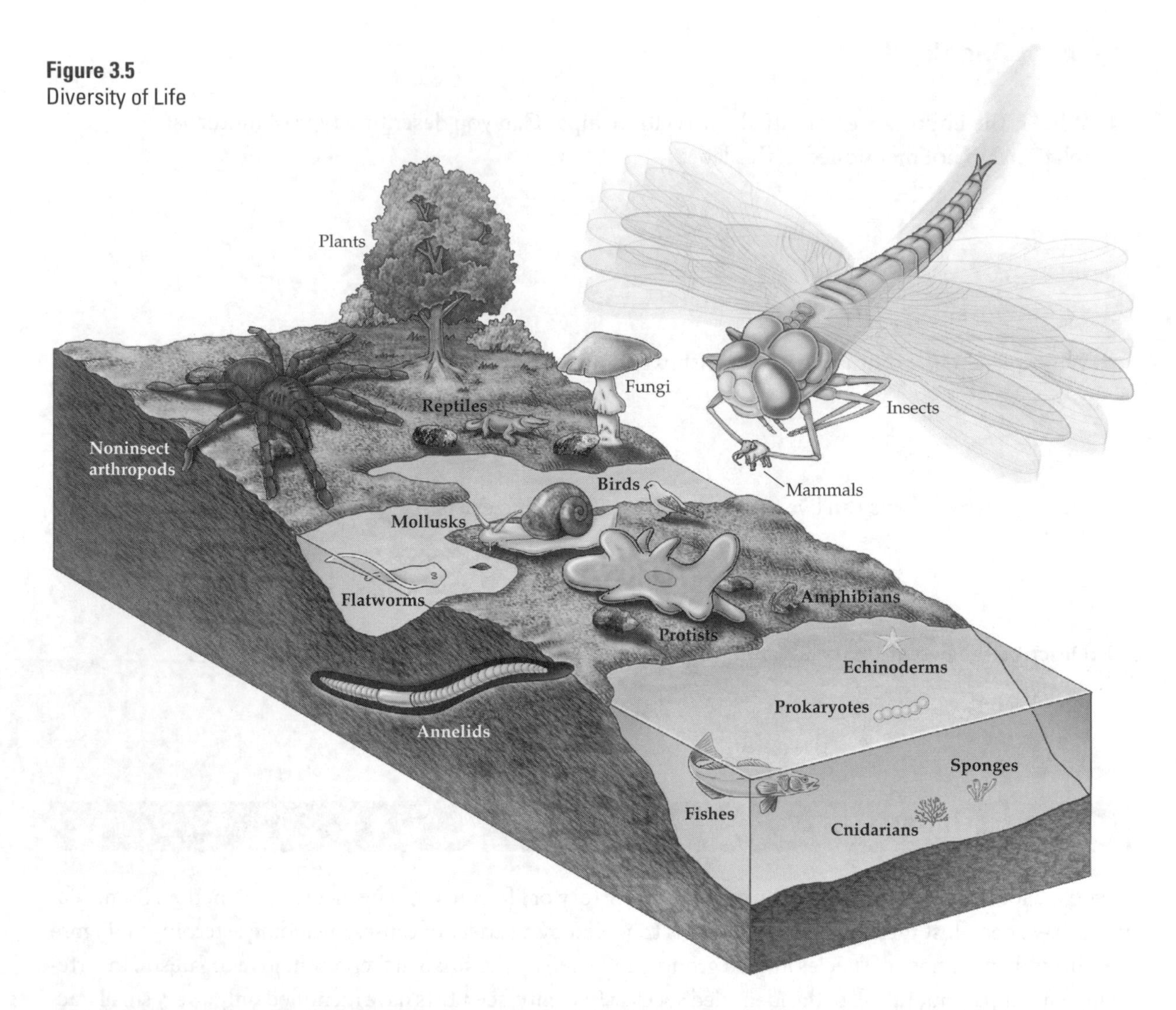

the incidence of disease. When a population that has a high level of genetic diversity is exposed to disease, there is a greater chance that some individuals will survive. For a population with low genetic diversity, all the individuals will have a much smaller chance of survival and therefore disease will have a greater impact.

Greater biodiversity equates to greater **stability** for all of the populations in a community. Imagine you are on safari in the Serengeti of Africa. Lions, being very effective predators, are often thought of as the "kings" of the animal world. They may prey on zebras, wildebeests, various bucks, and giraffes. These different types of prey animals represent species diversity. The number of individuals of each prey animal relative to all the other prey animals represents population diversity. If many prey species (species diversity) are equally distributed throughout the community (population diversity), then predators will have a variety of prey to choose from and will not decimate any single population. This variety increases the survival of both the predator because it has more than one food source, and the prey because it is not the only food source. If the genetic diversity of a particular pride of lions in the preceding example were very low, their ability to be successful as predators might be significantly affected by disease outbreaks.

In this activity you will examine a sample of land to assess biodiversity in a local region chosen by your instructor. Alternatively, this activity may be conducted by each student at home and then discussed at the next class meeting.

1. Obtain a clipboard, measuring sticks, and a field guide for the ecosystem you will examine.
2. At random, your instructor has chosen an area for you to examine. Using the measuring sticks, stand in the middle of your designated area and place your measuring sticks down to form a square on the ground.
3. Start from one corner of your box and work your way to the other corner. Count how many different species of organisms you can clearly see and identify.
4. Record the species in Table 3.2. If possible, identify how many individuals within each population are present in your area and record these numbers in the second column of the table. If your organism is too small to count the exact number of individuals present (as in the case of grasses, for example), make an estimate. For example, count how many grass blades per square inch and then multiply that number by the total area the grass covers. After you have completely surveyed your area, answer the questions in steps 5 through 12.

TABLE 3.2 BIODIVERSITY DATA SHEET

Species	Number of individuals

5. Describe the species diversity of your area.

6. Describe the population diversity of your area.

7. Describe your ecosystem.
   a. What is the time of year?

   b. How would you characterize the abiotic (nonliving) factors of your ecosystem, such as sun exposure, climate, rainfall/humidity, and so on?

   c. Does your area have human influence? If so, describe it.

8. Do you see any specialists? List them here, and explain what makes them specialists.

9. Do you see any generalists? List them here, and explain what makes them generalists.

10. How would you summarize the biodiversity of your area?

11. What factors have the most influence on the biodiversity of your area?

12. Name some pollutants that affect biodiversity.

Concept Check

1. Which are more likely to evolve in a rapidly changing environment—specialists or generalists?

2. How do populations within a community influence the survival, and possibly evolution, of another population?

3. Name three human activities that negatively affect biodiversity.

4. Biodiversity and ecosystem health are being severely tested by a burgeoning human population that encroaches on and transforms natural habitat. Think about how your own hometown has changed since you were a young child. Give two examples of encroachment on habitat in your hometown during your lifetime.

Key Terms

biodiversity (p. 3-1)
cell (p. 3-2)
decomposer (p. 3-8)
detritivore (p. 3-8)
detritus (p. 3-8)
DNA (p. 3-2)
energy (p. 3-3)
evolution (p. 3-1)
evolutionary tree (p. 3-5)
generalist (p. 3-13)
genetic diversity (p. 3-13)
growth and development (p. 3-3)
homeostasis (p. 3-3)
hyphae (p. 3-8)
lichen (p. 3-12)
lineage (p. 3-1)
mutualism (p. 3-10)
mycelium (p. 3-8)
parasitic (p. 3-10)
population (p. 3-2)
population diversity (p. 3-13)
reproduction (p. 3-2)
resilient (p. 3-13)
septum (p. 3-8)
shared derived feature (p. 3-5)
single common ancestor (p. 3-1)
specialist (p. 3-13)
species (p. 3-1)
species diversity (p. 3-13)
spore (p. 3-9)
stability (p. 3-14)
succession (p. 3-12)
vestigial structure (p. 3-1)
yeast (p. 3-10)

Review Questions

1. An organism's management of constant internal conditions through many regulatory channels is ________________.
2. Individuals within a given area that can interbreed are a(n) ________________.
3. The process by which all descendants of the single common ancestor came to be is known as ________________.
4. ________________ are passed down by a common ancestor to descendants in a closely related group.
5. The reciprocally beneficial association between two organisms is called ________________.
6. ________________ are species that can survive in only a very specific environment with specific needs.
7. ________________ are species that can survive under a variety of conditions and are relatively resilient in changing environments.
8. What methods do scientists use to determine whether two organisms may be related evolutionarily?

9. Describe an evolutionary tree and how systematists use shared derived features to determine how groups are related.

10. You encounter two animals. One requires a specific plant to feed on; the other can feed on at least five different plants. If a severe drought developed in the area, which animal would be best suited to survive, and why?

11. Describe the characteristics of life.

12. What are the three components of biodiversity?

13. How does biodiversity positively influence the stability of ecosystems?

14. Describe how bacteria and humans evolved from a common ancestor.

LAB 4 Common Ancestry and Invertebrates

Objectives

- Discuss evolutionary relatedness and recognize shared derived features.
- Observe living and preserved invertebrates.
- Compare the evolution of Porifera, Cnidaria, Platyhelminthes, Nematoda, Mollusca, Annelida, Arthropoda, and Echinodermata.
- Identify the evolutionary advantages of various physical structures for each phylum of invertebrates.

Introduction

Despite the common perception that humankind rules the biological world, it is the invertebrates that call the shots, and humans react. In number of species, insects—with an estimated 1 million species—dominate the world.

Although many invertebrates are harmful to humans in one way or another (there are more rats in New York City than humans, and more cockroaches than rats!), many others are beneficial. They provide food such as lobster, honey, and shrimp; or commercial products such as shells, silk, and pearls. Invertebrates are also important from an evolutionary perspective. Though they exhibit characteristics of all animals (multicellular, chemoheterotrophic lacking cell walls), invertebrates introduce evolutionary innovations that are important to the evolution of the complex bodies of vertebrates, including humans.

All life evolved from a common ancestor, and along the evolutionary timescale different groups branched off from others, forming their own **lineage** or line of descendants. Similar features define evolutionary relationships between groups: a feature that a common ancestor passes down to all of its descendants is a **shared derived feature**, clearly defining the descendants as a group of closely related organisms. Biologists use the term "shared derived feature" to refer to features that originate with the **most recent common ancestor** of a group. In Figure 4.1, each little rectangle between branches represents a shared derived feature that evolved in one group and is then shared by its descendants. For instance, the feature of "four limbs" evolved after the evolution of the fish and led to the evolution of the salamander. Salamanders, lizards, mice, chimpanzees, and humans *share* this *feature derived* from their common ancestor, which existed after fishes diverged from the evolutionary tree.

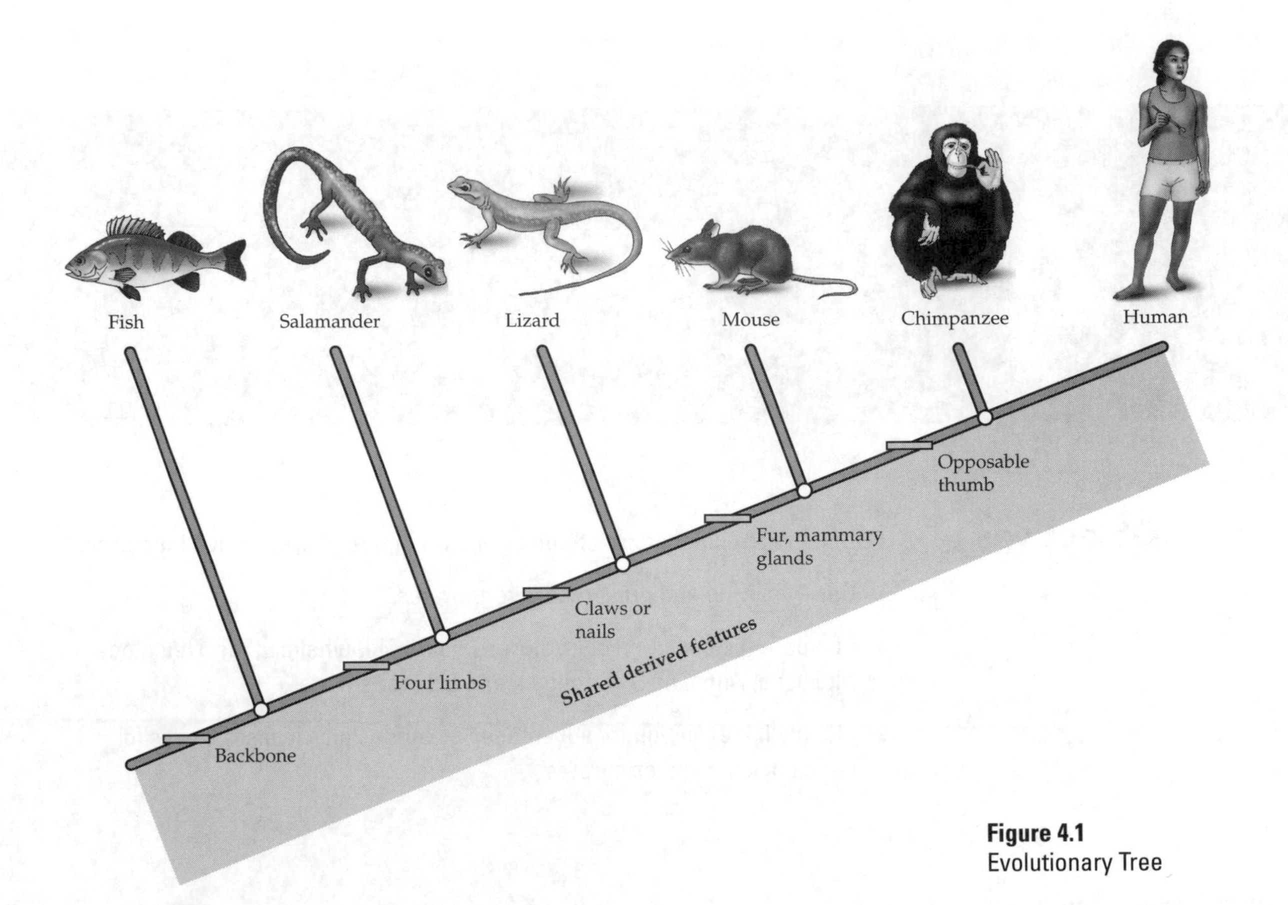

Figure 4.1
Evolutionary Tree

Humans and chimpanzees are set apart from other animals on the tree because they both have the shared derived feature of opposable thumbs. The opposable thumbs likely originated in the *most recent common ancestor* of chimps and humans, which also had opposable thumbs. The group containing chimps and humans has the shared derived trait of opposable thumbs; the group containing lizards, mice, chimps, and humans has the shared derived feature of claws and nails. Scientists draw **evolutionary trees** to help them assess the evolutionary relationships between particular groups. Many important evolutionary innovations (shared derived features) occurred during the group preceding vertebrates: the invertebrates.

Invertebrates all belong to the kingdom Animalia and are then classified more specifically into different phyla. Each phylum represents one or more evolutionary innovations that are not shared by the preceding phylum on an evolutionary tree, but that are inherited from a recent common ancestor that evolved more recently than the preceding phylum. Investigating the evolution of the common ancestors allows us to visualize the evolutionary steps that led to our complex bodies. All life—from prokaryotes to protists to plants to fungi to animals—evolved from a single common ancestor.

In this lab you will take a closer look at the phyla that precede vertebrates on the evolutionary timescale. Each activity will focus particularly on features that are significant from an evolutionary perspective, including shared derived features, in each of the major phyla of invertebrates leading to the evolution of human beings.

ACTIVITY 1 Phylum Porifera

Sponges, the simplest of animals, make up the oldest phylum of invertebrates, Porifera (Figure 4.2). They display the least complex body structure of animals by having no coordinated collection of cells (no tissue). Although sponges are only loose collections of cells, these cells form a collective body and some of the cells are specialized. This loose collection of cells doesn't display any symmetry or any consistent pattern of cells forming the body of an organism. Sponge cells work together to pull water into the body of the organism, filtering out food from the surrounding water.

Like many other invertebrates, sponges have the best of both worlds; they can reproduce asexually or sexually. They reproduce asexually through **budding**, cell division forming an outgrowth of the organism's body to create a completely new individual that eventually falls off of the parent and becomes an independent entity. **Asexual reproduction** creates offspring that are genetically identical to their parents—advantageous because the offspring are just as adapted to the environment as the parent is. Asexual reproduction is often fast, conserves energy, and does not require a mate to be found. The sexual reproduction of poriferans is characterized by the production of egg and sperm. **Sexual reproduction** is much more time- and energy-consuming, but the major benefit is the production of genetically diverse offspring. This advantage is important in unstable environments because it allows the offspring to be potentially better adapted than their parents.

All sponges are aquatic, but most are marine; only a few thrive in freshwater habitats. In this activity you will examine a sponge.

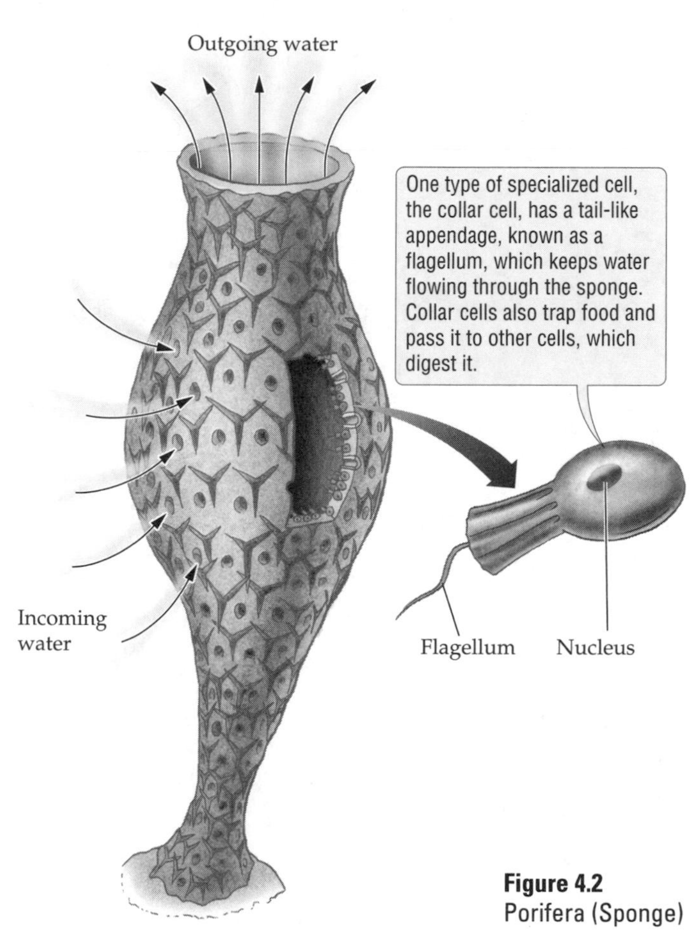

Figure 4.2
Porifera (Sponge)

1. Obtain either a living or a preserved specimen of a sponge, as provided by your instructor, along with a dissecting microscope.

2. Using the microscope, examine the sponge. Do these cells have any defining organization? Note the complex structure of the body, with its intricate crevices. Describe what you see.

3. In the circle below, sketch the sponge.

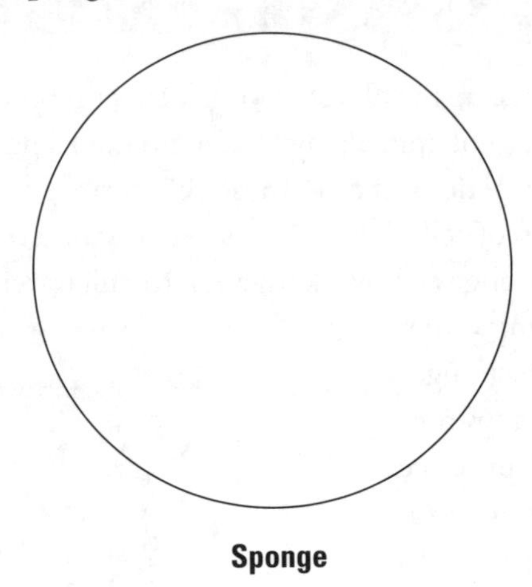

Sponge

Concept Check

1. Describe the advantages and disadvantages of asexual and sexual reproduction.

2. What are the evolutionary advantages of budding?

3. How might it be an advantage for a sponge to have unspecialized cells?

4. Given that poriferans only have specialized cells and no coordinated tissues, what types of organisms do you think sponges eat?

ACTIVITY 2 Phylum Cnidaria

Jellyfish, sea anemones, hydras, and corals are all part of the phylum Cnidaria, the first immediate descendant phylum from Porifera. This group of invertebrates is named after the Greek word for "nettle," a stinging plant found on land. It's an appropriate name, considering that the major feature of this phylum is the presence of stinging cells (Figure 4.3). These cells immobilize prey and protect the animal from predators. The cells of cnidarians are more organized than the cells of sponges. Cnidarians are the first phylum to have tissues, including specialized nervous tissues, musclelike tissues, and digestive tissues. **Tissues** are groups of cells that work together to perform a specialized function.

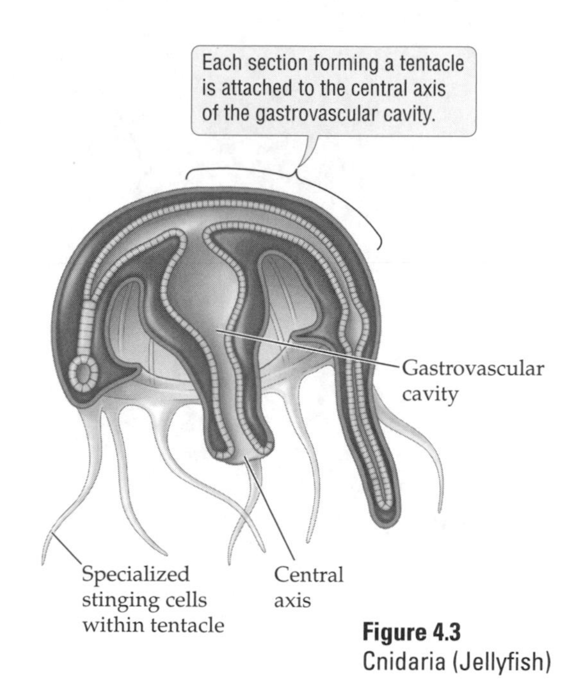

Figure 4.3
Cnidaria (Jellyfish)

Cnidarians are the first phylum to display symmetry in their body structure. The tissues of the cnidarian body are arranged in a circular repetitive pattern around a central axis, like pieces of a pie; this is **radial symmetry** (Figure 4.4). This is advantageous for sessile (nonmotile) organisms because they can obtain food from all around their body and deliver it to the central digestive cavity. The digestive tissues are located in this central cavity, called a **gastrovascular cavity**, a chamber that has one opening for both the mouth and the anus (Figure 4.3). Like sponges, cnidarians can reproduce asexually through budding or sexually. They exhibit two specific life stages: the **polyp** is the immature, stationary stage; the **medusa** is the adult, mobile stage. Ecologically, this phylum is extremely important. Corals lay down a vast network of calcium carbonate deposits, forming coral reefs around the world. These reefs provide a home to 30% to 50% of marine fish species. In this activity you will examine one type of cnidarian, a hydra.

1. Obtain either a living or a preserved specimen of a hydra, as provided by your instructor, along with a concave/depression slide, a pipette, a dissecting microscope, and a compound microscope.

2. Using the pipette, transfer the hydra onto the slide. Using the dissecting microscope, examine the hydra. Describe the organization of the cells.

3. Now transfer the slide to the compound microscope and examine it.

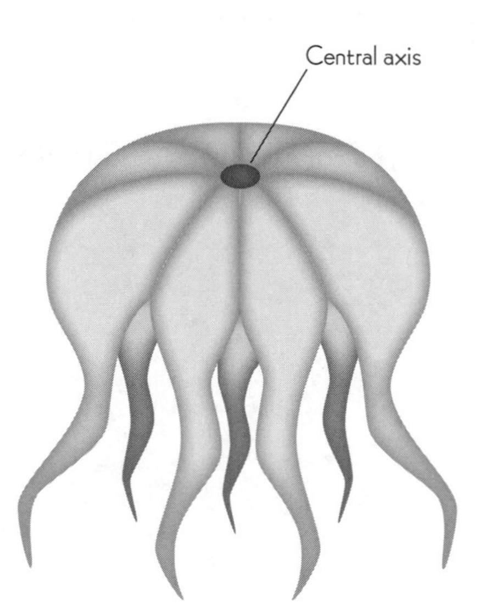

Figure 4.4
Radial Symmetry

4. In the circle below, sketch the hydra. Where are the hydra's digestive tissues are located? What structures contain the stinging cells? Indicate these features on your drawing.

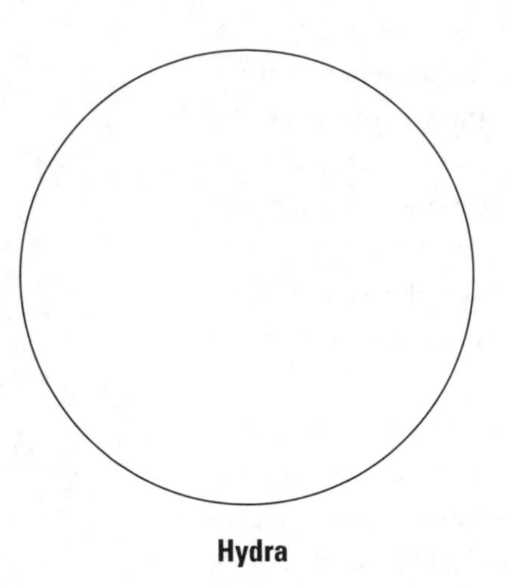

Hydra

Concept Check

1. What is the evolutionary advantage of having a gastrovascular cavity?

2. What new evolutionary features (shared derived features) are observed in these specimens?

3. What features do these specimens and the Porifera have in common?

4. How can you connect the evolution of a sessile stage in Cnidaria and the evolution of stinging cells?

ACTIVITY 3 Phylum Platyhelminthes

A group of simple wormlike animals called flatworms make up the phylum Platyhelminthes. Flatworms are the first phylum to have **organs** (different tissues with a defined boundary that work together to perform a special function), and complex **organ systems** (collections of organs working together to perform multiple complex tasks). The body pattern of flatworms shows bilateral symmetry for the first time (Figure 4.5). In **bilateral symmetry**, the left side of the body is a mirror image of the right side of the body, allowing for more complex organ systems with less energy input during development. Bilateral symmetry leads to the evolution of centralized organ systems, like the brain of the nervous system and the heart of the circulatory system in humans. Flatworms have a gastrovascular cavity and specialized sensory system, including sensory organs called **eyespots**. Light-sensitive cells in the eyespots detect light and send the message to a simple brain.

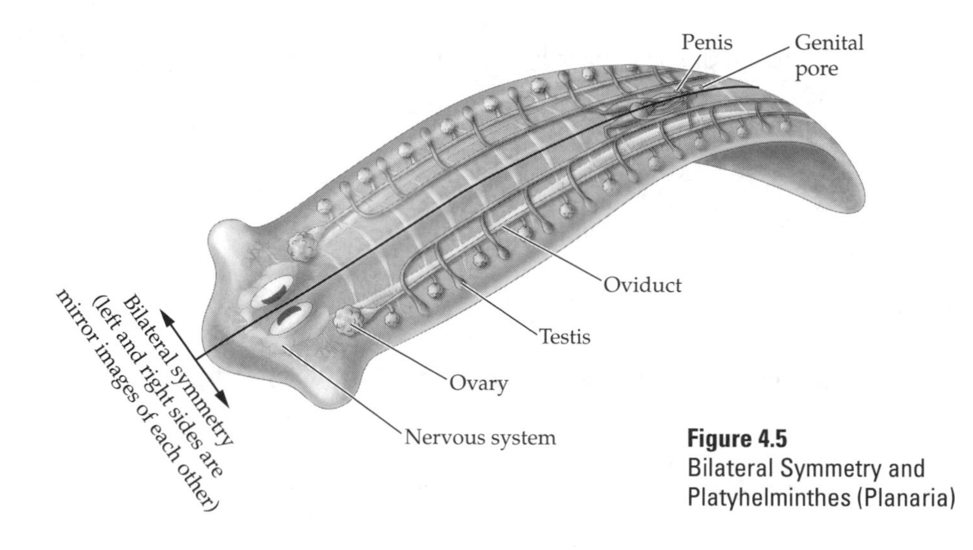

Figure 4.5
Bilateral Symmetry and Platyhelminthes (Planaria)

Many flatworms are **hermaphrodites** (having both male and female sexual organs) and reproduce sexually (Figure 4.5). Some also reproduce asexually, through regeneration, splitting down the center of an individual's body while generating the other half of the body. Flatworms live in a variety of habitats, both aquatic and terrestrial, and they can be parasitic. For instance, flukes cause the disease schistosomiasis, affecting the smooth muscle surrounding the urinary and digestive systems of humans. In this activity you will examine the flatworm *Planaria*.

1. Obtain either a living or a preserved specimen of a planaria, as provided by your instructor, along with a glass petri dish, a pipette, a razor blade, and a dissecting microscope.
2. Using the pipette, transfer the planaria into the petri dish. Using the microscope, examine the planaria. Can you locate the eyespots? Can you identify any internal organs, such as the gastrovascular cavity?

3. In the circle below, sketch the planaria and label the structures that you can identify.

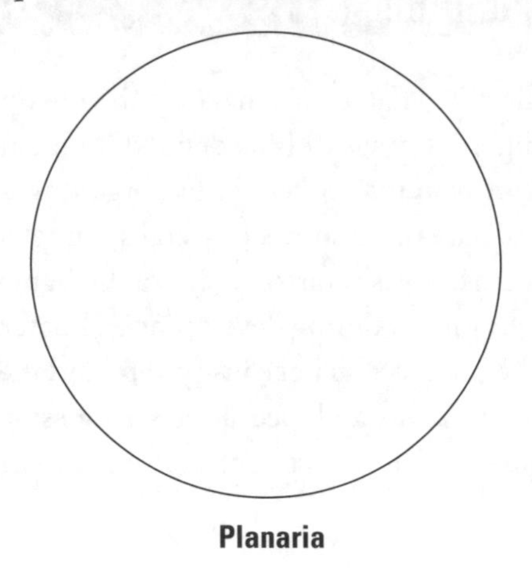

Planaria

4. Now use a razor blade to cut the planaria in half down the length of its body. Does the planaria die? What happens? Why?

Concept Check

1. Define "organ." Describe two organs and their functions within a flatworm.

2. What new evolutionary features (shared derived features) are observed in flatworms?

3. What features do these specimens and the Cnidaria have in common?

4. What is the evolutionary advantage of being hermaphroditic?

ACTIVITY 4 Phylum Nematoda

Nematoda is the next phylum to evolve after Platyhelminthes. Another group of worms, specifically roundworms, make up this phylum. Nematodes come in many varieties. They continue to display bilateral symmetry. The nematodes represent many significant evolutionary innovations. They are the first phylum to display a **complete digestive tract**, a continuous tube with two openings. Along with mollusks, annelids, and arthropods, nematodes develop their digestive tract starting with the mouth and are therefore named **protostomes** (meaning "first opening"). Two later groups, the echinoderms and chordates, are **deuterostomes** ("second opening"), which develop the anus first and then the mouth.

Another new feature seen in nematodes is a **partial body cavity**, also referred to as a **partial coelom** or **pseudocoelom**. A body cavity is an internal space lined with many tissues where organs reside. This space is filled with fluid to help protect and buffer organs from injury. The evolution of this feature also allows organs to develop independently of the body wall of the organism. Nematodes have a partial body cavity, so the fluid-filled cavity does not completely surround the tissues.

Nematodes inhabit all types of environments and range in size. They are the first phylum to have a protective outer layer called a **cuticle**. This structure allows nematodes to live in many environments and has promoted their survival through drastic global changes. Many are parasites of animals and plants. Hookworm is a nematode-caused disease that affects many people and their pets, causing damage to the digestive system and a slow loss of blood. Elephantiasis is a nematode-caused disease that infects the lymphatic system in humans, increasing fluid levels and causing swelling of tissues and thickening of the skin. Nematodes also attack plants, invading their root systems and their leaves, eventually killing the plants. In this activity you will examine a nematode called the vinegar eel.

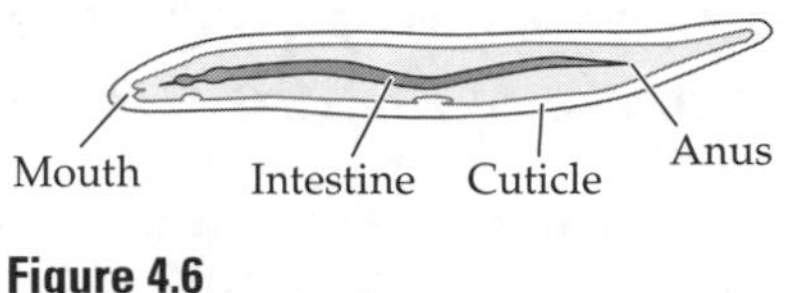

Figure 4.6
Nematoda (*Caenorhabditis elegans*)

1. Obtain either a living or a preserved specimen of vinegar eels, as provided by your instructor, along with a pipette, a concave/depression slide, a coverslip, and a compound microscope.
2. Using the pipette, transfer the vinegar eels onto the slide. Place the coverslip over the sample. Using the microscope, examine the slide. Can you see the outer protective layer?
3. Examine the eels. What is their behavior? Can you identify any structures such as the mouth or digestive tract?
4. In the circle below, sketch a vinegar eel.

Vinegar eel

Concept Check

1. How can nematodes survive in such a variety of environments?

2. What is the difference between a protostome and a deuterostome?

3. What new evolutionary features (shared derived feature) are observed in these specimens?

4. What features do these specimens and the Platyhelminthes have in common?

ACTIVITY 5 Phylum Mollusca

The phylum Mollusca was the first phylum to evolve a **complete body cavity**, or **complete coelom**. Mollusks retain protostome development with a complete digestive tract or an interior tube with two openings: a mouth and an anus. In addition, mollusks evolved an **open circulatory system**, or a group of vessels and a pumping organ to distribute nutrients and oxygen to organs. Typically, mollusks display bilateral symmetry.

The phylum Mollusca contains a hodgepodge of animals, such as **bivalves** (including oysters), **gastropods** (like snails), and **cephalopods** (for instance, squid and octopi). Although this range of animals exhibits a variety of physical characteristics, mollusks evolved three specialized features: **mantle**, **unitary foot**, and **radula**. All mollusks have a structure called the mantle, a fold of skinlike tissue. The mantle secretes a hardened material, normally forming the shell in bivalves and gastropods and the beaks (jaws) of the cephalopods. Aquatic mollusks evolved gills, and their terrestrial counterparts evolved lungs in a cavity of the mantle. Bivalves and gastropods use a unitary foot, a muscular structure, to move across their habitat. Finally, some mollusks feed using a scraping membrane called the radula.

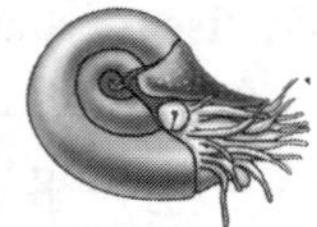

Figure 4.7
Mollusca (Nautilus)

The cephalopods are a unique group of mollusks. These creatures are considered the most intelligent invertebrates and even give some vertebrates a run for their money. They evolved very large brains with complex sensory organs, including very complex eyes. The variation of mollusks illustrates how each group evolved specific structures to deal with its environment. In this activity you will examine three different mollusks.

1. Obtain either living or preserved specimens of one bivalve, one gastropod, and one cephalopod, as provided by your instructor, along with a dissecting microscope.

2. Using the microscope, examine your specimens.

3. Use your observation of these specimens to complete Table 4.1, inferring from what you see the likely habitat and types of prey for each specimen.

TABLE 4.1 COMPARISON OF THREE MOLLUSKS

Specimen	Group in Mollusca	Specialized feature	Habitat	Type of prey

Concept Check

1. Other than the examples that you examined in this activity, give an example of a bivalve, a gastropod, and a cephalopod.

2. What new evolutionary features (shared derived feature) are observed in these specimens?

3. What do these specimens and the previous phylum, Nematoda, have in common?

4. What is the advantage of a complete body cavity?

ACTIVITY 6 Phylum Annelida

Another group of worms, the phylum Annelida, is composed of segmented worms. Annelida is the first phylum to evolve segmentation. **Segmentation** is an important evolutionary innovation providing a subdivision of repetitive units. Repetition of body parts enables autonomy while providing strength and mobility. Annelids evolved a **closed circulatory system**, a system with fluid (blood) that remains within a pumping structure and the vessels. Annelids maintain bilateral symmetry and protostome development. Earthworms are the most obvious annelid. Earthworms are ecologically important because they consume dead or dying organic matter, recycling nutrients back into the environment. In this activity you will examine an annelid.

Figure 4.8
Annelida (Earthworm)

1. Obtain either a living or a preserved specimen of an annelid, as provided by your instructor.
2. In the box below, sketch the annelid and label your drawing according to the specimen. How does an annelid move?

Annelid

Concept Check

1. Describe how the segmentation of an earthworm is beneficial. Can you name another animal that displays segmentation?

2. What new evolutionary features (shared derived features) are observed in these specimens?

3. What features do these specimens and the previous protostome phylum, Mollusca, have in common?

ACTIVITY 7 Phylum Arthropoda

The phylum Arthropoda is the largest, most successful group of animals and contains the most variety. Probably the best-known arthropod group is the insects, which includes grasshoppers, beetles, butterflies, ants, mosquitoes, and many others. Whereas prokaryotes dominate Earth in sheer number of individuals, insects dominate in number of species, having many more than any other group of organisms has (Figure 4.9).

Figure 4.9
Species Diversity of Life

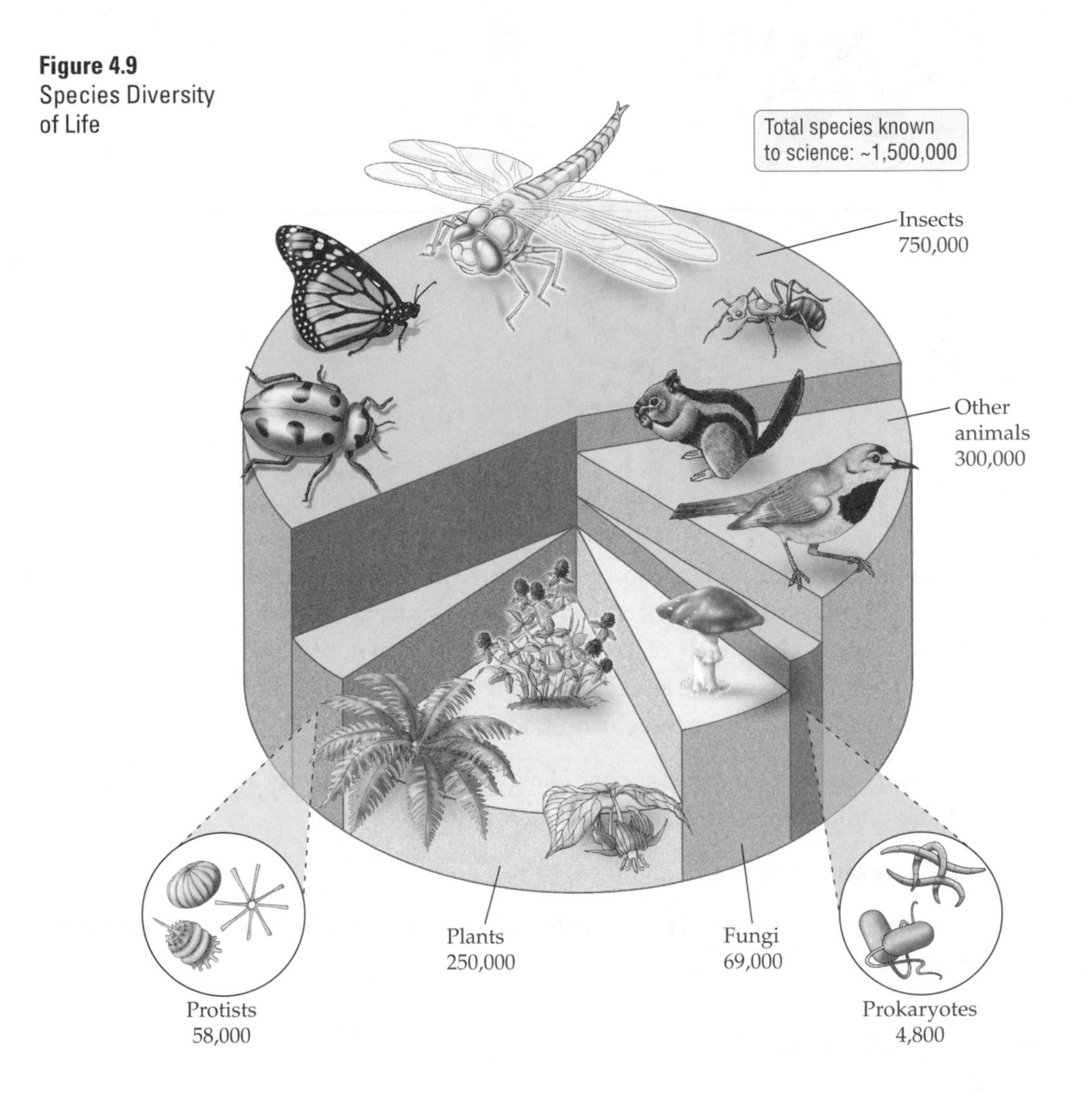

Arthropods still display bilateral symmetry with a complete coelom, complete digestive tract, and body segmentation (head, thorax, and abdomen). Unlike annelids, this group didn't evolve a closed circulatory system and retains an open one. They do have a hard outer skeleton called an **exoskeleton**, as well as **paired jointed appendages**. Arthropods exhibit a wide range of reproductive approaches, both asexual and sexual. Figure 4.10 shows some representative arthropods.

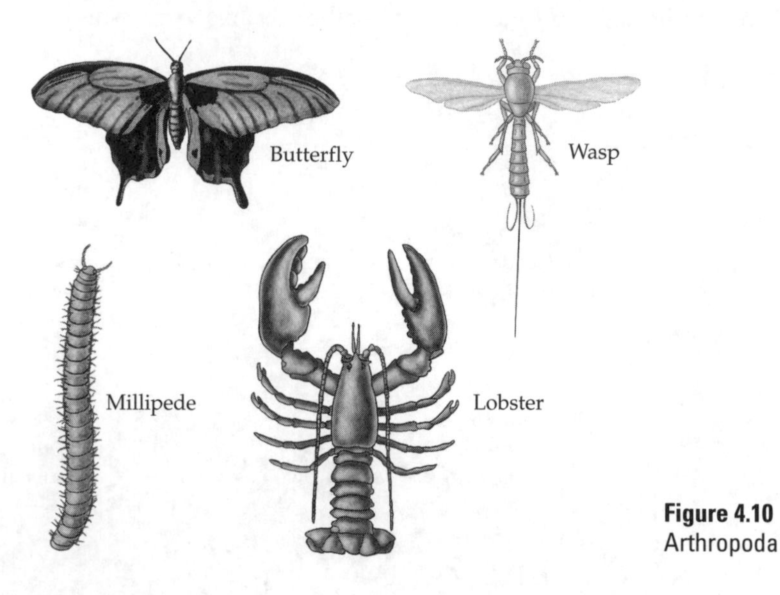

Figure 4.10
Arthropoda

Like the phylum Mollusca, Arthropoda is composed of three main groups: **chelicerates**, **crustaceans**, and **uniramians**. Chelicerates evolved a specialized front pair of appendages called the chelicerae and include spiders and ticks. Crustaceans include a variety of sea-dwelling animals that humans eat as food, including lobsters, shrimps, and crabs. Uniramians include insects, millipedes, and centipedes. In this activity you will examine one representative from each of these three arthropod groups.

1. Obtain either living or preserved specimens of one chelicerate, one crustacean, and one uniramian, as provided by your instructor, along with a dissecting microscope.
2. Using the microscope, examine your specimens.
3. Use your observation of these specimens to complete Table 4.2, inferring from what you see the likely habitat and types of prey for each specimen.

TABLE 4.2 COMPARISON OF THREE ARTHROPODS

Specimen	Group in Arthropoda	Number of appendages	Habitat	Type of prey

4. In Figure 4.10, label the exoskeleton, segmentation, and paired jointed appendages on each organism depicted.

Concept Check

1. What are the three common characteristics of arthropods?

2. What new evolutionary features (shared derived features) are observed in these specimens?

3. What features do these specimens and the previous phylum, Annelida, have in common?

4. What do you think are the advantages and disadvantages of having an exoskeleton?

ACTIVITY 8 Phylum Echinodermata

This final group of invertebrates includes sea urchins, sea stars, and sand dollars. The phylum Echinodermata is the first phylum to evolve deuterostome development, meaning that they produce their anus first, and the mouth forms later. This is the same early developmental pattern that humans undergo, making echinoderms more closely related to humans than to all other invertebrates. (It's hard to believe that a sea urchin is more like a human than like the intelligent octopus or squid!)

On the invertebrate evolutionary tree, protostomes and deuterostomes split after the phylum Platyhelminthes. Protostomes and deuterostomes share the features already present at this point in evolution. Their common ancestor had tissues, organs and organ systems, a complete digestive tract with a body cavity, and bilateral symmetry. Although all echinoderms display radial symmetry as adults, some, including the sea stars (or "starfish," as they are commonly called), are bilateral as embryos and develop into a radial pattern. They do not have brains or a sophisticated nervous system, but they do have a sensory organ called an eyespot. Many have tubelike projections to move around and attack food. Some echinoderms reproduce asexually through regeneration. As in plants, a portion of the body gives rise to a completely new individual. In this activity you will examine an echinoderm.

Figure 4.11 Echinodermata (Sea Star)

1. Obtain either a living or a preserved specimen of an echinoderm, as provided by your instructor.
2. In the box below, sketch the echinoderm and label your drawing according to the specimen. How does an echinoderm move?

Echinoderm

Concept Checks

1. Describe how echinoderms are more like humans than like squid.

2. What new evolutionary features (shared derived features) are observed in these specimens?

3. What features do these specimens and the previous phylum, Platyhelminthes (which represents the protostome/deuterostome evolutionary split), have in common?

ACTIVITY 9 Evolution of Invertebrates

In Activities 1 through 8 you examined many specimens across the kingdom Animalia. Scientists are still determining the relationships of these groups on the basis of their shared derived features. With the advent of new technology, this process has become more complicated. Instead of just using physical characteristics, scientists must combine this information with molecular data to discern **evolutionary relatedness** (a time line of shared ancestry between groups).

Evolutionary trees can be drawn from left to right (as in Figure 4.1), or from top to bottom, depending on which is the simpler way of depicting the relationships that a scientist wants to evaluate. In this activity you will label a tree that is drawn from top to bottom, The oldest groups are toward the trunk of the tree and the newly evolved groups are at the top of the tree. Note that evolutionary trees represent the relationships among groups; after branching from a main evolutionary line, distinct groups continue to evolve.

Some key features found in increasing complexity along the invertebrate evolutionary tree are specialized tissues (cells working together to perform a specialized function), organs (different tissues working together to perform specialized functions), and organ systems (two or more organs working together to perform multiple related tasks). These advances led to the evolution of a complete digestive tract, a complex nervous system, closed circulatory systems, and the urinary systems seen in vertebrates such as humans. Other characteristics, such as radial or bilateral symmetry and a body cavity, are shared among the various groups according to their lineage. Invertebrates provide an evolutionary timescale pinpointing these innovations as they evolved. In this activity you will identify a shared derived feature for each branch along the evolutionary tree of invertebrates.

1. Figure 4.12 shows an evolutionary tree of the phyla described in Activities 1 through 8. Each branch represents a different phylum that diverged from the main group of invertebrates. Use the example in Figure 4.12 of a shared derived feature for all animals (no cell wall) to help guide you through this activity.

2. From the list below, choose one shared derived feature for each branch shown in Figure 4.12. Write the features on the lines provided in the figure.

   Shared derived features:

   - bilateral symmetry
   - complete body cavity
   - complete digestive tract
   - deuterostome

- organ
- protostome
- radial symmetry
- segmentation
- tissue

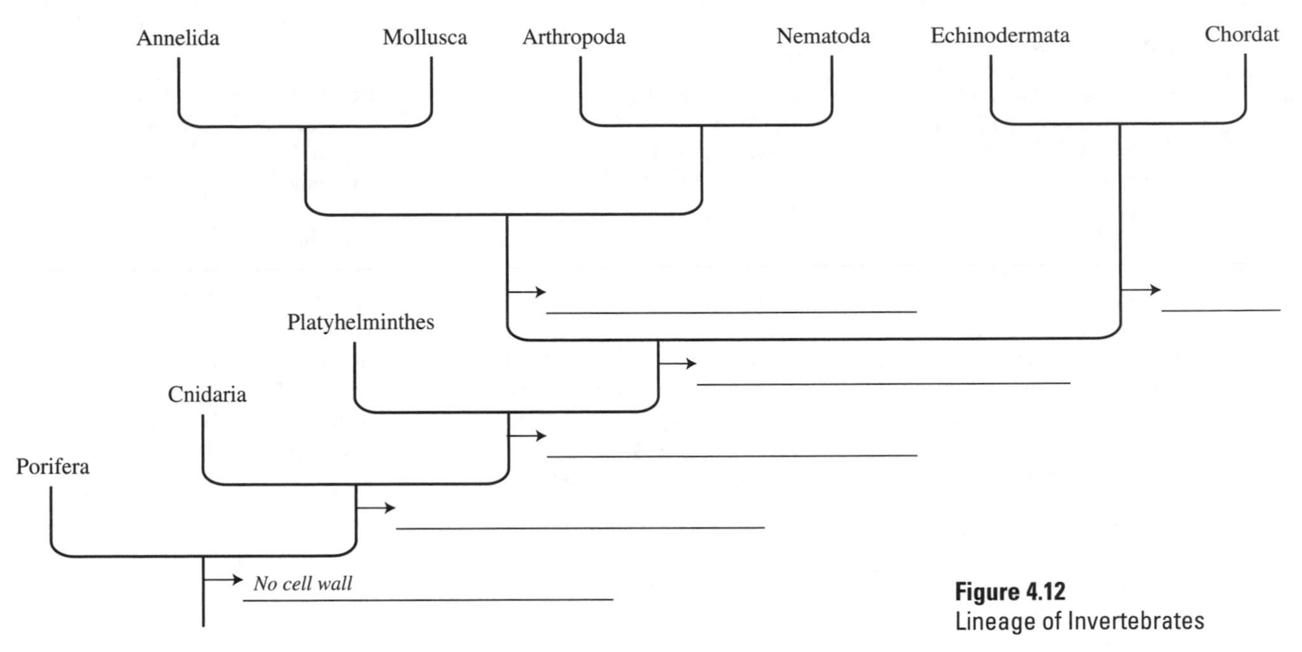

Figure 4.12
Lineage of Invertebrates

3. Some branches have more than one shared derived feature. For each branch, list all of the features that apply. Which branches have more than one shared derived feature?

Concept Check

1. What is a shared derived feature?

2 Define "evolutionary relatedness." How might the knowledge of evolutionary relatedness help scientists to learn more about human biology?

3. What features of invertebrates do humans share?

4. Describe how the branch showing nematodes, mollusks, annelids, and arthropods relates to the larger tree.

Key Terms

asexual reproduction (p. 4-3)
bilateral symmetry (p. 4-7)
bivalve (p. 4-10)
budding (p. 4-3)
cephalopod (p. 4-10)
chelicerate (p. 4-14)
closed circulatory system (p. 4-12)
complete body cavity (p. 4-10)
complete coelom (p. 4-10)
complete digestive tract (p. 4-9)
crustacean (p. 4-14)
cuticle (p. 4-9)
deuterostome (p. 4-9)
evolutionary relatedness (p. 4-16)
evolutionary tree (p. 4-2)
exoskeleton (p. 4-14)
eyespot (p. 4-7)
gastropod (p. 4-10)
gastrovascular cavity (p. 4-5)
hermaphrodite (p. 4-7)
lineage (p. 4-1)
mantle (p. 4-10)
medusa (p. 4-5)
most recent common ancestor (p. 4-1)
open circulatory system (p. 4-10)
organ (p. 4-7)
organ system (p. 4-7)
paired jointed appendages (p. 4-14)
partial body cavity (p. 4-9)
partial coelom (p. 4-9)
polyp (p. 4-5)
protostome (p. 4-9)
pseudocoelom (p. 4-9)
radial symmetry (p. 4-5)
radula (p. 4-10)
segmentation (p. 4-12)
sexual reproduction (p. 4-3)
shared derived feature (p. 4-1)
tissue (p. 4-5)
uniramian (p. 4-14)
unitary foot (p. 4-10)

Review Questions

1. Identify which phyla are deuterostomes and which are protostomes.

2. What types of asexual reproduction are mentioned in this lab? Describe each one.

3. How is segmentation of body structures advantageous? Which phyla display segmentation?

4. Which phyla have complete body cavities, or coeloms? What is the evolutionary advantage of this feature?

5. What is an evolutionary tree?

6. Name three shared characteristics of animals.

7. A coordinated collection of cells is called ________________.

8. ________________ are tissues working together to perform specific functions.

9. ________________ eat dead or dying organic matter.

10. The ________________ stage is the immature, stationary form and the ________________ is the adult, free-moving form of the Cnidaria.

11. Why is it difficult to determine which phylum is the most advanced/sophisticated group?

12. Describe the type of symmetry that each phylum you studied in this activity displays and then list the advantages of each.

13. Which phylum first evolved organs?

14. Which phylum was first to evolve segmentation?

LAB 5 Chemical Building Blocks and Nutrition

Objectives

- Explain the significance of the four macromolecules of life: nucleic acids, carbohydrates, proteins, and lipids.
- Use different tests to detect carbohydrates, proteins, and lipids in a variety of substances.
- Identify the macromolecules present in an unknown food.
- Consider the impact of food processing on the molecules you consume.
- Identify the components of a balanced diet, including vitamins and minerals.
- Interpret nutritional labels of food products to fully understand the components contributing to good health.

Introduction

Nutrients are the chemical building blocks required by an organism to grow and function. Many of the nutrients an organism needs to consume in relatively large amounts are **macromolecules**. These large molecules are formed from the covalent bonding of small **organic molecules** (each of which, by definition, contains at least one carbon atom covalently bonded to one or more hydrogen atoms). A **covalent bond** is the strongest chemical bond in which at least two atoms are linked by electron sharing. Macromolecules are too large to cross the plasma membrane of a cell. Therefore, when you ingest food, these macromolecules must be broken down so that your body can absorb their components.

Four main macromolecules are common to all life-forms: **nucleic acids**, **carbohydrates**, **proteins**, and **lipids**. Some nucleic acids, such as DNA and RNA, store and transport genetic information; others—most important, the molecule ATP—provide energy to cellular reactions. Carbohydrates are used primarily for direct or short-term energy storage; lipids provide long-term energy storage. Proteins assist the body in many different ways; for example, they form physical structures such as collagen in skin, they enable muscle contraction, and as enzymes they hasten chemical reactions vital to the functioning of cells.

TABLE 5.1 SOME IMPORTANT FUNCTIONAL GROUPS FOUND IN ORGANIC MOLECULES

Functional group	Formula	Ball-and-stick model
Amino group	$—NH_2$ —N(H)(H)	Bond to carbon atom; N, H, H
Carboxyl group	—COOH —C(=O)OH	C, O, O, H
Hydroxyl group	—OH	O, H
Phosphate group	$—PO_4$ $O—P(=O)(O^-)—O^-$	P, O, O, O^-, O^-

Macromolecules are often **polymers**, repetitive arrangements of small organic molecules called **monomers**. Covalent bonds link the smaller repeating units of monomers to form polymers. The properties of organic polymers are influenced by attached groups of atoms called **functional groups**. A functional group is a distinct arrangement of specific atoms, some examples of which are depicted in Table 5.1. The number and types of functional groups present in the macromolecule dictate the overall chemical properties and the roles that these molecules play within cells. Functional groups are especially important in protein structure and function.

Food ingested by human beings is composed of different chemical components, not necessarily containing all essential nutrients. For this reason, it is important to understand what you are consuming and the importance of all nutrients on a cellular level. Nucleic acids absorbed within the digestive system are relegated to their respective roles; the other macromolecules are digested, absorbed, and manufactured into macromolecules needed by your body. Vitamins and minerals—dietary nutrients required in smaller amounts—are also essential for the thousands of chemical reactions necessary to life. In this lab you will explore the composition and nutritional value of carbohydrates, lipids, and proteins, and how all nutrients play a role in your diet.

ACTIVITY 1 Carbohydrates

The common term "sugar" describes small carbohydrates such as **glucose**, **fructose**, and **sucrose**. These sweet molecules are very important to all life. Cells break down these sugars to harness the energy stored in the covalent bonds linking their atoms. Carbohydrates are sugars and any polymer of sugars. The smallest component (monomer) of carbohydrates is a **monosaccharide**, which means "one sugar" (Figure 5.1). A sugar formed by linking two monosaccharides is a **disaccharide** ("two sugars"), and any molecule larger than two is a **polysaccharide** ("many sugars"). All sugars are a specific combination of three elements—carbon, hydrogen, and oxygen—whose atoms are always in the ratio of 1:2:1, or CH_2O. The composition of carbohydrates gives this group of molecules its name: *carbo-* refers to the carbon; *hydrate* refers to water, which is made of oxygen and hydrogen.

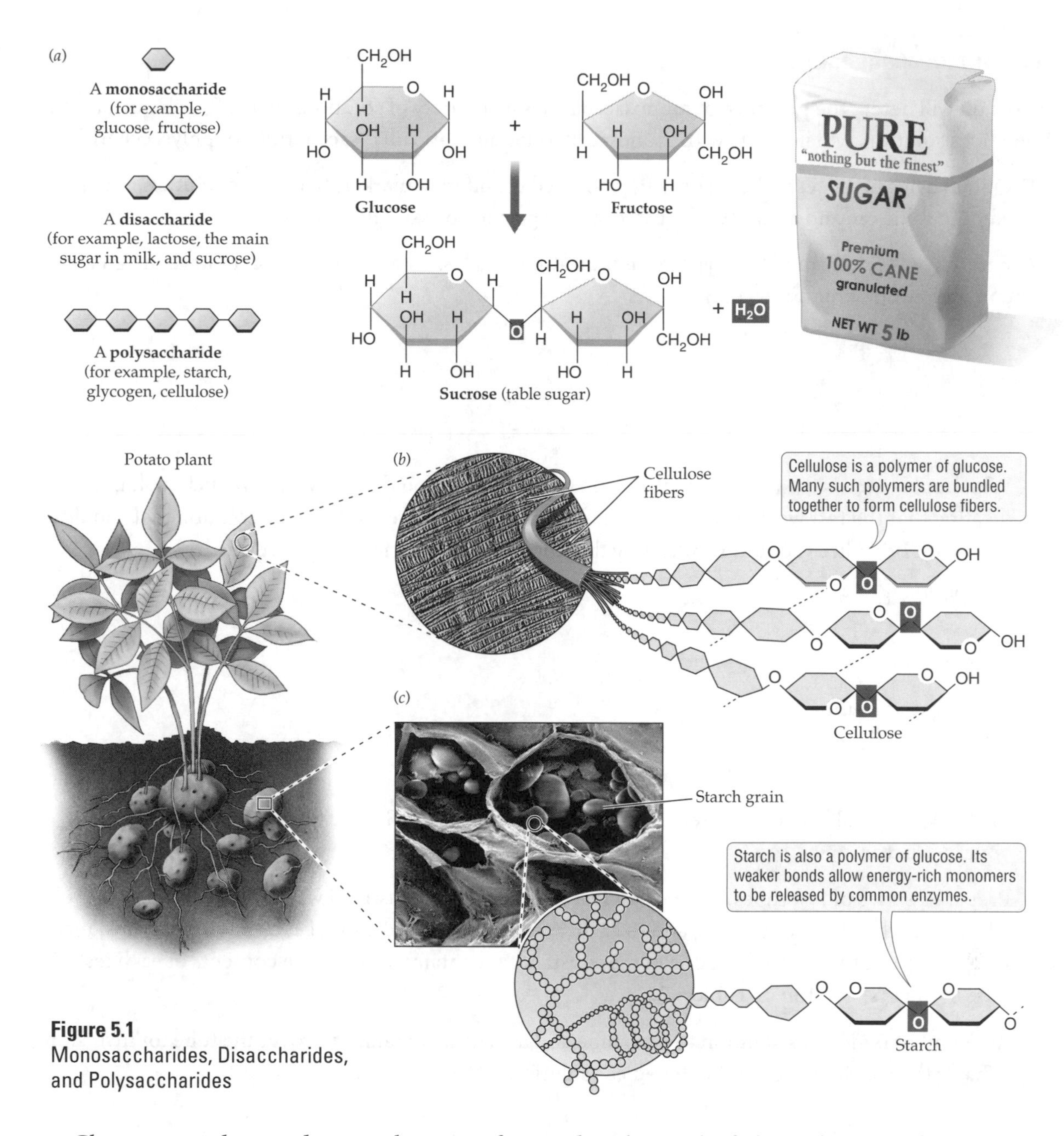

Figure 5.1
Monosaccharides, Disaccharides, and Polysaccharides

Glucose not only provides a quick source of energy but also can be fashioned into a polymer to provide structural support or short-term energy storage. Chitin and cellulose are both polysaccharides of glucose, found in the cell walls of fungi and plants, respectively. In plants, glucose molecules are linked to form the energy storage polymer called **starch**. Animals use **glycogen**, another polymer of glucose, to store energy. Some large polysaccharides, such as chitin and cellulose, cannot be broken down in our digestive system, but they add insoluble fiber to our diet, promoting intestinal health. The body must use energy to break down other polysaccharides (such as starch); the smaller monosaccharides are easily and quickly absorbed and used, or stored.

In this activity you will use two different tests to examine the presence and amount of monosaccharides in different substances and determine whether any of the substances contain the polysaccharide starch.

Test for Monosaccharides

Here you will detect the presence of monosaccharides in a variety of solutions by using a chemical called Benedict's solution, which reacts with monosaccharides but not with disaccharides or polysaccharides.

1. Obtain a hot plate, test tubes, a beaker, graduated cylinders, a marker, Benedict's solution, water, glucose solution, and a variety of substances, as specified by your instructor.
2. Before you start, formulate a hypothesis indicating which substance you expect to have the greatest concentration of monosaccharides.

   Hypothesis:

   Which substance is the positive control (the solution that should react with Benedict's solution, indicating that the test works) and which substance is the negative control (the solution that shouldn't react with Benedict's solution, indicating that the test is specific for monosaccharides).

   Positive control:

   Negative control:

3. Fill the beaker halfway with water and place it on the hot plate. Set the temperature as directed by your instructor.
4. Prepare a series of test tubes. To the first test tube add 1 ml of distilled water, to the second test tube add 1 ml of the glucose solution, and to each successive tube add 1 ml of each additional substance. Label each test tube with the name of the substance it contains. Record the contents of each test tube in the first column of Table 5.2.
5. Add 5 ml of Benedict's solution to each tube and mix, using Parafilm to cover the tube for inversion. What is the starting color of the Benedict's solution?
6. Place the test tubes in the beaker of water. Boil the tubes for 3 minutes to allow the Benedict's solution to react with the sugars.
7. The color indicates the amount of product—with green representing the least and red representing the most. In Table 5.2 (under "Monosaccharide test results"), rate your product on a scale of 0 to 4, depending on its final color:

   a. Blue = 0
   b. Green = 1
   c. Yellow = 2
   d. Orange = 3
   e. Red = 4

8. Which fluid contains the highest number of monosaccharides?

9. Which fluid contains the lowest number of monosaccharides?

10. Which fluid would you drink for a source of quick energy?

TABLE 5.2 ACTIVITY 1: CARBOHYDRATE TEST RESULTS

Test tube #	Name of substance	Monosaccharide test results	Starch test results (positive or negative)
1			
2			
3			
4			
5			
6			

Test for Starch

Now you will use an iodine solution to detect the presence of starch, a polysaccharide.

1. Obtain iodine, a dropper, and the substances provided by your instructor. What color is the iodine?

2. Before you start, formulate a hypothesis indicating which substance you expect to have the greatest concentration of starch.

   Hypothesis:

   Which substance is the positive control (the solution that should react with iodine, indicating the presence of starch) and which substance is the negative control (the solution that shouldn't react with iodine).

   Positive control:

   Negative control:

3. For each substance, place one drop of iodine on the item. If the drop turns black, the test is positive for starch. Record the results of the test in Table 5.2, under "Starch test results."

4. Which substances contain starch, a polysaccharide?

5. Which substances did you find to contain monosaccharides earlier in the activity?

6. Does it make a difference whether you ingest a monosaccharide or a polysaccharide? If so, how?

Concept Check

1. What type of solution contains sugar? What foods that you eat are high in carbohydrates?

2. Explain the general composition of carbohydrates. What nutritional role do carbohydrates play in humans?

3. What is the function of starch in plants? Why is starch a valuable nutrient in humans? What other molecule in humans is similar to starch?

ACTIVITY 2 Proteins

Proteins have many functions within the cell. They do pretty much everything from forming physical structures, to carrying oxygen to cells, to (as enzymes) speeding up reactions. The monomers of proteins are **amino acids**. All life uses amino acids to build proteins. Twenty different amino acids exist. For humans, only 12 of these can be manufactured within the body; the remaining 8 are considered **essential**, meaning that they need to be acquired from food. Like the 26 letters of the alphabet, the 20 different amino acids allow a vast array of possible arrangements of amino acids to produce millions of different proteins.

All amino acids are similar in composition, having a central carbon atom. Each carbon atom can form four bonds. In an amino acid, one of these bonds is to a hydrogen atom, a second bond is to an amino functional group (—NH_2), a third is to a carboxyl functional group (—COOH), and a fourth is to an **R group**, which determines the distinct chemical properties of each amino acid (Figure 5.2). These R groups vary in size, shape, and elemental composition. Amino acids covalently bonded to one another form large chains called **polypeptides** (Figure 5.3). Polypeptides range in size from hundreds to thousands of amino acids.

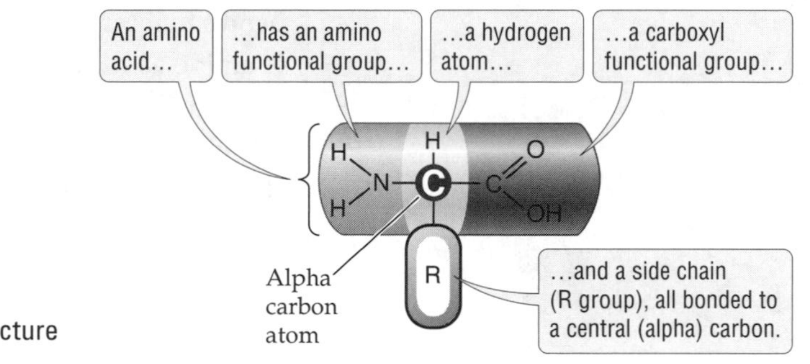

Figure 5.2
Amino Acid Structure

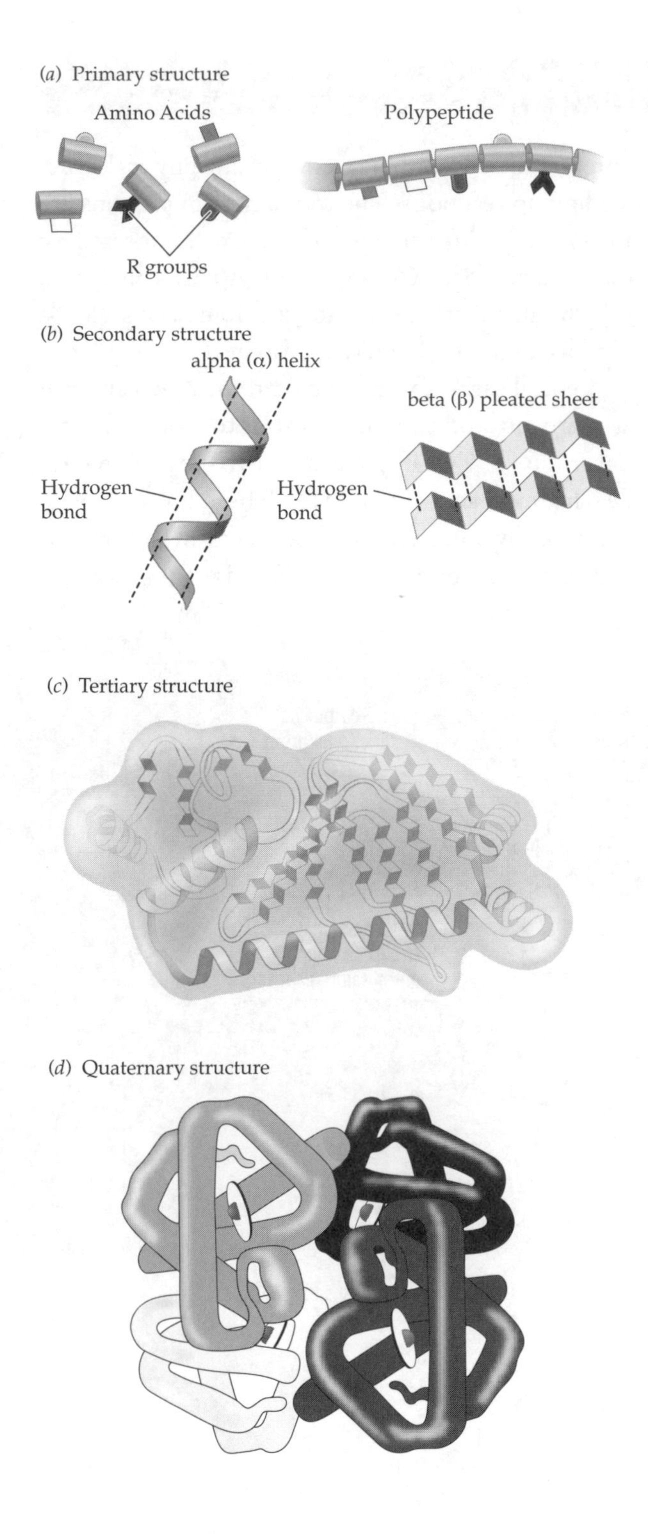

The structure of a protein is determined first by the sequence of amino acids, known as the **primary structure** (Figure 5.3). The polypeptide represents the primary structure of a protein. The amino acids within a polypeptide interact and can attract one another through a noncovalent bond called a **hydrogen bond**. This type of bond forms when a hydrogen atom with a partial positive charge interacts with a nearby atom with a partial negative charge. The formation of hydrogen bonds helps the polypeptide to fold into a specific pattern, such as an alpha helix or a beta pleated sheet, which is referred to as the protein's **secondary structure**. A polypeptide must be folded into a **tertiary structure**, or stable three-dimensional state, before it can function as a protein. The overall tertiary structure depends on interactions/bonds between R groups of amino acids. Certain conditions, such as high heat and extreme pH, break these bonds, unfolding the protein. **Denaturation** is the destruction of protein structure that results in loss of protein function. Many different proteins are formed by more than one polypeptide chain bound together, giving this type of protein another level of organization, called the **quaternary structure**.

In this activity you will test for the presence of proteins and determine conditions that cause proteins to become denatured.

Figure 5.3
The Four Levels of Protein Structure

Presence of Proteins

Here you will determine whether proteins are present in various different substances.

1. Obtain test tubes, a marker, Parafilm, Biuret solution, and a variety of substances provided by your instructor.
2. Before you start, formulate a hypothesis indicating which substance you expect to have the greatest concentration of proteins.

   Hypothesis:

   Which substance is the positive control (the solution that should react with Biuret solution, indicating the presence of protein), and which substance is the negative control (the solution that shouldn't react with Biuret solution).

   Positive control:

   Negative control:

3. Label one test tube for each substance you are instructed to test. Record the name of each substance in the first column of Table 5.3.
4. To each test tube, add 1 ml of the appropriate substance.
5. What is the starting color of the Biuret solution? Add 2 ml of Biuret solution to each test tube. Cover the tube with Parafilm and immediately invert to mix the contents.
6. If protein is present, the Biuret solution changes to a deep violet color. For each test tube, record any color change in Table 5.3, under "Results."

TABLE 5.3 ACTIVITY 2: TESTING FOR THE PRESENCE OF PROTEINS

Test tube #	Name of substance	Results (positive or negative for protein)
1		
2		
3		
4		
5		
6		

7. Which substances contain proteins?

8. Which substance would you consume to get the greatest amount of protein? The least amount?

Denaturation of Proteins

Now you will test to see what happens to proteins at different pH levels and at high temperature.

1. Obtain four eggs, four beakers, a marker, a hot plate, four stirring rods, gloves, and pH 3 and 10 solutions.
2. Label each beaker for the condition you're going to test:

   a. Beaker 1: normal conditions (plain distilled water)
   b. Beaker 2: low pH (pH 3)
   c. Beaker 3: high pH (pH 10)
   d. Beaker 4: high temperature

3. Before you start, formulate a hypothesis indicating which condition you expect to have the greatest effect on the egg.

   Hypothesis:

ALWAYS HANDLE EGGS CAREFULLY, BECAUSE THEY MIGHT BE CONTAMINATED WITH *SALMONELLA*. WASH YOUR HANDS AND GLASSWARE WITH SOAP AND WATER AFTER YOU HANDLE RAW EGGS. TO PREVENT CONTAMINATION, USE GLOVES WHILE HANDLING THE EGGS.

4. Crack open each egg and separate the egg white from the yolk. Put the yolk into the beaker designated by your instructor for yolk disposal. Pour one egg white into each of the four beakers.
5. Add 100 ml of distilled water to beakers 1 and 4. To beaker 2, add 100 ml of pH 3 solution; to beaker 3, add 100 ml of pH 10 solution.
6. Mix each solution with a separate stirring rod. Place beaker 4 on the hot plate and heat the solution to 100°C.

7. Once the temperature has reached 100°C, turn off the hot plate and analyze the differences among the egg whites in each condition. Describe the color and consistency of the solutions in the "Results" column of Table 5.4.

TABLE 5.4 ACTIVITY 2: TESTING THE EFFECTS OF pH AND TEMPERATURE ON PROTEINS

Condition tested	Results
Beaker 1: normal conditions	
Beaker 2: pH 3 (low)	
Beaker 3: pH 10 (high)	
Beaker 4: high temperature	

8. Which condition causes the most change?

9. What is changing in the solutions? How are the structures of proteins affected by different conditions?

10. How do the effects observed in this activity relate to how your digestive system functions to break down macromolecules?

Concept Check

1. Describe what happens when a protein is denatured. What conditions lead to denaturation?

2. What functions do proteins have within cells? How can their functions be inhibited?

3. What foods that you eat are high in proteins?

ACTIVITY 3 Lipids

Lipids are unlike the other three types of macromolecules in that they are **hydrophobic** (repelling water), while the other macromolecules are **hydrophilic** (attracting water). This characteristic is crucial to the many functional roles of lipids in a cell, such as providing a structural barrier that separates the inside of the cell from the environment outside. Lipids are three different groups of molecules: **fatty acids, phospholipids**, and **sterols**. In this activity we will focus on fatty acids—primarily long chains of carbon atoms. These carbon chains are bonded to hydrogen atoms, creating molecules called **hydrocarbons**, which can be arranged in many patterns, such as rings and chains. A hydrocarbon chain binds to a carboxyl group to form a fatty acid.

The type of fatty acid is determined by its length and the bonds within it, as illustrated in Figure 5.4. Fatty acids containing only single bonds between all of the carbon atoms are **saturated** fatty acids, which are straight hydrocarbon chains. This structural characteristic allows these molecules to pack together tightly, forming semisolids or solids at room temperature. If a fatty acid contains any double bond between two or more carbon atoms in the hydrocarbon chain, the chain will bend and is then called **unsaturated**. The presence of one or more double bonds reduces the number of hydrogen atoms in the molecule. Unsaturated fatty acids are bent and cannot lie tightly together in a solution at room temperature. Unsaturated fatty acids are typically produced in plants; animals form saturated fatty acids. There are exceptions to both generalizations. **Trans fatty acids**, commonly referred to as "trans fats," are a human invention—straight fatty acids created by the modification of unsaturated fats. In the nutritional labels on food products, the term "partially hydrogenated" indicates the presence of trans fats. Research shows that this form of fat is extremely unhealthy, increasing the risk of heart disease.

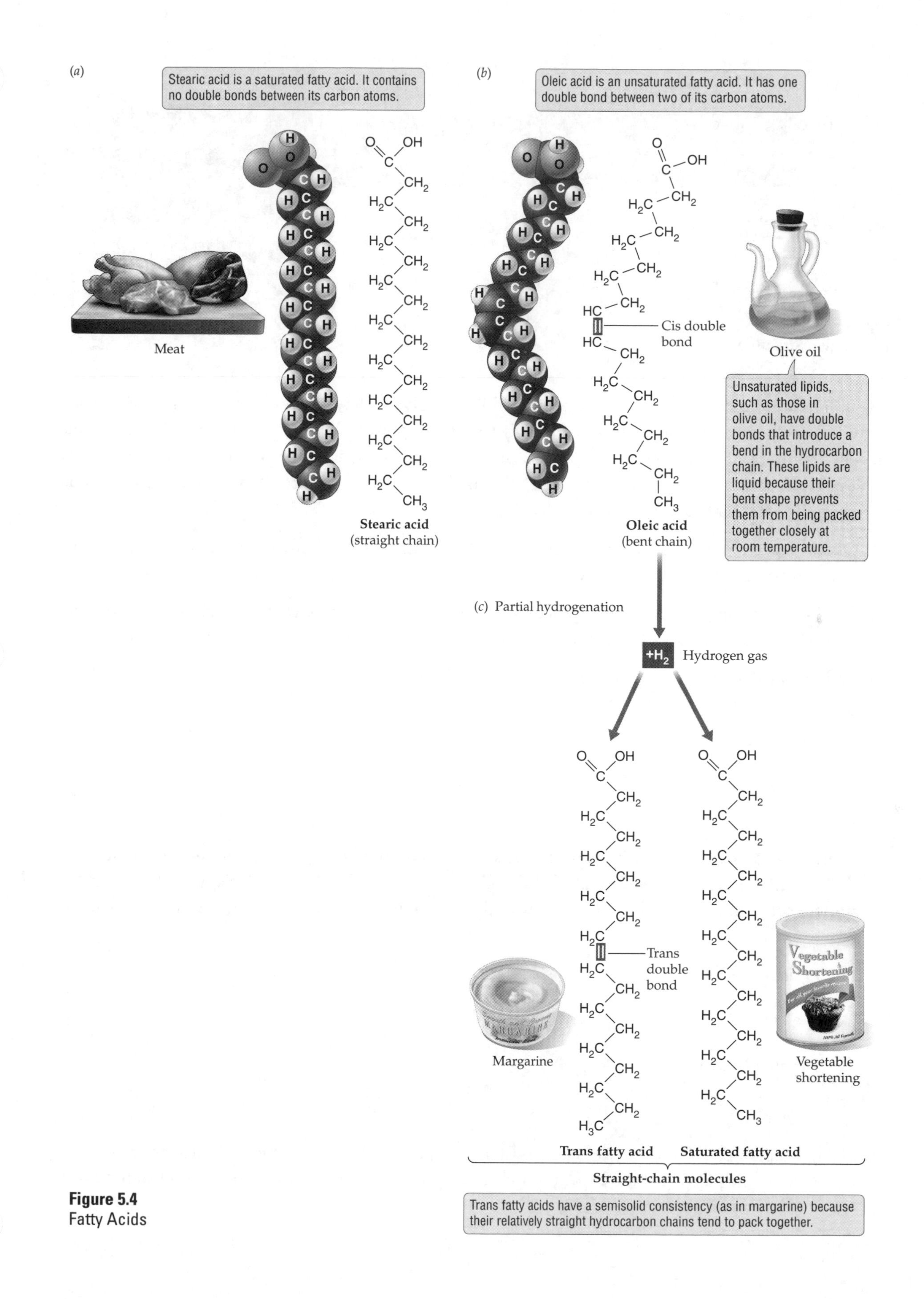

Figure 5.4
Fatty Acids

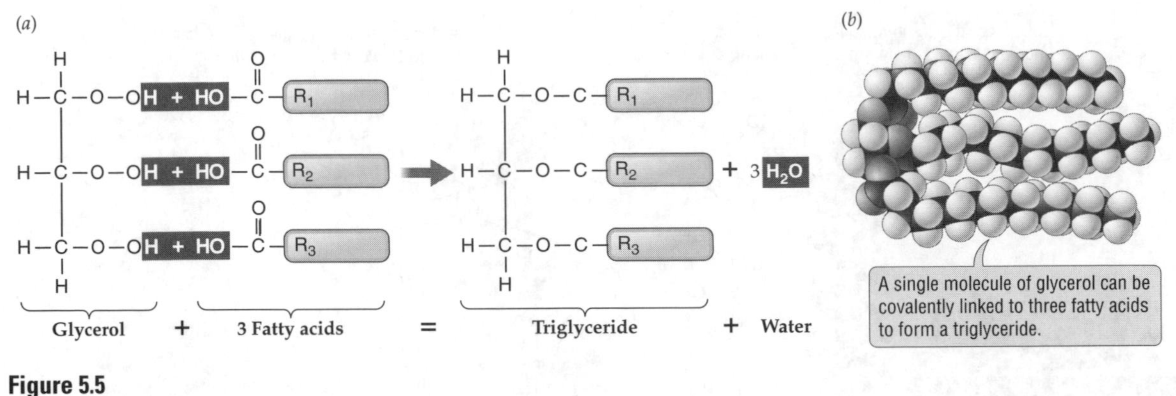

Figure 5.5
Triglyceride Structure

In humans, lipids are taken up and fashioned into **triglycerides**, compounds with three fatty acids bound to a glycerol molecule (Figure 5.5). Triglycerides are used as a long-term storage molecule in your fat cells. It is more efficient to store energy as fats instead of as carbohydrates or proteins because each gram of fat provides twice as much energy as these other molecules do. In this activity you will examine foods to determine whether they contain fats.

1. Obtain pipettes, a pencil, a circular stencil, filter (Whatman) paper, Sudan stain, and a variety of substances provided by your instructor.
2. On the Whatman paper, use the pencil to trace the outer edge of the circular stencil. Trace one circle for each of the substances. Label the circles for each substance that you are instructed to test.
3. Before you start the test, formulate a hypothesis indicating which substance you expect to have the greatest concentration of lipids.

   Hypothesis:

   Which substance is the positive control (the solution that should appear oily and react with the Sudan stain, indicating that the test works) and which substance is the negative control (the solution that shouldn't appear oily and react with the Sudan stain, indicating that the test is specific for fats).

   Positive control:

   Negative control:

4. Using a separate, clean pipette for each substance, apply one drop of the substance to the appropriate circle on the paper. Let the paper absorb the substance. Record the name of each substance in the first column of Table 5.5.

5. Examine the spots on your paper. Which ones look oily (appear shiny and don't dry) like the positive control? Record the results for each substance—either positive or negative for lipids—in Table 5.5, under "Whatman paper test results."

TABLE 5.5 ACTIVITY 3: LIPID TEST RESULTS

Name of substance	Whatman paper test results: positive or negative for lipids	Sudan stain test results: positive or negative for lipids

6. Use a stain to confirm your results. Sudan stain turns red when a fat is present or remains pink if no fat is present. Apply one drop of stain to each circle on your paper and wait 5 minutes.
7. Place the paper in distilled water for 2 minutes to rinse the excess stain from the spots.
8. Examine your positive and negative controls. What colors are they? Record these results in Table 5.5, under "Sudan stain test results."
9. Do the Sudan stain test results corroborate the paper test results?

10. Which substances, if any, differ in results between the two tests? What do you think causes these differences?

Concept Check

1. What is the difference between saturated and unsaturated fatty acids?

2. Which types of food are more likely to contain saturated fatty acids? Why?

3. Why are fats used as a long-term storage molecule?

4. What foods that you eat are high in unsaturated fats? saturated fats? trans fats?

ACTIVITY 4 Analyzing an Unknown

The common proverb "You are what you eat" holds a lot of weight and truth. With the advent of so many food-processing technologies, it is hard for people to truly understand what they are consuming. The more food is processed, the less it resembles the organisms that contribute to its composition. This activity will allow you to apply what you have learned about testing for macromolecules to analyze a processed food. Because most foods are solids when consumed, you may need to use a blender to macerate the tissue to release the macromolecules into water. In this activity you will perform the carbohydrate, protein, and lipid tests that you learned about in Activities 1 through 3.

1. Obtain a sample of an unknown food item provided by your instructor.
2. Test the food item for the presence of monosaccharides, starches, proteins, and lipids. For each test, write a summary of the procedure you used to detect the specific molecule and record your results in Table 5.6 for monosaccharides, Table 5.7 for starches, Table 5.8 for proteins, and Table 5.9 for lipids. Use your results to help you analyze the effect of food processing on the food item you're testing.

TABLE 5.6 ACTIVITY 4: MONOSACCHARIDE TEST
Summary of monosaccharide test:
Results of monosaccharide test:

TABLE 5.7 ACTIVITY 4: STARCH TEST

Summary of starch test:

Results of starch test:

TABLE 5.8 ACTIVITY 4: PROTEIN TEST

Summary of protein test:

Results of protein test:

TABLE 5.9 ACTIVITY 4: LIPID TEST

Summary of lipid test:

Results of lipid test:

3. Does your food item contain all types of macromolecules?

4. What organisms contributed to your food product?

5. Describe how your food product resembles its original source (chemical composition and physical characteristics).

Concept Check

1. How does food processing destroy or alter the organism that makes up the original food product?

2. Do processed foods contain a variety of nutritional ingredients? Do the amount and types of nutritional ingredients correlate to the degree of food processing?

ACTIVITY 5 Nutrition

The goal of **digestion** is to break down nutrients into a form that cells and the circulatory system can take up. Macromolecules are not the only nutrients that you ingest; **vitamins** and **minerals** are also necessary to life, in much smaller doses. Vitamins are small organic nutrients that regulate metabolic pathways. They come in two forms: water-soluble and fat-soluble (Table 5.10). Overconsumption of water-soluble vitamins is not a problem, because the body excretes these into the urine; but fat-soluble vitamins can be locked up in fat stores, and at very high levels they lead to medical problems. Minerals, the last group of essential nutrients, are solid substances often composed of one **element** (a pure substance made up of only one type of atom with a characteristic number of protons), or composed of two or more elements bonded

TABLE 5.10 VITAMINS NEEDED IN THE HUMAN DIET

Class	Vitamin	Main functions	Possible symptoms of deficiency and excess	Dietary sources
Water soluble	B vitamins: thiamin (B_1), riboflavin (B_2), niacin (nicotinamide), pyridoxine (B_6), pantothenic acid, folate (folic acid), cyanocobalamin(B_{12}), biotin	Act with enzymes to speed metabolic reactions, or act as raw materials for chemicals that do so. Work with enzymes to promote necessary biochemical reactions.	Deficiency: B vitamins act in concert; deficiency in one can cause symptoms related to deficiency in others. Deficiency diseases include pellagra and beriberi (damage to heart and muscles). *Excess: B_6 in excess can cause neurological damage.*	Folic acid, a B vitamin, is abundant in green vegetables, legumes, and whole grains. B_{12}, a B vitamin, is scarce in plant foods but abundant in milk, meat, fish, and poultry.
	Vitamin C (ascorbic acid)[a]	Assists in the maintenance of teeth, bones, and other tissues.	Deficiency: scurvy (teeth and bones degenerate), increased susceptibility to infection. *Excess: diarrhea, kidney stones with chronic overuse.*	Vitamin C is abundant in many fruits (e.g., kiwi, strawberry, citrus) and in many vegetables (e.g., bell peppers, broccoli, spinach).
Fat-soluble	Vitamin A (carotene)	Produces visual pigment needed for good eyesight; also used in making bone.	Deficiency: poor night vision; dry skin and hair. *Excess: nausea, vomiting, fragile bones.*	Carotene is responsible for the color of yellow and orange fruits and vegetables. It is converted to vitamin A within our bodies.

NOTE: The human body cannot make these essential vitamins, or else makes them in insufficient amounts, so it must get what it needs from food.

[a] Vitamin C is a vitamin only for primates (including humans) and a few other animals (such as guinea pigs, bats, some birds, and some fishes). Most other animals can make vitamin C as needed.

TABLE 5.10 VITAMINS NEEDED IN THE HUMAN DIET (continued)

Class	Vitamin	Main functions	Possible symptoms of deficiency and excess	Dietary sources
Fat soluble	Vitamin D	Promotes calcium absorption and bone formation.	Deficiency: poor formation of bones and teeth, irritability. *Excess: diarrhea and fatigue.*	Fish is the richest source of vitamin D; shellfish and egg yolks provide smaller quantities; fortified foods (such as milk, soy milk, and breakfast cereals) are important sources for most people.
	Vitamin E	Protects lipids in cell membranes and other cell components.	Deficiency: very rare. *Excess: heart problems.*	Vitamin E is abundant in nuts, vegetable oils, whole grains, and egg yolk.
	Vitamin K	Produces clotting agent in the blood.	Deficiency: prolonged bleeding, slow wound healing. *Excess: liver damage.*	Leafy green vegetables and some fruits (avocado, kiwi) are rich in vitamin K, which is also manufactured by intestinal bacteria.

to one another. Unlike macromolecules, many minerals are not formed by covalent bonds; rather they are bound by ionic bonds. In an **ionic bond**, one atom donates one or more electrons to another atom, creating positively and negatively charged ions. Minerals have a variety of roles within the body. Iron deficiency leads to anemia or low oxygen content in the bloodstream. Both deficiencies and overdoses of vitamins and minerals can lead to medical problems.

While the monomers of nucleic acids (nucleotides) are directly used to make other nucleic acids, the components of carbohydrates, fats, and proteins can provide cells and the body with energy. Fats and carbohydrates are the long- and short-term sources of energy for your body. In times of desperation, the body will degrade its own tissue formed of proteins to acquire energy only after depleting all of its stores of fats and carbohydrates. Remember that, gram for gram, fats provide more than twice the energy of proteins or carbohydrates. Biologists often describe the amount of energy provided by food in units called **kilocalories** (kcal), as in Table 5.11. Nutritional labels on food packages identify kilocalories as "Calories" (written with a capital "C," as in Figure 5.6).

TABLE 5.11 ENERGY AND BUILDING BLOCKS THAT NUTRIENTS PROVIDE TO ANIMALS

Nutrient	Absorbable units	Energy content of 1 gram (kcal)	Major use
Carbohydrates	Monosaccharides	4	Energy, building other macromolecules and cell structures
Fats	Fatty acids, monoglycerides	9	Energy storage, building other macromolecules and cell structures, especially cell membranes
Proteins	Amino acids, small peptides	4	Building other proteins, other organic molecules such as signaling substances

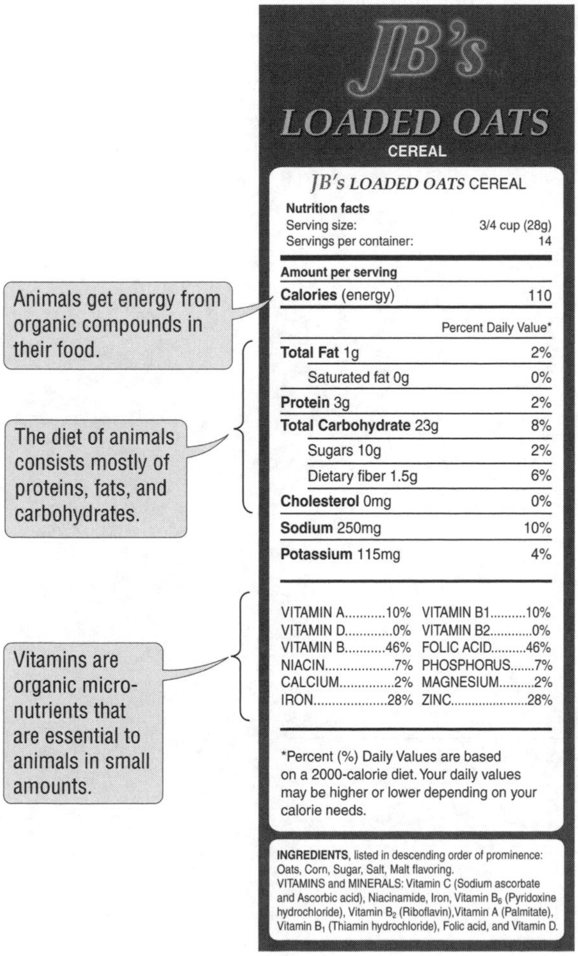

Figure 5.6
Sample Nutritional Label

There is a lot of flexibility in consuming a balanced diet, but problems do arise when not enough or the not the right type of food is consumed. In **undernourishment**, a person does not consume enough food to provide energy to the body. In **malnourishment**, a person does not consume the required types of molecules necessary for the body to function properly. A person can be **overnourished**, consuming more food than what is necessary to power the body, which is therefore stored as fat, but still be malnourished.

This activity will help you understand the food you eat. Food labels are often confusing, preventing most people from completely comprehending the impact of the food they ingest. In this activity you will be a food critic, but your critique will be based not on food taste or presentation, but on the nutritional value of the food.

Basic Nutritional Content

Here you will evaluate the basic characteristics and nutritional content of your food product.

1. Obtain a packaged food item containing the nutritional label dictated by the Food and Drug Administration (FDA).
2. What type of food product is your food item?
3. How many servings does the package contain?
4. What is the total size (in grams) of your food item?
5. Table 5.12 lists major components of food. Calculate the total amount of each component in your food item by multiplying the amount per serving size by the number of servings in the package. Enter each total in the table.

TABLE 5.12 ACTIVITY 5: NUTRITIONAL INFORMATION IN A SAMPLE FOOD ITEM

Component	Total amount
Calories (C)	
Calories from fat (C)	
Total fat (g)	
Saturated fat (g)	
Trans fat (g)	
Unsaturated fat (g)	
Sodium (mg)	
Potassium (mg)	
Total carbohydrates (g)	
Dietary fiber (g)	
Sugars (g)	
Protein (g)	

Lipids

Here you will evaluate the lipid content of your food product.

1. Calories are recommended in specific quantities from the different sources. Fat calories are only 30% of the recommended 2,000-calorie diet.

   a. Calculate the number of calories recommended from fat.

      $0.30 \times 2{,}000$ calories =

   b. What amount, in grams of fat, is this, if there are 9.9 calories in a gram?

2. Now put your food item in perspective.

   a. Calculate the number of fat calories you will consume if you eat the whole food product.

   b. Divide this number (the amount you have consumed) by the number of recommended calories you should receive from fat on a daily basis.

   c. Now multiply by 100 to get the percentage of fat calories that you would consume versus the amount recommended.

   d. Given this information based solely on fat, would this be a nutritional food product?

3. Does your ingredient list contain any partially hydrogenated oils or fats? These are actually trans fats. Does your food item claim to be free of trans fats?

Proteins

Here you will evaluate the protein content of your food product.

1. If you consumed the whole package of your food item, what percentage of recommended proteins would you be consuming? (An easy way to calculate this is to multiply the percentage of recommended daily allowances listed next to the amount of protein and multiply it by the number of servings in the package.)

2. Does all food contain the essential amino acids that humans need?

3. Can you think of any medical condition that requires limiting the amount of a specific protein consumed?

Carbohydrates

Here you will evaluate the carbohydrate content of your food product.

1. If you consumed the whole package of your food item, what percentage of recommended carbohydrates would you be consuming? (An easy way to calculate this is to multiply the percentage of daily allowances listed next to the amount of carbohydrates and multiply it by the number of servings in the package.)

2. Can you think of any medical condition that requires regulating the amount of carbohydrates consumed?

3. Why do athletes "carb up" before a big competition?

Vitamins and Minerals

Here you will evaluate the vitamin and mineral content of your food product.

1. List the water-soluble vitamins in your food item. For each vitamin you listed, is the amount in your food item more or less than the amount recommended by the FDA?

2. List the fat-soluble vitamins in your food item. For each vitamin you listed, is the amount in your food item more or less than the amount recommended by the FDA?

3. From your personal knowledge, what impact does sodium have on the human body? Can too much sodium in your diet disrupt the function of your body?

4. What other nutritional information would you consider while evaluating this food product?

5. Does the food item's packaging contain any slogans that attract a consumer to think the food item is nutritional? If so, do you think the advertisement is accurate?

Concept Check

1. If you were given only the grams of fat, protein, and carbohydrates in a food product, how could you determine the calories contributed by each?

2. What would you do to make nutritional labels clearer?

3. Explain the importance of a balanced diet.

4. What advertising slogans on food products do you find confusing, and why?

Key Terms

amino acid (p. 5-7)
carbohydrate (p. 5-1)
covalent bond (p. 5-1)
denaturation (p. 5-8)
digestion (p. 5-19)
disaccharide (p. 5-2)
essential (p. 5-7)
element (p. 5-19)
fatty acid (p. 5-12)
fructose (p. 5-2)
functional group (p. 5-2)
glucose (p. 5-2)
glycogen (p. 5-3)
hydrocarbon (p. 5-12)
hydrogen bond (p. 5-8)
hydrophilic (p. 5-12)
hydrophobic (p. 5-12)
ionic bond (p. 5-20)
kilocalorie (kcal) (p. 5-21)
lipid (p. 5-1)
macromolecule (p. 5-1)
malnourishment (p. 5-21)
mineral (p. 5-19)
monomer (p. 5-2)
monosaccharide (p. 5-2)
nucleic acid (p. 5-1)
nutrient (p. 5-1)
organic molecule (p. 5-1)
overnourishment (p. 5-21)
phospholipid (p. 5-12)
polymer (p. 5-2)
polypeptide (p. 5-7)
polysaccharide (p. 5-2)
primary structure (p. 5-8)
protein (p. 5-1)
quaternary structure (p. 5-8)
R group (p. 5-7)
saturated (p. 5-12)
secondary structure (p. 5-8)
starch (p. 5-3)
sterol (p. 5-12)
sucrose (p. 5-2)
tertiary structure (p. 5-8)
trans fatty acid (p. 5-12)
triglyceride (p. 5-14)
undernourishment (p. 5-21)
unsaturated (p. 5-12)
vitamin (p. 5-19)

Review Questions

1. Fill in the missing information in Table 5.13.

TABLE 5.13 REVIEW QUESTION 1

Type of macromolecule	Nucleic acids	Carbohydrates	Proteins	Lipids
Smaller organic components				
Name of one specific macromolecule				

2. Define "nutrient." What makes some nutrients essential?

3. ___________________ is the condition in which a person does not consume enough food to provide energy to the body.

4. ___________________ is the condition in which a person does not consume the required types of molecules necessary for the body to function properly.

5. How can a person be overnourished and malnourished at the same time?

6. How can just 20 amino acids generate the thousands of different proteins found in nature?

7. How many amino acids are essential?

8. ________________ is the lipid that humans use to store energy; ________________ is the carbohydrate that humans use to store energy.

9. ________________ is the most common type of monosaccharide.

10. Macromolecules are often composed of small repeating components called ________________.

11. The properties of organic molecules depend on ________________, clusters of atoms that impart specific chemical properties regardless of which molecule they are found in.

12. ________________ are fatty acids containing only single bonds between carbon atoms; ________________ are fatty acids containing one or more double bonds between carbon atoms.

13. How are proteins important to cell function?

14. Describe the four levels of organization of a protein.

15. What is the role of nucleic acids in your diet?

16. Name a medical condition caused or exacerbated by malnutrition.

LAB 6 Cell Structure, Internal Compartments, and Evolution

Objectives

- Describe the structural and functional differences between prokaryotic and eukaryotic cells.
- Examine and depict the variety of cells in all major domains and kingdoms.
- Identify cell structures and comprehend the evolutionary significance of membrane-enclosed organelles.
- Trace the cellular evolution of prokaryotes and eukaryotes.

Introduction

The **cell** remains the smallest and most basic unit of life. Cells can be independent organisms or just one of a trillion cells working together within a large organism. At the most general categorization, organisms are grouped according to cellular structure. **Prokaryotes** are simpler, smaller cells; **eukaryotes** are much larger and contain a more complex cellular structure. The domains **Bacteria** (including the familiar disease-causing bacteria) and **Archaea** (bacteria-like organisms that are best known for living in extreme environments) are prokaryotes. Prokaryotes are ancestors to the domain **Eukarya,** which is made up of the kingdoms **Protista** (a diverse group that includes amoebas and algae), **Plantae** (plants), **Fungi** (yeasts, mushrooms, and molds), and **Animalia** (animals). The evolutionary relatedness of these kingdoms is complex but leads to a greater understanding of their cell and organismal structure. In this lab you will explore prokaryotic cells and a representative of eukaryotic cells from each kingdom, noting the vast variety of structures and functions.

ACTIVITY 1 The First to Arrive: Prokaryotes

The term "prokaryote" (meaning "before the kernel") refers to the lack of a nucleus, the internal compartment housing the DNA of eukaryotic cells. The first domain emerging in the history of life is the Bacteria. Bacteria are ubiquitous, having intimate relationships—sometimes helpful (as a symbiont), sometimes harmful (as a disease-causing pathogen)—with almost all other forms of life. Bacteria are found in the soil, where they play an important role in decomposition and recycling essential nutrients. They exist in both fresh and salt water. They play an important role in the food industry, giving cheese, yogurt, and other foods their characteristic flavors and consistencies. The Archaea arose after the bacteria and live in very harsh environments, such as boiling-hot geysers, acidic waters, and hot sulfur vents.

Prokaryotes are single-celled and small, have little DNA with no nucleus, undergo a form of asexual reproduction called fission, and do not have membrane-enclosed organelles (Figure 6.1). The following structures are common to all prokaryotes:

- *Cell wall.* The outermost protective barrier.
- *Plasma membrane.* A selective gateway, letting in substances useful to a cell's growth and functioning, and keeping out substances that are harmful.
- *Cytosol.* The fluid within the plasma membrane, composed mainly of water.
- *Nucleoid region.* The area where DNA resides.
- *Ribosomes.* The organelles that create proteins.

Using these definitions, label the prokaryotic cell in Figure 6.1.

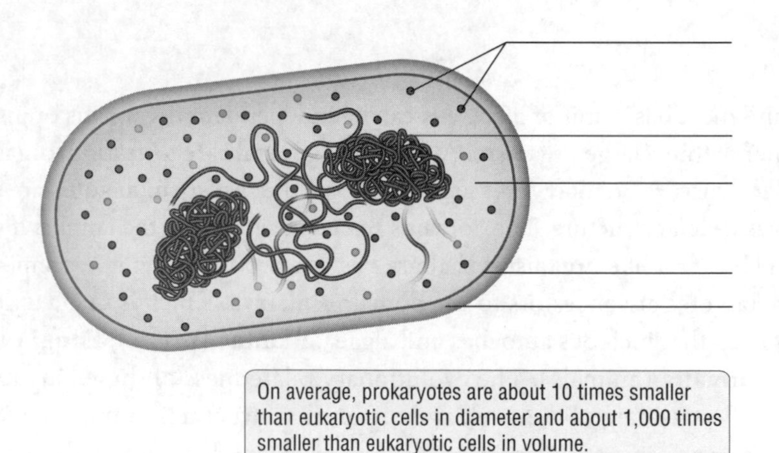

On average, prokaryotes are about 10 times smaller than eukaryotic cells in diameter and about 1,000 times smaller than eukaryotic cells in volume.

Figure 6.1
Prokaryotic Cell

Concept Check

1. Which domain is characterized by living in extreme environments?

2. Name four defining characteristics of prokaryotes.

3. What are two structures that provide protection for prokaryotes from the outside world?

ACTIVITY 2 Examining Prokaryotic Cells

Prokaryotes are quite variable in shape, including the following:

- *Spheres (called cocci [kock-eye], singular "coccus").* If the spheres remain attached to each other, they are called diplococci (pairs), streptococci (chains), or staphylococci (clusters).
- *Rods (called bacilli [buh-sill-eye], singular "bacillus").*
- *Corkscrews (called spirochetes [spy-roh-keets]).*

In this activity you will examine these bacterial shapes.

1. Look at the prepared slide of mixed bacteria. Be sure you can identify all three bacterial shapes.
2. In the circles below, sketch a sample of the bacteria—one of each shape—at the highest magnification.

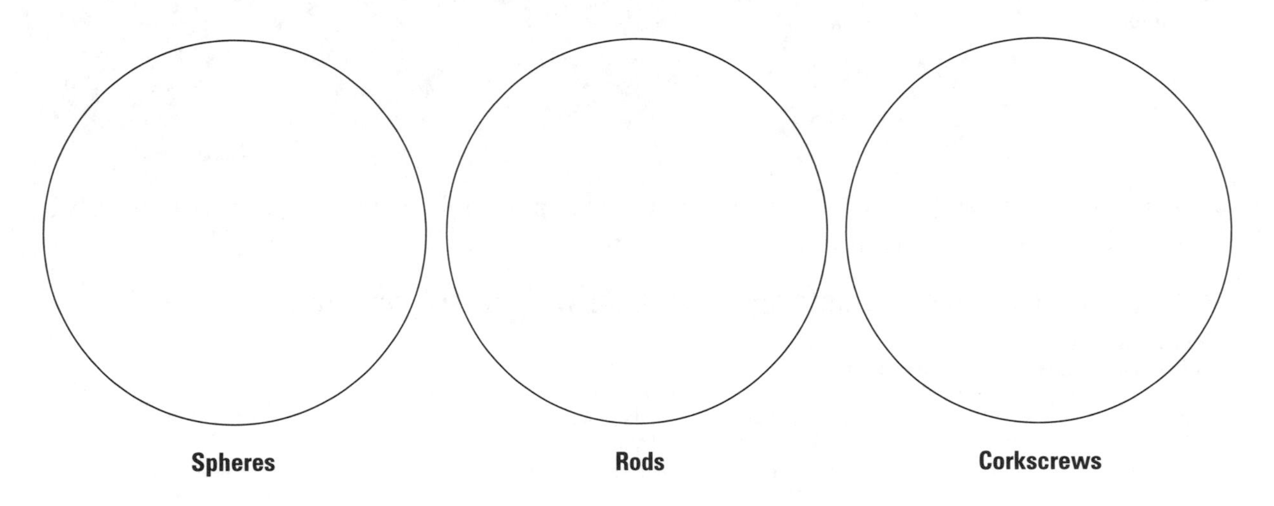

Spheres **Rods** **Corkscrews**

Concept Check

1. Name and describe the three main shapes of prokaryotes.

2. Why would prokaryotes evolve such a variety of shapes?

3. Why can prokaryotes survive in extreme conditions?

ACTIVITY 3 Eukaryotic Cells

Eukaryotic cells are as different as types of hairstyles, and they contain many types of specialized compartments called **organelles**. "Organelle" is an appropriate name because these internal compartments are like "little organs"; just as the heart, lungs, and other organs have unique functions in the human body, each organelle has specific duties in the life of the cell. Scientists believe that membrane-enclosed organelles evolved when primitive predatory eukaryotic cells engulfed prokaryotes. The engulfed cells then became specialized compartments within the larger cell. This evolutionary theory is called **endosymbiosis** (Figure 6.2).

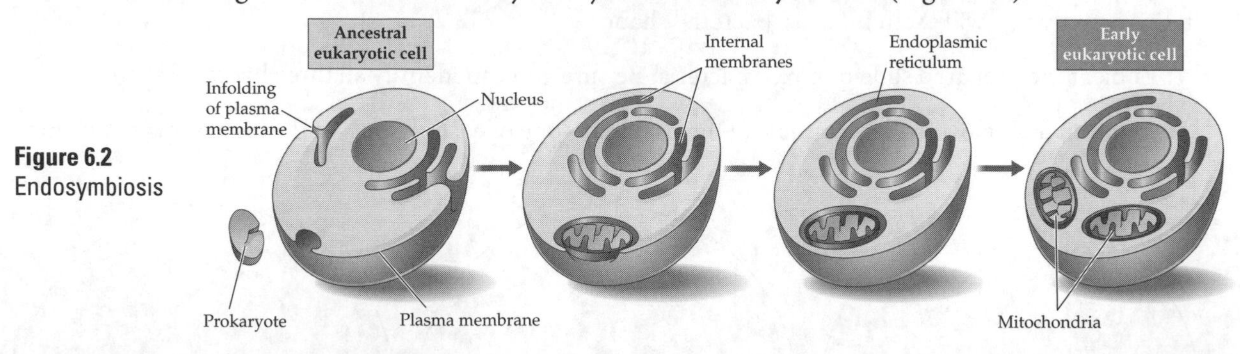

Figure 6.2 Endosymbiosis

Table 6.1 lists general and specialized structures and compartments found in eukaryotic cells. Using Figure 6.3 as a reference, match each structure to its function.

TABLE 6.1 EUKARYOTIC STRUCTURES AND COMPARTMENTS AND THEIR FUNCTIONS

Common structures and compartments	Function
A. Plasma membrane	____ Lipid-based outer boundary that functions as the gatekeeper of the cell. (Barrier of the cell dotted by proteins and other molecules.)
B. Cytosol	____ Structure that maintains and changes cell shape; acts as the rail system of the cell. (Scaffold of the cell.)
C. Ribosome	____ Membrane-enclosed organelle that houses DNA and the nucleolus. (Largest organelle; typically round.)
D. Nucleus	____ Membranous organelle that directs lipids and proteins made in the ER to their final destination. (A stack of membranous pockets.)
E. Rough endoplasmic reticulum (rough ER)	____ Small compartment that moves substances around the cytoplasm and into and out of the cell. (A transport bubble.)
F. Smooth endoplasmic reticulum (smooth ER)	____ Double-membrane organelle that produces energy by breaking down storage molecules. (An oval structure with a mazelike interior.)
G. Vesicle	____ Small organelle that is an essential component in protein manufacturing. (Small and round; found throughout the cell and on the rough endoplasmic reticulum.)
H. Golgi apparatus	____ Membrane network that extends from the nucleus, is studded with ribosomes, and functions to produce proteins. (An extension of the membrane surrounding the nucleus that has a rough surface.)
I. Mitochondrion	____ Membranous organelle that extends from the nucleus and is involved in lipid production and detoxification. (An extension of the membrane surrounding the nucleus that has a flat surface.)
J. Cytoskeleton	____ Water-based fluid that is composed of a multitude of free ions and molecules. (The liquid of the cell.)

Specialized structures and compartments	Function
K. Lysosome	____ Organelle that recycles macromolecules in animals. (Round structure containing enzymes.)
L. Central vacuole	____ Organelle that converts light energy into chemical energy in the form of sugar in plant cells. (Green structure with many compartments.)
M. Chloroplast	____ Organelle that handles recycling and maintains water balance in plants. (Large organelle that pushes other organelles toward the outer edge of the cell.)
N. Cell wall	____ Protective barrier that provides a rigid structure for plants, fungi, and some protists. (An additional outer boundary that is very rigid in shape.)

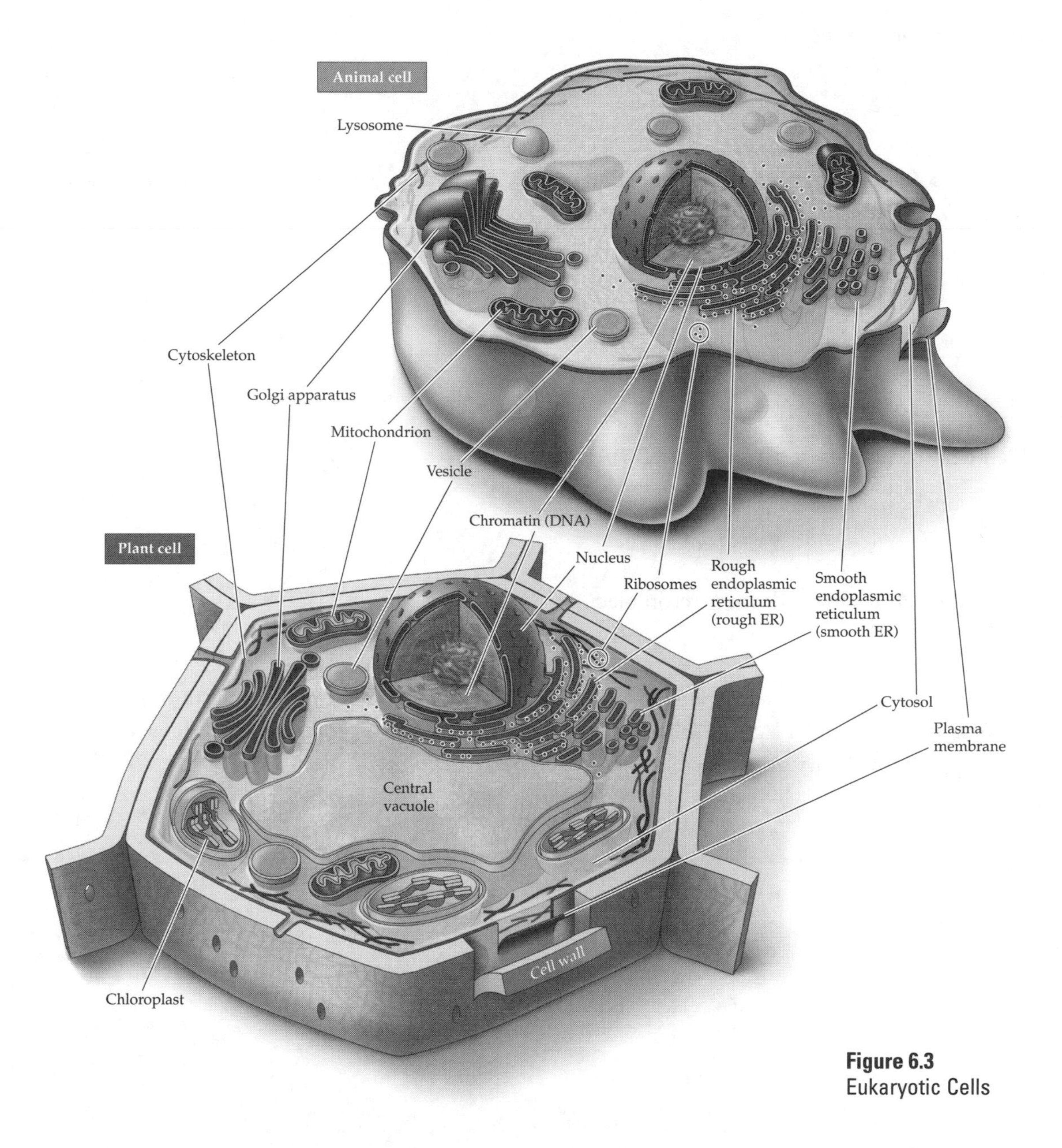

Figure 6.3
Eukaryotic Cells

Concept Check

1. Some cells need to change shape to function. Which structure gives a cell this flexibility?

2. What is an organelle?

3. Identify some characteristics of organelles that indicate they were once independent cells. Judging by the list in Table 6.1, which organelles are likely endosymbionts?

4. Name the structures involved in protein production.

5. Which structures break down macromolecules?

6. The prokaryotic plasma membrane and the internal membrane in both chloroplasts and mitochodria contain DNA (so each structure can produce its own proteins). How does this fact support the theory of endosymbiosis?

ACTIVITY 4 Kingdom Protista

The oldest kingdom of eukaryotes, Protista, consists mainly of single-celled organisms. This hard-to-define group has members diverse in size, shape, and lifestyle. Most protists are microscopic, ranging in size from about 5 to 500 μm. A few, such as the large single-celled aquatic protists called foraminiferans, are much larger—commonly visible to the naked eye. Protists include marine, freshwater, terrestrial, symbiotic, and pathogenic species; and some are larger than the smallest of the animals.

Much is unknown about the evolutionary relatedness of protists. Some are plantlike; for example, green algae can photosynthesize, and plants are thought to have evolved from green algae. There are also animal-like protists—such as the ciliates—that move and hunt for food. Still others—such as slime molds—are more like fungi. Protists demonstrate vast variation in the number and organization of their organelles; some lack organelles completely, while others have many organelles of a single type. Protists represent the first point in evolution that organisms reproduced sexually, but not all of them employ this strategy. Protists were traditionally grouped together into a kingdom before electron microscopy and molecular biology revealed their true diversity.

Even though we know that protists do not represent a true kingdom, it is useful to examine them together from the standpoint of the strategies and constraints of being single-celled. Remember that, within the limitations imposed by being single-celled, these creatures still must accomplish all of the basic life functions that animals and plants do.

In this activity you will look at examples of protists with a variety of cell structures, including cilia and flagella, contractile vacuoles, and other organelles. For each live specimen you will identify the organelles that you see.

NEVER PLACE ON THE MICROSCOPE A SLIDE THAT HAS EXCESS LIQUID. IF THERE IS TOO MUCH LIQUID, GENTLY APPLY THE CORNER OF A PAPER TOWEL TO THE AREA THAT'S WET AND LET THE TOWEL SOAK UP THE LIQUID.

Paramecium

Paramecium is a protist that uses cilia (tiny hairlike projections) to move in its habitat. Paramecia also use cilia to move food into a mouthlike structure called the oral groove, which forms food vacuoles. Lysosomes fuse with food vacuoles to break down the nutrients within the food.

1. Obtain a compound microscope, a microscope slide, a coverslip, and *Paramecium* culture.
2. Prepare a wet mount by placing one drop of *Paramecium* culture on the slide.
3. Add one small drop of nigrosin.
4. Slowly lower the coverslip onto the slide.
5. Put the slide on the microscope and let it sit for at least 1 minute to let the protists settle down.
6. Using the scanning lens, locate a paramecium. Increase the magnification to examine the paramecium in more detail. Examine the paramecium's external structures. Can you identify the oral groove and see the cilia moving?

7. Examine the paramecium's internal structures. Increase the magnification and find any star-shaped structures. Used by the paramecium to regulate water, these specialized organelles are called **contractile vacuoles.** Can you identify the nucleus? Are any other structures visible? In the circle below, sketch the paramecium and label any structures that you can identify.

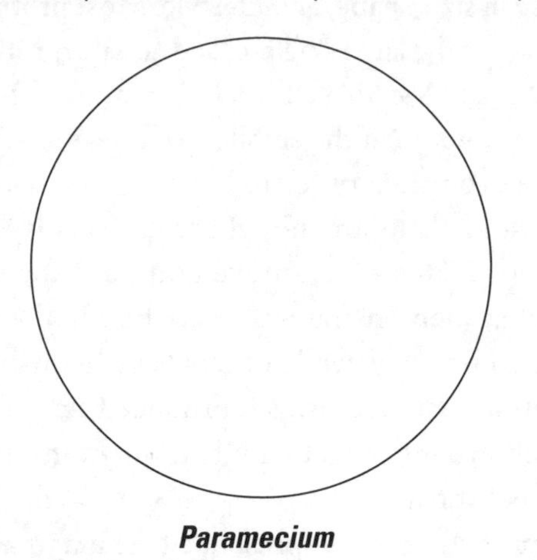

Paramecium

Volvox

Volvox, another type of protist, is not an individual cell but instead a colony of cells. An exterior layer of cells surrounds the colony, and daughter colonies are found inside the globe. *Volvox* lives in slow-moving freshwater systems.

1. Obtain a compound microscope, a microscope slide, a coverslip, and *Volvox* culture.
2. Place a drop of *Volvox* culture on a microscope slide.
3. Cover with a coverslip and examine.
4. What color is the *Volvox*? Does the color indicate anything about how this organism obtains food? Which organelle produces this coloring?

5. In the circle below, sketch the *Volvox*. Label any external structures (such as the external layer of cells) and the internal daughter colonies.

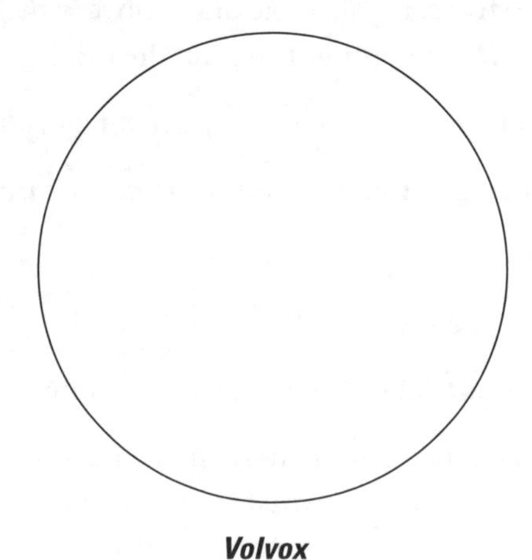

Volvox

Slime Molds

Slime molds were not always classified as protists. They appear more related to fungi, but molecular evidence has shown that they evolved much earlier than the fungi. Typically, they spend their lives as single-celled organisms, but through environmental cues they can form a collective or community structure.

1. Obtain a dissecting microscope, a microscope slide, and a culture plate of slime mold.
2. Examine the culture plate under the microscope. What is the texture of the slime mold?
3. In the first circle below, sketch the slime mold.
4. Use a swab to gently remove a sample of the slime mold from the culture plate and smear the sample on the microscope slide.
5. Now switch to a compound microscope and examine the cellular characteristics in the specimen on the slide. Can you discern any structures more clearly than with the dissecting microscope? In the second circle below, sketch and label any external structures (such as the plasma membrane) and internal structures (such as the nucleus) that you can identify.

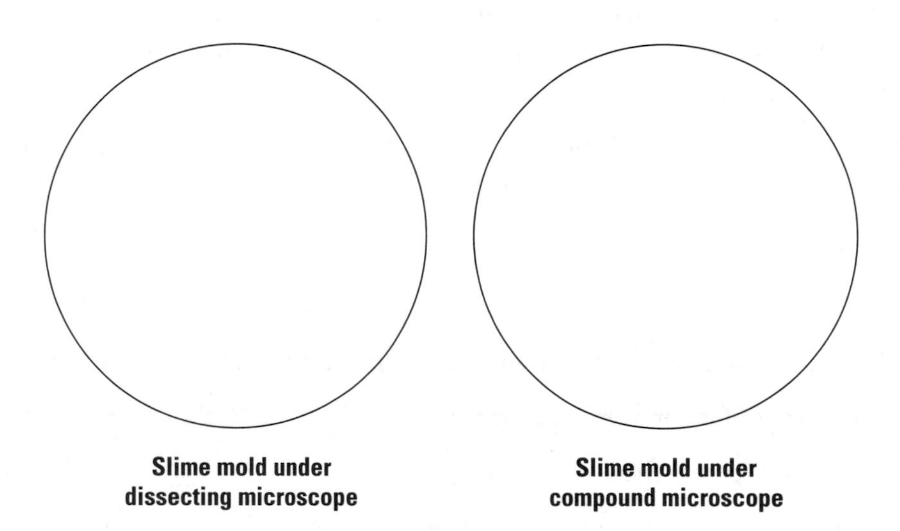

Slime mold under dissecting microscope

Slime mold under compound microscope

Concept Check

1. Which of the three protists you examined is most like an animal?
2. Which protist is most like a plant?

3. How does the paramecium propel itself around the slide?

4. Other than cellular structures, what evolutionary advance did protists achieve?

5. What organelles and structures were you able to identify, and what are their functions? Don't forget the most obvious ones.

ACTIVITY 5 Kingdom Plantae

Plants and animals have different physical structures, and their cells reflect the needs of their different kingdoms. Plant cells have three main structures that differentiate them from animal cells:

- *Cell wall.* The **cell wall** maintains water balance and protects the cell.
- *Central vacuole.* A large **central vacuole** fills most of the interior of the plant cell. It contains water dissolved with various nutrients and enzymes.
- *Chloroplasts.* Within the cytoplasm of photosynthetic cells lie the food-producing organelles called **chloroplasts**. Chloroplasts contain the green pigment chlorophyll, which absorbs light energy from the sun and transforms it into usable chemical energy.

In this activity you will examine two different plant cells under the compound microscope: the leaves of the common aquatic plant *Elodea* and the storage leaves of an onion. To view the cells more easily, you will have to make a wet mount.

Elodea

1. Obtain an *Elodea* leaf from the aquarium, and place it flat on a clean microscope slide.
2. Add one or two drops of water from the aquarium.
3. Slowly lower a coverslip onto the leaf.
4. Examine the leaf first under the scanning objective (lowest magnification), and then under higher magnification.

REMEMBER TO USE THE FINE FOCUS ONLY AFTER POSITIONING THE HIGHER-POWER OBJECTIVES.

5. Are the structures all in the same plane? Do you need to focus up and down to see clearly the different cells and structures?

6. Does the leaf contain more than one layer of cells?

7. In the two circles below, sketch a cell first at the lowest magnification and then at the highest magnification. Label the cell wall, central vacuole, and chloroplasts.

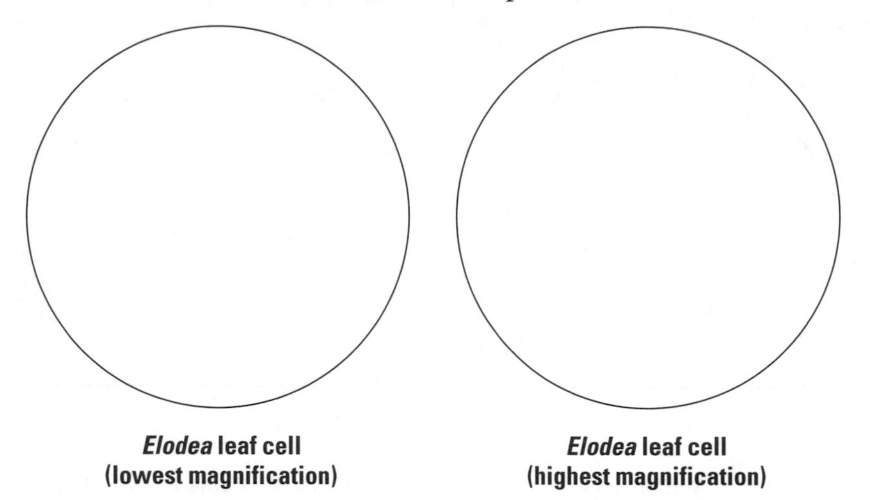

Elodea **leaf cell (lowest magnification)**

Elodea **leaf cell (highest magnification)**

Onion

1. Cut a small wedge (¼ inch thick) from an onion storage leaf. With forceps, remove the transparent membrane from the inner side of the section.
2. Place this membrane on a microscope slide, and try to keep the membrane flat with no folds.
3. Prepare a wet mount of this specimen by adding a drop of spring water to the slide.
4. Examine the specimen under the compound microscope on both low and high power, adjusting light intensity to improve contrast as necessary. In the first circle below, sketch what you see.
5. Can you identify any parts of the cell? Label any identifiable cellular structures in your drawing. What do you notice about the cell shape, size, and arrangement?

6. To improve the contrast of the organelles, remove the slide from the microscope, and then remove most of the water from the specimen by placing a piece of paper towel right up against the coverslip.
7. Tilt the slide so that one of the narrower edges is in the air while the opposite edge is still on the lab bench.
8. Add a drop of Lugol's solution (I_2KI, which is a solution of iodine and potassium iodide) above the higher edge of the coverslip and allow it to seep between the coverslip and the slide to stain the specimen. Make sure that no stain is on top of the coverslip.

9. Make additional observations under the microscope, at low and high power. Note the parts of the onion cells stained by the iodine. In the second circle below, sketch what you see. Remember to label any cell structures.

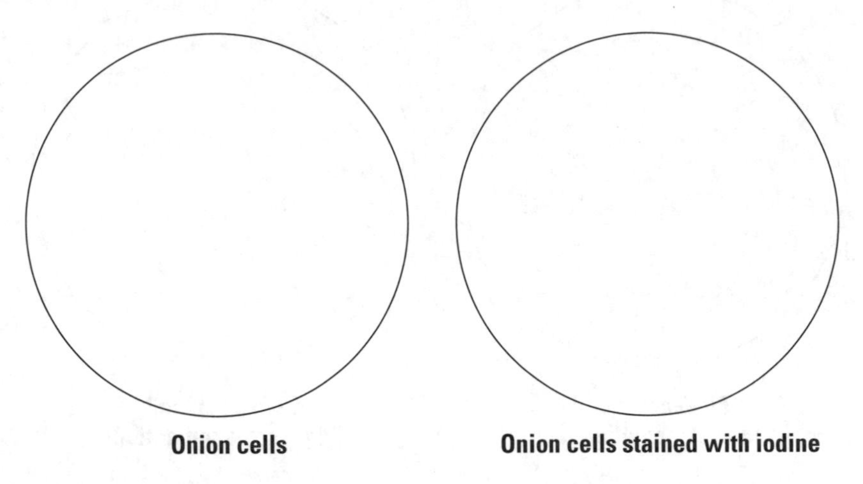

Onion cells **Onion cells stained with iodine**

10. Did the stain have any effect on your ability to see internal structures of the plant cells?

Concept Check

1. Name the three main plant cell structures that animal cells do not contain.

2. What is the main difference between the two types of plant cells?

3. What internal structures of the onion and *Elodea* cells were you able to identify?

4. Typically plant cells are green, but the onion cells are clear. Why are they not green?

ACTIVITY 6 Kingdom Fungi

For decades, the fungi—including yeasts, molds, and mushroom-producing organisms—were thought to be more closely related to plants than to animals. Unable to move and very unlike animals, these faceless organisms seem more akin to trees, shrubs, and mosses. Consequently, it came as a huge surprise when recent genetic studies showed that fungi are actually more closely related to animals, including people, than to plants. That is, fungi share a more recent common ancestor with animals than with plants. To put it in common, everyday terms, the mushrooms on a pizza are more closely related to us than they are to the green peppers sitting next to them.

Some fungi, such as **yeasts**, are single-celled. Beyond their great value in the food industry for alcohol, cheese, and bread production, yeasts are used in genetic and medical research for studying disease, therapy, and development. Yeasts typically reproduce asexually, by the process of fission. In this activity you will examine yeast cells.

1. Using a pipette, transfer a drop of yeast from the culture onto a depression slide, and place a coverslip over it.
2. Using a compound microscope, examine the yeast cells. In the circle below, sketch a yeast cell. Identify the cell wall and nucleus in your drawing.

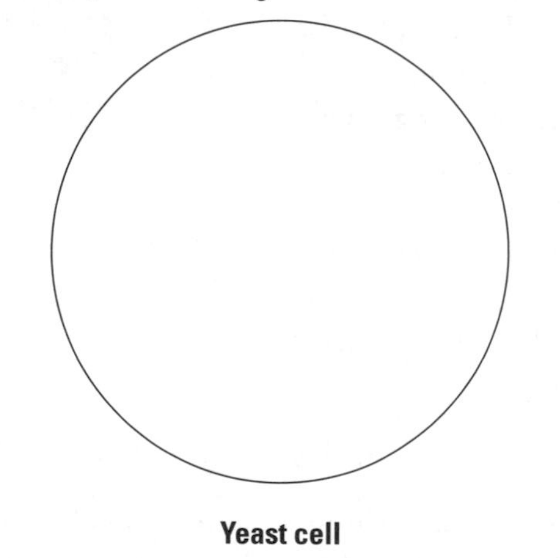

Yeast cell

3. Do you see any yeast cells undergoing reproduction?

Concept Check

1. What are some beneficial uses of fungi?
2. How are fungal cells physically like plant cells?
3. From an evolutionary perspective, how are fungal cells more closely related to animal cells?

ACTIVITY 7 Kingdom Animalia

Because they have no cell wall restraining their shape, animal cells come in a wider variety of shapes than do plant cells. They have one organelle that the plant cell does not have: the **lysosome**. The lysosome is the recycling center for the cell, so it takes the place of the plant cell's central vacuole. Some animal cells contain more organelles than others, depending on the function of the cell. All animals are multicellular, but some are simple (such as sponges) and others are more complex (like tigers).

Throughout evolution, plants and animals evolved similarities while remaining quite different. They both evolved **tissues**, collections of cells that carry out specific functions. Then, complex pressures led to the evolution of **organs**—structures composed of different tissues to carry out specialized functions. In this activity you will examine the epidermal cells composing your hair. Human hair is a specialized structure of the body that grows out of the tissue in skin, the body's largest organ (Figure 6.4). Hair is composed of epithelial cells that become hardened (or keratinized), and growth occurs from cells within the hair follicle. A hair originates in the follicle (located between layers of tissue called the dermis and hypodermis), then protrudes through the epidermis and finally penetrates the surface of skin. Staining helps outline the cells in the follicle and and their nuclei. In this activity you will examine a strand of your own hair.

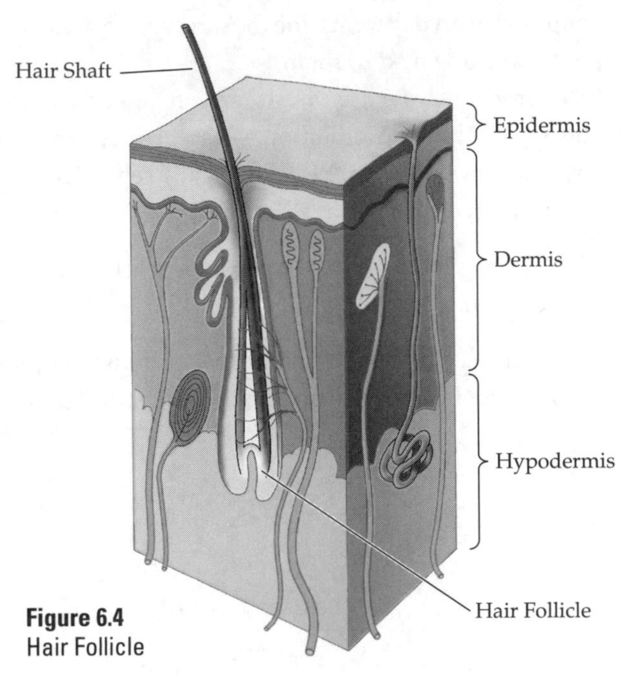

Figure 6.4
Hair Follicle

1. Obtain a microscope slide, a coverslip, transparent tape, and methylene blue.
2. Pull one hair firmly from your head, eyebrow, or arm. To make sure you've got the follicle, check that the end you removed from your skin has a white coating.
3. Fix the hair to a microscope slide with a small piece of transparent tape so that the follicle is in the middle of the slide. If necessary, cut the hair so that it's not falling off the slide.
4. Add two drops of methylene blue and gently place a coverslip on top.
5. Locate the follicle and the individual cells that comprise it. Are the cells distributed evenly throughout the follicle?
6. Locate an individual cell. In the circle, sketch the cell and label the nucleus and the plasma membrane.

Individual cell from hair follicle

Concept Check

1. What structures do plant cells have that animal cells do not? What organelle do animal cells have that plant cells do not?

2. Why do animal cells have such a variety of shapes?

3. What size are animal cells in comparison to plant cells?

4. How are animal cells different from fungal cells?

Key Terms

Animalia (p. 6-1)
Archaea (p. 6-1)
Bacteria (p. 6-1)
cell (p. 6-1)
cell wall (p. 6-10)
central vacuole (p. 6-10)
contractile vacuole (p. 6-8)
chloroplast (p. 6-10)
cytoskeleton (p. 6-4)
cytosol (p. 6-4)
endosymbiosis (p. 6-4)
Eukarya (p. 6-1)
eukaryote (p. 6-1)
Fungi (p. 6-1)
Golgi apparatus (p. 6-4)
lysosome (p. 6-14)
mitochondrion (p. 6-4)
nucleus (p. 6-4)
organ (p. 6-14)
organelle (p. 6-4)
Plantae (p. 6-1)
plasma membrane (p. 6-4)
prokaryote (p. 6-1)
Protista (p. 6-1)
ribosome (p. 6-4)
rough endoplasmic reticulum (rough ER) (p. 6-4)
smooth endoplasmic reticulum (smooth ER) (p. 6-4)
tissue (p. 6-14)
vesicle (p. 6-4)
yeast (p. 6-13)

Review Questions

1. The ________________ is the powerhouse of the cell.
2. The ________________ maintains the cell's shape and transports structures around the cell.
3. ________________ and ________________ are the two domains of prokaryotic cells.
4. Prokaryotes house their DNA in the ________________ region.
5. Lipids are produced and detoxified in the ________________.
6. In plants, the ________________ breaks down macromolecules; in animals, the ________________ performs this function.
7. ________________ and ________________ are two structures shared by all cells—both prokaryotic and eukaryotic.
8. ________________ is the evolutionary hypothesis that some organelles are remnants of prokaryotic cells.
9. ________________ is a kingdom that should be separated into many different kingdoms, but it remains just one because most of the organisms it contains are single-celled.
10. Describe all the organelles involved in protein production.

11. If the liver is in charge of detoxifying the bloodstream, what organelle would be found in great concentration in liver cells compared to a skin cell?

12. The heart works constantly and therefore needs a lot of energy. Which organelle would be found in greatest concentration in heart cells?

13. Why are plant cells so rigid?

14. Why is there so much variety in the protist kingdom?

15. Describe the main differences between prokaryotes and eukaryotes.

LAB 7 Molecular Movement across Membranes

Objectives

- Simulate the function of the plasma membrane and explain its importance to the survival of a cell.
- Observe the processes of osmosis and diffusion.
- Describe the dynamic relationship of two solutions separated by a membrane in terms of isotonic, hypertonic, and hypotonic conditions.
- Investigate how different cells react to a variety of environments.
- Discuss the role of cellular structures involved in molecular regulation.

Introduction

The plasma membrane surrounding the cell monitors the exchange of material into and out of the cell's fluid (the cytosol). The plasma (cell) membrane is a **semipermeable membrane**, meaning it allows certain types of molecules across, but not others. The movement of molecules is either passive or active. In **passive transport**, molecules move from a higher concentration to a lower concentration without the input of energy. This natural process describes molecules moving down their **concentration gradient**, from areas of abundance to areas of scarcity. **Active transport**, in contrast, requires energy and moves molecules against or up their concentration gradient.

Diffusion is a type of passive movement of molecules from an area of high concentration to an area of low concentration. Molecules tend to diffuse toward equal distribution. When molecules are evenly distributed, the net movement is zero and the solution is at **equilibrium**. Solutions within and surrounding cells never reach equilibrium, because cells are constantly creating and using molecules. Diffusion moves molecules into and out of the cell according to their concentration gradient. If the molecule that diffuses across a semipermeable membrane is water, the process is called **osmosis**. Cells sustain themselves through the passive transport of molecules by diffusion and osmosis, combined with the active transport of molecules that the cell uses in high concentrations. In this lab you will investigate the passive processes of diffusion and osmosis.

ACTIVITY 1 Diffusion: The Movement of Molecules

You experience diffusion on a daily basis. When someone sprays perfume on one side of a room, the molecules will eventually be distributed across the room, and you will smell the perfume on the other side. Colored, powdered drink mix provides another example of diffusion. When you pour the powdered drink mix into water, diffusion occurs constantly until equilibrium is reached and the particles are evenly distributed in the water (Figure 7.1). A molecule moves in a straight line until it bumps into another molecule. After hitting the second molecule, the first molecule changes its path until it hits yet another molecule. Diffusion distributes molecules throughout the space to achieve equal distribution throughout the space (equilibrium). Equilibrium does not prevent molecules from moving—it simply describes the state at which molecules are evenly distributed.

All organisms require diffusion to function. Diffusion allows nutrients to enter cells and allows wastes to be expelled and carried away through the bloodstream. The eye offers a good example of diffusion. Blood vessels do not reach all the cells composing the human eye. Instead, a fluid called the vitreous humor fills the cavity of the eye. Molecules diffuse through the vitreous humor, providing these cells with the nutrients retrieved from the bloodstream and removing waste from the cells to the bloodstream. This process also dictates the movement of molecules of a solid as it dissolves into a liquid. In this activity you will examine diffusion and the effects of temperature on this process.

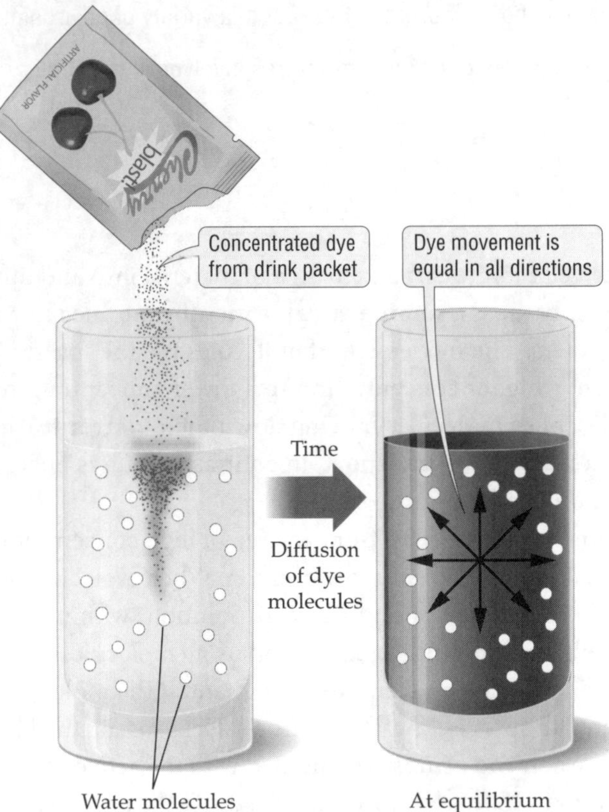

Figure 7.1
Diffusion of a Solid in a Liquid

1. Obtain two petri dishes, two rulers, a timer, and a few crystals of potassium permanganate.
2. Place each petri dish over a ruler as diagrammed in Figure 7.2.

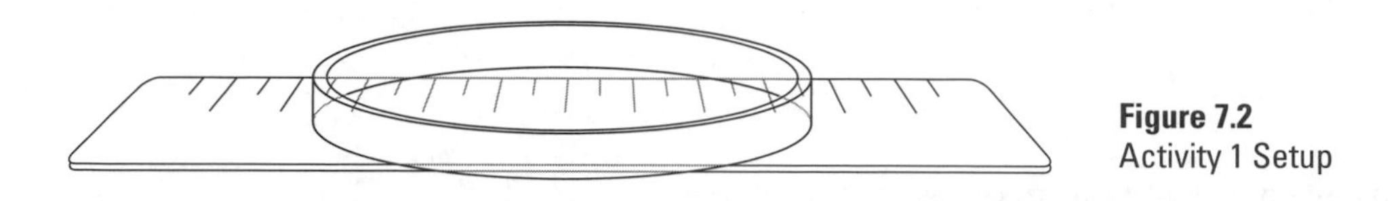

Figure 7.2
Activity 1 Setup

3. Cut two 1-inch-square pieces of paper. Using tweezers, place one crystal of potassium permanganate on each piece of paper. You will use the paper to accurately add the crystal to the center of each petri dish.
4. Pour room-temperature water into the left dish and 70°C water into the right dish.
5. Prepare the timer and start it while simultaneously using the paper to add one crystal to the center of each dish, directly above the ruler. Record in Table 7.1 (under "Starting position") the point on the ruler where the crystal was placed.
6. After 10 minutes, record in Table 7.1 (under "Ending position") the point on the ruler to which the purple color (potassium permanganate molecules) has moved.

TABLE 7.1 ACTIVITY 1: DATA ON DIFFUSION

Condition	Starting position (mm)	Ending position (mm)	Distance traveled (mm)	Speed of diffusion (mm/h)
Room temperature				
70°C				

7. How fast has the solid moved through the liquid? Subtract the ending position you recorded for each condition from the starting position, and record this number in the "Distance moved" column.
8. The number you recorded for distance traveled pertains to a time period of 10 minutes. To determine the speed of diffusion in millimeters per hour, multiply the distance traveled (third column) by 6 (10 minutes × 6 = 60 minutes, or 1 hour) and record this rate in the column labeled "Speed of diffusion."
9. Is there a difference between the speed of diffusion at each temperature?
10. How does temperature affect the diffusion of potassium permanganate?

Concept Check

1. The area of an injury such as a twisted ankle normally gets warm as the body reacts to damage to the tissue. Why would the body increase the temperature in an injured area?

2. Why is diffusion considered a form of passive transport?

3. What effect does warmer water have on molecules moving within the solution?

ACTIVITY 2 Diffusion across a Selectively Permeable Membrane

Water makes up approximately 80% of each cell and nearly the same percentage of the human body. This makes water the solvent of life. A **solution** is composed of a liquid called a **solvent,** and any molecule dissolved in that solvent is called a **solute.** Any solution in which water is the solvent is called an **aqueous solution.** The amount of solute contained in a solvent defines its concentration. A concentration can be expressed as a percentage (for example, in a 10% solution the solvent is 10% solute).

In this activity you will examine whether the size of a molecule affects its diffusion. You will be using an iodine solution, which is brownish yellow; and starch, a storage molecule for plants. When iodine and starch come into contact, the solution turns black, indicating the presence of starch. You will also use dialysis tubing to simulate the movement of molecules across a semipermeable membrane. Dialysis tubing is a flattened cylinder of a selectively permeable substance. It is an artificial plasma membrane.

The plasma membrane of a cell contains a double layer of phospholipids (Figure 7.3). The phosphate groups are hydrophilic (water-loving) molecules, so they face outward (hydrophilic heads). The lipids are hydrophobic (water-hating) molecules, so they make up the interior of the membrane (hydrophobic tails). The plasma membrane is also dotted with proteins, cholesterol, and glycoproteins/glycolipids. This variety of molecules bestows on plasma membranes their mosaic characteristic. Some molecules that are small or similar in composition to the plasma membrane cross freely by passive diffusion. Other molecules require channel or carrier proteins, which perforate the plasma membrane allowing specific molecules to diffuse freely across the membrane. Still other molecules are needed at higher concentrations inside the cell and must be pumped by carrier proteins against their concentration gradients by active transport.

Like the phospholipid layer of the plasma membrane, dialysis tubing is made up of primarily a selective molecular barrier. In the case of dialysis tubing, this barrier is composed of cellulose, the sugar found within plant cell walls. Like a cell membrane, the dialysis tubing allows only certain molecules to cross. You will use this simulated membrane to mimic diffusion in a cell.

1. Obtain two test tubes, two beakers, dialysis tubing, a ruler, two elastic bands, iodine solution, and 1% starch solution.

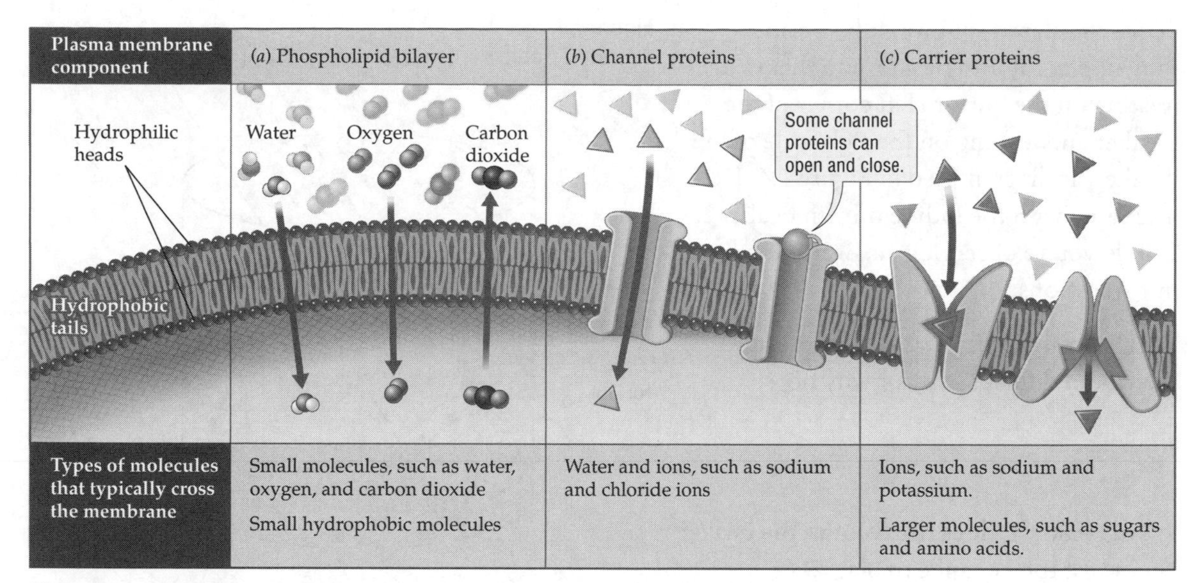

Figure 7.3 The Plasma Membrane Controls What Enters and Leaves the Cell

2. Using the ruler, mark each test tube halfway up the tube. Label one tube "starch" and the other tube "iodine," and fill each tube to the midpoint with the appropriate solution.

3. Cut the dialysis tubing so that it will cover the top of the test tube, with some overlap. Wet the dialysis tubing in distilled water to make it more flexible. Using the elastic bands, fix the dialysis tubing over each opening of the test tubes.

4. Turn each test tube upside down over a beaker to ensure that the dialysis tubing is fixed securely. Is any solution dripping? If so adjust the tubing to prevent any leaks.

5. Now fill beaker 1 with 250 ml of iodine, and beaker 2 with 250 ml of 1% starch solution. To make any change in the solutions easier to see, place both beakers on a sheet of white paper.

6. You are testing for diffusion. Remember that when starch and iodine make contact the solution turns black. You are going to add the starch test tube to beaker 1 and the iodine test tube to beaker 2. Before starting the activity, make predictions by answering the following questions:
   a. *Prediction 1:* In beaker 1, will the starch diffuse from the test tube into the beaker?

   b. *Prediction 2:* In beaker 1, will the iodine diffuse from the beaker into the test tube?

   c. *Prediction 3:* In beaker 2, will the starch diffuse from the beaker into the test tube?

   d. *Prediction 4:* In beaker 2, will the iodine diffuse from the test tube into the beaker?

7. Turn the starch solution tube upside down; place it into beaker 1 so that it is resting on the bottom left corner of the beaker and leaning on the right side of the beaker, as diagrammed in Figure 7.4. Do the same with the iodine tube in beaker 2. While you're observing, answer the following questions:

   a. Did you predict that both molecules would diffuse? Why or why not?

   b. In beaker 1, does the color in the beaker or test tube change to black?

   c. In beaker 2, does the color in the beaker or test tube change to black?

   d. Did both molecules diffuse? What enables you to determine which molecule diffused?

   e. Would increasing the concentration of starch or iodine to 10% affect the rate of diffusion?

   f. How does this activity relate to diffusion in your body? Do you think there are molecules that can or cannot diffuse into your cells? What characteristics of these molecules would prevent them from entering a cell?

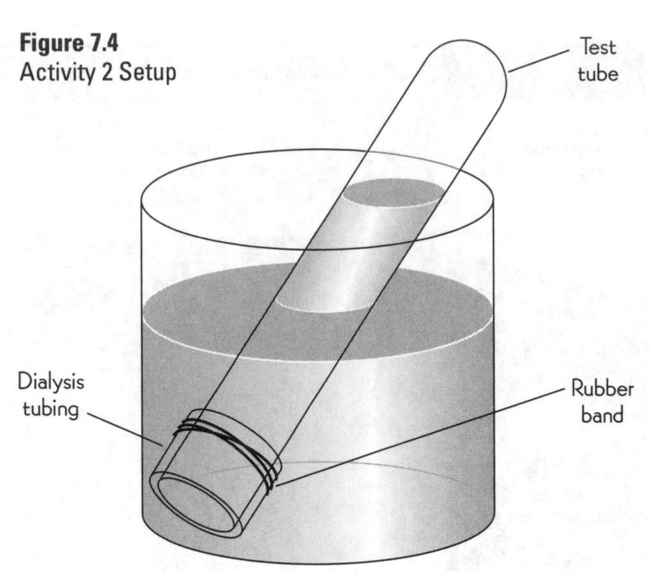

Figure 7.4
Activity 2 Setup

Concept Check

1. Define "solution" and "concentration."

2. What characteristics of a molecule prevent or allow it to enter a cell?

3. Why is water the solvent of life?

4. Why is the cell membrane so important to the life of the cell?

ACTIVITY 3 Observing Osmosis

Cells need a sufficient amount of water in their cytosol to function. Osmosis, the diffusion of water across a semipermeable membrane, is based on the concentrations of solutes inside and outside the cell. Three terms are used to describe the relationship of solutions separated by a membrane. In this activity you will use these terms to describe the solution within a cell and the solution surrounding a cell.

Isotonic describes the condition in which equal concentrations of solutes are found on the inside and outside of a cell (Figure 7.5). Diffusion and osmosis still take place, but the net movement of molecules is equal, so the concentrations on either side of the plasma membrane remain the same. The terms "hypotonic" and "hypertonic" describe the relationship when concentrations are not equal. If the inside of the cell has a low solute concentration (**hypotonic**), then the outside of the cell has a high solute concentration (**hypertonic**). The root of these words, *tonic,* refers to the solute. The prefix determines how much of the solute is present, so in the case of *hyper* think of hyperactivity, or having too much energy.

Figure 7.5
Effects of External Cellular Environment

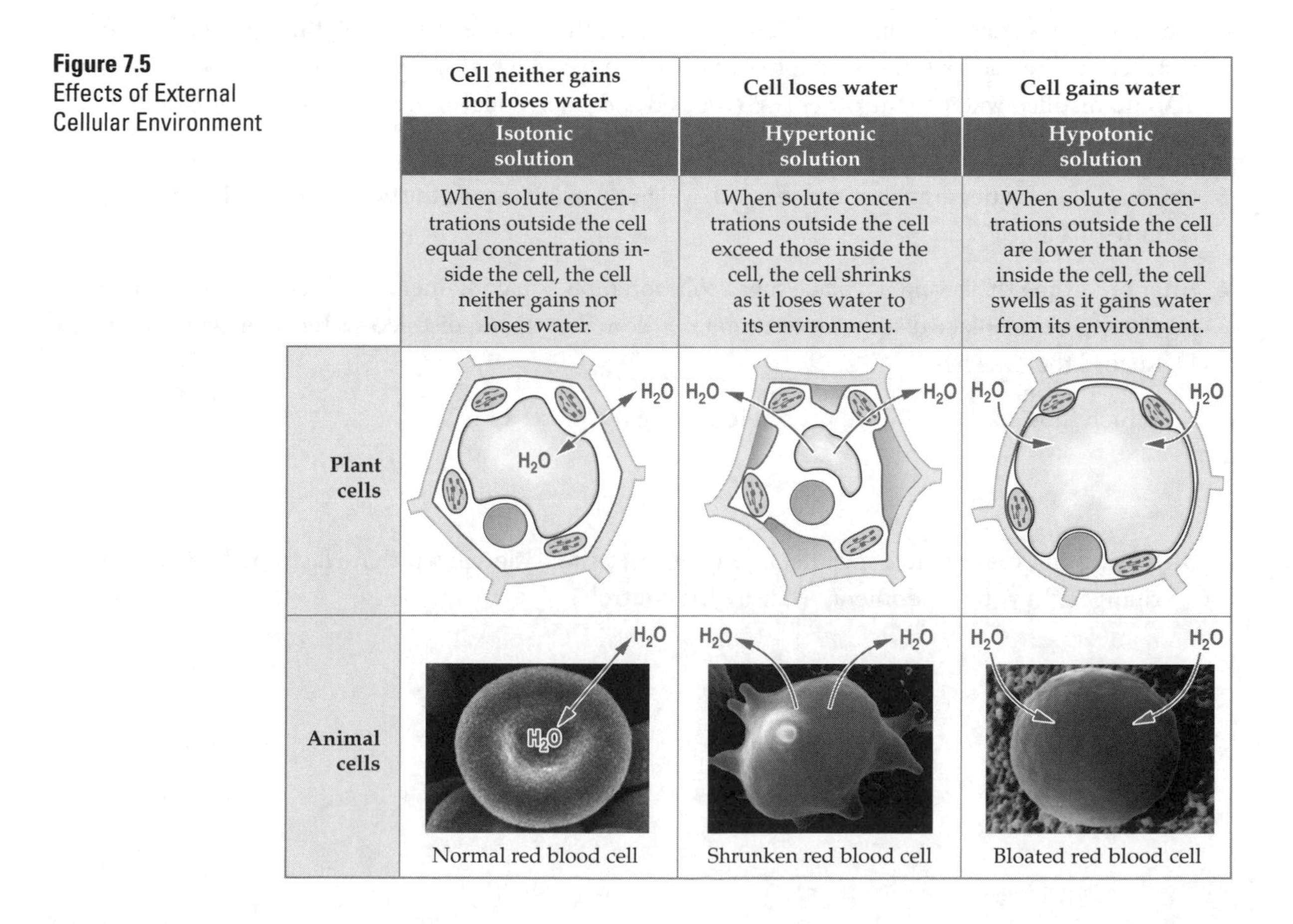

Remember that these terms describe the relationship between two solutions, so if one solution is hypertonic then the other has to be hypotonic. Similarly, if the solution inside a cell is isotonic, the outside solution must be isotonic, or equal in concentration of solutes to the solution within the cell.

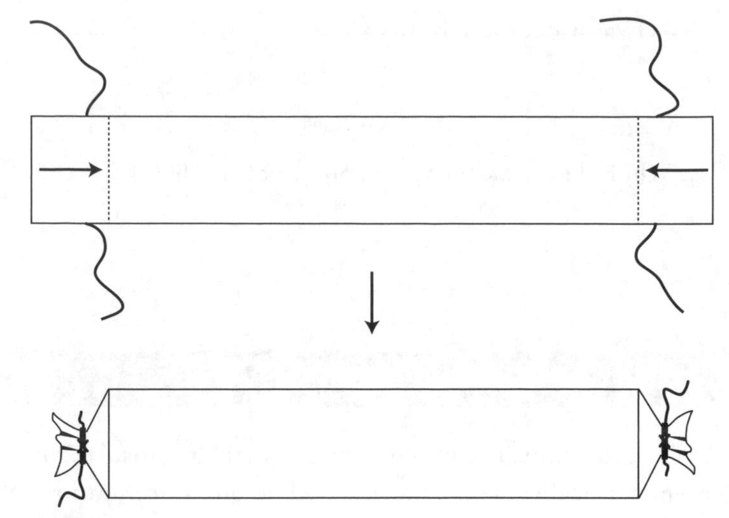

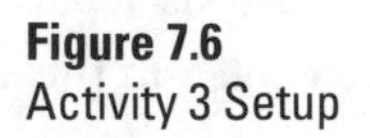

Figure 7.6
Activity 3 Setup

1. Obtain four glass containers, four sections of dialysis tubing, eight pieces of string/clips, 10% sugar solution, and distilled water.
2. Label the four glass containers A through D. Fill containers A and B with distilled water, and C and D with sugar solution.
3. The dialysis tubing is cut from one continuous length of tube, so both ends need to be sealed in order for you to make your "cell." The goal is to fill the tubing with a solution, so first close one end of the tubing by folding it along the dotted line as indicated in Figure 7.6. Next, tie the string around the area with several secure knots, or, if you are provided with clips, simply fasten the clip around this area. Do this for all of your tubes.
4. Now the tubes are ready for their solutions. Measure the volume of each solution specified by your instructor. Then add sugar solution to two of your tubes in equal amounts (these will be tubes A and C), and distilled water to the other two (tubes B and D) in equal amounts. Be sure to secure the open end of each tube immediately after you fill it.
5. Once all of the tubes are filled and secured, weigh each and record the weight in the "Initial weight" column of Table 7.2.
6. After recording their weights, place sugar-solution tube A in container A and sugar-solution tube C into container C. Place distilled-water tube B in container B and distilled-water tube D in container D. Record the time here: ___________________.
   a. Which tube's weight do you predict to change the most? Why?

   b. During an experiment, a scientist uses controls, a simulation in which no factor is deliberately changed. In your experiment, what are the controls?

c. What is the importance of having a control?

d. Will molecules move into and out of the controls?

e. What do you predict will move into or out of each tube?

7. After the time period designated by your instructor, remove the tubes and describe their condition. Are they limp or full?

   Tube A:

   Tube B:

   Tube C:

   Tube D:

8. Blot off any excess solution from the surface of the tube and record the weight of each tube in the "Final weight" column of Table 7.2. Is there any change? Did you expect the weights to stay the same?

TABLE 7.2 ACTIVITY 3: DATA ON OSMOSIS

Solution in container	Solution in tube	Initial weight of tube (g)	Final weight of tube (g)	Weight change (g) (+ or –)
A: Distilled water	A: Sugar solution			
B: Distilled water	B: Distilled water			
C: Sugar solution	C: Sugar solution			
D: Sugar solution	D: Distilled water			

9. Determine the weight change by subtracting the initial weight from the final weight, and record this number in the "Weight change" column in Table 7.2.

10. Which tube had the greatest change in weight?

11. Describe each setup (solution in bowl: solution in tube) using the terms "isotonic," "hypertonic," and "hypotonic."

    Setup A:

    Setup B:

    Setup C:

    Setup D:

12. Describe a scenario in which your cells are surrounded by a hypotonic or hypertonic environment.

Concept Check

1. When does water move into a cell?

2. What is the difference between hypertonic and hypotonic?

3. How does dialysis tubing simulate the plasma membrane of a cell?

ACTIVITY 4 Osmosis in Plant Tissue

Cellular needs are dictated by the function of the cell, so the ideal environment for different types of cells varies. Remember that cell function requires energy and constant exchange of molecules through the passive processes of diffusion and osmosis, as well as active transport. The passive process of osmosis quickly moves water into or out of a cell according to the concentration of solutes in the surrounding solution.

When surrounded by a hypotonic solution, most animal cells burst from too much water flowing into the cell. Plant cells don't burst as easily, because the cell wall prevents too much water from flowing inside the cell. But increased water volume within all cell types can also move molecules that are normally in close proximity so that they're very far apart, making them unable to react, and inhibiting cellular activity.

When surrounded by a hypertonic solution, both plant and animal cells shrink as water moves out of the cell. Cells can shrink to a point of dying because low water volume within the cell forms a thick fluid, preventing normal cellular functions. In this activity you will explore the effects of hypotonic and hypertonic solutions on a large portion of plant tissue (Figure 7.5).

1. Obtain a potato, a cork borer, 10% salt solution, and distilled water. (A cork borer is a long metal cylinder with a sharpened end to either poke holes or make a cylinder out of a soft substance.) Use the cork borer to make a cylinder of potato.

2. Cut the potato cylinder into two pieces of the lengths specified by your lab instructor. Record the length and diameter of each cylinder under "Initial measurements" in Table 7.3.

3. Place one cylinder in a solution of distilled water and the other in the salt solution.

4. Measure the length and diameter of the cylinders at the two points in time designated by your instructor, and record these measurements, along with the times, in Table 7.3. Have the cylinders changed in length?

TABLE 7.3 ACTIVITY 4: DATA ON OSMOSIS IN PLANT CELLS

	Initial measurements (cm)		Measurements (cm) at time _______		Measurements (cm) at time _______	
Solution	**Length**	**Diameter**	**Length**	**Diameter**	**Length**	**Diameter**
Distilled water						
Salt solution						

5. Did the salt move into or out of the cells?

6. The swelling of body tissues—**edema**—is caused by a variety of problems, but all relate to the osmosis of water from the bloodstream into the surrounding tissue. What problems do you think can cause edema?

Concept Check

1. What cellular structure(s) in plants enable them to withstand the pressure of osmosis?

2. Why do cells die from water loss?

3. How does a surrounding hypotonic solution affect cellular function?

ACTIVITY 5 The Cellular Effects of Osmosis

In this activity you will examine how osmosis affects both plant and animal cells.

Plant Cells

Elodea is a common aquatic plant often used in science labs because it has very large cells with clear cellular structures. The leaves have two layers of cells; only one layer is typically visible at any given time under a microscope.

1. Obtain three clean microscope slides and prepare them as follows. For each slide, use only one leaf picked directly from a fresh *Elodea* stem and a couple of drops of each solution:

   Preparation A: Elodea leaf + water from an aquarium

   Preparation B: Elodea leaf + distilled water

   Preparation C: Elodea leaf + 10% salt solution

2. Place a coverslip over each leaf. Let the leaves sit in their solutions for 10 minutes.
3. Observe the leaves, using the compound microscope with the high-power objective in place.
4. In the three circles below, sketch your observations. Have any of the cells burst open?

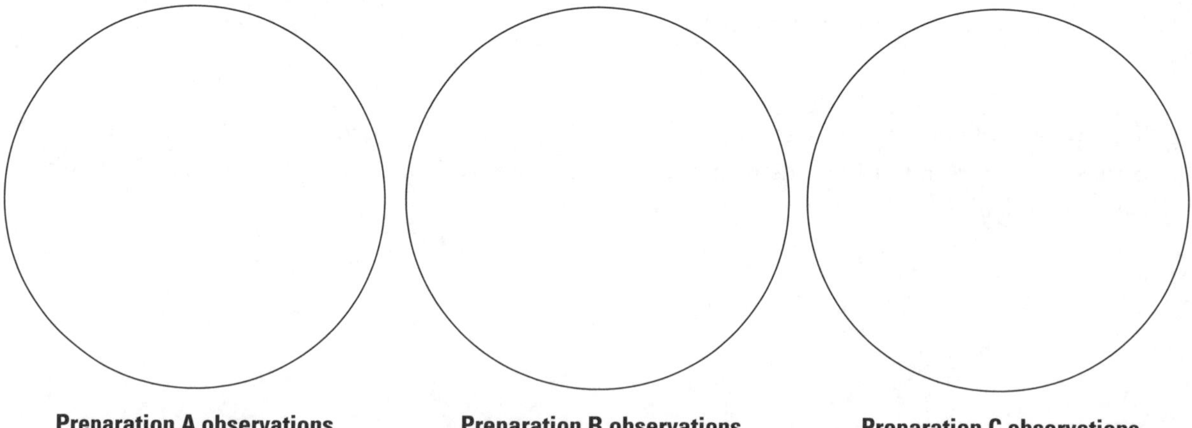

Preparation A observations **Preparation B observations** **Preparation C observations**

5. To preparation C (with the salt solution), add distilled water to the edge of the coverslip. Place a paper towel on the edge of the coverslip to soak up the excess water. Continue until the salt solution has been completely replaced with distilled water. Wait 5 minutes and observe. What has happened to the cells and cellular structures?

6. Use a single term to describe the condition of the cells on each of the microscope slides.

   Preparation A:

   Preparation B:

   Preparation C:

7. Under which condition did you observe the chloroplasts moving around the central vacuole?

8. Why would moving chloroplasts around the cell be helpful for a plant?

9. Describe the activity and physical structure of the chloroplast and central vacuole under each condition.

Animal Cells

Now that you have investigated the effects of these solutions on plant cells, how do you think animal cells act under the same conditions? In this activity you will examine red blood cells under the same conditions that you used for the *Elodea* leaf.

1. Obtain three clean microscope slides and prepare them as follows. For each slide, use only one drop of blood cells from the stock provided to you by your instructor.

   Preparation A: red blood cells alone

   Preparation B: red blood cells + distilled water

   Preparation C: red blood cells + 10% salt solution

2. Place a coverslip over the red blood cells. Let the red blood cells sit in their solutions for 10 minutes.
3. Observe the red blood cells using the compound microscope with the high-power objective in place. Can you identify any cellular structures?
4. In the three circles below, sketch your observations. Are any of the cells dead?

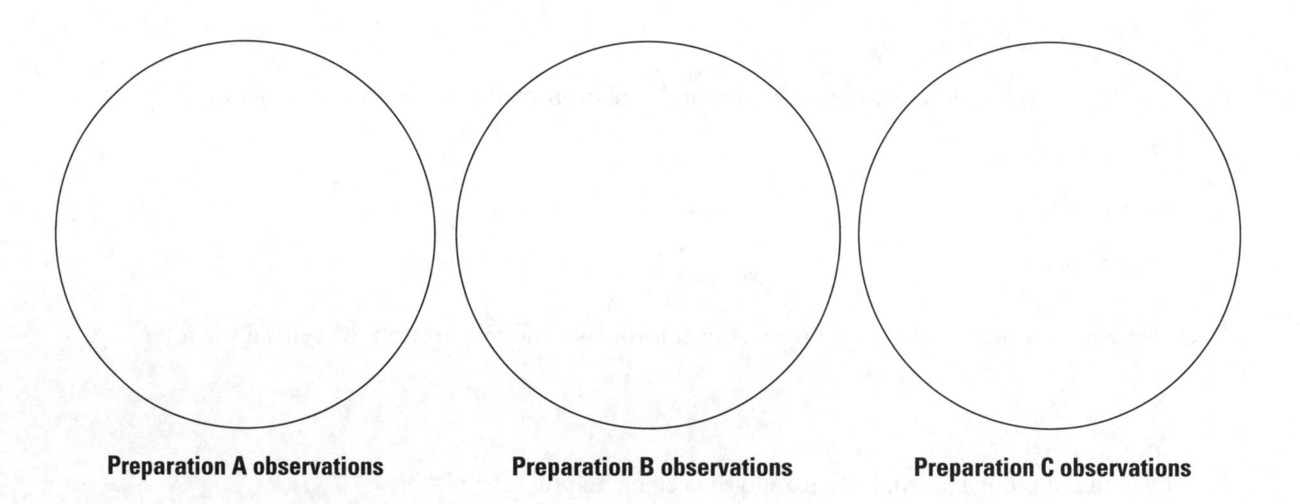

5. To preparation C (with the salt solution), add distilled water to the edge of the coverslip. Place a paper towel on the edge of the coverslip to soak up the excess water. Continue until the salt solution has been completely replaced with distilled water.
6. Wait 5 minutes and observe. What has happened to the cells and cellular structures?
7. What condition would be the most favorable for cellular function? Why?
8. What condition would be the least favorable for cellular function? Why?

Concept Check

1. Does any one of the slides show an isotonic relationship? If yes, which one?

2. Why is it common practice to place lettuce and celery in tap water before serving?

3. Why do grass plants die if they are exposed to too much fertilizer or salt?

Key Terms

active transport (p. 7-1)
aqueous solution (p. 7-4)
concentration gradient (p. 7-1)
diffusion (p. 7-1)
edema (p. 7-11)
equilibrium (p. 7-1)
hypertonic (p. 7-7)
hypotonic (p. 7-7)
isotonic (p. 7-7)
osmosis (p. 7-1)
passive transport (p. 7-1)
semipermeable membrane (p. 7-1)
solute (p. 7-4)
solution (p. 7-4)
solvent (p. 7-4)

Review Questions

1. Describe the process of diffusion.

2. Describe the process of osmosis.

3. Freshwater fish are surrounded by a hypotonic solution, and in the opposite scenario, marine fish are surrounded by a hypertonic solution. How can you explain the fact that freshwater fish don't drown as a result of water rushing into their cells and marine fish don't dehydrate as a result of water rushing out of their cells?

4. Provide one example of osmosis in humans.

5. Provide one example of diffusion in humans.

6. How does temperature affect the processes of osmosis and diffusion?

7. A solution of high solute concentration is described as __________________; a solution with low solute concentration is __________________.

8. When solutions on either side of a membrane have equal solute concentration, their relationship is described as __________________.

9. A solution is composed of molecules called __________________ dissolved in a liquid called a __________________.

10. __________________ is the barrier of all cells that allows certain substances across while preventing others from leaving or coming into the cell.

11. Describe the characteristics of a semipermeable membrane.

12. A(n) __________________ solution is any solution in which water is the solvent.

13. Why do plant cells function better in hypotonic conditions than in hypertonic conditions?

14. Describe what would happen to an animal if it were placed into a hypotonic, isotonic, or hypertonic solution.

15. Define "concentration."

16. Why does drinking too much salt water dehydrate your body?

LAB 8 Energy and Enzymes

Objectives

- Recognize the importance of enzymes in metabolic chemical reactions.
- Explain the concept of activation energy and how enzymes differ from heat in speeding up chemical reactions.
- Describe the function and significance of substrates.
- Determine the optimal concentration of an enzyme.
- Test how variables such as temperature and pH affect enzymes and their functions.

Introduction

Life would not exist without the thousands of chemical reactions that occur inside cells everyday. Collectively, these reactions within living cells are known as **metabolism**. In a chemical reaction, **reactants** undergo a chemical change to produce **products**. These reactions are not isolated, discrete events; typically, a product from one chemical reaction becomes the reactant for another reaction. In cells, this chain effect is called a **metabolic pathway.** Metabolic pathways usually build up or break down large molecules called macromolecules, such as proteins and carbohydrates.

A chemical reaction can be expressed as follows:

$$A + B \rightarrow C + D$$

where A and B are the reactants and C and D are the products of the reaction. Some metabolic reactions are **anabolic,** requiring energy to create complex molecules out of smaller compounds. Others are **catabolic,** breaking down complex molecules to release energy (Figure 8.1). Both types of enzymatic reactions merely speed up the rate of reaction; these reactions would occur naturally, but too slowly to allow a cell or organism to function properly.

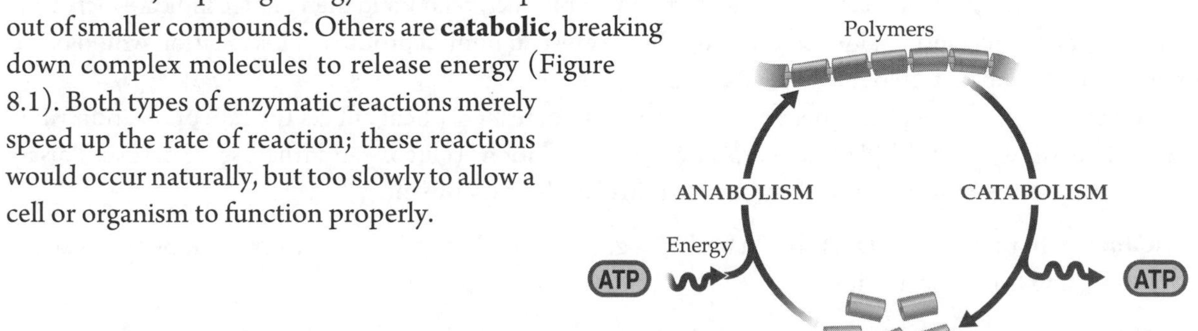

Figure 8.1
Anabolic and Catabolic Reactions

The minimum amount of energy required for a reaction to occur is called the **activation energy** of the reaction. In order for a chemical reaction to take place, the reactants must collide with enough frequency and force and in the proper orientation for molecules to separate and atoms to recombine. Scientists sometimes refer to the meeting of these conditions as "overcoming the **activation barrier**."

Several conditions affect the rate at which a reaction occurs. Heat can speed up the movement of atoms or molecules—when the energy of these atoms or molecules equals the activation energy, then the reaction will proceed. Heat will speed up almost all chemical reactions, but cells could not survive if all chemical reactions were hastened. Specific chemicals called **catalysts** also speed up chemical reactions. Unlike heat, catalysts increase the rate of a reaction by lowering the activation energy so that reactants can cross the activation barrier. Cells produce biological catalysts called **enzymes**. Enzymes are specific: normally an enzyme can bind to only one or very few reactants, called the **substrate(s)** of that enzyme. Enzyme–substrate pairs depend on the chemical properties, size, shape, and overall geometry of the substrate and the enzyme. Every enzyme has an **active site,** like a little pocket, where its substrate physically fits into it.

The reactions that occur within cells are specific and must proceed at specific times in order for the cell to function. Given the specificity of the enzyme–substrate relationship, one enzyme will catalyze only a few, and often only one, of the thousands of reactions that could take place in a cell—and this is why enzymes are so important to cell function. Luckily, enzymes remain chemically unaltered when the reaction that they catalyze is complete—and they are used again and again in the body to catalyze specific reactions in cells.

In this lab you will explore how heat and enzymes speed up catabolic reactions, and how different factors, including temperature, affect the enzyme activity.

ACTIVITY 1 Effect of Heat on Reaction Rate

Heat is a direct input of energy into a reaction. It causes atoms and molecules to move faster, creating a greater likelihood that reactants will encounter each other in the right orientation for the reaction to occur. Because it acts indiscriminately, heat does not initiate most cellular processes.

In this activity you will examine the process of heating a mixture to increase the rate of reaction. The reaction involves Benedict's solution, which contains cupric hydroxide—$Cu(OH)_2$—an indicator that turns blue when dissolved in water. Cupric hydroxide reacts with *simple sugars* (monosaccharides) such as glucose and fructose, forming cuprous oxide (Cu_2O):

$$\underset{\text{(Cupric hydroxide)}}{Cu(OH)_2} + \text{monosaccharide} \rightarrow \underset{\text{(Cuprous oxide)}}{Cu_2O}$$

The color produced during this reaction indicates the amount of product, cuprous oxide (Cu_2O): blue indicates no product, green indicates a minimal amount of product, yellow and then orange indicate increasing amounts of product, and finally, red indicates the greatest amount of product. This reaction would occur on its own, but application of heat speeds up the process.

In this activity you will use this Benedict's solution to test how heat affects the rate of reaction when cupric hydroxide [$Cu(OH)_2$] is added to simple sugar solutions (glucose and fructose). Sucrose, a disaccharide and not a simple sugar, will act as the control in this experiment.

1. Obtain a hot plate, nine test tubes, a beaker, a graduated cylinder, a marker, Parafilm, Benedict's solution, and three sugar solutions.

2. Fill the beaker halfway with water and place it on the hot plate. Set the temperature according to your instructor's directions.

3. Label three test tubes "glucose," three "fructose," and three "sucrose." Then label the three glucose tubes as follows: one with "RT" for room temperature, one with "37°C," and one with "100°C." Do the same for the fructose and glucose tubes.
4. Add 5 ml of Benedict's solution to each test tube. Is the resulting solution bright blue? If not, tell your instructor, because any other color indicates contamination.
5. Add 5 ml of the appropriate sugar solution to each test tube. Stretch a piece of Parafilm (much like Saran wrap, except stretchy) over the top of the tube and invert several times. Incubate (let your tube sit) at the temperature labeled on the test tube for 5 minutes. Do you notice any change in color?
6. Rate your product on a scale of 0 to 4, depending on its final color:
   a. Blue = 0
   b. Green = 1
   c. Yellow = 2
   d. Orange = 3
   e. Red = 4

   For each test tube, record the rating in the "Results" column of Table 8.1.

TABLE 8.1 ACTIVITY 1: DATA FOR THE EFFECT OF HEAT ON REACTION RATE

Test tube number	Test tube contents	Temperature	Results (rating of product based on color)
1	5 ml glucose 5 ml Benedict's solution	Room temperature (RT)	
2		37°C	
3		100°C	
4	5 ml fructose 5 ml Benedict's solution	Room temperature (RT)	
5		37°C	
6		100°C	
7	5 ml sucrose 5 ml Benedict's solution	Room temperature (RT)	
8		37°C	
9		100°C	

7. Compare your results.
   a. Why is sucrose a useful solution to test in this experiment?

   b. Which temperature produced the most product? Why?

   c. Why wasn't the same amount of product produced at lower temperatures?

   d. Which solution formed no product? Why?

Concept Check

1. How does heat speed up a reaction?

2. How is heat different from an enzyme in speeding up a reaction?

3. Can you think of any commonplace situations in which heat is used to facilitate a chemical reaction?

4. Some organisms undergo a state of torpor, where their metabolism slows and their core body temperature is lowered. Torpor allows an organism to conserve energy, usually during periods of extreme weather when food is scarce. Why would metabolism and body temperature slow down simultaneously? In what type of environment do you think most organisms that undergo torpor live?

ACTIVITY 2 Enzymes and Substrates

An enzyme speeds up a reaction by binding to a substrate(s) to produce a product(s) (Figure 8.2).

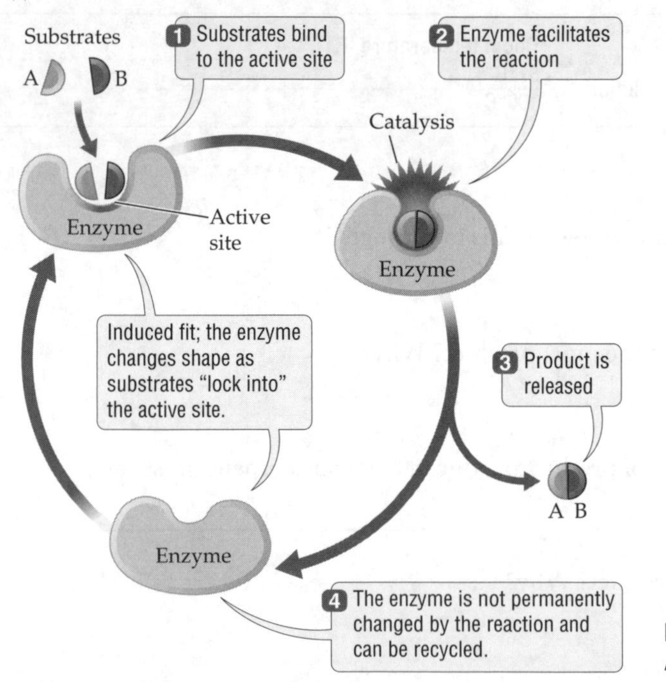

Figure 8.2
An Enzymatic Reaction

In this activity you will test different substances to determine which ones are substrates for the enzyme invertase. Extracted from yeast cells, **invertase** breaks down the disaccharide sucrose (table sugar) into the monosaccharides glucose and fructose.

$$\underset{\text{(Sucrose)}}{C_{12}H_{22}O_{11}} + \underset{\text{(Water)}}{H_2O} \xrightarrow{\text{Invertase}} \underset{\text{(Glucose)}}{C_6H_{12}O_6} + \underset{\text{(Fructose)}}{C_6H_{12}O_6}$$

Invertase is the common name for the enzyme **sucrase,** which is present in your small intestines. So you will be investigating enzymes in your own body! All the enzymes involved in digestion are involved in catabolic reactions. When you take in food, it must be broken down (releasing energy) from large molecules into smaller ones that are more easily absorbed by your body.

In this activity you will test for the substrates of invertase. After setting up the enzymatic reaction, you will use Benedict's solution as an indicator to determine which substances are substrates of invertase, by detecting the presence of simple sugars.

1. Obtain a hot plate, six test tubes, a beaker, eight graduated cylinders, a marker, Parafilm, Benedict's solution, 1% invertase solution, and six test solutions provided by your instructor.
2. Fill the beaker halfway with water and place it on the hot plate. Set the temperature according to your instructor's directions.
3. You will be adding three different solutions to each test tube: invertase solution, Benedict's solution, and one of the test solutions. Label each tube with a number (1–6) and one of the six substances, and record in Table 8.2 (under "Test tube contents") the substance that you will put in each tube.
4. Add 1 ml of each substance to the appropriate tube.
5. Add 1 ml of invertase solution to each tube, cover with Parafilm, and immediately invert. Let the tubes sit for 10 minutes.
6. Add 5 ml of Benedict's solution to each tube, cover with Parafilm, and mix again.
7. Place the test tubes in the beaker of water. Boil the tubes for 3 minutes, to allow Benedict's solution to react with the sugars.
8. Rate your product on a scale of 0 to 4, depending on its final color. Remember that blue indicates no product, and each subsequent color indicates increasing amounts of product.
   a. Blue = 0
   b. Green = 1
   c. Yellow = 2
   d. Orange = 3
   e. Red = 4

   For each test tube, record the rating in the "Results" column of Table 8.2.

TABLE 8.2 ACTIVITY 2: ENZYMES AND SUBSTRATES		
Test tube number	**Test tube contents**	**Results (rating of product based on color)**
1	1 ml ______________ 1 ml invertase 5 ml Benedict's solution	
2	1 ml ______________ 1 ml invertase 5 ml Benedict's solution	
3	1 ml ______________ 1 ml invertase 5 ml Benedict's solution	
4	1 ml ______________ 1 ml invertase 5 ml Benedict's solution	
5	1 ml ______________ 1 ml invertase 5 ml Benedict's solution	
6	1 ml ______________ 1 ml invertase 5 ml Benedict's solution	

9. Which tubes contain a substrate of invertase?

10. Which tubes do not contain a substrate of invertase?

11. Did you test a control? Which substance would that be?

Concept Check

1. What characteristics of an enzyme and a substrate enable them to bind together?

2. What role do substrates play in enzyme specificity and function?

3. In this activity you used an enzyme to break down a larger molecule into two smaller molecules. What type of enzymatic reaction is this?

4. Knowing that enzymes are very specific, is your body likely to produce many or few enzymes in breaking down the food that you consume?

ACTIVITY 3 Enzyme Concentration

One enzyme molecule is required to react with each molecule of a substrate. Because enzymes remain unchanged after a reaction, an enzyme can be used over and over again. An optimal **concentration** of an enzyme is the amount of enzyme relative to the other contents in a solution that allows the reaction to occur at maximum speed. If not enough of the required enzyme is present, the reaction may proceed too slowly to be efficient. If too much enzyme is present, the extra protein production wastes the resources of a cell.

In this activity you will examine the effects of enzyme concentration on the rate of reaction by measuring the effect of varying enzyme concentrations on the amount of product produced when invertase reacts with sucrose.

1. Obtain a hot plate, five test tubes, a beaker, seven graduated cylinders, a marker, Parafilm, Benedict's solution, 1% sucrose solution, and five invertase solutions (0%, 2%, 5%, 8%, and 10%).
2. Fill the beaker halfway with water and place it on the hot plate. Set the temperature according to your instructor's directions.
3. Label each test tube with a percentage and a solution name (for example, "10% invertase"), as shown in Table 8.3. To each tube, add 5 ml of the appropriate invertase solution.
4. Add 2 ml of sucrose to each tube, cover with Parafilm, and immediately invert. Let the tubes sit for 10 minutes.
5. Add 5 ml of Benedict's solution to each tube, cover with Parafilm, and mix again.
6. Place the test tubes in the beaker of water. Boil the tubes for 3 minutes, to allow Benedict's solution to react with the sugars.

7. Rate your product on a scale of 0 to 4, depending on its final color:

   a. Blue = 0
   b. Green = 1
   c. Yellow = 2
   d. Orange = 3
   e. Red = 4

   For each test tube, record the rating in the "Results" column of Table 8.3.

TABLE 8.3 ACTIVITY 3: ENZYME CONCENTRATION

Test tube number	Enzyme concentration	Results (rating of product based on color)
1	0% invertase	
2	0.1% invertase	
3	0.5% invertase	
4	1% invertase	
5	5% invertase	

8. Plot the results in Figure 8.3. Do you see any trend?

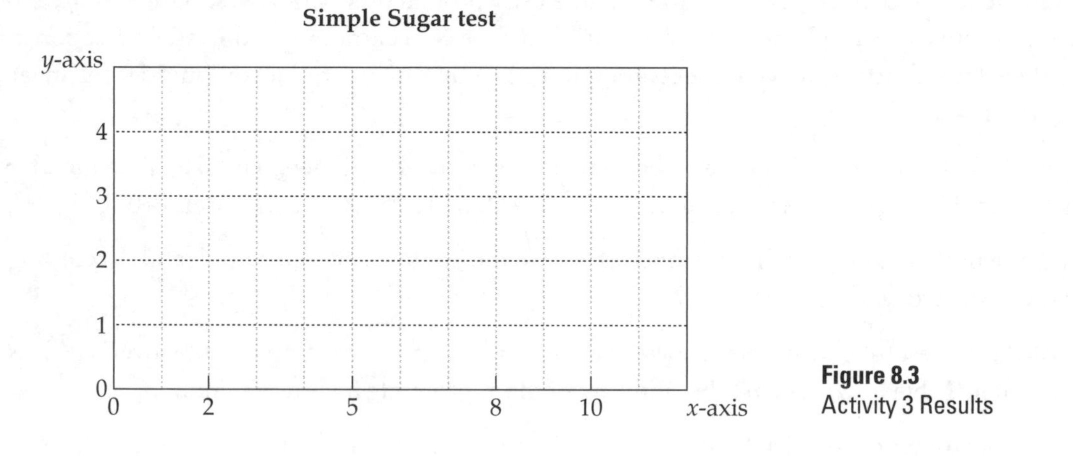

Figure 8.3
Activity 3 Results

9. What is the optimal enzyme concentration (the lowest enzyme concentration that will yield the maximum amount of product) in this reaction?

10. Which tube(s) showed no reaction? Why?

11. What is the purpose of having a control for each experiment?

Concept Check

1. Does your body produce the same amount of all enzymes? Why or why not?

2. How does enzyme concentration limit a reaction?

3. Would substrate concentration limit a reaction in the same way?

ACTIVITY 4 Effect of Temperature on Enzyme Activity

Temperature is a measure of heat energy. As more heat is applied, atoms and molecules move faster. Consequently, enzymes and their substrates collide more frequently at higher temperatures. In general, raising the temperature by only 10°C tends to increase the rate of reaction by two to three times. Each enzyme has a range of effective temperatures at which it operates. For most enzymes, high temperatures cause the structure of the enzyme to break down, or **denature,** rendering it ineffective. Lower temperatures, however, slow down the movement of molecules and atoms, reducing the product of the reaction by reducing the occurrence of collisions. In this activity the variable will be the temperature of the reaction. Both the enzyme and the reactant concentration will remain unchanged.

1. Obtain a hot plate, five test tubes, a beaker, three graduated cylinders, a marker, Parafilm, Benedict's solution, 1% sucrose solution, and 0.5% invertase solution.
2. Fill the beaker halfway with water and place it on the hot plate. Set the temperature according to your instructor's directions.
3. Your instructor has incubated the enzyme (0.5% invertase) and substrate (1% sucrose) solutions for you, at the temperatures listed in Table 8.4. Label the five test tubes according to those temperatures.
4. Add 2 ml each of invertase solution and sucrose solution to each tube. Once both solutions have been added, cover the tube with Parafilm and immediately invert. Let the test tubes sit at the designated temperature for 10 minutes.
5. Add 5 ml of Benedict's solution to each tube, cover with Parafilm, and mix again.
6. Place the test tubes in the beaker of water. Boil the tubes for 3 minutes, to allow Benedict's solution to react with the sugars.
7. Rate your product on a scale of 0 to 4, depending on its final color:

   a. Blue = 0
   b. Green = 1
   c. Yellow = 2
   d. Orange = 3
   e. Red = 4

   For each test tube, record the rating in the "Results" column of Table 8.4.

TABLE 8.4 ACTIVITY 4: DATA FOR THE EFFECT OF TEMPERATURE ON ENZYME ACTIVITY

Test tube number	Temperature	Results (rating of product based on color)
1	0°C	
2	Room temperature (RT)	
3	37°C	
4	70°C	
5	100°C	

8. Plot the results in Figure 8.4. Do you see any trend?

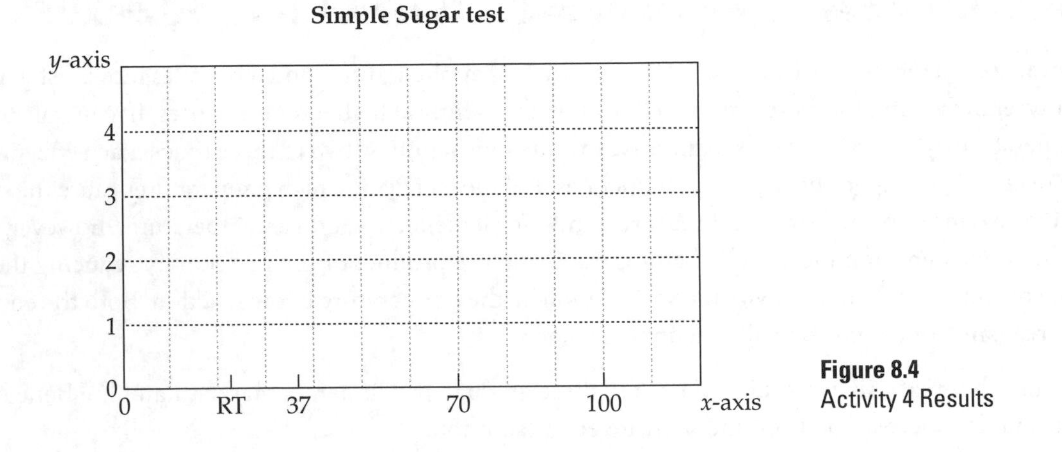

Figure 8.4
Activity 4 Results

9. During this activity, only five temperatures were tested, which is not a sufficient number of different temperatures to determine the precise optimal temperature for invertase activity. However, the temperature producing the greatest amount of product is likely close to the optimal temperature. What temperature yielded the most product?

Concept Check

1. What is the general relationship between temperature and enzyme activity?

2. Would you expect all enzymes to have the same optimal temperature that invertase has? Why or why not?

3. Do you think the enzymes of your digestive system all have the same optimal temperature?

ACTIVITY 5 Effect of pH on Enzyme Activity

pH is a measurement of hydrogen ions (H^+) in a solution. The presence of hydrogen ions influences the chemical reactions within a cell. The pH of a solution can inhibit or speed up reactions by affecting the geometry of an enzyme.

The pH scale is an easy way of expressing the concentration of hydrogen ions. A pH value of 1 represents the greatest concentration of hydrogen ions; pH 14, the lowest concentration; and pH 7, the middle ground and, therefore, the neutral pH (Figure 8.5). Most cellular activity occurs most efficiently at pH 7. Each unit of the pH scale represents a 10-fold increase or decrease in hydrogen ion concentration in a solution. Anything below pH 7 is considered **acidic;** anything above pH 7 is **basic.** Your body uses acidic conditions within your digestive, reproductive, and urinary systems to fight off infection by other organisms. Disease-causing microbial cells may enter the body through the digestive, reproductive, or urinary systems; the acidic conditions of these environments break down cellular structures preventing the microbes from functioning and reproducing. Although extreme conditions are not ideal for most proteins, some proteins can still function in very acidic or basic conditions, demonstrating that every enzyme has optimal and functional pH ranges, just as an enzyme has optimal and functional temperature ranges. In this activity you will determine the optimal pH for the enzyme invertase.

1. Obtain a hot plate, four test tubes, a beaker, seven graduated cylinders, a marker, Parafilm, and seven solutions (1% sucrose, 0.5% invertase, four pH solutions, and Benedict's solution).

2. Fill the beaker halfway with water and place it on the hot plate. Set the temperature according to your instructor's directions.

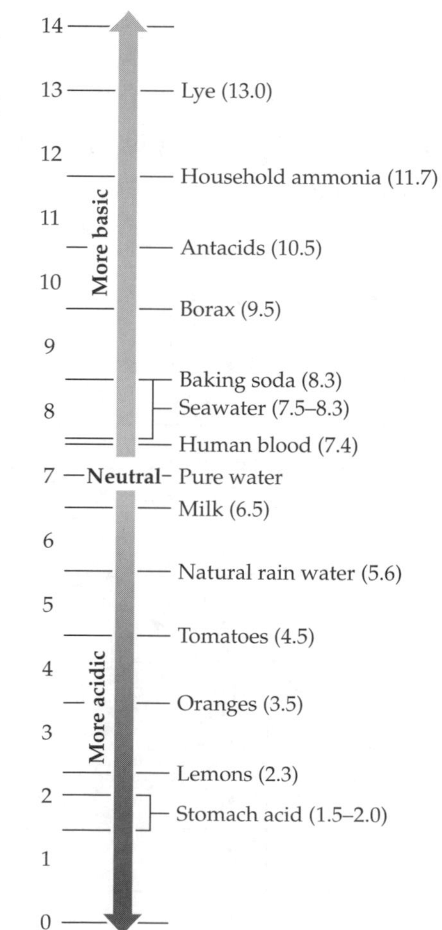

Figure 8.5
pH Scale

3. Label the four test tubes according to the different pH levels listed in Table 8.5. Then add 5 ml of each pH solution to the appropriate tube.

4. Formulate a hypothesis and prediction of the outcome of this test.

   Hypothesis:

   Prediction:

5. Add 1 ml each of invertase solution and sucrose solution to each tube. Cover with Parafilm and immediately invert. Let the test tubes sit at room temperature for 10 minutes.

6. Add 5 ml of Benedict's solution to each tube, cover with Parafilm, and mix again.

7. Place the test tubes in the beaker of water. Boil the tubes for 3 minutes, to allow Benedict's solution to react with the sugars.

8. Rate your product on a scale of 0 to 4, depending on its final color:

   a. Blue = 0
   b. Green = 1
   c. Yellow = 2
   d. Orange = 3
   e. Red = 4

   For each test tube, record the rating in the "Results" column of Table 8.5.

TABLE 8.5 ACTIVITY 5: THE EFFECT OF pH ON ENZYME ACTIVITY

Test tube number	pH	Results (rating of product based on color)
1	3.0	
2	5.0	
3	7.0	
4	10.0	

9. Plot the results in Figure 8.6. Do you see any trend?

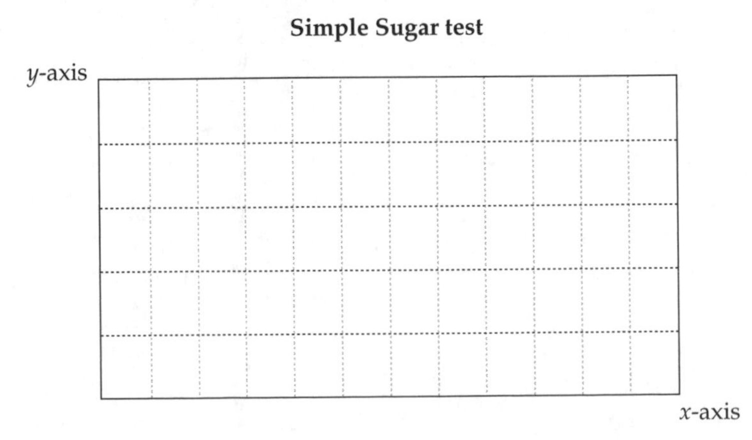

Figure 8.6
Activity 5 Results

10. What is the optimal pH for invertase? Does this finding support or reject your hypothesis?

Concept Check

1. What is pH? How can pH prevent an enzyme from functioning?

2. Where do you think invertase is more likely to catalyze a reaction: within the juices of your stomach (pH 2.0) or in the small intestine (pH 8.0)?

3. Is the optimal pH the same for all enzymes? Why or why not?

4. Why do different organs of your body have different pH?

ACTIVITY 6 Application of Enzymes

Enzymes may appear to be obscure molecules, small and unrelated to your everyday life. However, they are not just the tiny molecules enabling your cells and body to function; they also make life enjoyable! Enzymes enable humans to create many food products, one of which is cheese. A solution called **rennet** (a combination of enzymes extracted from the gut of an animal) produces the clumps of proteins called curds. Rennet contains an enzyme, **rennin**, that breaks down the protein **casein** in milk, allowing casein molecules to stick together to form curds. Casein gives milk its white color and is the reason that cheese is also white.

The process of making cheese combines all of the concepts you examined in Activities 1 through 6. In cheese making, pH, temperature, and enzyme concentration all affect the amount of product (cheese) produced. This activity is designed to optimize these conditions. Have fun and consider these conditions as you make your cheese.

NOTE: ALL THE SUPPLIES USED IN THIS ACTIVITY ARE FOR CHEESE MAKING ONLY. TO ENSURE THAT THE CHEESE PRODUCED WILL BE EDIBLE, THESE SUPPLIES HAVE NEVER BEEN USED FOR ANY OTHER PURPOSE. THERE IS NO REQUIREMENT TO EAT THE CHEESE; THE LAB IS INTENDED TO SHOW YOU HOW CHEESE IS MADE. BUT FEEL FREE TO TRY IT! YOUR INSTRUCTOR WILL PROVIDE PLASTIC ZIPPER BAGS, IF YOU WOULD LIKE TO TAKE THE CHEESE WITH YOU.

30-Minute Mozzarella

1. Obtain two stainless steel bowls, a hot plate, milk, two disposable pipettes, two disposable cups, citric acid, liquid rennet, spring water, a strainer, two sheets of cheesecloth, a spoon, salt, and plastic zipper bags.
2. Pour 2.5 liters of milk into the stainless steel bowl. Place the bowl on the hot plate and apply heat.
3. Be careful not to overheat the milk. Slowly bring the temperature of the milk to 13°C. Why should you pay careful attention to the temperature?
4. While waiting for the milk to warm up, prepare the citric acid solution. Using a clean pipette, transfer 3 grams of citric acid into a clean disposable cup and add cool, nonchlorinated water (bottled spring water) to the fill line. Stir to dissolve the citric acid. How does an acid affect the milk?
5. When the milk reaches 13°C, stir it and add the citric acid solution. Continue to mix and increase the temperature of your hot plate. Heat the milk to 31°C. Do you notice any textural difference in the milk? Any clumps?
6. While waiting for the milk to warm up, prepare the rennet solution. Using a clean pipette, transfer 1 ml of liquid rennet into a clean disposable cup and add cool, nonchlorinated water (bottled spring water) to the fill line.
7. When the milk reaches 31°C, add the rennet solution and continue stirring with an up-and-down motion. Increase the temperature of the hot plate and heat the milk to 41°C. Why don't you just increase the temperature to boil the milk?
8. When the milk reaches 41°C, the curds should be pulling away from the sides of the pot. Turn off the heat. Large curds will appear and begin to separate from the whey (the clear, greenish liquid).
9. Set up the strainer with two full sheets of cheesecloth covering the inside of the strainer, and place the strainer in another stainless steel bowl. Then slowly pour the mixture into the strainer, capturing the curds in the cheesecloth and the whey in the stainless steel bowl.
10. Use the cheesecloth to gather the curds together and gently press them to squeeze out as much whey as possible. Remove the curds from the cheesecloth and place the curds into the whey in the bowl.
11. Place the bowl back on the hot plate and heat to 41°C for 10 minutes. The curds will be very warm at this point, so use the spoon to scoop the curd back into the strainer and let cool for 3 minutes. With the spoon, press the curds into a ball until cool.
12. If you want a harder cheese, place the curd ball back into the warm whey. Why would adding the curds back into the hot whey make the cheese harder?
13. If you are satisfied with the cheese consistency, add some salt, folding it into the cheese.

14. Voilà! You have mozzarella. You can divide the cheese into small separate balls for each student at the bench and place them into plastic zipper bags to take home.

Concept Check

1. Why would an animal have an enzyme in its stomach that breaks down proteins?

2. Do you have an enzyme similar to rennin in your stomach? What other conditions of your stomach would potentially help or prevent the formation of curds?

3. How do you think rennin was discovered?

Key Terms

acidic (p. 8-11)
activation barrier (p. 8-2)
activation energy (p. 8-2)
active site (p. 8-2)
anabolic (p. 8-1)
basic (p. 8-11)
casein (p. 8-13)
catabolic (p. 8-1)
catalyst (p. 8-2)
concentration (p. 8-7)
denaturation (p. 8-9)
enzyme (p. 8-2)
invertase (p. 8-5)
metabolic pathway (p. 8-1)
metabolism (p. 8-1)
pH (p. 8-11)
product (p. 8-1)
reactant (p. 8-1)
rennet (p. 8-13)
rennin (p. 8-13)
substrate (p. 8-2)
sucrase (p. 8-5)
temperature (p. 8-9)

Review Questions

1. The minimum amount of energy required for a reaction to occur is called the ________________.
2. Metabolic pathways usually build large molecules through ________________ reactions, or they break down large molecules through ________________ reactions.
3. The collective reactions within a cell or an organism are known as ________________.
4. A(n) ________________ is a biological catalyst.
5. The speed at which a reaction occurs is the ________________.
6. Groups of enzymes usually catalyze multiple steps in a chain of chemical reactions known as a(n) ________________.
7. Describe how a substrate reacts with an enzyme.
8. How are enzymes specific for a substrate?
9. How does heat speed up a reaction?
10. Why are enzymatic reactions important in the context of your digestive system?
11. Can you think of a medical condition that might be caused by lack of an enzyme?
12. Describe the range of affects that temperature has on an enzyme.
13. Why does pH affect an enzymatic reaction?

14. Describe how enzyme and substrate concentration may affect the rate of a reaction.

15. Can you think of any commercial product that contains or uses enzymes?

16. Many pills have an outer coating made of dense fatlike substances to help ensure that they will be processed in the small intestine. Why would the fatlike substance protect the pill from all the other structures of the digestive system?

LAB 9

Photosynthesis and Cellular Respiration

Objectives

- Describe the role of carbon dioxide in photosynthesis.
- Identify the pigments present in plant tissue that enable plants to capture light energy.
- Determine the effects of different wavelengths of the visible light spectrum on photosynthesis.
- Describe the chloroplast, its components, and their functions.
- Observe cellular respiration and fermentation in yeast cells, and compare and contrast the two processes.
- Identify the cellular compartments used during cellular respiration and fermentation.
- Explain why photosynthesis and cellular respiration are complementary processes, and map each process as it occurs within a cell.

Introduction

The complex chemical reactions and structures that allow living organisms to exist require a substantial amount of energy. Ultimately, the primary source of energy is the sun, powering the producers of the world. **Producers** such as plants capture energy from the sun and convert this solar energy into chemical energy through **photosynthesis. Consumers** acquire their energy through digestion of the chemical components of other organisms. Both consumers and producers release energy stored in the bonds of large molecules during **cellular respiration**. Photosynthesis builds up large molecules, storing energy, through **anabolic** reactions; cellular respiration breaks down large molecules, releasing energy, through **catabolic** reactions (Figure 9.1).

The light reactions and the Calvin cycle of photosynthesis occur within the chloroplasts of plant cells. During the **light reactions**, the energy from sunlight is harnessed by pigments to create energy carriers and release oxygen. During the **Calvin cycle**, energy carriers such as ATP and NADPH fuel anabolic reactions, forming sugar. Overall, photosynthesis uses carbon dioxide, water, and light energy to produce glucose and oxygen.

The opposite process—cellular respiration—breaks down large molecules to release stored energy in catabolic reactions. Cellular respiration is composed of three major steps: **glycolysis**, the **Krebs cycle**, and **oxidative phosphorylation**. During these steps, glucose is slowly broken down, releasing carbon dioxide and creating ATP. This process is **aerobic**, meaning that it uses oxygen. Cellular respiration produces water as a by-product.

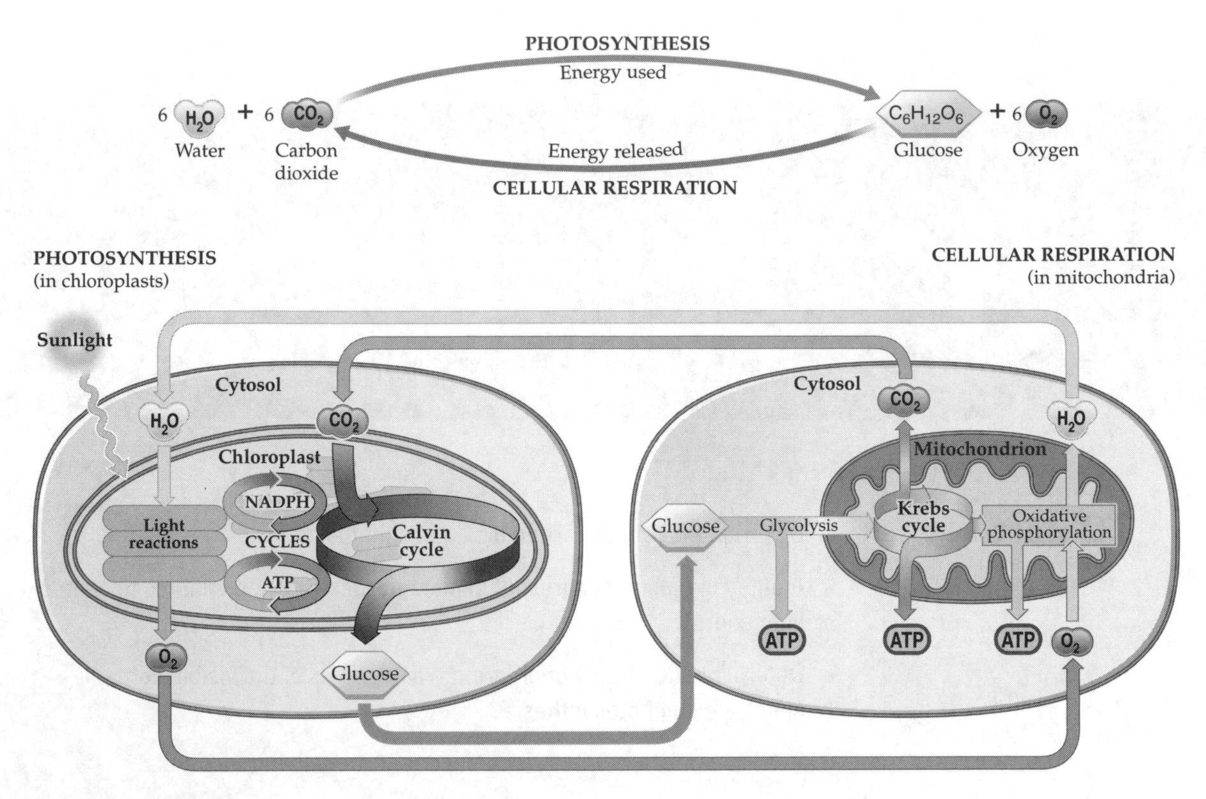

Figure 9.1
Overview of Photosynthesis and Cellular Respiration

When oxygen is absent or in low supply, organisms including humans can produce ATP through glycolysis, without using oxygen. This process is **anaerobic** and is called **fermentation**. There are many types of fermentation, including the familiar **alcoholic fermentation** and **lactic acid fermentation**. When oxygen is available, most eukaryotes use cellular respiration rather than fermentation to satisfy their need for relatively large quantities of ATP. In this lab you will explore the complex processes of photosynthesis, cellular respiration, and fermentation, allowing you to appreciate their significance to all life.

ACTIVITY 1 Role of CO_2 in Photosynthesis

Carbon dioxide provides producers with the carbon atoms necessary to form glucose during photosynthesis. Most plants obtain carbon dioxide through structural pores in their leaves called **stomata** (singular "stoma"). Stomata consist of two specialized cells called guard cells, which open and close to allow gas exchange. The process of absorbing carbon dioxide and incorporating its atoms into an organic molecule is called **carbon fixation**, the first step of the Calvin cycle. This process exemplifies the importance of producers: incorporating carbon into an organic molecule makes atmospheric carbon available to all life on Earth.

During the light reactions, enzymes speed up chemical reactions, forming the energy carriers **ATP** and **NADPH**. NADPH is an electron carrier; ATP is the short-term energy storage molecule that fuels chemical reactions in all cells of life. NADPH and ATP provide energy for the reactions of the Calvin Cycle, which forms sugar from carbon dioxide (see Figure 9.1).

In this activity you will use a pH indicator, bromthymol blue, to detect the amount of carbon dioxide present. Bromthymol is green or yellow in acidic conditions and turns blue in neutral or basic conditions. High carbon dioxide levels in water increase the levels of carbonic acid and turn the indicator green or yellow, indicating acidity. In contrast, if carbon dioxide is removed from the water by a plant performing photosynthesis, the pH will become more basic and turn blue. In this activity you will detect carbon dioxide use by an aquatic plant, *Elodea*, in different conditions.

1. Obtain a beaker, spring water, bromthymol blue, a straw, *Elodea*, a ruler, and four test tubes.
2. Fill the beaker with 150 ml of spring water. (*Elodea* is an aquatic plant; spring water is used because tap water contains many additives not found in spring water or in the natural water of *Elodea*'s environment.)
3. Add 10 drops of bromthymol blue to the beaker. If the solution in the beaker is blue, use the straw to blow air into the solution until it turns and remains green or yellow. DO NOT drink this solution! Only blow into it. The CO_2 in your breath will increase the amount of carbon dioxide and therefore increase the carbonic acid concentration. If the solution is already green or yellow after the bromthymol blue has been added, there is no need to blow air into it.
4. Obtain two sprigs of *Elodea* and, using a ruler, measure and cut each sprig to 10 cm in length. Label your test tubes 1 through 4 and place one piece into tube 1 and the other into tube 2, leaving tubes 3 and 4 empty.
5. Fill tubes 1 and 2 with the bromthymol blue to directly above the sprigs of *Elodea*. Fill tubes 3 and 4 to approximately the same level, making sure the level of bromthymol blue is approximately the same in each test tube. What color is the fluid in each tube? Record these colors in Table 9.1, under "Initial Color."
6. Seal the top of each of the four tubes with a stopper. Place tubes 1 and 3 under a light source. Place tubes 2 and 4 in a dark place designated by your laboratory instructor.
7. After 1 hour, record the final color of each solution in Table 9.1, under "Final Color."

TABLE 9.1 ACTIVITY 1: DETECTION OF CO_2 REMOVAL DUE TO PHOTOSYNTHESIS

Test tube number	Test tube conditions	Initial color	Final color
1	*Elodea*/light		
2	*Elodea*/dark		
3	No *Elodea*/light		
4	No *Elodea*/dark		

a. Which tubes indicate a reduction of carbon dioxide? What color are these tubes?

b. What are the control tubes in this experiment? What do these tubes indicate?

Concept Check

1. How does bromthymol blue indicate carbon dioxide use and therefore photosynthetic activity?

2. Why is carbon fixation so important to the Calvin Cycle?

3. Why are producers so important to the existence of all life?

ACTIVITY 2 Photosynthetic Pigments

Plants and algae are the major photosynthetic organisms on Earth. **Chloroplasts** are the organelles in plant cells that convert solar energy into usable chemical energy for the rest of life on Earth. A chloroplast is a manufacturing facility with distinct compartments and regions performing specific functions (Figure 9.2). A

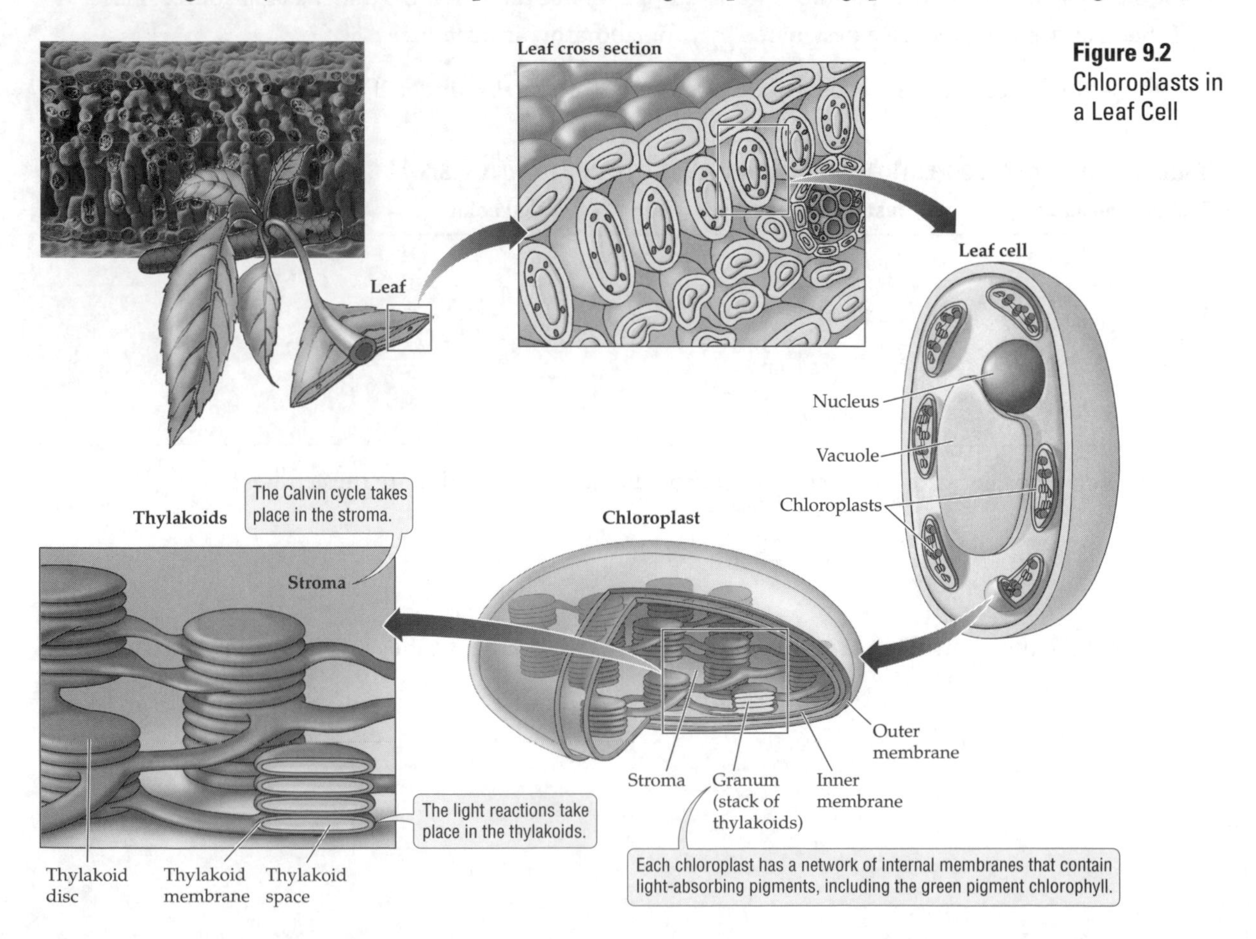

Figure 9.2 Chloroplasts in a Leaf Cell

double membrane within this organelle forms two compartments: the **intermembrane space** between the **inner** and **outer membranes** and the **stroma**, enclosed by the inner membrane. Inside the stroma is a system of interconnected membrane sacs called **thylakoids**.

The light reactions occur within thylakoid membranes, which contain pigments arranged in antenna complexes. **Pigments** are any molecule capable of absorbing light energy. In photosynthesis, pigments convert light energy into chemical energy. The **antenna complexes** are concentrated regions of pigments. The main pigment in these complexes is **chlorophyll *a***. Both chlorophyll *a* and **accessory pigments**, such as chlorophyll *b*, carotenes, and xanthophylls, absorb light energy. The accessory pigments do not directly convert this energy into chemical energy; instead, they pass it on to chlorophyll *a*, which is the only pigment that converts light energy into chemical energy, by exciting electrons. These electrons go on to power the formation of sugar in the Calvin cycle. The electrons lost in chlorophyll *a* are replaced in a reaction that breaks down water into the oxygen gas we breathe.

Plant leaves have different colors depending on the types and concentrations of pigments they contain. During the fall, the beautiful colors exhibited by deciduous plants result from the breakdown of pigments as the leaves die.

In this activity you will use **chromatography** to identify the pigments present in a plant extract. An organic solvent moves the pigments up the chromatography paper according to the geometry and chemical properties of each pigment. The smaller and less polar the molecule, the more soluble it is and the faster the molecule travels. The ratio of the distance traveled by the pigments to the distance traveled by the solvent identifies the pigment. This relationship is called the **ratio factor** or **Rf value**.

1. Obtain a strip of chromatography paper, a large test tube and a stopper with a hook/paper clip attached, a pencil, a ruler, a wax pencil, a paintbrush, and plant extract.

TO PREVENT OILS ON YOUR SKIN FROM BEING ABSORBED BY THE CHROMATOGRAPHY PAPER, HOLD IT BY THE EDGES OR USE GLOVES WHILE HANDLING IT.

2. With your pencil, mark a line 2 cm from the end of the chromatography strip. Using the paintbrush, apply a stripe of plant extract over the pencil mark, and then blow on the stripe to dry it. Repeat 10 times.

KEEP THE PLANT EXTRACT CONTAINER CLOSED WHEN NOT IN USE. THE SOLVENT IS ACETONE, AND THE FUMES SHOULD NOT BE INHALED.

3. This step should be performed within a fume hood or a well-ventilated area. Using the hook/paper clip on the test tube stopper, attach the paper strip to the stopper (Figure 9.3). Secure the stopper in the test tube so that the strip of paper hangs down inside the tube with the pencil stripe toward the bottom. Measure from the bottom of the test tube to 1 cm below your plant extract line (which should be 1 cm above the end of the chromatography strip). Mark the test tube at this point with a wax pencil.

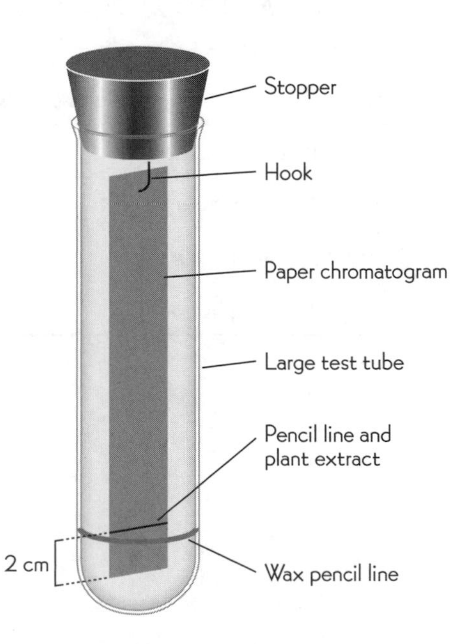

Figure 9.3
Activity 2: Position of Paper Chromatogram within Large Test Tube

4. Remove the stopper and fill the test tube to the wax pencil line with the organic solvent provided by your instructor.

HANDLE THIS TOXIC SUBSTANCE CAREFULLY!

5. Carefully place the test tube in the rack provided. Do not disturb the tube as the solvent makes its way up the paper. Do you see the solvent moving up the paper?
6. Remove the paper when the solvent is within 2 cm of the top. Using a pencil and a ruler, draw a line along the solvent front. The solvent front is the maximum distance the acetone and pigments travel up the paper. Do you see several bands of color? These bands represent the pigments within the plant extract (Figure 9.4).

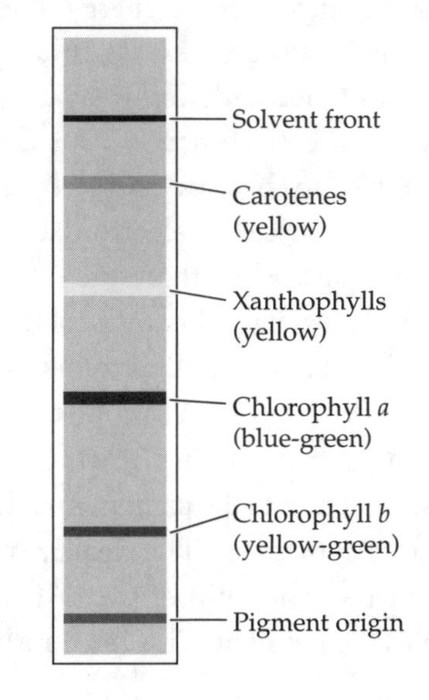

Figure 9.4
Chromatogram Depicting Photosynthetic Pigments

7. Measure the distance the solvent has traveled and record this value here.

8. Using Figure 9.4 for reference, identify the color of each band and record it in Table 9.2.
9. Measure the distance traveled by each pigment and record these values in Table 9.2. Start with the pigment closest to the solvent front.

TABLE 9.2 ACTIVITY 2: PHOTOSYNTHETIC PIGMENTS PRESENT IN A PLANT EXTRACT

Pigment	Color of band	Distance traveled (mm)	Rf value: distance traveled by pigment ÷ distance traveled by solvent
Pigment 1: carotenes			
Pigment 2: xanthophylls			
Pigment 3: chlorophyll *a*			
Pigment 4: chlorophyll *b*			

10. Using the formula provided in Table 9.2, calculate the Rf value for each pigment and record it in the table. Can you have an Rf value greater than 1? Why or why not?

11. What does a small Rf value tell you about the characteristics of the moving molecules?

12. Why is chlorophyll *b* closest to the pigment origin? Why is the carotene line closest to the top of the paper?

Concept Check

1. What role do pigments play in photosynthesis?

2. Where do the light reactions occur?

3. Why do plants have different-colored leaves?

ACTIVITY 3 Role of Light in Photosynthesis

Chloroplasts in plant cells convert solar energy into usable chemical energy for the rest of life on Earth. The pigments in chloroplasts use solar energy, which is composed of **photons**, discrete packets of energy that have specific wavelengths. The different wavelengths of light equate to the amount of energy within that light. White light is the rainbow of visible light from violet to red, known as the **visible light spectrum.**

The objects around you appear to have specific colors because the pigments present in them absorb certain colors/wavelengths of visible light and reflect the colors you see. If an object is black, its pigments absorb all the colors of the visible spectrum; if it is white, it contains no pigments and reflects all the colors of the visible light spectrum. Table 9.3 displays the range of wavelengths for each color in the visible light spectrum that humans are able to detect. **Wavelength** ultimately describes the energy of the photons within any particular wavelength (color). As the wavelength increases, the amount of energy decreases. Wavelengths of light are measured in nanometers, abbreviated nm (1 cm = 10 million nm).

TABLE 9.3 WAVELENGTHS OF VISIBLE LIGHT

Color	Wavelength (nanometers, nm)
Violet	380–435
Indigo	435–460
Blue	460–495
Green	495–570
Yellow	570–590
Orange	590–620
Red	620–750

NOTE: White light is composed of violet, indigo, blue, green, yellow, orange, and red.

Unlike photosynthetic organisms, most objects on Earth have pigments and simply radiate the energy absorbed by those pigments as heat. Photosynthetic organisms instead convert light energy into chemical energy. Photosynthetic pigments in the chloroplast absorb light at specific wavelengths, converting it into chemical energy through the excitation of electrons. The energy from these electrons is eventually incorporated through the energy carrier molecules ATP and NADPH into sugar molecules. This is the process of photosynthesis. It hinges on the presence of pigments and their ability to absorb specific wavelengths of light.

In this activity you will examine the relationship between wavelengths of light (as represented by the colors of visible light spectrum) and photosynthesis. You will isolate wavelengths of the visible light spectrum to analyze their specific effects on photosynthetic production.

1. Obtain an *Elodea* sprig; spring water with sodium bicarbonate (a pinch equivalent to 1%); a wax pencil; a test tube rack; a large beaker: three sheets of filter paper (blue, red, and green); aluminum foil; a lamp with a 75-watt halogen bulb; and six test tubes, each with a stopper that has a 1-ml pipette threaded through the stopper hole.

2. Label each test tube with a number (1–6). Cover tube 2 with blue filter paper, tube 3 with red filter paper, tube 4 with green filter paper, and tube 5 with aluminum foil. Keep tubes 1 and 6 uncovered.

3. Using fresh *Elodea* leaves, cut six 7.5-cm pieces—one for each test tube. Place one piece of *Elodea* leaf, cut side up, into each tube.

4. Add spring water with bicarbonate to tubes 1–5 and tap water to tube 6. Tap water contains very little or no carbon dioxide, but spring water with bicarbonate provides extra carbon dioxide, so it should increase the photosynthetic product—in this case, oxygen. The water should cover the *Elodea* leaf in each tube.

5. Close each test tube with a rubber stopper containing a 1-ml pipette, so that the water fills the pipette about 2.5 cm above the top of the stopper. If the water level is not high enough, add more water to the tube as needed.

6. Using the wax pencil, mark the water level on each pipette. In Table 9.4 (under "Starting water level"), record the amount indicated by the graduations on each pipette (for example, 1.2 ml).

7. Place all the test tubes in the rack provided. Between the rack and the lamp, place a large beaker containing tap water. The lamp will produce a lot of heat as it emits light, and you do not want that heat to affect the water level in the test tubes, so the water in the beaker will absorb the heat while allowing the light to pass through to the tubes (Figure 9.5).

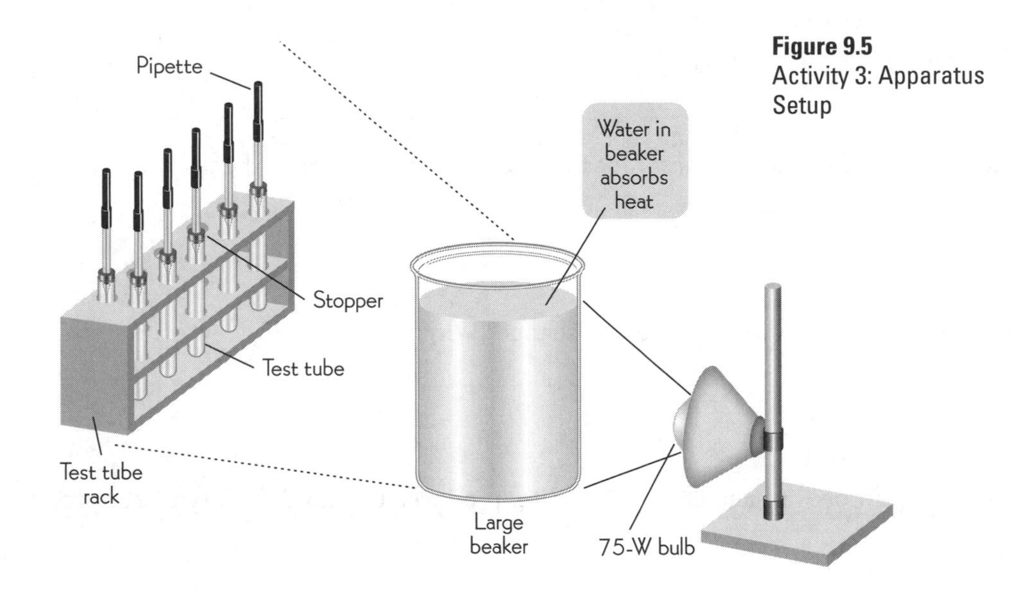

Figure 9.5
Activity 3: Apparatus Setup

8. Turn on the lamp and start timing the reaction. Do not touch the setup, as any movement will skew the results.

9. After 10 minutes, in Table 9.4 record the water level indicated on each pipette.
10. After 20 minutes, in Table 9.4 record the water level indicated on each pipette.
11. Calculate the difference between the volume at the beginning time and the volume at the ending time, and record the value in Table 9.4 (under "Total change").

TABLE 9.4 ACTIVITY 3: PHOTOSYNTHESIS AT DIFFERENT WAVELENGTHS OF LIGHT

Test tube number	Water	Lighting conditions	Starting water level (ml)	Water level after 10 minutes (ml)	Water level after 20 minutes (ml)	Total change (ml)
1	Spring water with sodium bicarbonate	White light				
2	Spring water with sodium bicarbonate	Red light				
3	Spring water with sodium bicarbonate	Blue light				
4	Spring water with sodium bicarbonate	Green light				
5	Spring water with sodium bicarbonate	No light				
6	Tap water	White light				

12. What is being collected in the test tube that causes the water level in the pipette to change?

13. Which conditions provided the greatest change? Why?

14. What is the purpose of each filter? Which filter produced the most photosynthetic product?

15. Sodium bicarbonate provides a greater amount of carbon dioxide for the plant to perform photosynthesis. Is there a difference between test tubes 1 and 6?

Concept Check

1. Why does a black car feel hot on a hot, sunny day?

2. Why do most plants have green leaves?

3. Why do photosynthetic pigments absorb only certain wavelengths of light?

ACTIVITY 4 Cellular Respiration

While photosynthesis harnesses energy from carbon in the atmosphere, the complementary reactions in cellular respiration transform this harnessed energy into a usable form. Almost all eukaryotes use cellular respiration. The three steps of cellular respiration are glycolysis, the Krebs cycle, and oxidative phosphorylation.

The first stage of cellular respiration, glycolysis, occurs in the cytosol of all cells. Glycolysis breaks the molecule glucose into two identical molecules called **pyruvate**. During this process, ATP and an electron carrier (NADH) are formed (Figure 9.6). The final stages of cellular respiration occur in the **mitochondrion**. Pyruvate is moved into the mitochondrion and prepared to enter the Krebs cycle. During this preparation, pyruvate is converted into acetyl CoA, releasing carbon dioxide and producing an electron carrier (NADH). The Krebs cycle slowly breaks down acetyl CoA, releasing carbon dioxide, ATP, and an electron carrier (NADH).

All of the electron carriers from each step, including glycolysis, move to the inner compartment of the mitochondrion (see Figure 9.6). Within this membrane, a series of enzymes accepts electrons dropped off by the electron carriers. The enzyme-catalyzed transfer of electrons provides energy to create ATP; this is oxidative phosphorylation. Finally, at the end of oxidative phosphorylation, oxygen accepts the electrons and forms water. In this activity you will examine cellular respiration in yeast cells, a type of fungi.

1. Obtain a yeast culture, a compound microscope, a microscope slide, a coverslip, and methylene blue.

2. Place one drop of yeast culture on the slide. Directly in the middle of the yeast, add one drop of methylene blue. Slowly place the coverslip on top of the solutions.

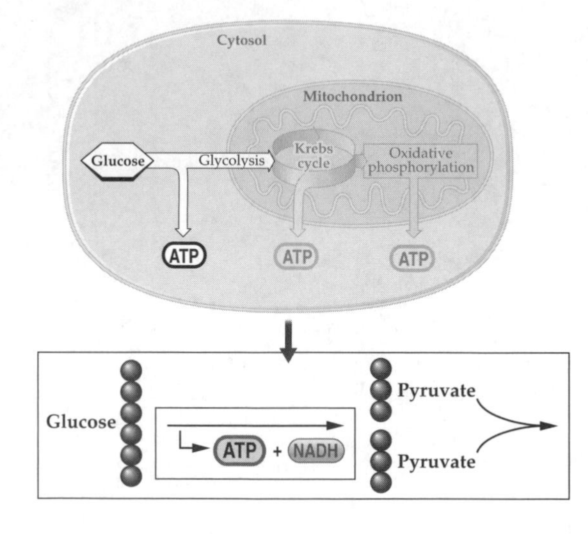

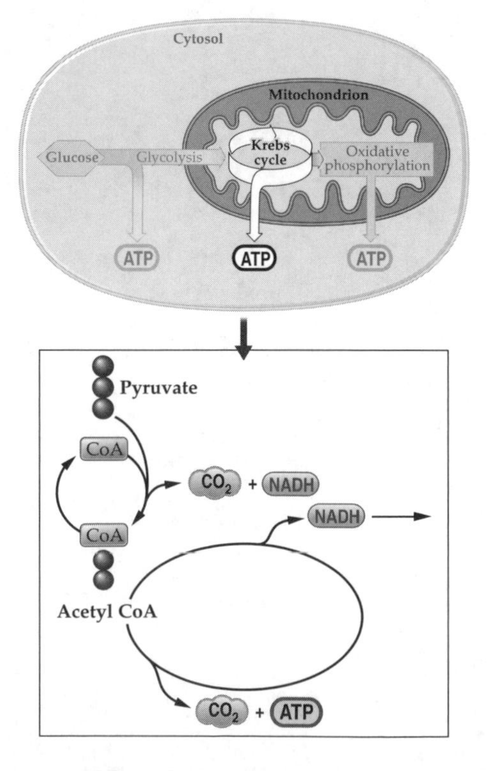

3. Using the compound microscope, examine the slide, starting with the lowest magnification and working up to the highest magnification. Do you notice any differences between cells?

4. Methylene blue acts as an indicator of cellular respiration. It absorbs ions and electrons (that are ultimately accepted by oxygen) produced during cellular respiration, becoming colorless in the process. Do you see any cells that are colorless? In the two circles below, sketch an example of one yeast cell that is undergoing cellular respiration and one that is not.

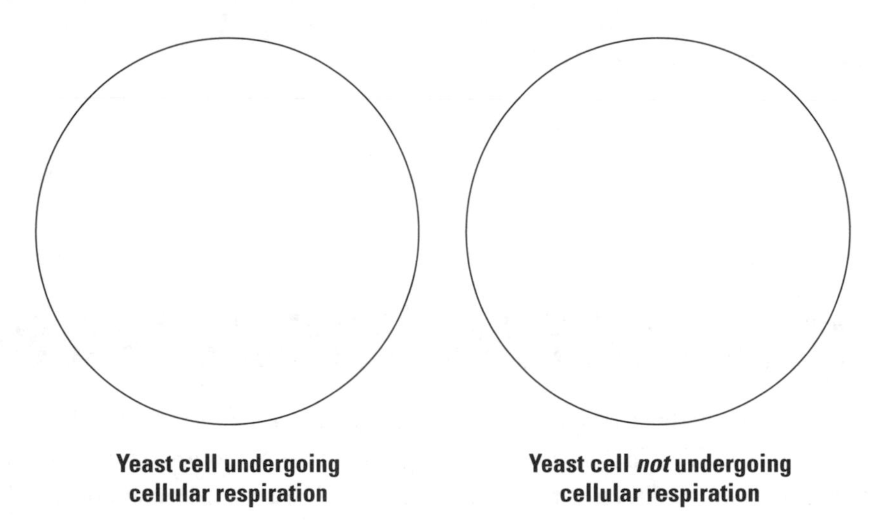

Yeast cell undergoing cellular respiration | **Yeast cell *not* undergoing cellular respiration**

5. Can you identify any cellular structures? Label them in your drawings.

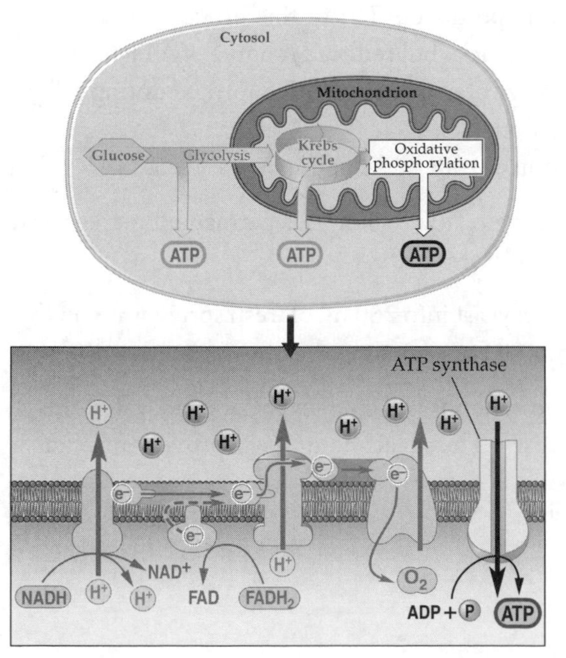

Figure 9.6
Cellular Respiration

Concept Check

1. Which step of cellular respiration requires oxygen?

2. At what stage of cellular respiration is carbon dioxide produced?

3. How are yeast cells like human cells?

ACTIVITY 5 Fermentation

An absence of oxygen in the early atmosphere forced the first life-forms to create ATP through anaerobic processes. The production of ATP in the cytosol by glycolysis does not require oxygen, so it is an anaerobic process. To sustain glycolysis, NADH must be recycled into NAD^+, a main reactant of glycolysis. The reactions recycling NAD^+ and processing pyruvate are called fermentation.

There are many different types of fermentation, but all of them have the same purpose: to break down pyruvate to regenerate NAD^+ so that it can be used again during glycolysis. Alcoholic fermentation recycles NAD^+ by producing ethanol and carbon dioxide (Figure 9.7). Many bacteria and yeast undergo alcoholic fermentation. Humans use lactic acid fermentation to quickly produce ATP; in this process, lactic acid is formed instead of carbon dioxide and ethanol. Lactic acid is a quick (but reduced) source of ATP for a high-intensity, short-term activity such as a sprint; in contrast, the greater amount of ATP formed during cellular respiration fuels long-term activity, such as running a marathon.

In this activity you will examine the production of carbon dioxide and alcohol in yeast cells.

1. Obtain a yeast packet, spring water, a 500-ml beaker, a pipette, four balloons, a permanent marker, and sugar solution. Your instructor will have also set up a 37°C water bath.

2. Create a fresh culture of yeast cells by pouring a packet of yeast into 250 ml of fresh spring water in the beaker. Stir with the pipette for a minute. Let the culture sit for 5 minutes.

3. Set up four balloons as directed by your instructor. Two balloons will contain culture and water; the other two will contain culture and sugar solution (glucose). The sugar solution will provide an extra source of glucose to be used during fermentation.

Figure 9.7
Fermentation

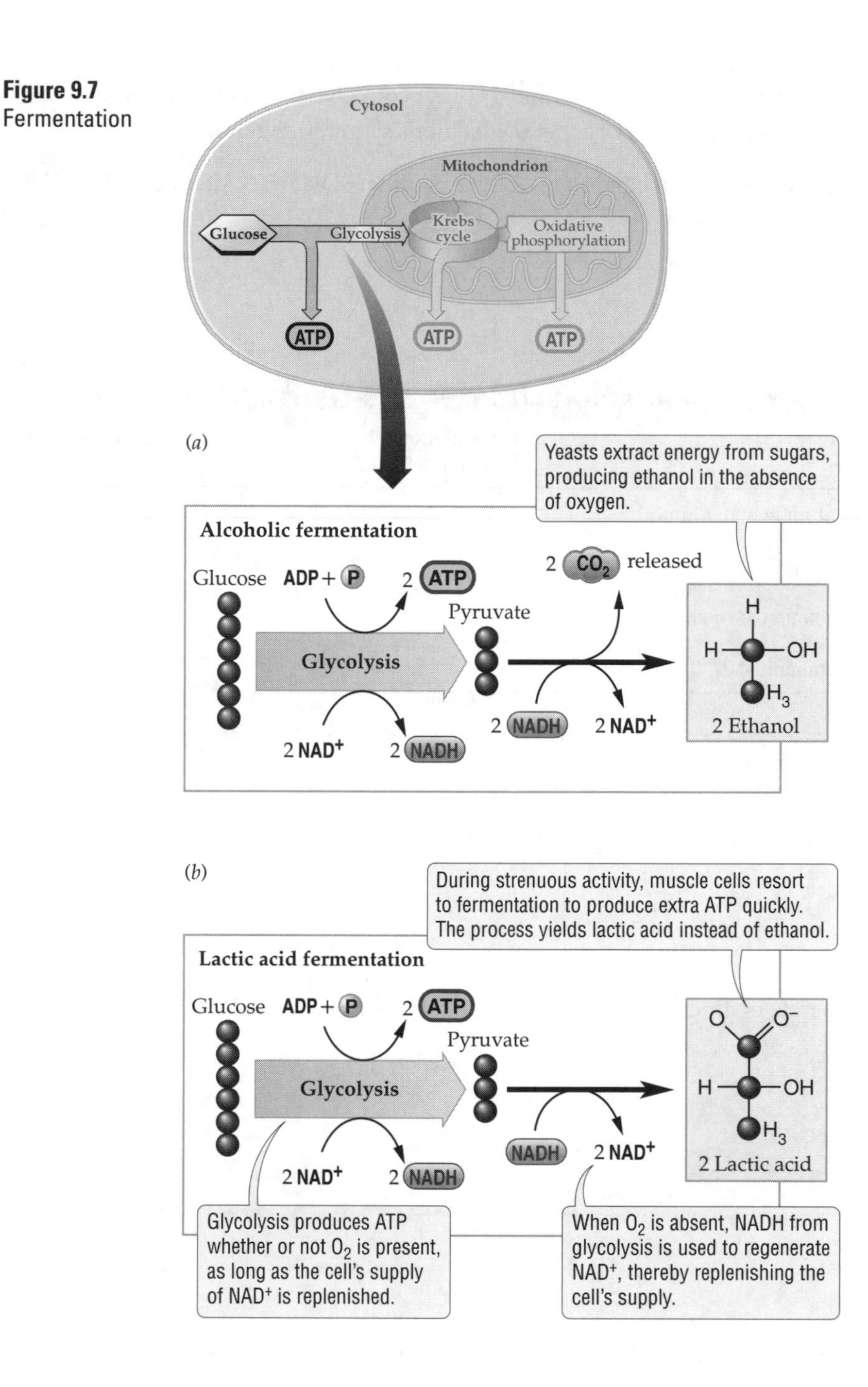

4. Using the permanent marker, label the balloons as follows: "Yeast and water at 37°C," "Yeast and water at room temperature," "Yeast and sugar solution at 37°C," "Yeast and sugar solution at room temperature," as indicated in Table 9.5.

5. For each of the four balloons, remove as much air as possible and tie a knot sealing the balloon. Hold up the balloon by the knot and measure the diameter of the solution inside. Record the diameter in Table 9.5, under "Diameter before incubation."

6. Place one balloon that contains water and one that contains sugar in the water bath at 37°C, and leave the other two at room temperature. Let the balloons sit for 30 minutes.
7. Record the diameter of each balloon in Table 9.5, under "Diameter after incubation." Why have the balloon diameters changed?

TABLE 9.5 ACTIVITY 4: FERMENTATION RATES OF YEAST CELLS AT DIFFERENT TEMPERATURES

Balloon	Circumference before incubation (cm)	Circumference after incubation (cm)
Yeast and water at room temperature		
Yeast and water at 37°C		
Yeast and sugar solution at room temperature		
Yeast and sugar solution at 37°C		

8. Which balloon's diameter changed the most?
9. What is causing the balloon to expand?
10. Did the presence of glucose increase the product of fermentation?

Concept Check

1. What is the goal of fermentation?
2. How do fermentation and cellular respiration differ? How are they similar?
3. Why do organisms like humans use fermentation instead of cellular respiration at certain times?

ACTIVITY 6 Overview of Photosynthesis and Cellular Respiration

The relationship between photosynthesis and cellular respiration is complementary. One builds up the molecules of life (photosynthesis); the other breaks down the molecules of life (cellular respiration). These processes go hand in hand, and in plant cells they literally occur within the same cells. Most plant cells contain mitochondria, and some contain chloroplasts as well. The mesophyll cells of plants contain both mitochondria and chloroplasts (Figure 9.8). The following activity will illustrate the reactants and products of the stages of photosynthesis and cellular respiration. It is important to remember that all the various life-forms that exist on Earth use some form of cellular respiration or fermentation to produce ATP, but only select groups can perform photosynthesis. In this activity you will follow photosynthesis and cellular respiration within a plant cell. You will locate each cellular component and map key steps in the general metabolic pathway of each process.

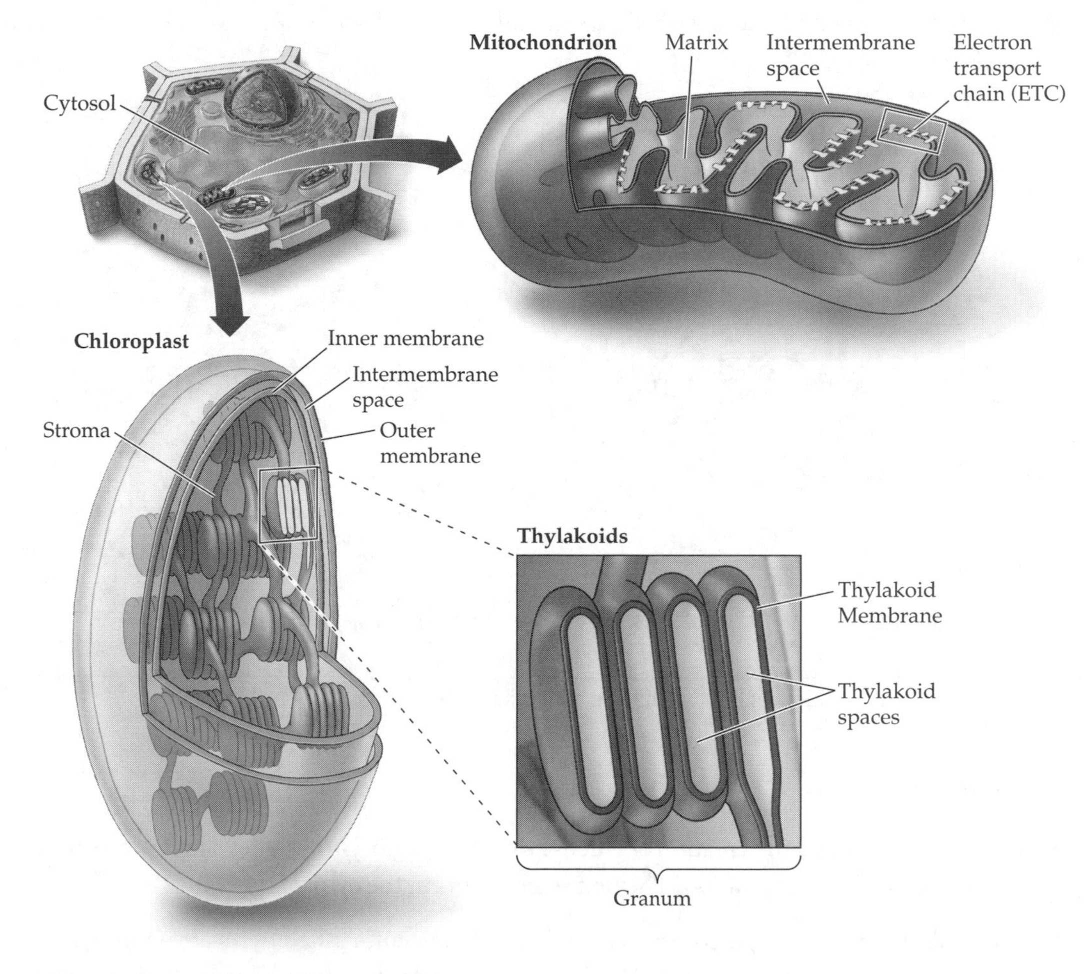

Figure 9.8
Locations of Photosynthesis and Cellular Respiration

Photosynthesis

1. Examine the chloroplast in Figure 9.8.

2. The chloroplast is composed of the intermembrane space between the inner and outer membranes, the stroma enclosed by the inner membrane, and thylakoids within the stroma. Find these compartments in Figure 9.8.

3. Photosynthesis is composed of light reactions and the Calvin cycle. The light reactions occur within the thylakoid membranes, where pigments are located.

4. The light reactions use sunlight and water, form oxygen and release electrons, and ultimately produce ATP and NADPH. ATP and NADPH then move on to the Calvin cycle within the stroma. Add the labels "Water," "Oxygen," "ATP," and "NADPH" where appropriate in boxes 1 through 4 of Figure 9.9.

5. In the Calvin cycle, six carbon dioxide molecules, along with ATP and NADPH molecules, are used to create one glucose molecule. Add the labels "CO_2" and "Glucose" where appropriate in boxes 5 and 6 of Figure 9.9.

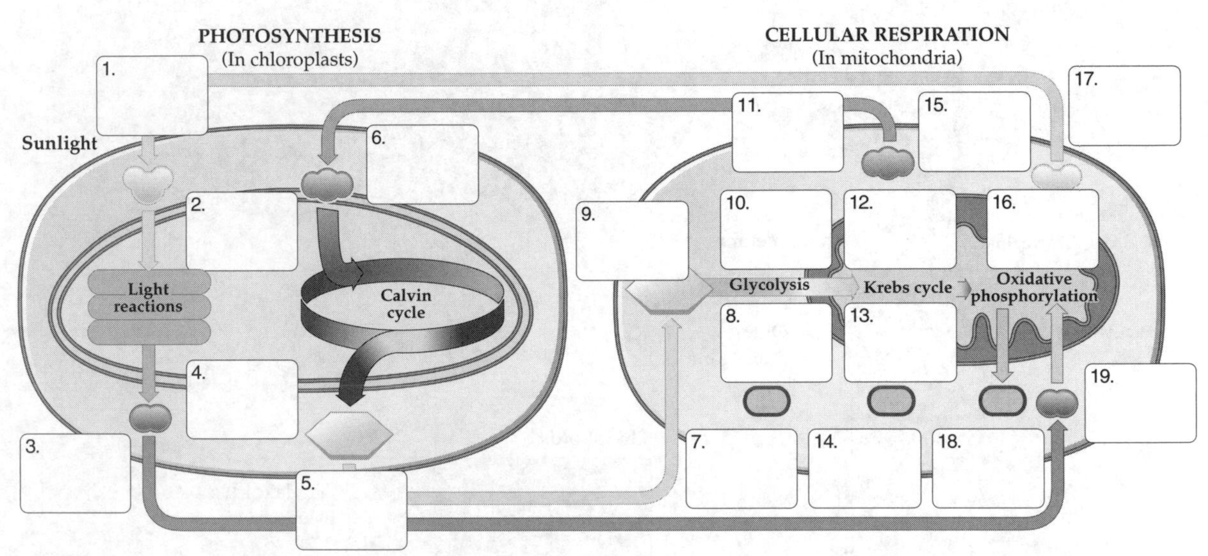

Figure 9.9
Locations of Molecules and Products in Photosynthesis and Cellular Respiration

Cellular Respiration

1. Now that you have mapped the creation of glucose through photosynthesis, let's follow the breakdown of glucose through cellular respiration. The first stage occurs in the cytosol.

2. Glycolysis breaks the molecule glucose into two identical molecules called pyruvate. During this process, ATP and an electron carrier (NADH) are formed. Add the labels "Glucose," "Pyruvate," "ATP," and "NADH" where appropriate in boxes 7 through 10 of Figure 9.9.

3. The final stages of cellular respiration occur within the mitochondrion. Examine the mitochondrion in Figure 9.8.

4. Pyruvate is processed in the mitochondrion. First, pyruvate is converted into acetyl CoA, releasing carbon dioxide and producing an electron carrier (NADH). Add the labels "Acetyl CoA," "CO_2," and "NADH" where appropriate in boxes 11 through 13 of Figure 9.9.

5. The Krebs cycle starts with acetyl CoA releasing carbon dioxide, ATP, and an electron carrier (NADH). Add the labels "CO_2," "ATP," and "NADH" where appropriate in boxes 14 through 16 of Figure 9.9.
6. Oxidative phosphorylation uses the electrons deposited by electron carriers (NADH). These electrons are used by enzymes within the membrane of the inner compartment of the mitochondrion. The process powers the production of three ATP molecules for the electrons deposited by every NADH. Add the labels "ATP," "Oxygen," and "Water" where appropriate in boxes 17 through 19 in Figure 9.9.

Concept Check

1. What is the source of carbon used to make glucose in the Calvin Cycle?

2. Where are NADH molecules formed during cellular respiration?

3. Where are most ATP molecules formed during cellular respiration?

4. In one sentence, describe the purpose of photosynthesis.

5. In one sentence, describe the purpose of cellular respiration.

6. How are photosynthesis and cellular respiration complimentary processes in a plant cell?

7. How are photosynthesis and cellular respiration complementary processes in a community of organisms?

Key Terms

accessory pigment (p. 9-5)
aerobic (p. 9-1)
alcoholic fermentation (p. 9-2)
anabolic (p. 9-1)
anaerobic (p. 9-2)
antenna complex (p. 9-5)
ATP (p. 9-2)
Calvin cycle (p. 9-1)
carbon fixation (p. 9-2)
catabolic (p. 9-1)
cellular respiration (p. 9-1)
chlorophyll *a* (p. 9-5)
chloroplast (p. 9-4)
chromatography (p. 9-5)
consumer (p. 9-1)
fermentation (p. 9-2)
glycolysis (p. 9-1)
inner membrane (p. 9-5)
intermembrane space (p. 9-5)
Krebs cycle (p. 9-1)
lactic acid fermentation (p. 9-2)
light reactions (p. 9-1)
mitochondrion (p. 9-12)
NADPH (p. 9-2)
outer membrane (p. 9-5)
oxidative phosphorylation (p. 9-1)
photon (p. 9-8)
photosynthesis (p. 9-1)
pigment (p. 9-5)
producer (p. 9-1)
pyruvate (p. 9-12)
ratio factor (p. 9-5)
Rf value (p. 9-5)
stoma (p. 9-2)
stroma (p. 9-5)
thylakoid (p. 9-5)
visible light spectrum (p. 9-8)
wavelength (p. 9-8)

Review Questions

1. ________________ are the organelles in plant cells that convert solar energy into usable chemical energy for the rest of life on Earth.
2. The process of absorbing carbon dioxide and incorporating the atoms from this gas into an organic molecule is called ________________.
3. A(n) ________________ is any molecule capable of absorbing light energy.
4. The reactions recycling the reactants of glycolysis are called ________________.
5. ________________ are discrete packets of light energy.
6. ________________ consist of two specialized cells called guard cells, which open and close to allow gas exchange.
7. ________________ is the organelle where photosynthesis occurs; ________________ is the organelle where the majority of cellular respiration occurs.
8. What are the three steps of cellular respiration?
9. Explain the relationship between cellular respiration and photosynthesis.
10. Muscle cells require quick sources of ATP provided by fermentation. What product of fermentation causes the soreness you feel after intense exercise?

11. Yeast used in bread making undergoes alcohol fermentation. This process produces alcohol, yet bread clearly does not contain alcohol. What happens to the alcohol?

12. Explain the light reactions and the Calvin cycle of photosynthesis.

13. What is the function of light during photosynthesis?

14. The final stage of cellular respiration is called oxidative phosphorylation. From your understanding about the reactions of this stage, why is this an appropriate name?

15. What is the visible light spectrum?

16. Describe the difference between pigments in your clothing and pigments in chloroplasts.

LAB **10**

Mitotic Cell Division, Meiosis, and Mendelian Genetics

Objectives

- Trace the cell cycle through the phases that make up a cell's lifetime.
- Explain the significance of two types of cell division: mitosis and meiosis.
- Observe mitotic cell division in animal and plant cells.
- Simulate meiosis and show how independent assortment and crossing-over contribute to genetic variation.
- Use a Punnett square to display parental gametes and probable offspring in a one-trait cross.
- Discuss how the drawing of a Punnett square relates to Mendel's law of segregation.
- Simulate mating between two parents to demonstrate a connection between Mendel's law of independent assortment and genetic variation.

Introduction

One of life's defining characteristics is growth. Growth may be an increase in the size or the number of cells in an organism. The **cell cycle** describes a series of distinct stages in the life of a cell that is capable of dividing. Multicellular organisms use mitotic cell division—**mitosis**—to create two genetically identical cells from one parent cell during stages of development in which much growth is needed, or to replace dead or dying cells throughout life. Figure 10.1 depicts the life of a cell from its creation and growth to its production of daughter cells by means of mitotic cell division. The length of each phase of the cell cycle depends on the organism, cell type, and stage of life. Checkpoints are built into the cell cycle to ensure that the process is occurring correctly. Sometimes these checkpoints fail, mainly because of DNA damage, causing inappropriate uncontrolled cell division known as **cancer**.

Other cells within sexually reproducing organisms specialize in producing **gametes** (sex cells) through the process of **meiosis**. Both meiosis and mitotic cell division undergo DNA replication, but whereas mitosis creates two daughter cells that are identical to the parent cell, meiosis creates four daughter cells (gametes) that each contain half of the chromosomes found in the parent cell. By creating four genetically different gametes, meiosis contributes to genetic variation. Gametes from two different individuals then randomly come together during fertilization to produce an individual cell called a **zygote** that then undergoes mitotic cell division.

Mitosis and meiosis depend on the duplication of **chromosomes**, DNA molecules packaged tightly with proteins and containing many genes that dictate the organism's characteristics (traits). **Mendelian genetics** describes the general processes by which traits are passed on to a gamete and the likelihood that an offspring will inherit those traits. In this lab you will examine mitotic cell division, simulate meiosis, and explore the laws of inheritance put forth by Gregor Mendel, a pioneer in understanding the inheritance of characteristics.

ACTIVITY 1 The Cell Cycle and the Structure of Chromosomes

The cell cycle has two main phases (Figure 10.1): interphase and cell division. Most cells that are destined to divide spend 90 percent or more of their lifetime in **interphase**, taking in nutrients, making proteins, performing other day-to-day functions, and finally preparing for cell division (see Figure 10.1). In mitotic cell division, a parent cell divides into two identical daughter cells. In meiosis (meiotic cell division), a parent cell divides twice, resulting in four gametes that have half the chromosomes of the parent cell. During sex that results in pregnancy, one gamete from a female will pair with one gamete from a male to conceive a child with genetic information from both parents. Both types of cell division—mitosis and meiosis—undergo interphase.

Interphase is divided into three special phases: G_1, S, and G_2. The **G_1 phase** is the first phase of the cell cycle, where normal cellular function occurs and proteins prepare the cell for division. Cells destined for cell division exit the G_1 phase to enter the **S phase**, which is characterized by DNA replication. The last phase of interphase is the **G_2 phase**, a period of cell growth in which regulatory cell proteins further prepare the cell for division. Both the G_1 and G_2 phases of interphase are very important checkpoints for either mitotic or meiotic cell division, ensuring that the cell is ready to divide and properly prepared. (Some cells never divide, and some divide very infrequently. Cells in a nondividing state are in the **G_0 phase**, which is separate from interphase, as illustrated in Figure 10.1.)

In this activity you will follow the steps of interphase and learn more about the structure of chromosomes and duplicated chromosomes.

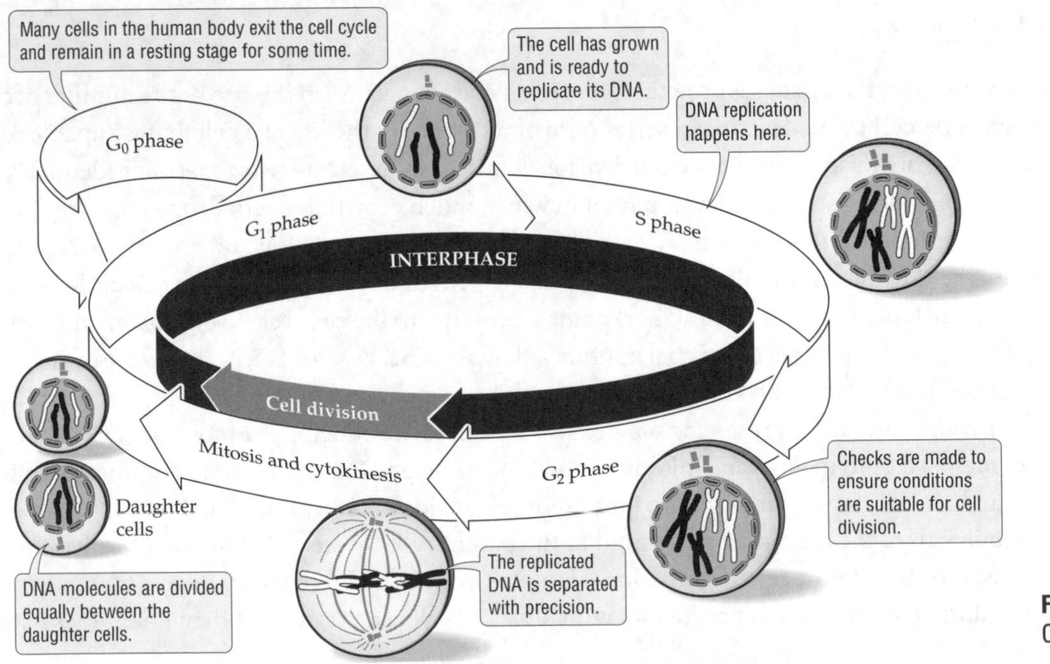

Figure 10.1
Cell Cycle

1. Mitotic cell division is for generic cells (**somatic cells**)—cells that are not gametes and do not produce gametes. What is an example of a somatic cell? What two types of gametes do humans produce?

2. DNA is not simply a twisted ladder of two nucleotide chains. It contains proteins defining its structure in a package called a chromosome. Examine Figure 10.1 and describe how DNA within the nucleus of a cell changes from the end of the G_1 phase to the end of the S phase.

3. When DNA condenses, each chromosome replicated during the S phase pairs with its identical duplicate to form an X-like structure. The two paired identical chromosomes are called **sister chromatids**. The right-hand side of the X-like structure represents one chromatid, and the left side represents the identical DNA structure. Examine Figure 10.2, which illustrates this process. The two chromatids are shaded differently to make this structure easier to understand. Are both chromatids the same size?

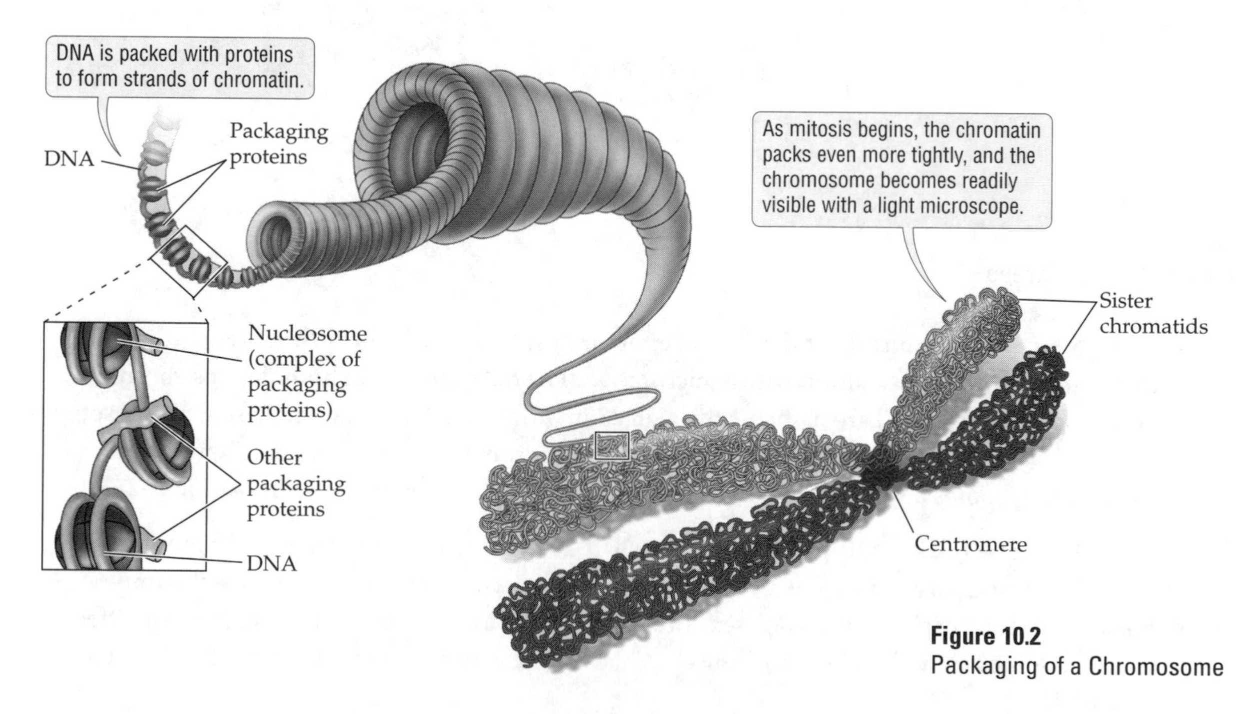

Figure 10.2
Packaging of a Chromosome

4. Notice that the sister chromatids contain one area that is significantly more constricted than the other regions; this constricted area is called the **centromere**. The centromere is an important structure during mitosis. Locate the centromere of the sister chromatids in Figure 10.2. The thicker portions of the chromatids above and below the centromere are typically different lengths. Identify the long arm and the short arm of each chromatid in Figure 10.2.

5. The nucleus of a human cell has 46 chromosomes (two sets of 23). One set is maternal (inherited from the mother); the other is paternal (inherited from the father). Those that have the same shape and contain the same genes are referred to as a **homologous pair**, or a set of homologous chromosomes, as depicted in Figure 10.3. Using the figure for reference, describe the composition of each homologous pair.

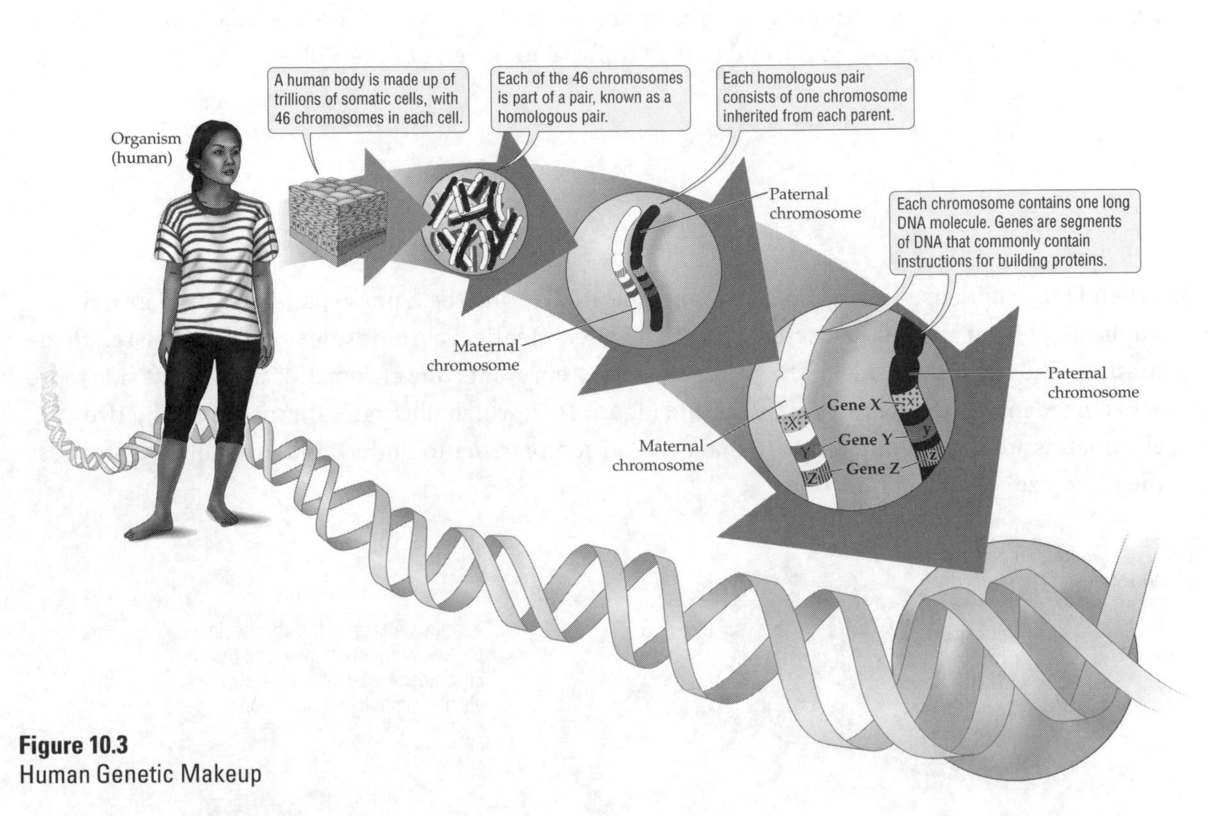

Figure 10.3
Human Genetic Makeup

6. When DNA condenses further at the start of either mitosis or meiosis, all of the sister chromatids within a nucleus can be visualized with a microscope. The different sets of homologous chromosomes (homologous pairs) are named 1 through 23 in order from largest to smallest. Chromosomes 1 through 22 are autosomes, nonsex-determining chromosomes of the individual. Examine the human chromosomes depicted in Figure 10.4 and identify the **autosomal chromosomes**.

7. Figure 10.4 shows the **karyotype** of a somatic cell. It depicts each pair of condensed homologous chromosomes of an organism. The twenty-third homologous pair consists of the **sex chromosomes**. Females have two of the same sex chromosome (**X chromosome**); males have two different sex chromosomes (X and **Y chromosomes**). What sex is the person whose karyotype is depicted in Figure 10.4?

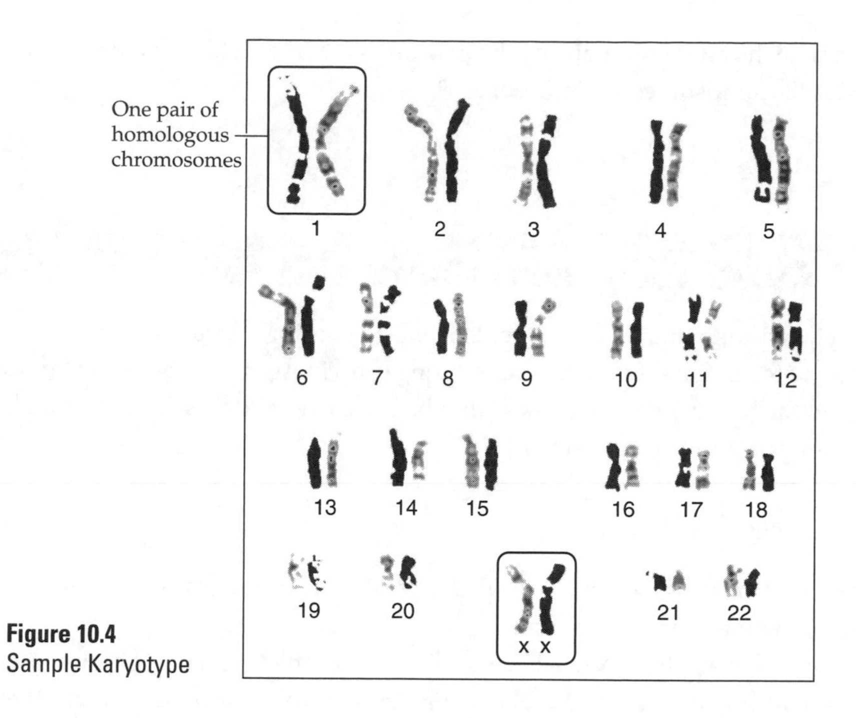

Figure 10.4
Sample Karyotype

Concept Check

1. Describe the three phases of interphase.

2. What is the chromosomal difference between males and females?

3. Define "homologous pair."

4. What are sister chromatids?

5. What information can a karyotype tell you about an organism?

6. How many homologous pairs of chromosomes are in the standard human karyotype? How many individual chromosomes are represented?

ACTIVITY 2 Mitosis and Cytokinesis

Mitotic cell division requires the replication of DNA during interphase, the separation of DNA during mitosis, and the formation of two new cells during cytokinesis. In this activity you will examine the stages of mitosis and the process of cytokinesis, and then observe mitosis in animal and plant cells, using slides of an onion root tip and a whitefish blastula.

Mitosis and Cytokinesis

Four stages of mitosis occur after DNA has already been replicated during interphase: prophase, metaphase, anaphase, and telophase.

Prophase ("prep the DNA") occurs directly after the G_2 phase of interphase. In early prophase, the sister chromatids condense further, becoming visible with a microscope; and the cytoskeletal structures called **centrosomes** move toward opposite sides (poles) of the cell, sprouting microtubules (Figure 10.5). These microtubules form a cage of fibers, the **mitotic spindle**, which later aligns the chromosomes and pulls them apart. In late prophase, the **nuclear envelope** (the membrane surrounding DNA) breaks down (see Figure 10.5). Microtubules from the mitotic spindle attach to proteins (called **kinetochores**) surrounding the centromeres of sister chromatids.

In **metaphase** ("meet in the middle"), the microtubules align the sister chromatids in the middle of the cell equidistant from each centrosome, as illustrated in Figure 10.5. The sister chromatids face opposite poles in preparation to be separated during anaphase. The alignment is along an arbitrary line called the **metaphase plate**.

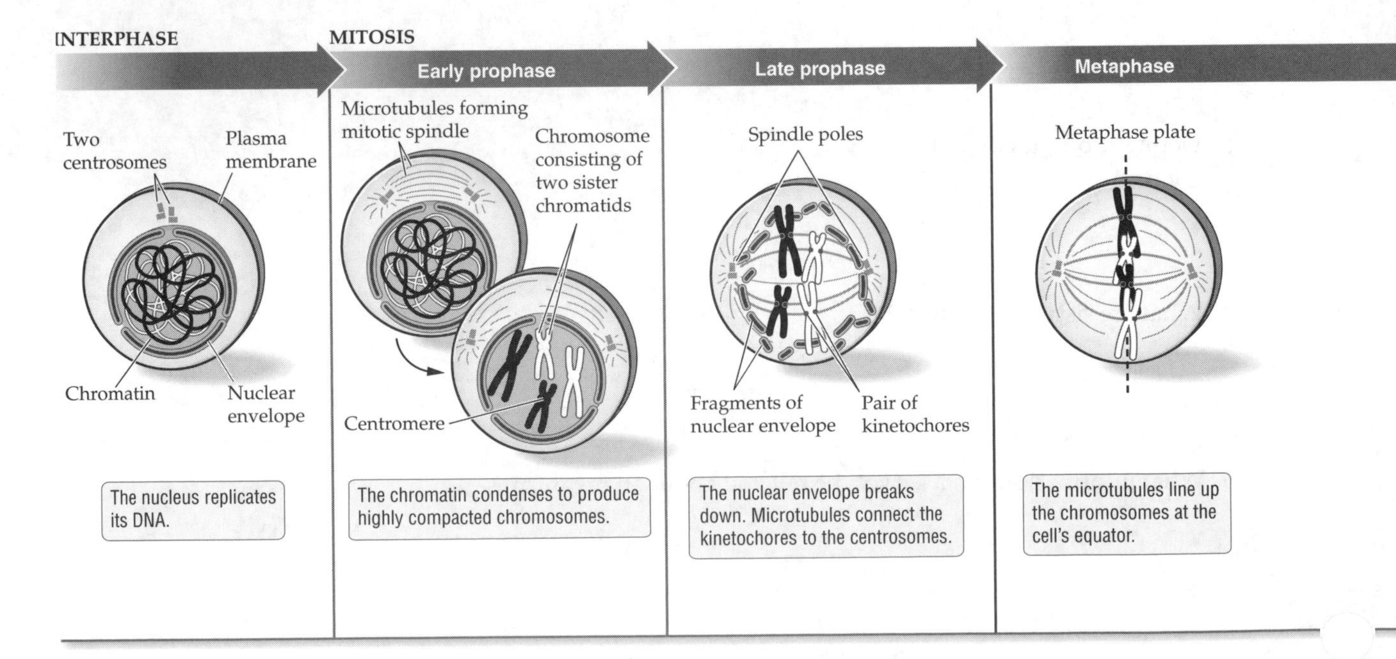

In **anaphase** ("pull sister chromatids apart"), the duplicated DNA is separated. The microtubules attached to the kinetochores shorten, pulling the sister chromatids toward the centrosomes located on opposite sides of the cell.

Telophase ("one cell becomes two") is the final stage of mitosis. As depicted in Figure 10.5, it occurs simultaneously with the final step of cell division, **cytokinesis**. As soon as sister chromatids reach opposite sides of the cell, telophase starts as a nuclear envelope forms around each set of chromosomes. The chromosomes are decondensed, and the parent cell starts to split.

Use the information above and Figure 10.5 to answer questions about mitotic cell division.

1. What stage of mitosis separates the sister chromatids?

2. During what stage can you visualize DNA?

3. What does mitotic cell division produce?

4. How are chromatids aligned during metaphase, and what is the purpose of their alignment?

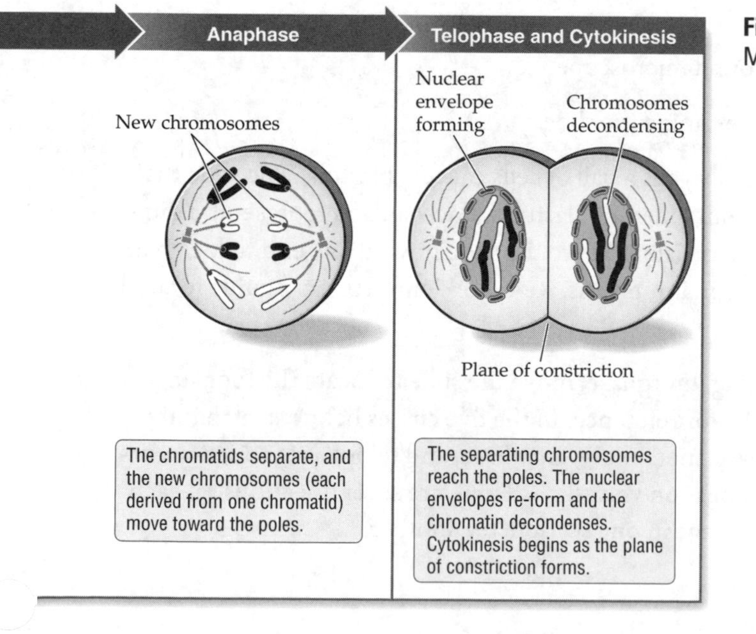

Figure 10.5
Mitotic Cell Division in Animals

Animal and plant cells divide differently during cytokinesis. Animal cells cinch the plasma membrane inward along the metaphase plate, forming a **plane of constriction**. A ring of actin microfilaments contracts, pulling the plasma membrane inward until all sides meet in the middle, fusing to form two new cells. A plant cell, by contrast, cannot form two cells by the same process, because of the rigidity of its cell wall. Instead a plant cell positions vesicles with components of both the plasma membrane and the cell wall along the metaphase plate. These vesicles fuse, forming a **cell plate** along the metaphase plate, as illustrated in Figure 10.6.

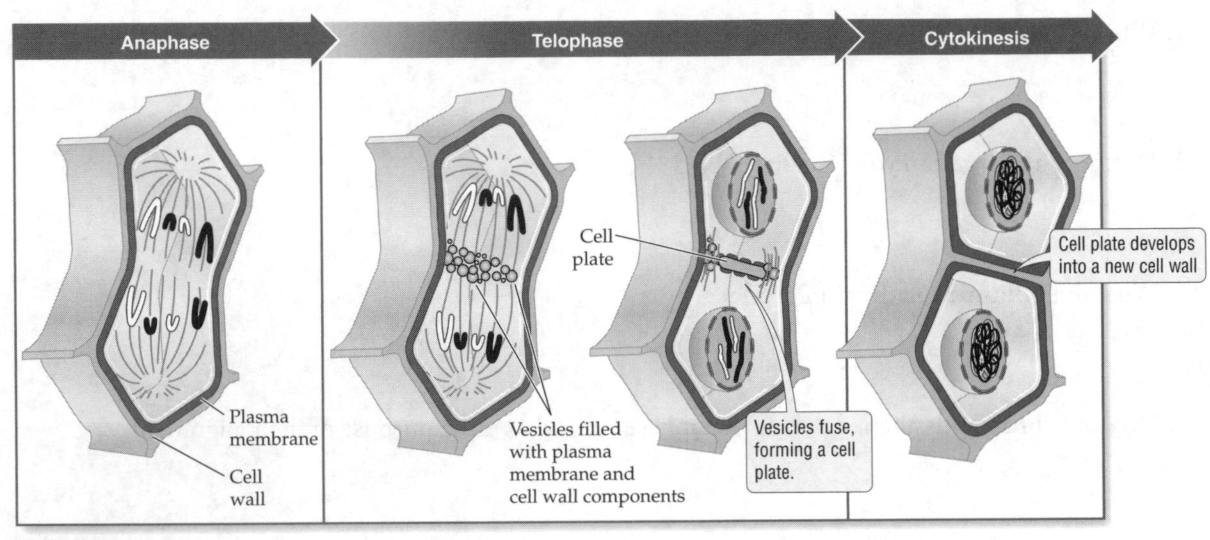

Figure 10.6
Mitotic Cell Division in Plants

Although the details of the process may vary from species to species, in most cases the stages are readily recognizable under the microscope because chromosomes are heavily condensed.

Observing Mitotic Cell Division in Animals

1. Obtain a whitefish blastula slide and a compound microscope.
2. Using the microscope starting on low power, examine the slide.
3. During the early stages of development, the embryo is a ball of cells called a blastula. The slide that you are examining contains several sections from different blastulas. Each section represents a different step during this process of coordinated cell division. Move the slide toward one end to locate interphase, where DNA is replicated in preparation for mitosis. Refer to Figure 10.5 as a guide for each stage.
4. Once you have located the section representing interphase, move the slide to locate the four stages of mitotic cell division, again using Figure 10.5 for reference. In the five circles below, sketch interphase and the four stages of mitosis—prophase, metaphase, anaphase, and telophase (which occurs simultaneously with cytokinesis)—as they appear on your slide. There are several sections on each slide; if you cannot locate a particular mitotic stage in one section, try another.

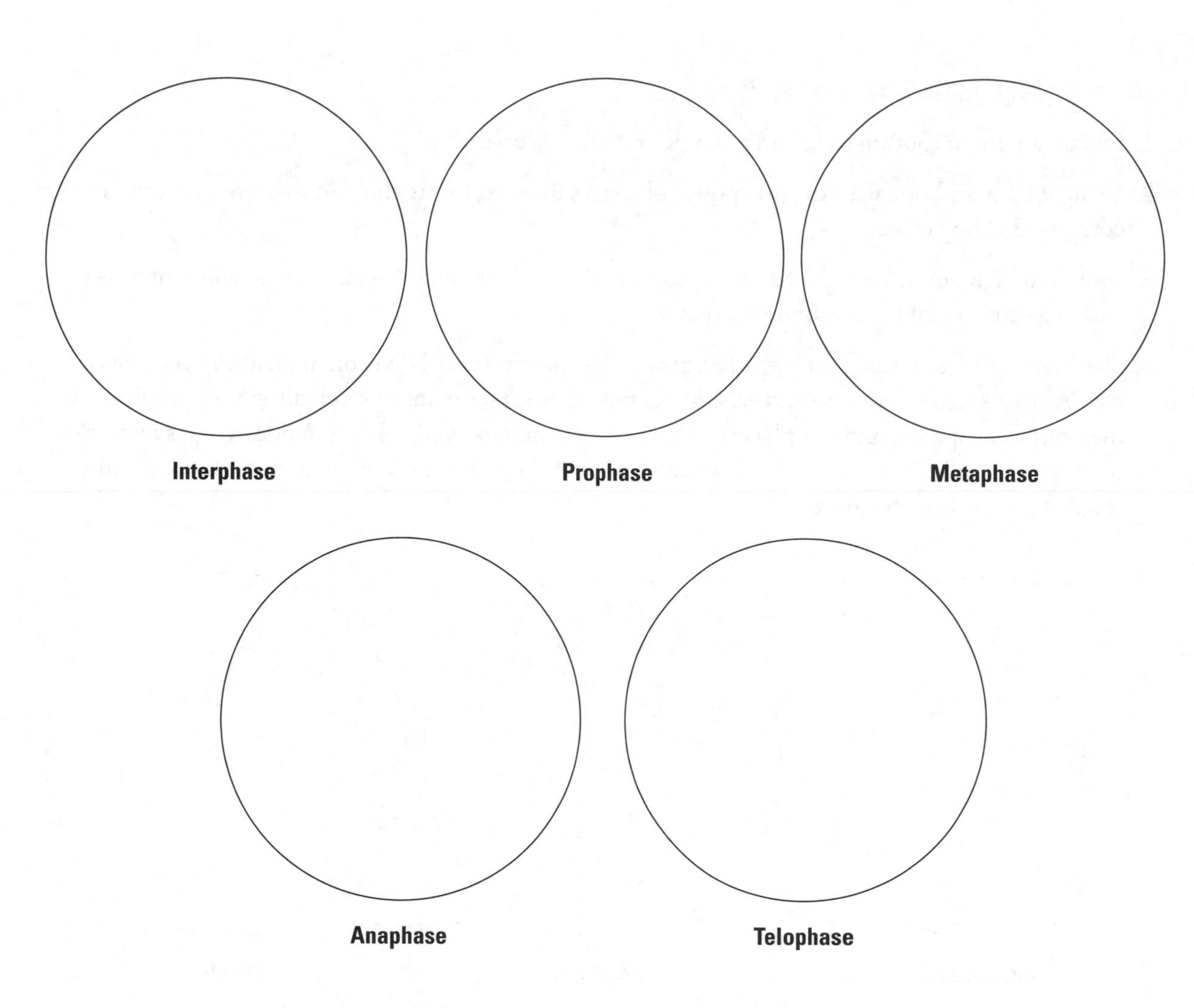

5. The centrosomes are too small to identify, but by metaphase you should be able to identify the different poles of the cell, where the centrosomes are located. Where they are visible under the microscope, label the poles in your drawings.

6. In which stage(s) is the mitotic spindle visible? Label this structure in your drawings.

7. Can you see the plasma membrane pulling inward during telophase? Why is this movement important?

Observing Mitotic Cell Division in Plants

1. Obtain an onion root tip slide and a compound microscope.
2. Using the microscope with the low-power objective in place, locate the portion of the root tip that contains dividing cells.
3. Switch to high power and identify each stage of mitosis. (The slide contains many different cells within a single plant tip at different stages of mitosis.)
4. Use Figure 10.6 as a guide for identifying the stages of mitotic cell division in plants. In the five circles below, sketch interphase and each of the following stages of mitotic cell division—prophase, metaphase, anaphase, and telophase (which occurs simultaneously with cytokinesis)—as they appear on your slide. There are several sections on each slide; if you cannot locate a particular mitotic stage in one section, try another.

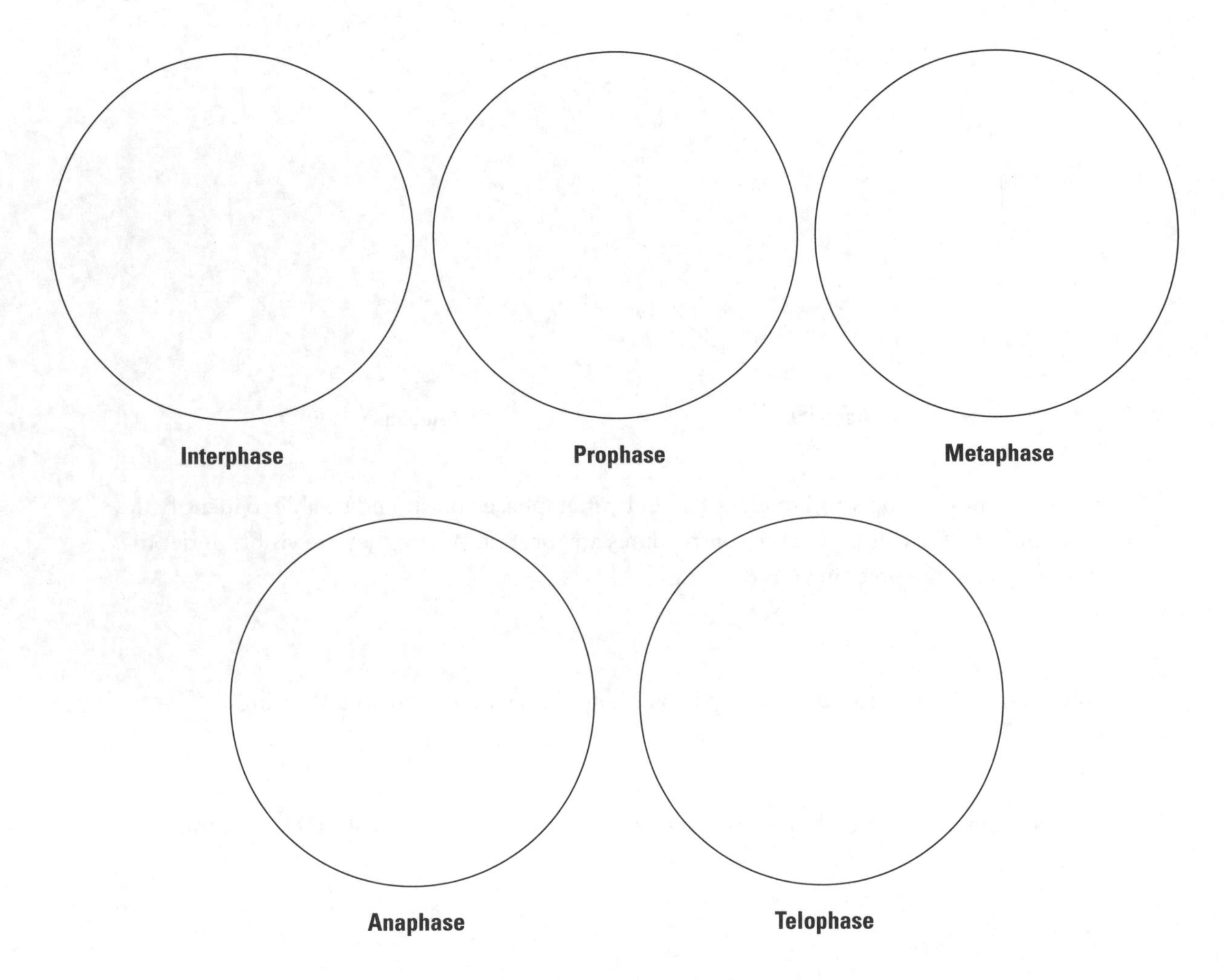

5. Is which stage(s) is the mitotic spindle visible? Label this structure in your drawings.
6. In the telophase drawing, label the cell plate of cytokinesis.
7. Why do you think plant and animal mitotic phases are so similar?

Concept Check

1. What structures are responsible for moving chromosomes around the cell during mitosis?

2. Explain how cell division creates two daughter cells that are identical to the parent cell. Why is this an important function in a plant or an animal?

3. Why is cytokinesis in plants different from cytokinesis in animals?

ACTIVITY 3 Meiosis

Meiosis, a second type of cell division, produces gametes. Meiosis is different from mitotic cell division in three ways: it produces genetically different cells, results in four cells instead of two, and reduces the amount of DNA in the cell. Somatic cells contain a double set of chromosomes (one set of 23 from the mother and one set of 23 from the father). Somatic cells are **diploid** cells, named as such because they have two complete sets of homologous chromosomes. Diploid cells are typically represented as $2n$ (where n is the number of chromosomes). During meiosis, a diploid cell duplicates its DNA and then divides twice instead of just once. This process reduces the amount of DNA in the parent cell by half, yielding four daughter cells, each with just one set of chromosomes (n). These are called **haploid** cells.

As with mitotic cell division, cells that undergo meiosis first replicate DNA during interphase and then enter the first stage of cell division. In meiosis, cell division begins with **meiosis I**. When the chromosomes condense during prophase I of meiosis I, not only the sister chromatids, but also the pairs of homologous chromosomes, come together. Examine Figure 10.7 to see how the chromosomes arrange themselves. The structure of the duplicated maternal and paternal homologous chromosomes is known as a **bivalent**. The bivalents meet on the metaphase plate in metaphase I, and then the homologous chromosomes separate during anaphase I. The result is two cells, each containing one set (n) of duplicated sister chromatids.

As Figure 10.7 shows, the second cell division, **meiosis II**, separates the duplicate sister chromatids to form haploid daughter cells, each with one copy of half of the number of chromosomes in the parent cell. Each daughter cell, or gamete, receives one copy of one member of the homologous pair of the parent cell. In this activity you will simulate meiosis and discover two ways that meiosis creates genetic diversity in offspring: through crossing-over and independent assortment.

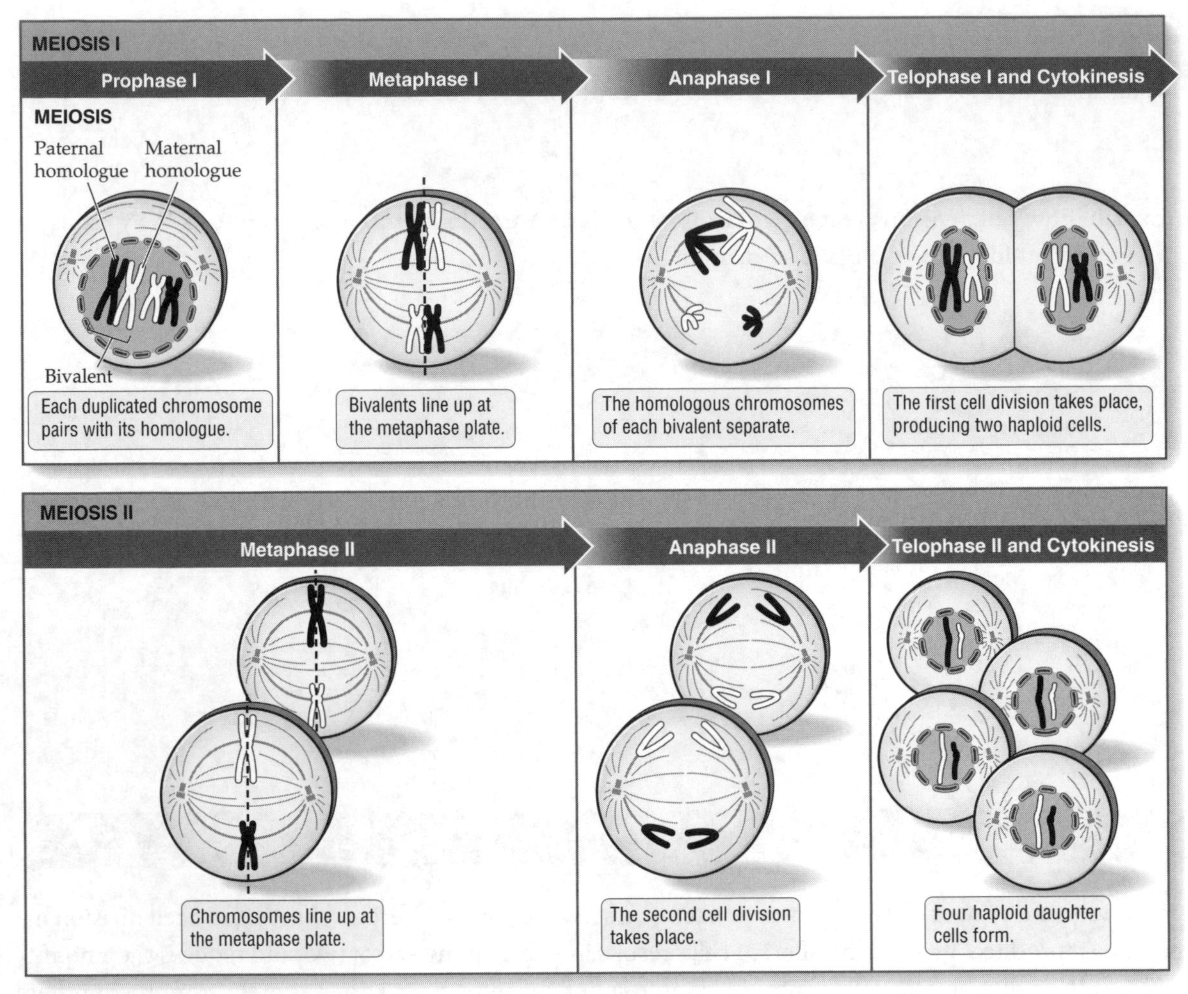

Figure 10.7
Meiosis I and Meiosis II

Interphase

1. Your instructor will provide you with four chromosomes made out of plastic interlocking building blocks. Each chromosome is labeled either "1" (longer chromosome) or "2" (shorter chromosome) and is either red, indicating maternal, or blue for paternal. The maternal chromosome 1 is homologous to the paternal chromosome 1, and the same is true of each chromosome 2. These two homologous pairs that have been created for you represent two homologous pairs in a diploid cell, *before* DNA is duplicated during the S phase of interphase.

2. What are the similarities between homologous chromosomes?

3. Your instructor will also provide the additional building blocks of the hypothetical organism's chromosomes. Mimic DNA replication by building a duplicate of both maternal and paternal chromosomes 1 and 2, adding the correct color and size blocks in the same order as the original chromosomes. This duplication ends interphase and now the cell is ready for meiosis I.

NOTE: A CHROMOSOME IS A DNA MOLECULE WRAPPED AROUND PROTEINS WITH SPECIFIC GENES. A "CHROMATID" IS EQUIVALENT TO A CHROMOSOME, BUT THIS TERM INDICATES THAT THE CHROMOSOME HAS ALREADY BEEN REPLICATED DURING INTERPHASE, AND THEREFORE THERE IS ANOTHER IDENTICAL CHROMATID. THE TWO IDENTICAL CHROMATIDS ARE REFERRED TO AS SISTER CHROMATIDS, WHICH FORM THE X-LIKE STRUCTURE DURING PROPHASE (SEE FIGURE 10.7).

Meiosis I

1. Examine Figure 10.7 to see how the chromosomes condense during prophase and come together. As illustrated in the figure, place the two identical maternal chromatids of chromosome 1 (the original and the one you built) side by side. Use a rubber band to cinch the chromatids together, imitating the centromeres of the pair during meiosis and uniting sister chromatids. Do this for each pair of identical chromatids. At the end you should have four pairs of sister chromatids, each pair composed of chromatids of the same size and color.

2. Now, using the illustration of prophase I in Figure 10.7 for reference, bring together the maternal and paternal homologues of each chromosome. Remember: homologous chromosomes are the same size, with the same genes located along the chromosomes. Therefore, for chromosome 1 you should be bringing together the longest red (maternal) sister chromatids and the longest blue (paternal) sister chromatids. Now, four long chromatids (two red and two blue) should be in a row forming a homologous bivalent (see Figure 10.7, prophase I). Create a second bivalent with the sister chromatids of chromosome 2. Your bivalents should resemble those in Figure 10.7.

3. How many homologues are in each bivalent?

4. How many chromatids are in one homologue (either maternal or paternal)?

5. Once the bivalents have formed, the proximity of the two homologues in each bivalent creates the potential for paternal and maternal chromatids to exchange a portion of DNA, a process called **crossing-over**. Homologous chromosomes have similar sequences of DNA, allowing these chromosomes to exchange pieces of DNA. Using Figure 10.8 for reference, mimic this process by removing two red blocks from one maternal chromatid and two blue blocks from one paternal chromatid within the bivalent of chromosome 1. Replace the missing two blocks from the maternal chromatid with the blocks that you took from the paternal chromatid, and vice versa. You started with a bivalent with two completely red chromatids next to two completely blue chromatids.

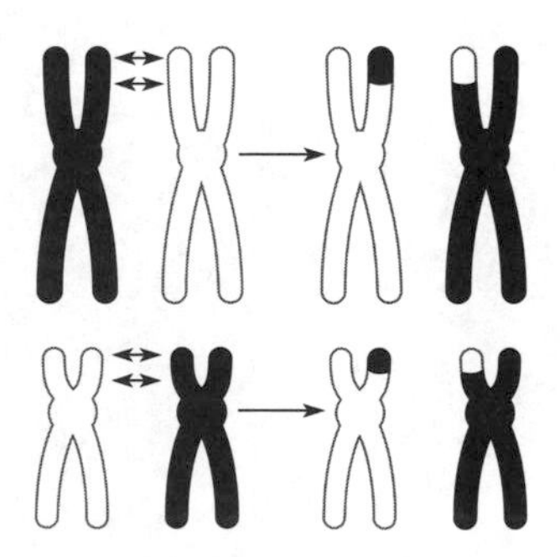

Figure 10.8
Crossing-Over

Now you have a bivalent with one completely red chromatid, next to a red chromatid with a blue top, next to a blue chromatid with a red top, next to a totally blue chromatid. Repeat this process for the bivalent of chromosome 2. You have just performed crossing-over!

6. Are any of the chromatids genetically identical at this point?

7. In the four circles below, sketch the sister chromatids of each homologue, using red pencils to distinguish the maternal chromatids and blue to distinguish the paternal chromatids. Label each circle according to its chromosome (either 1 or 2) and homologue (either maternal or paternal).

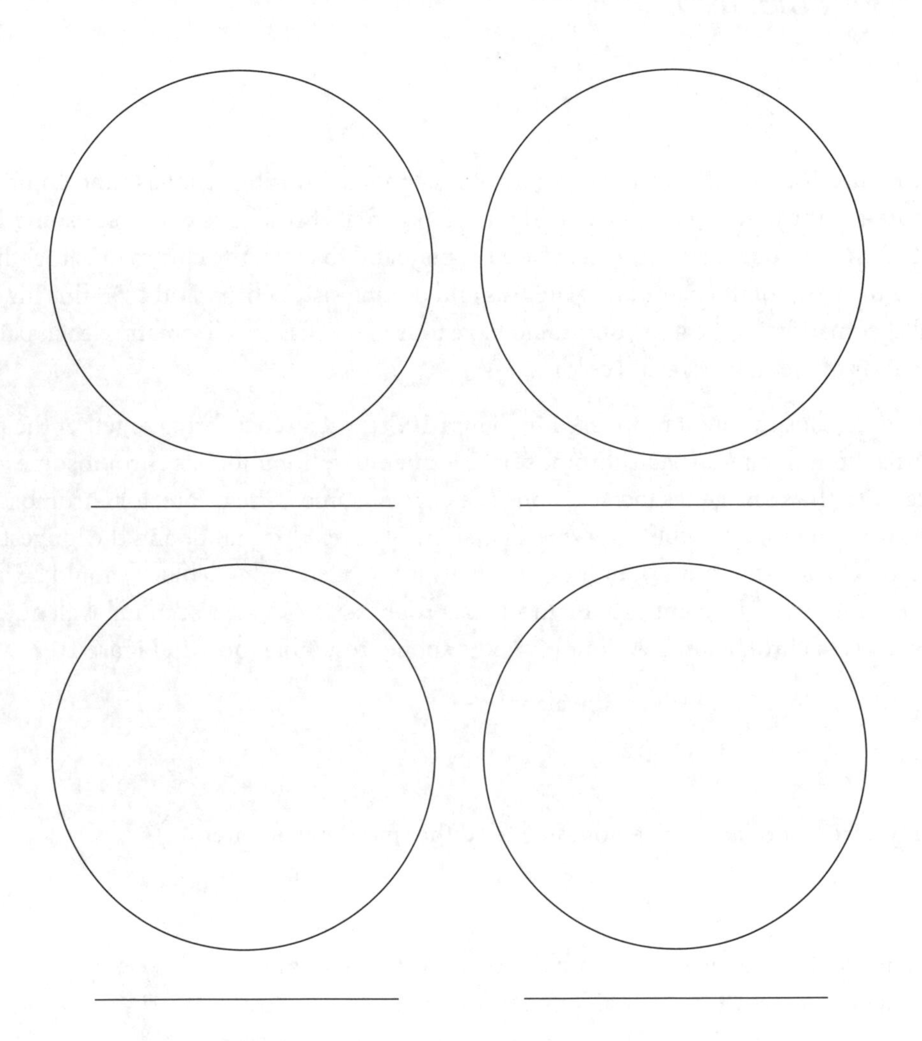

8. Now the DNA is ready for metaphase I. Use Figures 10.7 and 10.9 to guide you through this process. Imagining that your lab bench is a cell, create a metaphase plate by laying a piece of tape on your lab bench extending away from you, partitioning the bench into right and left sections, as illustrated by the dashed arrow in Figure 10.9. The tape represents the metaphase plate. Still using your hypothetical organism, simulate metaphase I by bringing the chromosome 1 bivalent onto the metaphase plate (tape) so that the maternal sister chromatids are to your left, as shown in Figure 10.9.

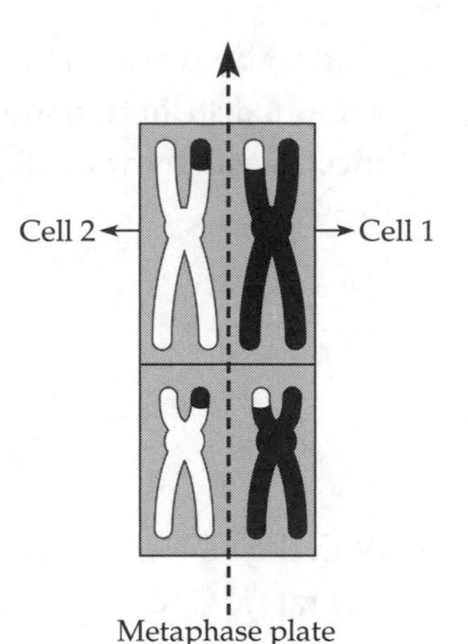

Figure 10.9
Meiosis I Simulation Setup

9. When the bivalents migrate to the metaphase plate, the orientation of each homologue is independent of all others. For instance, the maternal sister chromatids within bivalent 1 can face either the right or the left side of the cell. Now align the chromosome 2 bivalent along the metaphase plate. This represents one possible combination of how the bivalents can align along the metaphase plate during metaphase I. How many combinations are possible? In the four circles below, sketch all the possible combinations.

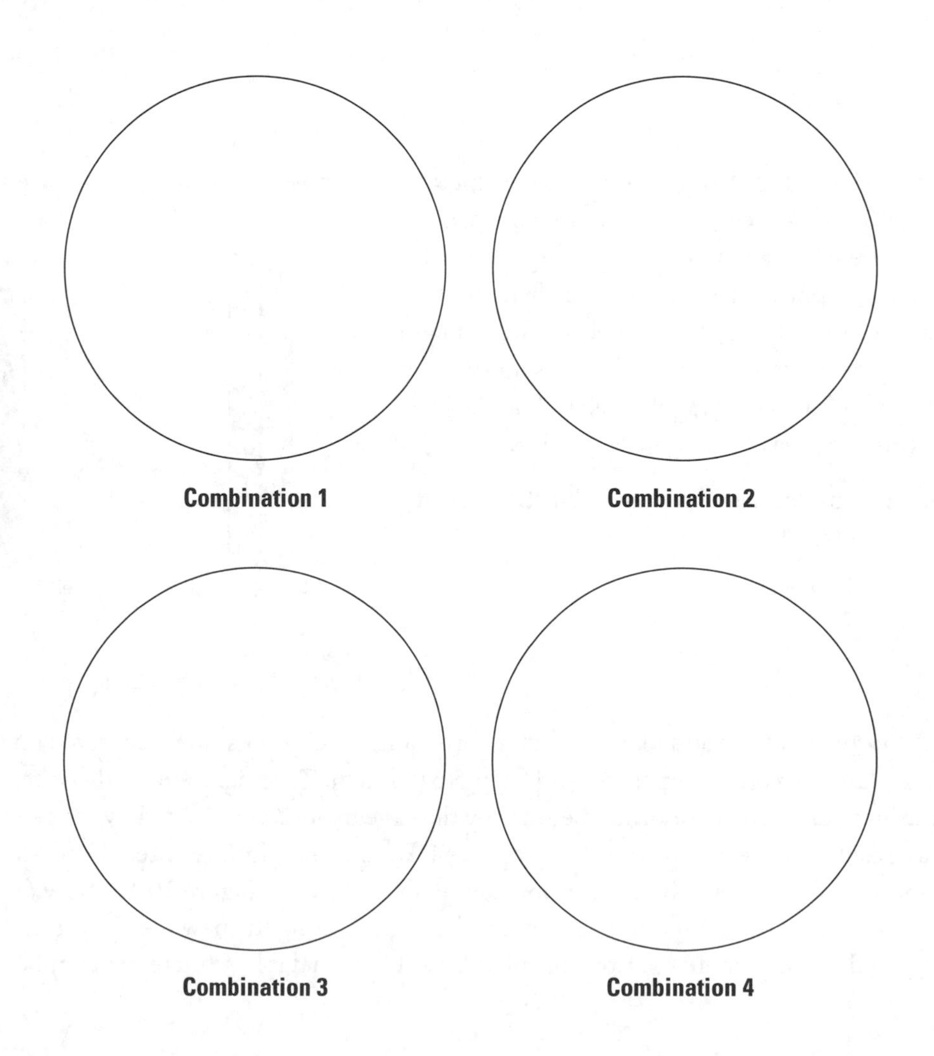

Combination 1 **Combination 2**

Combination 3 **Combination 4**

10. The random orientation of each pair of maternal and paternal homologues and their subsequent random distribution into gametes is called independent assortment. Using Figure 10.7 as a reference, explain how mixing maternal and paternal DNA when creating gametes contributes to genetic diversity.

11. With the bivalents lined up on the metaphase plate, the homologous chromosomes are ready to be separated. Choose one combination and align the bivalents along the metaphase plate. Simulate anaphase I by separating the homologous chromosomes of bivalent 1. Move the homologues facing the right and the left sides of the lab bench to their respective sides, using Figure 10.9 as a guide. Then separate the homologues of bivalent 2 in the same way. You have simulated meiosis I, the separation of homologous chromosomes into two daughter cells (the right and left sides of the lab bench representing the two new cells).

12. Examine Figure 10.7 to see how the cell completes meiosis I by re-forming the nucleus in telophase I and completing cytokinesis. How many chromosomes are present in each new cell?

Meiosis II

1. Referring to Figure 10.10, now create two new metaphase plates in the two cells formed during meiosis I. On your lab bench, place a new piece of tape parallel to the original metaphase plate in each cell, dividing the cell in half, as illustrated by the dashed arrows in Figure 10.10. These new pieces of tape will represent the new metaphase plates in each daughter cell (left and right) formed during meiosis I.

2. How are the homologues of prophase I distributed at the start of meiosis II?

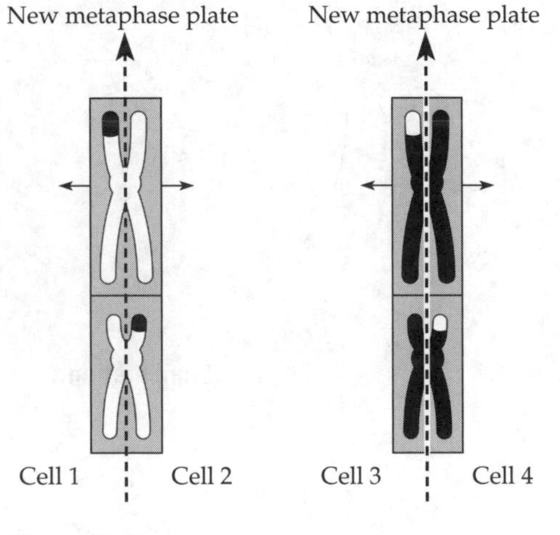

Figure 10.10
Meiosis II Simulation Setup

3. Start in the right-hand cell and place the sister chromatids of chromosome 1 centered on the new metaphase plate with one sister chromatid facing to the right. The alignment of the sister chromatids during meiosis II is important because by the time meiosis II occurs, they are genetically different, as a result of crossing-over during prophase I. Do the same for the sister chromatids of chromosome 2, placing these below chromosome 1, as depicted in Figure 10.10. Now do the same for the left cell. All sister chromatids should now be aligned along the new metaphase plate. You have simulated the alignment of sister chromatids along the metaphase plate in metaphase II. How

many possible resulting combinations are there in the right-hand cell undergoing meiosis II? Are the same combinations possible in the left cell? In the four circles below, sketch all the possible combinations for the left cell.

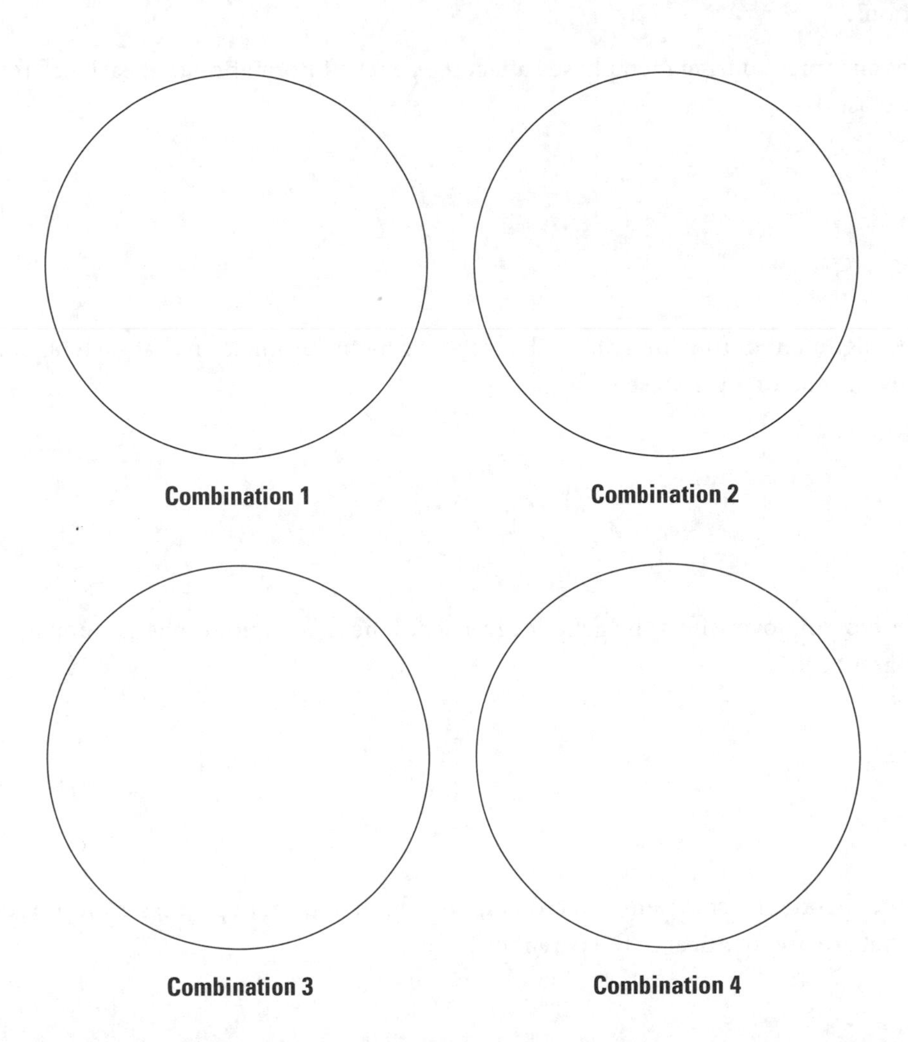

4. Choose one alignment. Remove the rubber band holding the sister chromatids together so that you can separate the sister chromatids of your hypothetical organism in both the left and right cells, as in anaphase II of Figure 10.7. Do this by moving the sister chromatids toward their respective poles (to the far right and left of each new metaphase plate, as in Figure 10.10). This separation represents the creation of four gametes (telophase II and cytokinesis, as shown in Figure 10.7). How many chromosomes are in each gamete?

5. How many chromosomes did your original parent cell have?

6. Meiosis II creates four daughter cells, each haploid. Are any of these gametes genetically identical?

A Closer Look at Genetic Variation

Take another look at your drawings of the different ways that bivalents can line up along the metaphase plate in metaphase I. Using your building blocks simulation and Figure 10.7 for reference, answer the following questions.

1. How does alignment during metaphase I affect the genetic information that each cell receives at the start of meiosis II?

2. Why does alignment during metaphase II affect the genetic information that each haploid cell receives by the end of metaphase II?

3. How does crossing-over affect the genetic diversity of the four haploid cells as compared to the original parent cell?

4. Why are independent assortment and crossing-over both important processes for producing gametes that are not identical to the parent cell?

5. To understand the genetic variation generated by independent assortment in the context of humans, again consider just the possible alignments of homologues during meiosis I. The number of potential chromosome combinations resulting from meiosis I is the possible number of orientations of the homologues during metaphase I (left or right = 2), raised to the number of homologous pairs in a cell (23 in humans): 2^{23} in human meiosis. Assuming that your hypothetical organism has only two homologous pairs in each of its cells, in how many different ways can homologues be combined in your hypothetical organism?

Concept Check

1. How many potential chromosome combinations could result from meiosis I in a diploid organism with six homologous pairs of chromosomes? (*Hint:* Possible number of orientations, raised to the number of homologous chromosomes).

2. In one sentence, explain how crossing-over generates genetic diversity. In one sentence, explain how independent assortment generates genetic diversity.

3. What gets separated during meiosis I? What gets separated during meiosis II?

ACTIVITY 4 The Law of Segregation and Drawing Punnett Squares

Gregor Mendel led science to a greater understanding of the molecules influencing the physical and behavioral characteristics of life. Although he did not have the advances of molecular biology to elucidate his knowledge, Mendel succeeded in discovering some basic generalizations of the laws of inheritance, through genetic experiments involving pea plants.

Mendel understood through his studies that inherited characteristics (genetic traits) are passed on from parent to offspring. These traits are dictated by information stored in DNA. Mendel also intuited from his experiments the existence of genes from each parent that separate and end up in different gametes. After Mendel's lifetime, it was discovered that a **gene** is a section of DNA (located on a chromosome) that governs one or more genetic traits. Genes have variations called **alleles**; for instance, the gene controlling the color of a pea plant's flower has alleles coding for different colors: purple and white. Although Mendel did not use the term "allele," he understood conceptually that in combining the alleles of two parents, each allele could influence the resulting color of a pea plant's flower. More specifically, Mendel's **law of segregation** indicates that alleles separate and end up in different gametes.

A **Punnett square** is a useful tool for representing the different combinations of alleles that could result from fertilization. In this activity you will learn more about using Punnett squares and describing the possible relationships between single inherited traits.

1. A diploid organism such as a pea plant inherits a double set of chromosomes—one set from each parent, with one allele from each parent controlling the same genetic trait, such as flower color. For instance, the pea plant can inherit a purple allele from one parent and a white allele from the other parent. The pair of alleles for a gene that an organism contains constitute its **genotype**. The

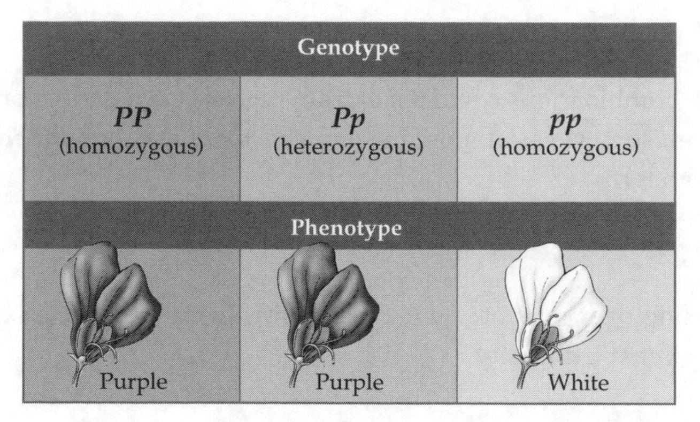

Figure 10.11
Genotype and Phenotype

relationship between these alleles influences the outward expression of a genotype, which is the **phenotype** of the gene. Examine Figure 10.11 for an example of genotypes and phenotypes, corresponding to flower color in pea plants.

2. Which process describes the separation of alleles?

3. An individual who carries two of the same allele for a given gene is said to have a **homozygous genotype**, or to be a **homozygote**. On the other hand, an individual who has two different alleles for a given gene is a **heterozygote**, or has a **heterozygous genotype** (see Figure 10.11). Alleles influence one another's expression. An allele that prevents another allele from being expressed is said to be **dominant**; the allele whose expression is prevented is said to be **recessive**. Dominant alleles are represented by a capital letter; recessive alleles, by a lowercase letter. A recessive phenotype is expressed only when two recessive alleles are present. In Figure 10.11, which phenotypes rely on the presence of a dominant allele in the genotype?

4. In Figure 10.11, which phenotype results from the expression of a recessive allele?

5. Punnett squares are designed to help you determine the possible genotypes and phenotypes that could result from combining the gametes of two specific parents. A Punnett square enables you to display the genotype of parental gametes and determine the possible genotypes of offspring. Take a look at Figure 10.12. The parents in this example were the offspring of a previous genetic cross of parents *PP* and *pp*. In a genetic cross, the parental generation is called the **P generation**. The P generation cross results in the **F_1 generation** of *Pp* offspring. In the genetic cross depicted in Figure 10.12, two individuals from the F_1 generation are mated to produce the **F_2 generation**. The gametes and associated alleles for the F_1 parents are shown outside the box. Which alleles are coming from the father in this mating? Which alleles are coming from the mother?

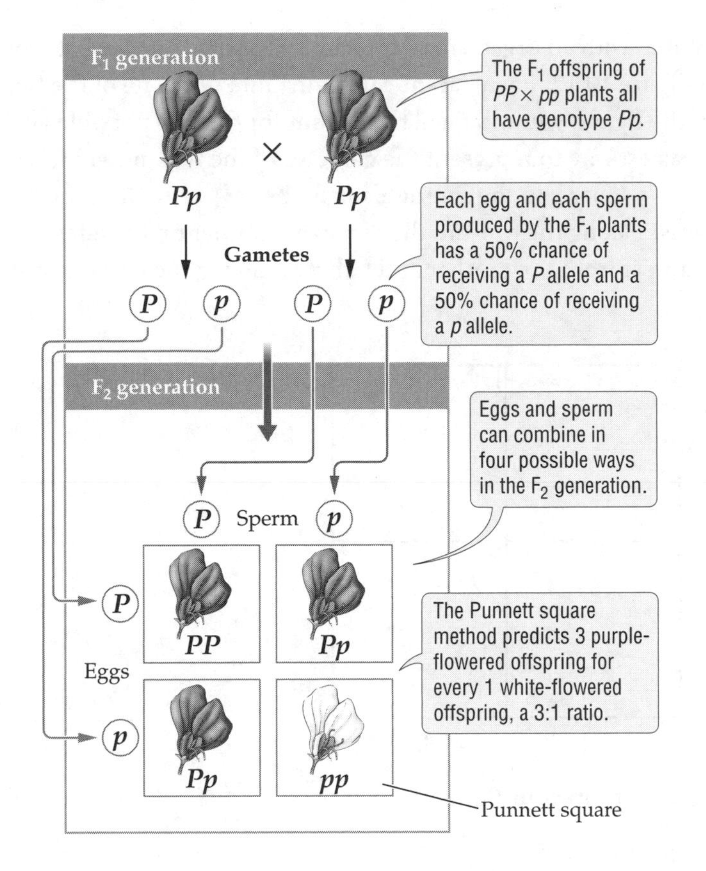

Figure 10.12
Using a Punnett Square

6. The possible allelic combinations of the F_2 generation are shown inside the boxes of the Punnett square. What is the likelihood of a white-flowered offspring? Why is this offspring the least likely result of a cross between the F_1 plants?

7. Imagine that you have a hypothetical organism with the genotype of *Bb* representing eye color, where brown (*B*) is the dominant allele and blue (*b*) is the recessive allele. What is the phenotype of your hypothetical organism?

8. Meiosis duplicates the two alleles of a parent cell, creating four gametes—two gametes with one allele, and two with the other. What are the possible alleles contained by the gametes of your hypothetical organism?

   Gamete 1:

   Gamete 2:

   Gamete 3:

   Gamete 4:

9. If your hypothetical organism mated with an organism of the same genotype (*Bb*), what would be the possible offspring? Refer to Figure 10.12. Each parent can contribute only one of the two possible alleles to a gamete. Just as in the figure, you will need to account for the two possible alleles contributed from each parent. The easiest way to represent the gametes of the two individuals and their possible offspring is to use a Punnett square. In the Punnett square below, write the gametes for the father of your hypothetical organism along the top and those for the mother on the left side. Now fill in the boxes by combining the gamete from the mother and father, showing the possible outcome for each combination.

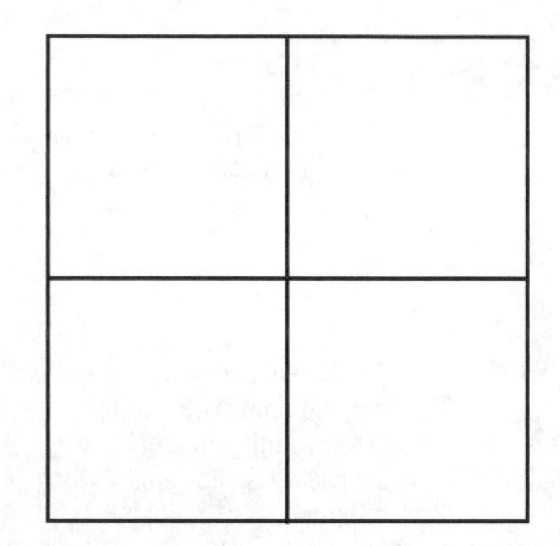

10. What is the phenotype of each genotype in your Punnett square?

    Genotype 1: *Phenotype 1:*

    Genotype 2: *Phenotype 2:*

    Genotype 3: *Phenotype 3:*

    Genotype 4: *Phenotype 4:*

    Which genotypes are homozygous? Which genotypes are heterozygous?

11. Drawing a Punnett square enables you to determine the probability that offspring will inherit specific allelic combinations for a gene. In your genetic cross, there are four possible outcomes. We determine the probability that any one of these genotypes will appear in the offspring by dividing the number of times a specific genotype appears in the Punnett square by the total of all possible outcomes (four for a single trait, as represented by the four boxes of the Punnett square). For instance, your hypothetical genetic cross can result in the genotype *bb*, but the probability of an offspring with this genotype is only ¼, or 25 percent. What is the probability of an offspring for each of the other genotypes?

    Probability of genotype BB*:*

    Probability of genotype Bb*:*

Concept Check

1. What is a variation of a gene called?

2. What term describes an individual with two copies of the same allele? What term describes an individual with two different alleles for a given phenotype?

3. What determines the outward expression of a gene?

4. How are Punnett squares useful in visualizing the possible offspring of two parents?

5. How does the drawing of a Punnett square reflect Mendel's law of segregation?

ACTIVITY 5 Genetic Diversity and Mendel's Law of Independent Assortment

Many of Mendel's experiments tracked two different genes controlling two different traits. For example, many of his experiments with pea plants tested whether the traits of seed shape and seed color were inherited together or independently. Mendel found that when gametes form, the separation of alleles of one gene is independent of the separation of alleles of other genes. This discovery is called Mendel's **law of independent assortment.** Note that this law applies only to genes that do not travel together on the same chromosome during meiosis.

In this activity you will use a new hypothetical organism to illustrate how genes on different chromosomes sort independently and how this independent assortment contributes to genetic diversity. You have a diploid organism with two chromosomes. Therefore, each cell has a pair of homologues for chromosome 1 and a pair of homologues for chromosomes 2, represented in this activity by sticks.

1. Your instructor will provide you with four sticks—two long and two short—and a beaker. The long sticks represent chromosome 1 (the big X's in Figure 10.7); the shorter sticks, chromosome 2 (the little X's in Figure 10.7). This simulation simplifies the independent assortment of chromosomes during meiosis, with each stick representing one homologue in a bivalent. Your four sticks will form only two gametes instead of four. How many bivalents are represented by your four sticks?

2. Your instructor has color-coded a gene on each chromosome—one color for a gene along chromosome 1 and another color for a gene along chromosome 2. If the gene on your chromosome has a thick band (1 cm), then the allele is dominant; if the band is thin, the allele is recessive. What is the genotype of your hypothetical organism? Use the capital first letter of the color to represent the dominant allele and the lowercase letter to represent the recessive allele (for example, "B" for dominant and "b" for recessive). Record your genotypes here.

   Genotype of chromosome 1 (longer sticks):

   Genotype of chromosome 2 (shorter sticks):

3. It is sometimes difficult to determine the possible combinations of alleles when you're examining two traits, so let's begin with an example. Consider the following two traits: one for eye color and another for ear color. *B* codes for brown eye color, and *b* is the recessive blue-color allele. The dominant allele *Y* codes for yellow ear color, and the recessive allele *y* codes for white ear color. This hypothetical organism is heterozygous for eye and ear color (*BbYy*). Since independent assortment allows either allele for the eye color gene to potentially be in the same gamete as either allele for the ear color gene, meiosis can produce four possible gametes: *BY*, *By*, *bY*, and *by*. Using this example for reference, what are the two possible alleles (one for each gene) in the gametes produced by your stick organism? Record them in Table 10.1, under "Possible gametic alleles."

4. Now that you have determined the gametes that your hypothetical organism could generate during meiosis, place your sticks in a beaker, shake them up, and toss them out onto your lab bench.

5. Simulate the independent assortment of chromosomes 1 and 2. Separate the homologous chromosomes, moving one of each pair (long stick and short stick) to the left and one to the right, according to the way they happen to fall. You have now simulated the end of meiosis, creating gametes. You

have formed two gametes, each consisting of one long and one short stick. (If you end up with two long sticks in one gamete, and two short ones in the other, you have made a gametic goof.)

6. How does the random alignment of sticks in the toss represent the random alignment of homologues in metaphase I (see Figure 10.7)?

7. Which alleles do each of your gametes contain? In Table 10.1, under "Resulting gametic alleles," for each combination of alleles that was created, make a mark in the row corresponding to that gamete.

8. Toss the sticks 19 more times and add marks for the gametes resulting from each toss to the rows under "Resulting gametic alleles" in Table 10.1.

9. Count your marks for each row to determine how many times each combination of alleles appeared as a result of your tosses. Record this total under "Experiment totals of resulting alleles" in Table 10.1. Now obtain experiment totals from each group in your class. Record the class totals in the last column of Table 10.1.

TABLE 10.1 POSSIBLE GENOTYPES RESULTING FROM RANDOM CROSSES IN A HYPOTHETICAL ORGANISM

Possible gametic alleles	Resulting gametic alleles	Experiment totals of resulting alleles	Class totals of resulting alleles

10. How do your totals compare to the class totals?

11. Why do the combined class data give a better representation of the likelihood of different genotypes?

Concept Check

1. How does this activity illustrate genetic diversity resulting from the independent assortment of chromosomes?

2. How does this activity help to explain the differences and similarities between siblings?

3. How does this activity show how offspring of sexually reproducing organisms can be better adapted to their environment than their parents were?

Key Terms

allele (p. 10-19)
anaphase (p. 10-7)
autosomal chromosome (p. 10-4)
bivalent (p. 10-11)
cancer (p. 10-1)
cell cycle (p. 10-1)
cell plate (p. 10-8)
centromere (p. 10-3)
centrosome (p. 10-6)
chromosome (p. 10-2)
crossing-over (p. 10-13)
cytokinesis (p. 10-7)
diploid (p. 10-11)
dominant (p. 10-20)
F_1 generation (p. 10-20)
F_2 generation (p. 10-20)
G_0 phase (p. 10-2)
G_1 phase (p. 10-2)
G_2 phase (p. 10-2)
gamete (p. 10-1)
gene (p. 10-19)
genotype (p. 10-19)
haploid (p. 10-11)
heterozygote (p. 10-20)
heterozygous genotype (p. 10-20)
homologous pair (p. 10-4)
homozygote (p. 10-20)
homozygous genotype (p. 10-20)
interphase (p. 10-2)
karyotype (p. 10-4)
kinetochore (p. 10-6)
law of independent assortment (p. 10-24)
law of segregation (p. 10-19)
meiosis (p. 10-1)
meiosis I (p. 10-11)
meiosis II (p. 10-11)
Mendelian genetics (p. 10-2)
metaphase (p. 10-6)
metaphase plate (p. 10-6)
mitosis (p. 10-1)
mitotic spindle (p. 10-6)
nuclear envelope (p. 10-6)
P generation (p. 10-20)
phenotype (p. 10-20)
plane of constriction (p. 10-8)
prophase (p. 10-6)
Punnett square (p. 10-19)
recessive (p. 10-20)
S phase (p. 10-2)
sex chromosome (p. 10-4)
sister chromatids (p. 10-3)
somatic cell (p. 10-3)
telophase (p. 10-7)
X chromosome (p. 10-4)
Y chromosome (p. 10-4)
zygote (p. 10-1)

Review Questions

1. For each of the following sets of parents, determine the possible gametes, the genotypes of possible offspring, and the proportions of possible phenotypes. Use the Punnett square provided to display the answer, place the gametes for parent 1 on the top outside and parent 2 on the left outside of the box.

   a. *Ss* × *SS*

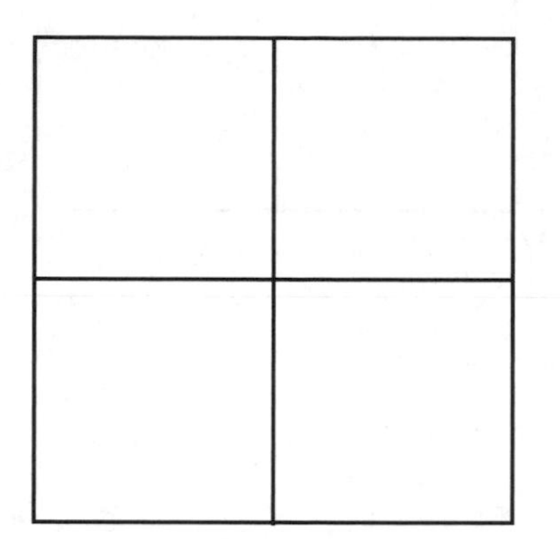

   Phenotype 1: Probability of phenotype:

   b. *ss* × *Ss*

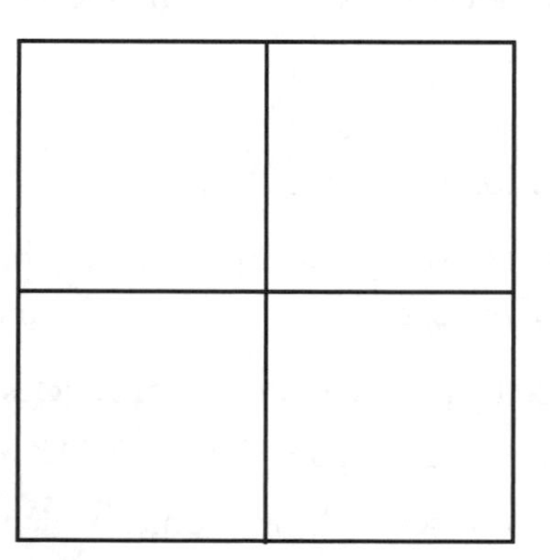

   Phenotype 1: Probability of phenotype:

   Phenotype 2: Probability of phenotype:

c. $Ss \times Ss$

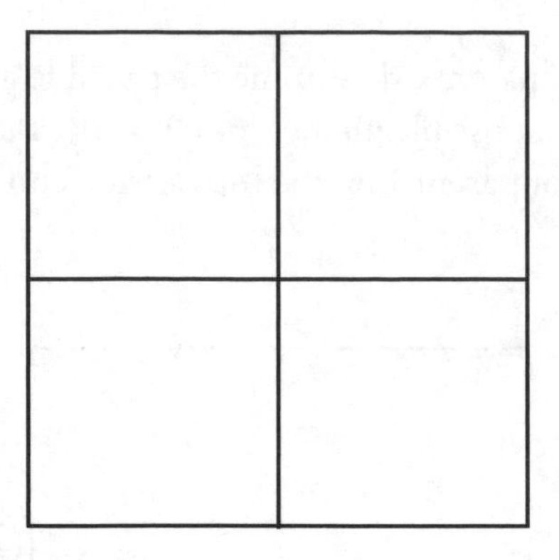

Phenotype 1: Probability of phenotype:

Phenotype 2: Probability of phenotype:

2. ________________ are a pair of identical chromatids resulting from DNA duplication.
3. Uncontrolled cell division is ________________.
4. An area along a chromatid that is more constricted than other regions is called the ________________.
5. Microtubules attach to special proteins called ________________, which surround a chromosome during cell division to pull apart replicated DNA.
6. A(n) ________________ depicts each pair of homologous chromosomes in a cell after DNA duplication and condensation of the chromosomes within a somatic cell.
7. Plant cytokinesis is characterized by the accumulation of vesicles along the metaphase plate, which then fuse to form a(n) ________________, creating two new daughter cells.
8. Meiosis produces cells with only one set of chromosomes, which are referred to as ________________ (n); mitotic cell division produces cells with two sets of chromosomes, which are referred to as ________________ ($2n$).
9. The structure of duplicated maternal and paternal homologues that form during meiosis I is known as a(n) ________________.
10. ________________ is the stage of the cell cycle in which cells never divide.
11. Explain the difference between the law of segregation and the law of independent assortment.

12. What two processes during meiosis generate genetic variation?

13. What is a section of DNA that governs physical and behavioral traits called?

14. What is the difference between plant and animal cytokinesis?

15. What terms describe the two possible relationships between two alleles of the same gene?

16. What is the difference between meiosis I and meiosis II?

17. Describe the differences and similarities between mitotic cell division and meiosis.

18. Why do cells undergo cell division?

19. Punnett squares represent the separation of alleles during ___________________.

Human Genetics

Objectives

- Describe the chromosomal theory of inheritance and explain why this theory enables us to analyze inheritance more effectively than does Mendelian genetics alone.
- Identify phenotypes of physical characteristics and determine the possible genotypes causing these phenotypes.
- Describe genetic linkage and sex linkage, and how they affect inheritance patterns.
- Use a Punnett square to demonstrate the production of gametes and determine the probability of genotypes and phenotypes in offspring.
- Analyze pedigrees for both autosomal and sex-linked traits.

Introduction

Gregor Mendel made great strides toward the understanding of how we come to be who we are. Another intuitive scientist, August Weismann, extended our knowledge of inheritance by suggesting that hereditary information is contained within chromosomes. This concept is now called the **chromosome theory of inheritance**. The discoveries of molecular biology confirm the chromosome theory of inheritance, uncovering the many molecules that influence gene expression and revealing relationships between alleles that were not clearly elucidated by Mendel.

According to the chromosome theory of inheritance, a **chromosome** contains specific genes within a DNA molecule wrapped around proteins. A **gene** is a stretch of DNA that affects a particular genetic trait. Figure 11.1, a **karyotype**, depicts the 23 pairs of human chromosomes. Each pair consists of one maternal and one paternal chromosome of the same shape and size and containing the same genes. These pairs are called **homologous chromosomes** (**homologues**). Homologues 1 through 22 are **autosomes**, and the last pair of chromosomes—chromosome 23, the **sex chromosome**—determines whether an offspring will be male or female. Homologous chromosomes contain **alleles** (variations) for the same genes. A mother and father can pass on the same allele for a gene (homozygous) or different alleles for a gene (heterozygous).

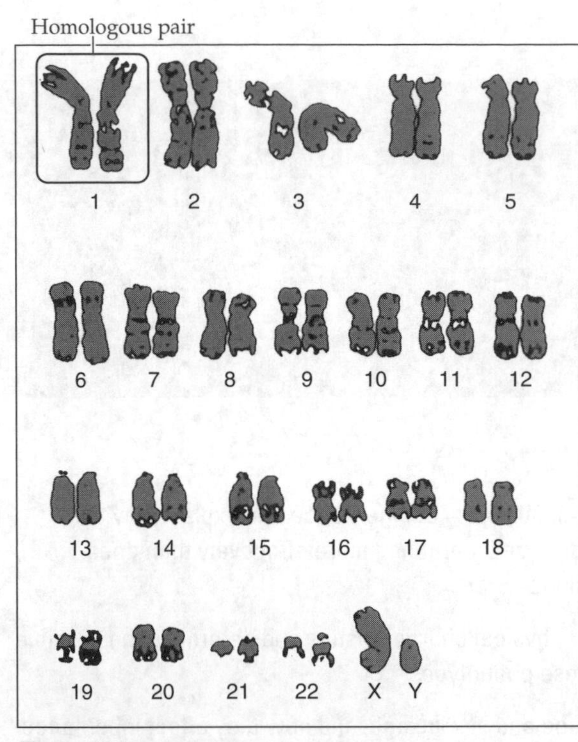

Figure 11.1
Human Karyotype

Human chromosomes range in size, but on average each chromosome contains approximately 1,000 genes. The chromosome theory of inheritance has important implications for our ability to analyze the inheritance of these genes. For instance, genes located on the same chromosome and positioned close to one another will often be inherited together as **linked traits**, so they will not be sorted independently into gametes as Mendelian genetics predicts. The theory also helps us understand the role of sex chromosomes in inheritance. Some genes controlling **genetic disorders** are located on sex chromosomes, affecting how they will be inherited and expressed differently in male and female offspring.

In this lab you will investigate the general patterns of inheritance for nonlinked traits, explore the chromosome theory of inheritance, and analyze family histories to assess the likelihood that various disorders will be passed from one generation to the next.

ACTIVITY 1 Human Phenotypes

Mendel determined important laws of inheritance crucial to our understanding of how inheritance works. Like all other humans, you, as a diploid organism, have two alleles for every gene—one inherited from your mother and one from your father. The relationship of these copies influences your development, behavior, physical appearance, and abilities. As Mendel suggested, if you have two of the same allele for a given gene, you are **homozygous** for the trait encoded by that gene and you will express that genetic trait. If you have two different alleles for a given gene, then you are **heterozygous**. Many alleles display a dominant/recessive relationship, meaning that one allele (**dominant**) can suppress the expression of the other allele (**recessive**). A heterozygote is a **carrier** of a recessive allele and can pass the recessive allele on to offspring without exhibiting the trait specified by that allele.

In this activity you will examine your own genetic makeup and simulate reproduction with a labmate to illustrate the general patterns of inheritance for autosomal, nonlinked traits controlled by one gene (as articulated by Mendel).

NOTE: AS YOU WORK THROUGH THE STEPS OF THIS ACTIVITY, IT IS IMPORTANT TO REMEMBER THAT GENES DO NOT JUST INFLUENCE HOW YOU LOOK, AND THAT THE GENERAL LAWS OF INHERITANCE ALLOW FOR GREAT VARIATION AND COMPLEXITY IN THE GENETIC COMPOSITION OF OFFSPRING.

Determining Your Genetic Makeup

1. Obtain three pieces of paper: one untreated (the control), one treated with phenylthiocarbamide (PTC), and one treated with thiourea.

2. Recall that your **genotype** describes the two alleles that you possess for a given gene (for example, *bb* for blue eyes), and that your **phenotype** describes the outward expression of a genotype for a given gene (for example, blue eyes). In Table 11.1, under "Your phenotype," record the *phenotype* that you express for each trait (except the last two).

3. The last two traits in Table 11.1 are the ability to taste specific chemicals. Such ability is determined by the expression of receptors in the taste buds that detect these chemicals. To discern your phenotype for these traits, first taste the control paper by touching this paper to the tip of your tongue. Then test the two additional pieces of paper, each treated with one of the following chemicals: phenylthiocarbamide (PTC) or thiourea. For each paper, if you taste anything other than what you tasted with the control, you are a "taster" for the chemical on that paper. The ability to taste either chemical is controlled by a dominant allele. Record your phenotype for each of these traits in Table 11.1.

DO NOT CHEW THE PAPERS. JUST TASTE THEM WITH THE TIP OF YOUR TONGUE.

4. Next, determine which *genotype* you have for each trait and record the information in Table 11.1, under "Your genotype."

REMEMBER: IF YOU EXPRESS THE DOMINANT PHENOTYPE, YOU CANNOT KNOW WHETHER YOU HAVE A HOMOZYGOUS OR HETEROZYGOUS GENOTYPE FOR THAT TRAIT. TO MAKE THINGS SIMPLE, DESIGNATE HALF OF YOUR DOMINANT TRAITS AS A HOMOZYGOUS PAIR OF ALLELES AND GIVE THE OTHER HALF OF YOUR DOMINANT TRAITS HETEROZYGOUS ALLELES.

TABLE 11.1 INDIVIDUAL GENETIC CHART

Trait	Genotypes	Your phenotype	Your genotype	Total class phenotypes	Total class genotypes
Free-hanging earlobe	*CC* or *Cc*				
Attached earlobe	*cc*				
Tongue roller	*DD* or *Dd*				
Nonroller	*dd*				
Widow's peak	*EE* or *Ee*				
No widow's peak	*ee*				
Freckles	*FF* or *Ff*				
No freckles	*ff*				
Dimples	*HH* or *Hh*				
No dimples	*hh*				
Right-handedness	*LL* or *Ll*				
Left-handedness	*ll*				
Normal thumb	*JJ* or *Jj*				
Hitchhiker's thumb	*jj*				
2nd finger shorter than 4th	*MM* or *Mm*				
2nd finger not shorter than 4th	*mm*				
PTC taster	*PP* or *Pp*				
PTC nontaster	*pp*				
Thiourea taster	*TT* or *Tt*				
Thiourea nontaster	*tt*				

5. Report to your instructor the genotypes and phenotypes that you logged in Table 11.1 so that the frequency of expression for each trait can be determined for the class as a whole. Record the totals for each phenotype and genotype for the entire class in the last two columns of Table 11.1. What phenotypes are most common in your class?

Simulating Reproduction

1. Obtain ten small pieces of colored paper (pink if you're female; blue if you're male) and a brown paper bag.
2. Choose a labmate to simulate creating an offspring.
3. On the basis of your phenotypes and those of your labmate, as each of you determined in the preceding part of this activity and recorded in Table 11.1, predict the traits that the offspring of your partnership will have. Record your predictions in Table 11.2, under "SIMULATION 1: Predicted phenotype 1."
4. For each of the ten traits listed in Table 11.2, write a single allele on each side of one of the small pieces of colored paper to represent your genotype, while your labmate does the same with his/her

TABLE 11.2 PREDICTED TRAITS OF OFFSPRING WITH LABMATE

	SIMULATION 1			SIMULATION 2		
Trait	**Predicted phenotype 1**	**Offspring 1 genotype**	**Offspring 1 phenotype**	**Predicted phenotype 2**	**Offspring 2 genotype**	**Offspring 2 phenotype**
Free-hanging earlobe (*C*) Attached earlobe (*c*)						
Tongue roller (*D*) Nonroller (*d*)						
Widow's peak (*E*) No widow's peak (*e*)						
Freckles (*F*) No freckles (*f*)						
Dimples (*H*) No dimples (*h*)						
Right-handedness (*L*) Left-handedness (*l*)						
Normal thumb (*J*) Hitchhiker's thumb (*j*)						
2nd finger shorter than 4th (*M*) 2nd finger not shorter than 4th (*m*)						
PTC taster (*P*) PTC nontaster (*p*)						
Thiourea taster (*T*) Thiourea nontaster (*t*)						

pieces of paper for his/her genotypes. For example, if you have attached earlobes, write a lowercase "c" on both sides of the paper chip that codes for earlobe attachment. But if you have free-hanging earlobes, write either "C" on one side of the paper and "c" on the other side, or "C" on both sides, depending on whether you specified the trait as homozygous or heterozygous. *Hint:* Underline the capital "C" to distinguish it from the lowercase "c."

5. Place both your pieces of paper indicating genotypes and your partners' pieces of paper together in the bag. Shake the bag and carefully dump the pieces of paper onto your lab bench.
6. The alleles facing up represent the genotypes of your offspring. Find the two alleles for each trait and record the genotypes of your offspring in Table 11.2, under "SIMULATION 1: Offspring 1 genotype."
7. Determine the phenotype for each genotype and record it in Table 11.2 under "SIMULATION 1: Offspring 1 phenotype." Compare your predicted phenotypes to the actual phenotypes. How many times were you correct?
8. Repeat steps 3 through 7 with another labmate to get a sense of the heritability of your personal traits. Record your predictions and results for this second simulation under "SIMULATION 2" in Table 11.2.
9. Do the offspring from your two simulations closely resemble each other? How can you explain the existence of both resemblance and variation?

Concept Check

1. Using your knowledge of how dominant and recessive alleles generally function, explain why some phenotypes are more common in a population than others.
2. How does what you observed in this activity explain the variation found among siblings?
3. Can you think of human characteristics other than physical appearance and taste that are influenced by genetics?
4. What sort of genotype must a carrier have?

5. In order for a carrier to have an offspring that displays the recessive phenotype, must the mate of the carrier be homozygous or heterozygous?

ACTIVITY 2 Genetic Linkage

Many studies since the time of Mendel, especially Thomas Hunt Morgan's investigations of fruit flies, indicate that some genes are inherited together and do not follow the law of independent assortment. The main reason Mendel did not detect this type of inheritance was that he believed genes were individual units. In this activity, you will examine how Mendel developed the concept of independent assortment and how some genetic traits are linked such that they are inherited together.

1. According to Mendel, if two traits are examined from generation to generation, a specific pattern of inheritance is normally expressed. At the time he did his work, Mendel did not know about chromosomes, and he did not know that many genes reside within each chromosome. He therefore considered each gene to be an entity or unit separate from all other genes and the inheritance of one gene to be independent of other genes. For instance, Figure 11.2 examines the possible gametes produced during meiosis for two independent genes as Mendel described. Each level of the figure represents a cell division during meiosis. What is the genotype of the individual who is producing gametes in this figure (for both traits)?

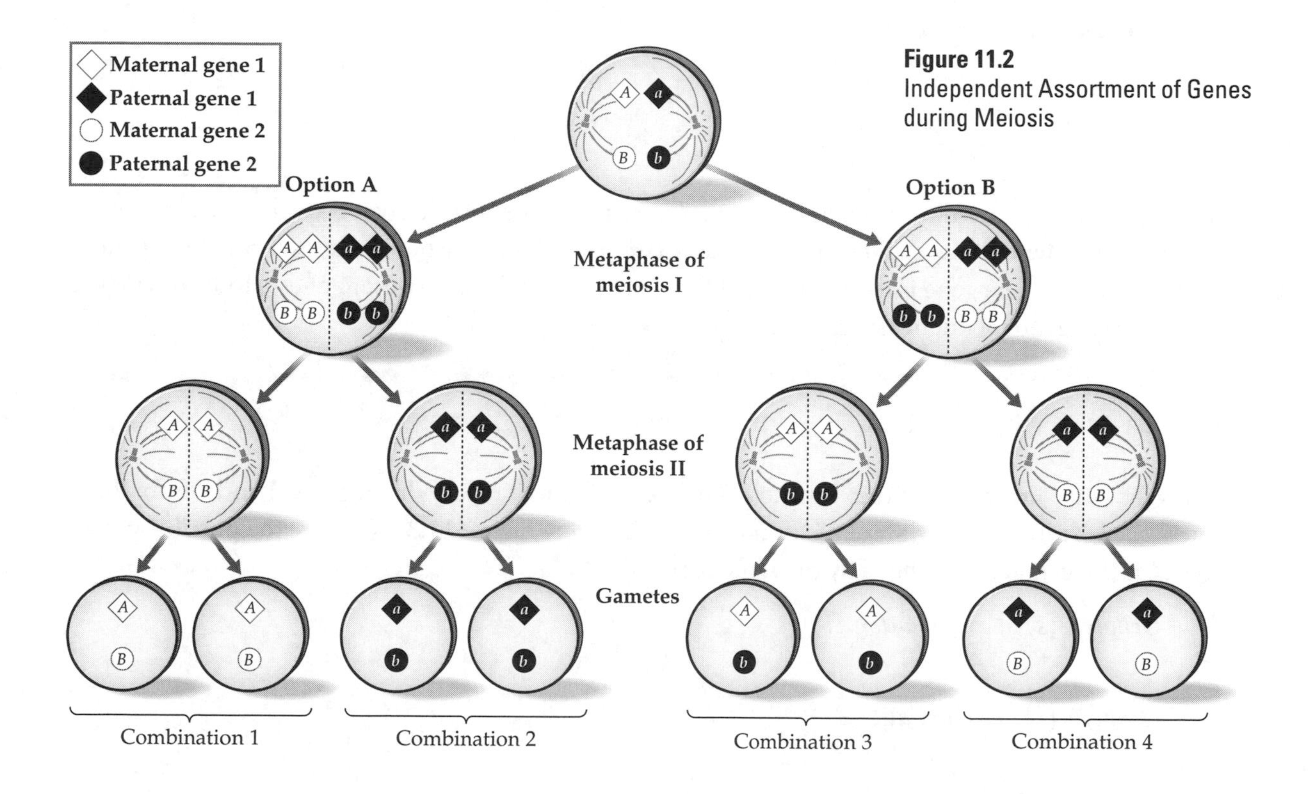

Figure 11.2
Independent Assortment of Genes during Meiosis

2. In Figure 11.2, gene 1 is represented by the *A* or *a* allele, and gene 2 is represented by the *B* or *b* allele. In the results of option A (combinations 1 and 2), the maternal genes from the parent cell end up together in two of the four gametes produced. The same occurs for the paternal genes. Using the figure for reference, why do the four gametes produced in option A contain the genes that they contain?

3. How does metaphase of meiosis I affect metaphase of meiosis II?

4. What are the genotypes of the four gametes that could result from option A?

   Genotype(s) of combination 1:

   Genotype(s) of combination 2:

5. In option B, the alleles for gene 2 align differently from those for gene 1, allowing maternal and paternal alleles to be distributed together in new combinations. What are the genotypes of the four gametes that could result from option B?

   Genotype(s) of combination 3:

   Genotype(s) of combination 4:

6. All four combinations illustrated in Figure 11.2 are equally possible during production of a gamete. Remember: Mendel's **law of independent assortment** states that when gametes form, the separation of alleles of one gene is independent of the separation of alleles of other genes. How does Figure 11.2 illustrate independent assortment?

7. Now let's take into account the chromosome theory of inheritance. What if genes 1 and 2 from Figure 11.2 were located on different chromosomes? Examine Figure 11.3. What are the possible genotypes of the gametes produced by the same parent cell?

   Genotype(s) of combination 1:

   Genotype(s) of combination 2:

   Genotype(s) of combination 3:

   Genotype(s) of combination 4:

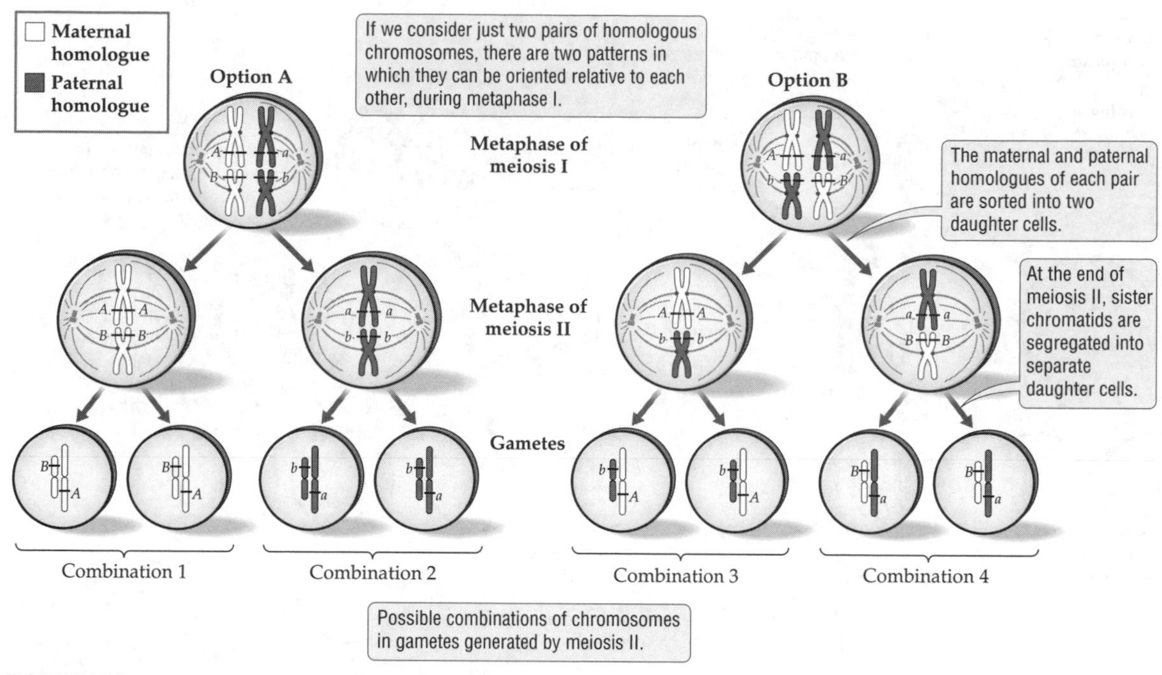

FIGURE 11.3
Independent Assortment of Chromosomes during Meiosis

8. Do the genes in Figure 11.3 sort independently, as in Figure 11.2?

9. Genes that are located close to one another on the same chromosome tend to be passed on together and are therefore considered **genetically linked**. Let's see what happens when the two genes from Figure 11.3 travel together on the same chromosome during meiosis. Examine Figure 11.4. In options A and B, we start with the same possible arrangements of chromosomes during metaphase of meiosis I.

   a. What are the genotypes of the four gametes that could result from option A?

      Genotype(s) of combination 1:

      Genotype(s) of combination 2:

   b. What are the genotypes of the four gametes that could result from option B?

      Genotype(s) of combination 3:

      Genotype(s) of combination 4:

10. When genes are sorted independently (Figures 11.2 and 11.3), what are the genotypes of the possible gametes?

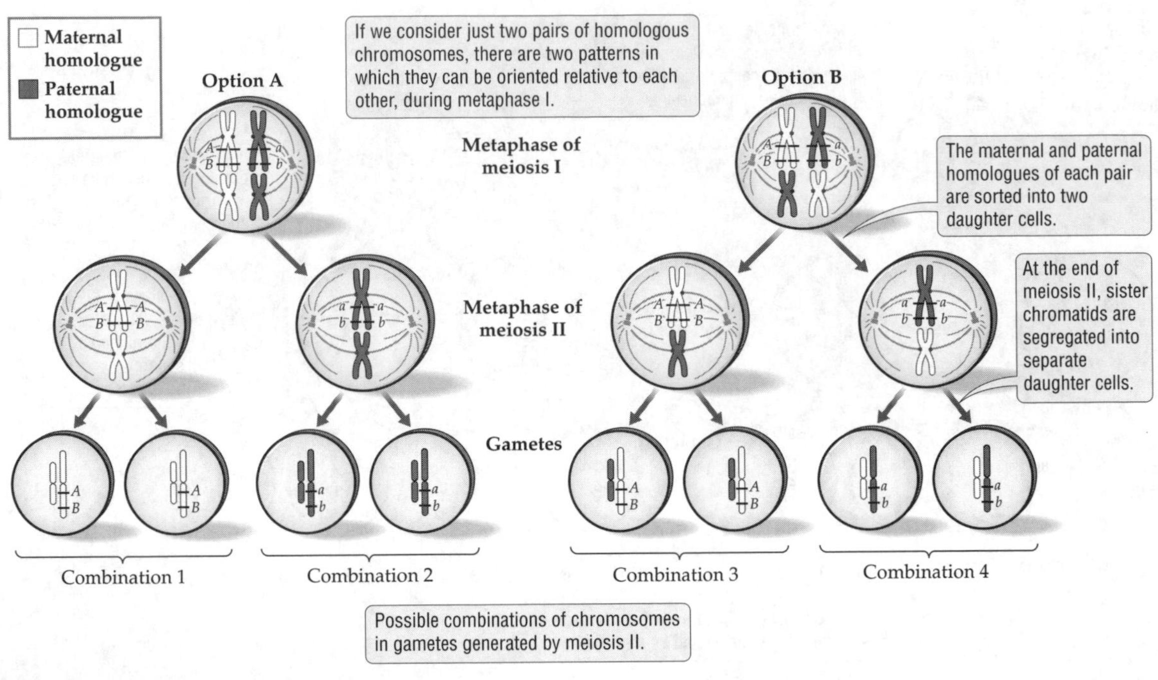

FIGURE 11.4
Genetic Linkage Alters the Outcome of Meiosis

11. When genes are linked (Figure 11.4), what are the genotypes of the possible gametes?

12. Does genetic linkage (linked genes) allow for more or less variation in the genotypes of offspring?

13. A process called **crossing-over** reduces genetic linkage by shuffling parts of chromosomes during meiosis. In crossing-over, segments of one homologue are physically exchanged with segments of a different homologue during meiosis. Examine Figure 11.5, which illustrates this concept.

    a. How does crossing-over disrupt the linkage between genes?

    b. How does crossing-over increase genetic diversity?

14. When Mendel performed a two-trait genetic cross in which the genes were on the same chromosome, he observed the variations in offspring that one would expect on the basis of independent assortment. How might crossing-over explain the diversity that Mendel saw in the offspring?

FIGURE 11.5
Crossing-Over

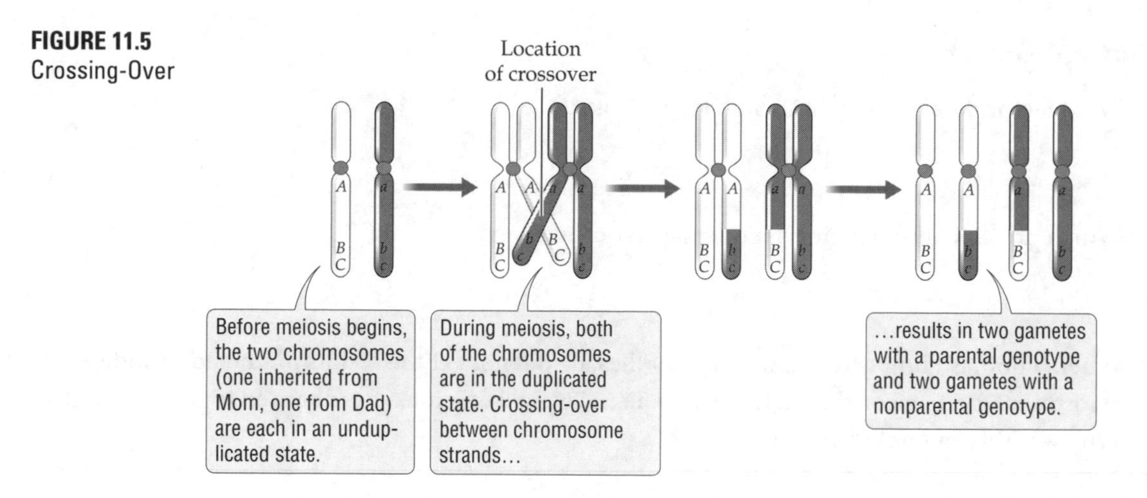

15. Examine Figure 11.6. Each gene resides at a specific location on a chromosome; this position is called a **locus** (plural "loci"). A mother and father can pass on the same allele for a gene or different alleles at the same locus. Using what you know about crossing-over, explain why it is likely that genes positioned close to one another on the same chromosome will remain genetically linked in spite of crossing-over (whereas genes far from one another on the same chromosome are likely to be separated by crossing-over).

FIGURE 11.6
Location of Genes on Chromosomes

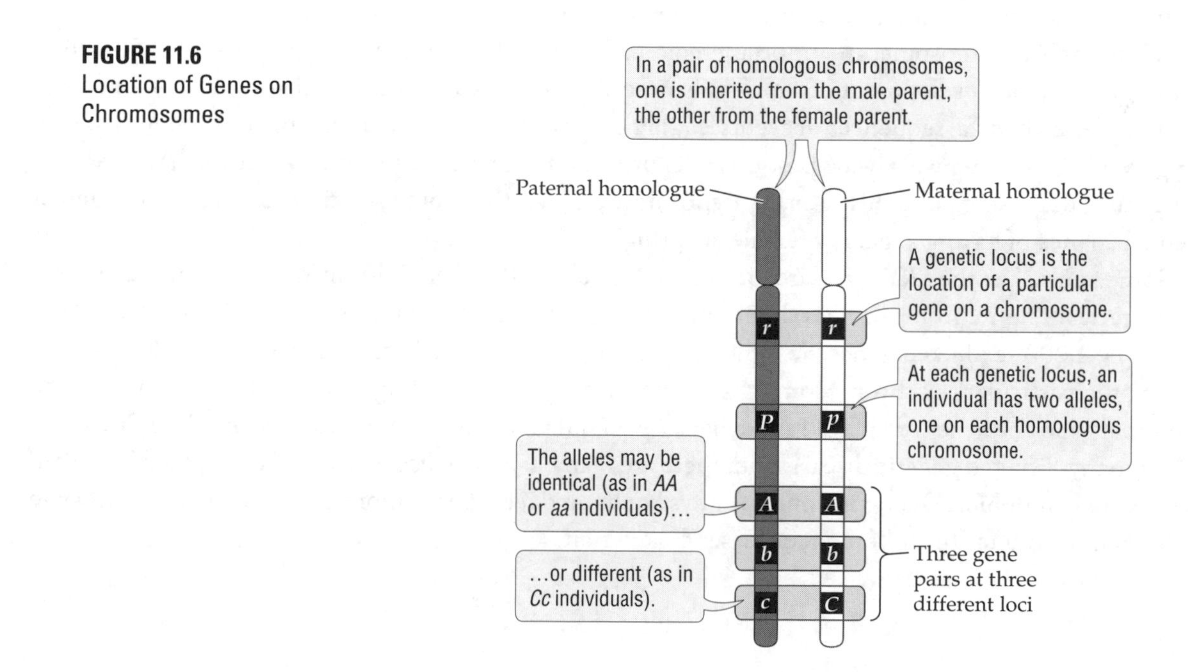

Concept Check

1. What factor determines whether genes are linked?

2. Which process during meiosis reduces genetic linkage?

3. When Thomas Hunt Morgan mated fruit flies, he did not get the results predicted by independent assortment, and he concluded that the genes he was testing must be inherited together as a group. Why was this an intelligent conclusion?

ACTIVITY 3 Sex-Linked Genes

Every human has 22 pairs of autosomal chromosomes (or nonsex-determining chromosomes). The twenty-third pair of chromosomes consists of the sex chromosomes. Figure 11.1 displays a karyotype of human chromosomes in a male somatic cell. Each male receives one **Y chromosome** from the sperm of his father and one X chromosome from the egg of his mother. Each female receives two **X chromosomes**, one from the sperm donated by her father and one from the egg donated by her mother.

Along the Y chromosome, an important region called ***SRY*** (*s*ex-determining *r*egion of the *Y* chromosome), initiates the development of a male offspring. All embryos are female until a switch is flipped and the *SRY* gene causes the sequences of events leading to the production of male characteristics. As depicted in Figure 11.7, a mother will always pass on an X chromosome to her offspring. A father will have a 50 percent chance of passing either his X or his Y chromosome, and therefore each time mating occurs there is an equal chance of having a male or female offspring.

Some genes, like the *SRY* gene, are located exclusively on the X or Y chromosome and are called **sex-linked genes.** The Y chromosome is significantly smaller than the X chromosome, which contains 92 percent of the 1,200 sex-linked genes. Sex-linked genes are also specifically referred to as X- or Y-linked, indicating the particular sex chromosome that contains the gene. To distinguish a sex-linked gene from an autosomal gene, the gene is displayed as a superscript of the sex chromosome—for example, Y^S or X^S. There are not many Y-linked genetic diseases, but there are many X-linked diseases. A few examples of X-linked diseases are hemophilia, Duchenne muscular dystrophy, and red–green color blindness. In this activity you will learn how to trace the inheritance of a sex-linked trait.

FIGURE 11.7
Distribution of Sex Chromosomes

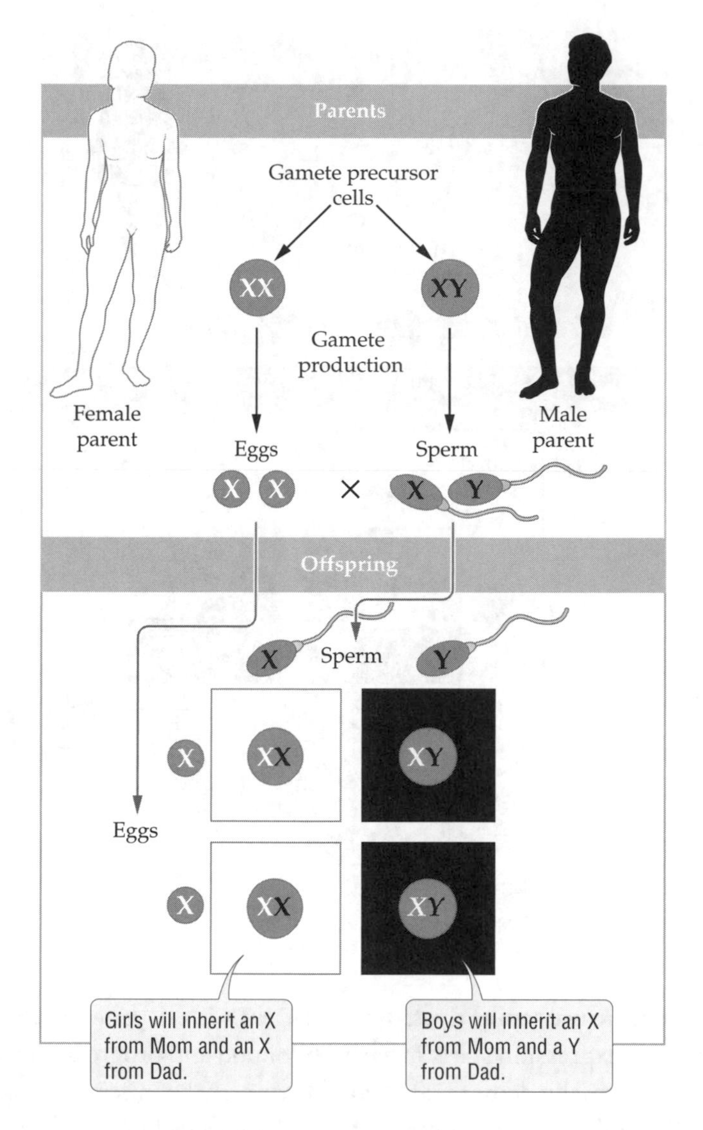

1. X-linked diseases are commonly recessive and are displayed more often in males than in females. Males express these recessive diseases more often because they do not have another X chromosome to mask the effects of the recessive allele. Females inherit two alleles of the gene, giving them a higher chance of inheriting another allele to mask the effects of the disease allele. What percentage of the offspring from the mating depicted in Figure 11.8 would you expect to express an X-linked recessive disorder if the mother is a carrier and the father does not have the recessive allele?

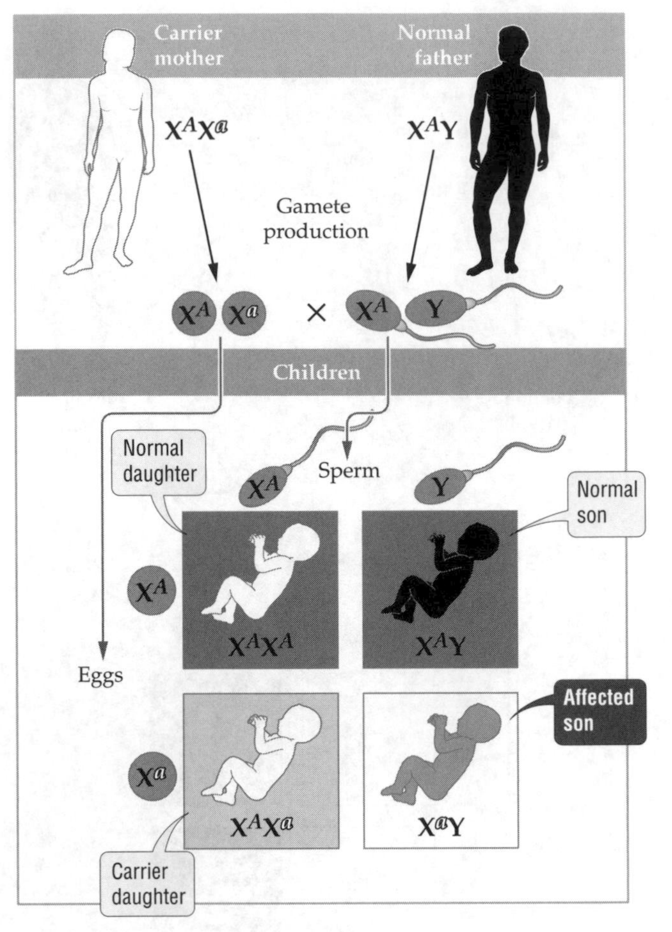

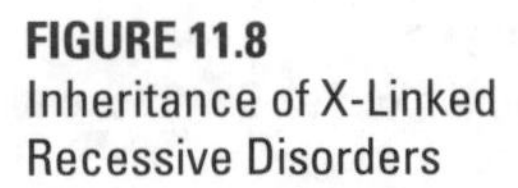

FIGURE 11.8
Inheritance of X-Linked Recessive Disorders

2. **Duchenne muscular dystrophy**, or **DMD**, is a degenerative disorder affecting voluntary muscle control. Expression starts in early childhood and progresses to affect the lungs and heart. Victims of DMD normally do not live beyond their early twenties. Consider the case in which a father has DMD while the mother carries no alleles for DMD. Determine the genotype of each parent and then depict the offspring in the Punnett square below. DMD is an X-linked recessive disease. Indicate the recessive allele as X^d and the dominant allele as X^D.

a. According to your Punnett square, what proportion of the male offspring will display the disease?

b. What proportion of the female offspring will display the disease?

c. Are any of the offspring carriers?

d. Do any offspring *not* have the DMD allele?

3. Now consider an alternate scenario, in which the mother exhibits Duchenne muscular dystrophy and the father does not. Determine the genotype of each parent and then depict the offspring in the Punnett square below.

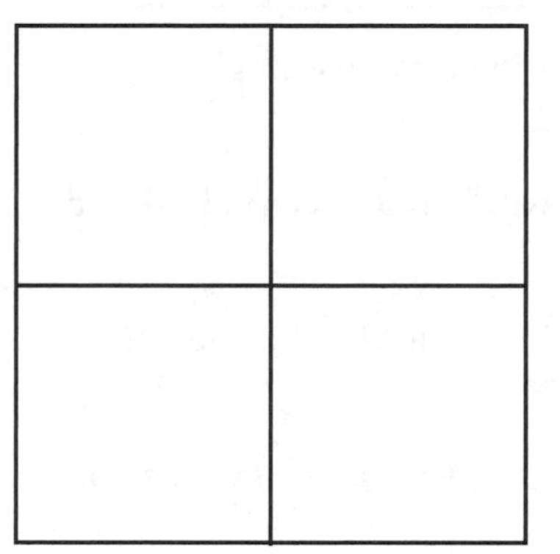

a. What proportion of the sons will display the disease?

b. What proportion of the daughters will display the disease?

c. Are any of the offspring carriers?

d. Do any offspring *not* have the DMD allele?

4. Red–green color blindness is also a recessive, X-linked condition. A red–green color-blind father and a color-seeing mother (whose father was red–green color-blind) have children. Determine the genotype of each parent and then depict the offspring in the Punnett square below. Indicate the recessive allele as X^b and the dominant allele as X^B.

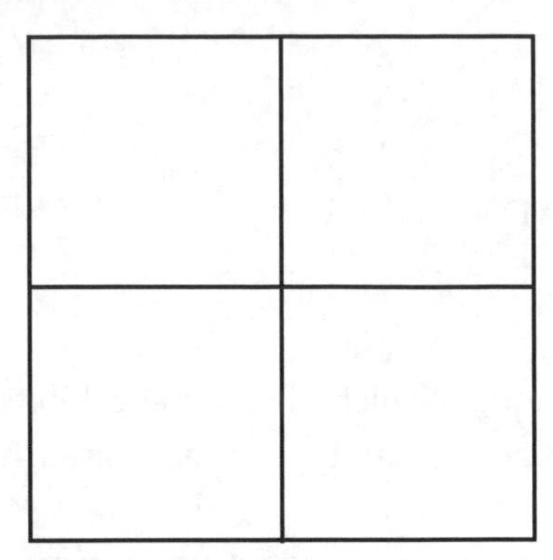

a. What proportion of the sons will be color-seeing?

b. What proportion of the sons will be red–green color-blind?

c. What proportion of the daughters will be color-seeing?

d. What proportion of the daughters will be red–green color-blind?

Concept Check

1. Which parent determines the sex of the offspring?

2. How is sex linkage different from genetic linkage as described in Activity 2?

3. Why do males display an X-linked recessive disorder more often than females do?

4. How would sex linkage affect the predicted outcome of a Mendelian cross?

ACTIVITY 4 Autosomal Disease Traced through a Pedigree

In order for alleles to remain in the population, the individuals carrying or expressing them must reproduce. Therefore, it is necessary for an allele's survival in the human population that individuals possessing the allele not only reach reproductive age but also have the ability to produce offspring. An allele that does not allow an individual to reproduce will eventually be eliminated from the population. As with regular inherited traits, most disease-causing alleles reside on autosomal chromosomes. Some diseases result from inheritance of a dominant allele and are called **dominant autosomal disorders**. Others, like that depicted in Figure 11.9, are expressed only if two recessive alleles are inherited and are referred to as **recessive autosomal disorders**.

The history of genetic inheritance is very important for tracking genetic diseases. A **pedigree** is a chart (similar to a family tree) that exhibits the phenotype for one specific trait among family members over two or more generations. Like the one illustrated in Figure 11.10, pedigrees trace the members of a family, representing each female as a circle, each male as a square, and each person who expresses the disease (affected phenotype) with a darkened circle or square. By examining these diagrams, we can interpret how a phenotype is inherited and potentially determine the underlying genotypes of individual family members. This type of diagram is a powerful tool for families exhibiting disease-related phenotypes. In this activity you will use a family pedigree to track the history of Huntington disease, a dominant autosomal disorder.

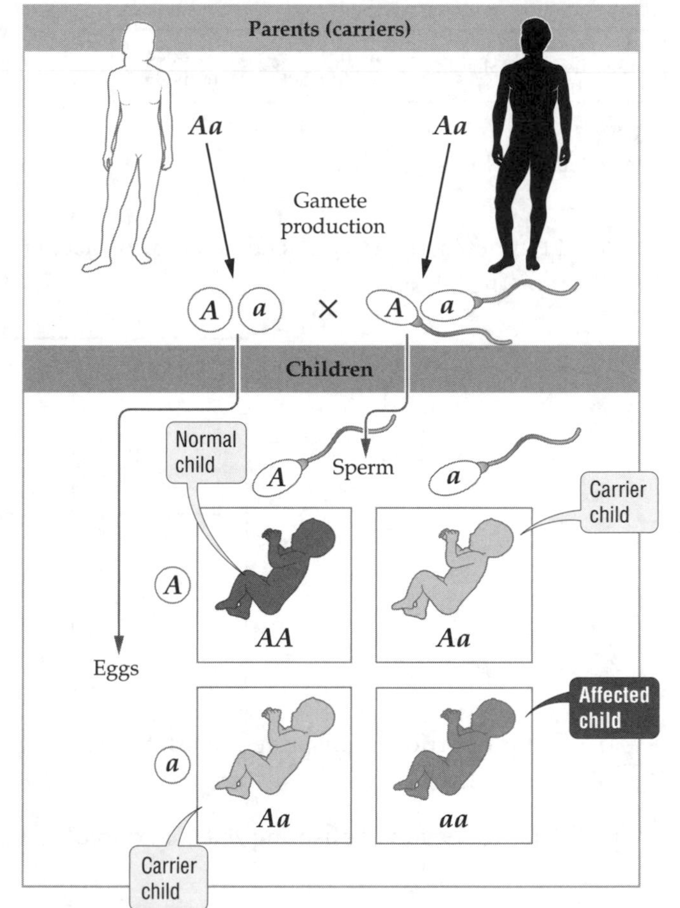

FIGURE 11.9
Inheritance of Autosomal Recessive Disorders

1. **Huntington disease** is a degenerative neurological disorder causing loss of voluntary movement. Affected individuals don't show symptoms of the disease until their early thirties or later. Because it is a dominant autosomal disorder, an individual needs to inherit only one dominant allele to express the disease. The recessive allele does not confer the disease but also does not protect against the disease. Examine Figure 11.10, which displays the phenotype of Huntington disease across three generations of a family. First, considering how Huntington disease is inherited, determine the genotypes of the mother (M), father (P), and their three offspring (1, 2, and 3). Use *A* to indicate the dominant allele and *a* for the recessive allele. Record the genotypes on the lines provided in the pedigree.

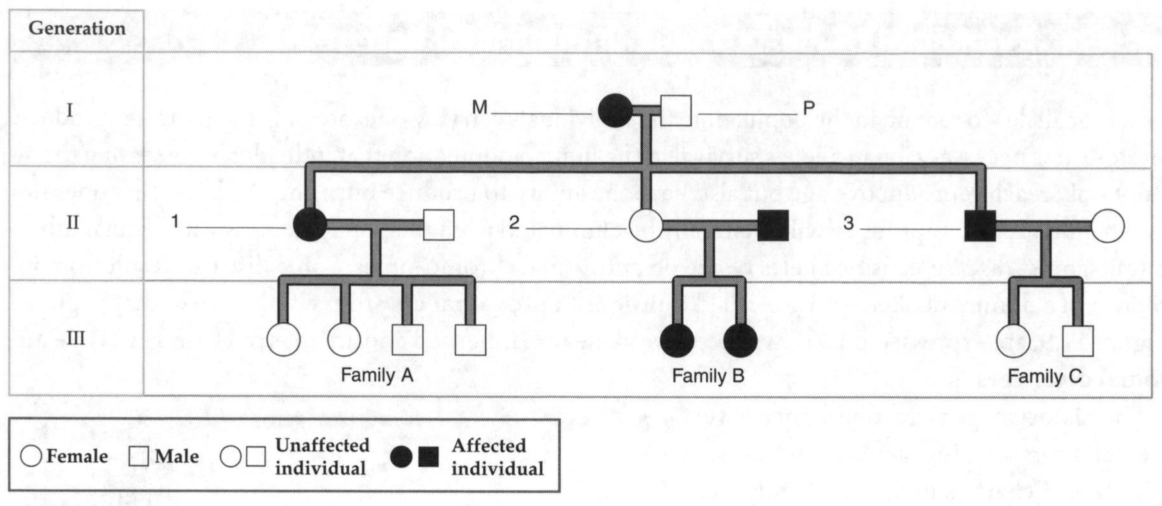

FIGURE 11.10
Family Pedigree: Huntington Disease

2. Use a Punnett square to depict this scenario, placing the paternal alleles on top and the maternal alleles on the left side.

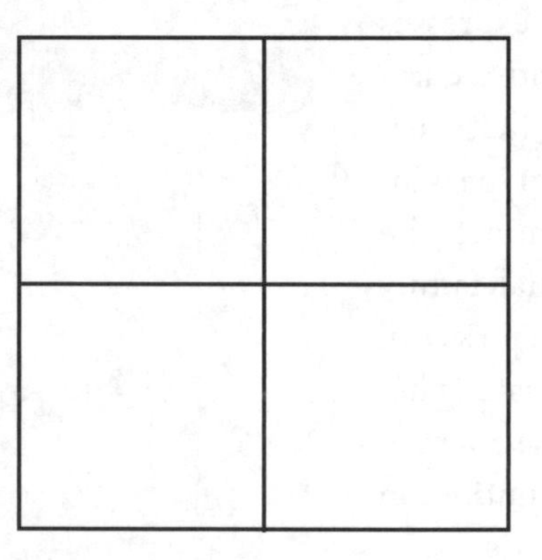

   a. What proportion of offspring will exhibit Huntington disease?

   b. What proportion of offspring will *not* exhibit Huntington disease?

3. Which genotype(s) will exhibit Huntington disease? Which genotype(s) will not exhibit Huntington disease?

4. Judging by the offspring depicted as generation III in Figure 11.10, can you determine the genotype of each generation II offspring's mate? If yes, write the answer below; if no, explain why.

   Offspring 1 mate:

   Offspring 2 mate:

   Offspring 3 mate:

5. Which families in Figure 11.10 will definitely *not* pass on Huntington disease to generation IV? Which families still have the potential of passing on Huntington disease?

6. Will all future generations of family B have Huntington's disease? Why or why not?

Concept Check

1. Why is a pedigree a valuable tool?

2. In a dominant autosomal genetic disorder, which genotypes would display the disorder?

3. In a recessive autosomal genetic disorder, which genotypes would cause expression of the disorder?

ACTIVITY 5 Sex-Linked Disease Traced through a Pedigree

Pedigrees can be used to trace any trait and its expression in a family. Sex-linked diseases show very specific patterns of inheritance and expression. Pedigrees displaying a sex-linked trait are often easy to detect. X-linked recessive genetic diseases are the most common because females can be carriers without expressing the diseases. A male that inherits an X-linked recessive disease expresses the disease because he does not have another X chromosome with a complementary allele to protect against the disease. A female can inherit a recessive allele from her mother and a dominant (protective) allele from her father. The only way a female can express a recessive X-linked genetic disease is if both her mother and her father pass on a recessive allele. This means that her father must express the disease.

In this activity you will analyze a pedigree that follows the inheritance of hemophilia, an X-linked recessive disorder.

1. **Hemophilia** is a disorder in which the blood does not clot properly. A hemophiliac is at risk of uncontrolled bleeding when a blood vessel is injured. Because it is an X-linked recessive disorder, only a female who receives two recessive alleles or a male who receives one recessive allele will exhibit the disease. Any individual who receives a dominant allele will not have hemophilia. Examine Figure 11.11, which displays the phenotype of hemophilia across three generations of a family. Judging by how hemophilia is inherited, determine the genotypes of the mother (M), father (P), and their four offspring (1, 2, 3, and 4). Use X^H to indicate the dominant allele and X^h for the recessive allele. Remember that this disease is displayed only if two recessive alleles are inherited in a female and one allele in a male. The dominant allele is protective, resulting in a normal individual. Record the genotypes in the lines provided in the pedigree.

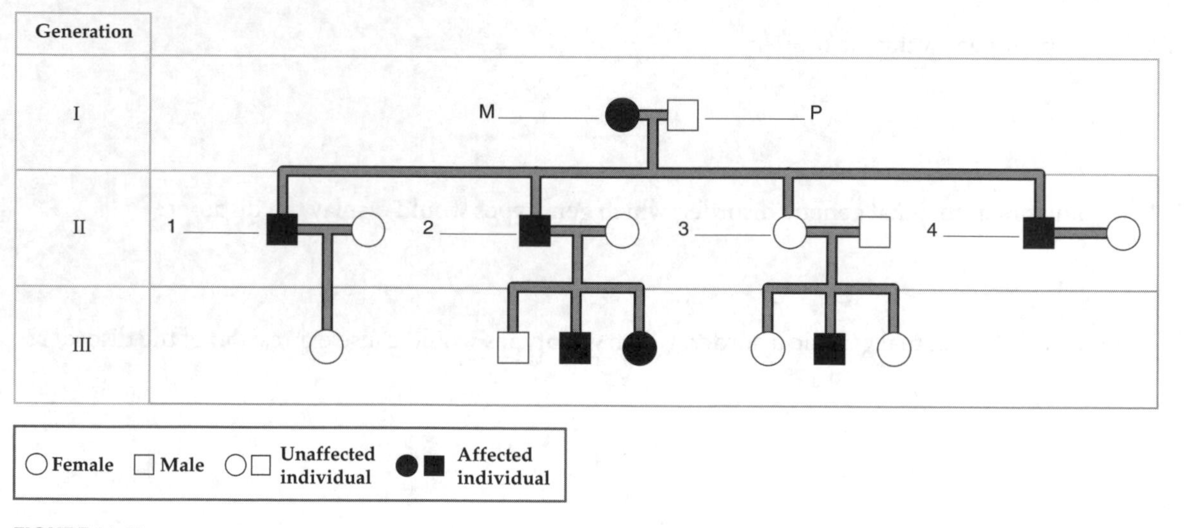

FIGURE 11.11
Family Pedigree: Hemophilia

2. Use a Punnett square to depict this scenario, placing the paternal alleles on top and the maternal alleles on the left side.

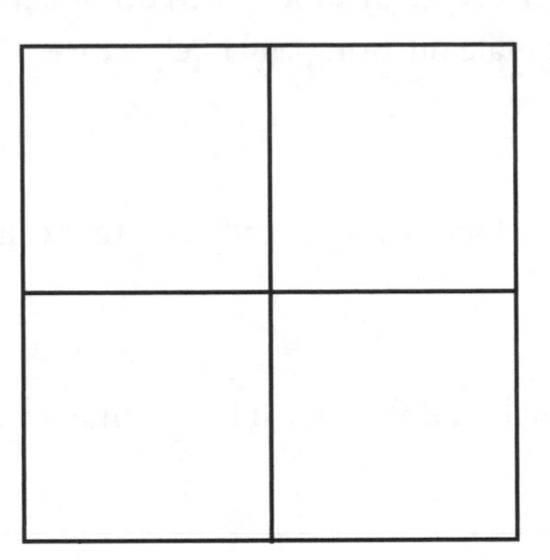

a. What proportion of offspring will exhibit hemophilia?

b. What proportion of offspring will *not* exhibit hemophilia?

3. What allele does a father who has hemophilia pass on to his male offspring? his female offspring?

4. What allele does a mother who has hemophilia pass on to her male offspring? her female offspring?

5. Judging by the offspring depicted as generation III in Figure 11.11, can you determine the genotype of each generation II offspring's mate? If yes write it below; if no, explain why.

Offspring 1 mate:

Offspring 2 mate:

Offspring 3 mate:

Offspring 4 mate:

6. Can two parents who do not express an X-linked recessive disease have an offspring who does express that disease?

7. What can you definitively state about the potential offspring of generation II's offspring 4 in Figure 11.11?

Concept Check

1. In a pedigree describing the phenotypes of an X-linked dominant disease, if a mother expresses the disease, what proportion of her male offspring will likely express the disease?

2. For X-linked recessive diseases, which sex is the only one that can be classified as a carrier?

3. If a disease is Y-linked, does it make a difference if it is dominant or recessive? Why or why not?

Key Terms

allele (p. 11-1)
autosome (p. 11-1)
carrier (p. 11-2)
chromosome theory of inheritance (p. 11-1)
chromosome (p. 11-1)
crossing-over (p. 11-10)
dominant (p. 11-2)
dominant autosomal disorder (p. 11-17)
Duchenne muscular dystrophy (DMD) (p. 11-14)
gene (p. 11-1)
genetic disorder (p. 11-2)
genetic linkage (p. 11-9)
genotype (p. 11-3)
hemophilia (p. 11-20)
heterozygous (p. 11-2)
homologous chromosomes (p. 11-1)
homologues (p. 11-1)
homozygous (p. 11-2)
Huntington disease (p. 11-17)
karyotype (p. 11-1)
law of independent assortment (p. 11-8)
linked traits (p. 11-2)
locus (p. 11-11)
pedigree (p. 11-17)
phenotype (p. 11-3)
recessive (p. 11-2)
recessive autosomal disorder (p. 11-17)
sex chromosome (p. 11-1)
sex-linked gene (p. 11-12)
SRY (p. 11-12)
X chromosome (p. 11-12)
Y chromosome (p. 11-12)

Review Questions

1. What are genes and what do they do?

2. ________________ is the theory describing how hereditary information is located within chromosomes.

3. ________________ is a stretch of DNA affecting a genetic trait.

4. The specific location of a gene on a chromosome is called a(n) ________________.

5. An individual possessing a recessive allele but not displaying the corresponding phenotype, because the dominant allele masks the recessive phenotype, is considered a(n) ________________.

6. Genetic loci within the same chromosome in close proximity tend to be passed on together and are considered genetically ________________.

7. A(n) ________________ is a chart similar to a family tree that shows genetic relationships among family members over two or more generations of a family's history.

8. Why do genetic diseases persist in the human population?

9. Why are dominant genetic disorders that produce serious medical effects less common then recessive genetic disorders?

10. If a father displays a dominant X-linked trait, what is the likelihood that his daughters will display that trait? Sons?

11. What information does a pedigree provide?

12. How is sex determined in humans?

13. If two carriers for a recessive autosomal disorder mate, what proportion of their offspring will be likely to inherit the disorder? What proportion will be carriers? Diagram the potential results of the mating in the Punnett square below.

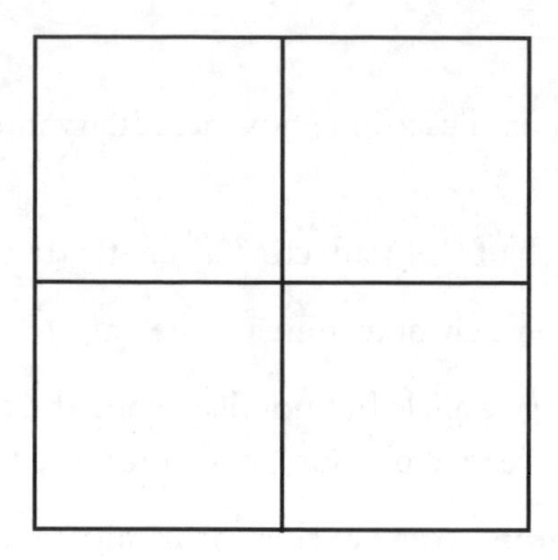

14. If an individual with a dominant autosomal disorder mated with a person who was homozygous recessive, what proportion of their offspring would be likely to inherit the disorder? Would any of the offspring be considered carriers?

15. How are genes located on sex chromosomes expressed and inherited differently from those located on autosomes?

16. Why don't linked genes follow Mendel's law of independent assortment?

17. How does the chromosome theory of inheritance change the way we analyze genetic inheritance?

LAB **12**

DNA and DNA Technology

Objectives

- Describe the structure of a DNA molecule and demonstrate DNA replication.
- Isolate DNA from human cheek cells.
- Copy DNA by PCR.
- Demonstrate transcription and translation.
- Separate DNA fragments by gel electrophoresis.
- Use restriction enzymes to make DNA fragments.
- Interpret profile results of DNA fingerprinting.
- Consider the application of DNA technology to real-world scenarios.

Introduction

DNA (**deoxyribonucleic acid**) is the molecule of inheritance, used by every organism to store information. It encodes the molecules of life that run individual cells and, consequently, entire organisms. Molecules of DNA are packaged into **chromosomes** in the nucleus of a cell. DNA is composed of two strands of nucleotides twisted around each other and has two different types of regions: **coding regions** (called **genes**) and **noncoding regions**.

Your genetic makeup is more complex than one molecule of DNA. Think of your genetic makeup as a library containing two reference books—books that are not permitted to leave the library. The nucleus of each cell is the library, and the DNA from your mother and from your father are the two reference books. Each book has 23 chapters, one for each of the 23 chromosomes you inherit from each parent. These chapters are not just letters on a page; they have page numbers, order, and structure, just as a chromosome is not just a molecule of DNA, but one molecule of DNA plus proteins that regulate its structure and function (Figure 12.1). Each chapter is filled with lots of paragraphs, but only specific sentences throughout the chapter provide the key concepts of the chapter; the genes (coding regions) within the chromosome are these key sentences. The filler sentences of the chapter are like the noncoding regions of DNA. Both the filler and the key sentences are composed of letters, just as both genes and noncoding regions of DNA are made of nucleotides.

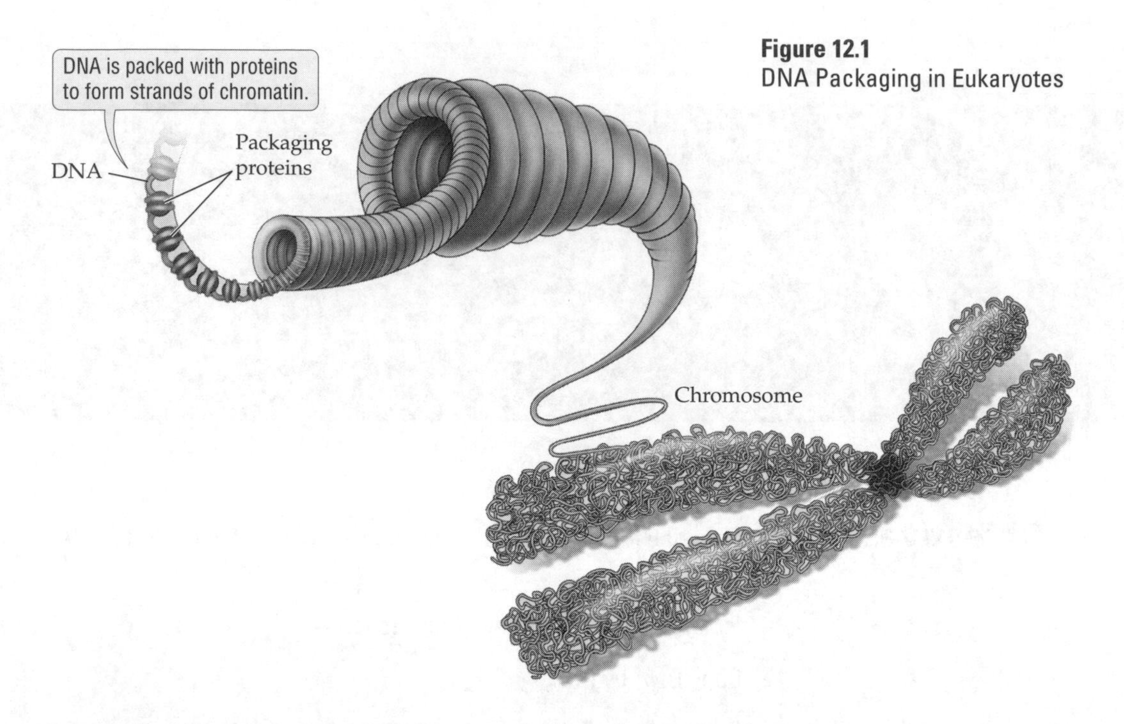

Figure 12.1
DNA Packaging in Eukaryotes

A process called transcription copies instructions stored in DNA, which dictates the production of proteins. Imagine that you visit the library in hopes of learning how to build a bicycle. Again, think of a reference book in a library: just as only a photocopy can be used to carry information from the reference book (such as instructions for building a bicycle) out of the library, only a copy of DNA can be used for making proteins that the body needs. During transcription, DNA is copied to **RNA** (**ribonucleic acid**), and RNA travels to the cytoplasm. Three specialized types of RNA—messenger, transfer, and ribosomal RNA—then function in unison to build a protein during a process called translation.

During translation, messenger RNA (mRNA) is like your photocopy: it contains the information (sequence) for putting the parts (amino acids) of a protein together—like the instructions for building a bicycle. Transfer RNAs (tRNAs) carry the amino acids (parts) to be strung together, as the parts of a bicycle would need to be gathered for assembly. And finally, ribosomal RNAs (rRNAs) make up ribosomes that work to read the mRNA (instructions) and bring the tRNAs (carrying the parts) together. In this analogy of translation, you are like the ribosome: you read the photocopy (mRNA) that has the instructions and put the parts (tRNAs carrying amino acids) together to build a bicycle (protein). The overall role of genes is to produce a functional protein, just as, in our analogy, the overall function of your instructions is to produce a functional bicycle.

Discovering the many secrets of our cells and bodies is a never-ending process, but scientists can use a variety of techniques to manipulate and examine DNA, RNA, and proteins. Molecular research has created many techniques used in a variety of applications, from identifying patients with genetic disorders to creating new plant species or identifying a criminal. DNA technology offers many opportunities to improve our understanding of life; in this lab you will learn about some important techniques commonly used in laboratories.

ACTIVITY 1 Deoxyribonucleic Acid

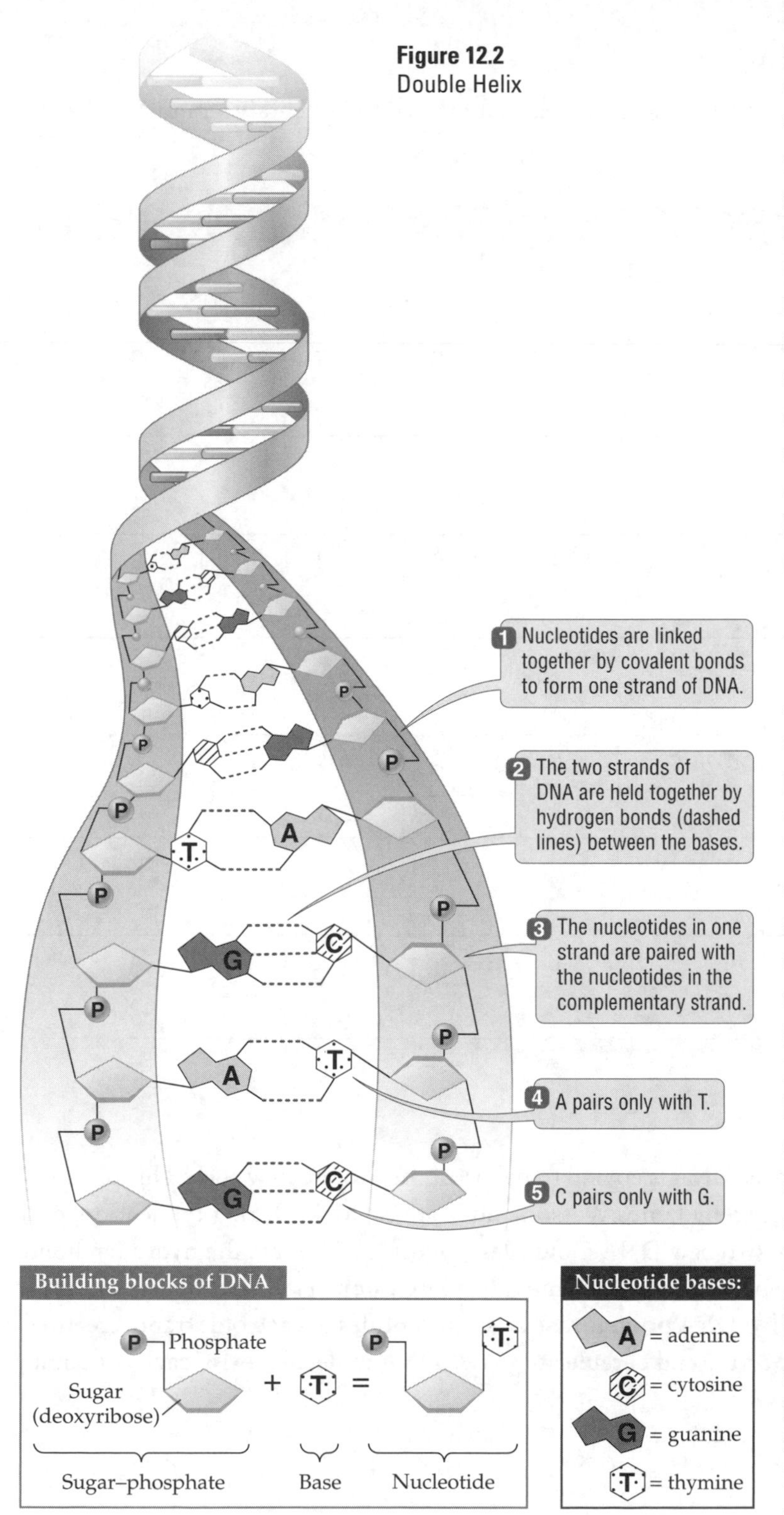

Figure 12.2
Double Helix

DNA is a polymer composed of four different **nucleotides**. Each nucleotide contains a phosphate group, a sugar, and a nitrogenous (nitrogen-containing) base. The phosphate group and sugar are the same among nucleotides, but the nitrogenous base can be one of four bases and determines the type of nucleotide. The four nucleotides that comprise DNA are **adenine** (A), **thymine** (T), **cytosine** (C), and **guanine** (G) (see Figure 12.2). In this activity you will explore how these nucleotides combine to make the DNA molecule and then take a look at DNA replication.

The Double Helix

A DNA molecule contains two strands of nucleotides that run antiparallel (in opposite directions) to each other like railroad tracks. The "rails" of each strand are the sugar and phosphate groups; the inside ties are the bases from each strand bound together by hydrogen bonds. These two strands twist around each other into a **double helix** (Figure 12.2). Each base has one complementary base. Adenine binds only to thymine and cytosine binds only to guanine, so the two strands of the double helix are held together by hydrogen bonds between adenine in one strand and thymine in the other, and by bonds between cytosine in one strand and guanine in the other. These binding patterns are called the **base-pairing rules**. Base-pairing rules are useful in DNA analysis because a scientist who knows the sequence of one strand can deduce the sequence of the other, **complementary strand**.

Here you will determine the sequence of nucleotides for the second (complementary) strand within a DNA molecule. Use the following example of complementary strands as a guide.

Strand 1: GGTACTACGAGTATC

Strand 2: CCATGATGCTCATAG

1. For each example shown in Table 12.1, indicate the bases for the complementary strand 2 in the second row.

TABLE 12.1 COMPLEMENTARY STRANDS

Example 1:

Strand 1	G	T	G	G	C	A	T	A	G	C	T	T	A	G	C
Strand 2															

Example 2:

Strand 1	T	C	C	G	A	T	A	C	C	A	T	G	C	G	A
Strand 2															

Example 3:

Strand 1	A	G	C	C	C	T	G	T	A	T	A	G	C	A	A
Strand 2															

Example 4:

Strand 1	C	G	A	A	T	C	C	G	T	A	G	T	C	A	T
Strand 2															

DNA Replication

Every time a cell divides during mitosis, DNA is copied to be distributed to the new cells. This process is called **DNA replication**, first described by James Watson and Francis Crick. During replication, each strand is used as a template to create two new DNA molecules (Figure 12.3). First, the hydrogen bonds holding the two strands together are broken by an enzyme called **DNA helicase**, enabling the strands to unwind and separate. An enzyme called **DNA polymerase** adds nucleotides to each old strand, creating a new complementary strand of DNA. At the end of replication, two DNA molecules exist, each containing an old strand and a new strand of DNA.

Here you will replicate a molecule of DNA.

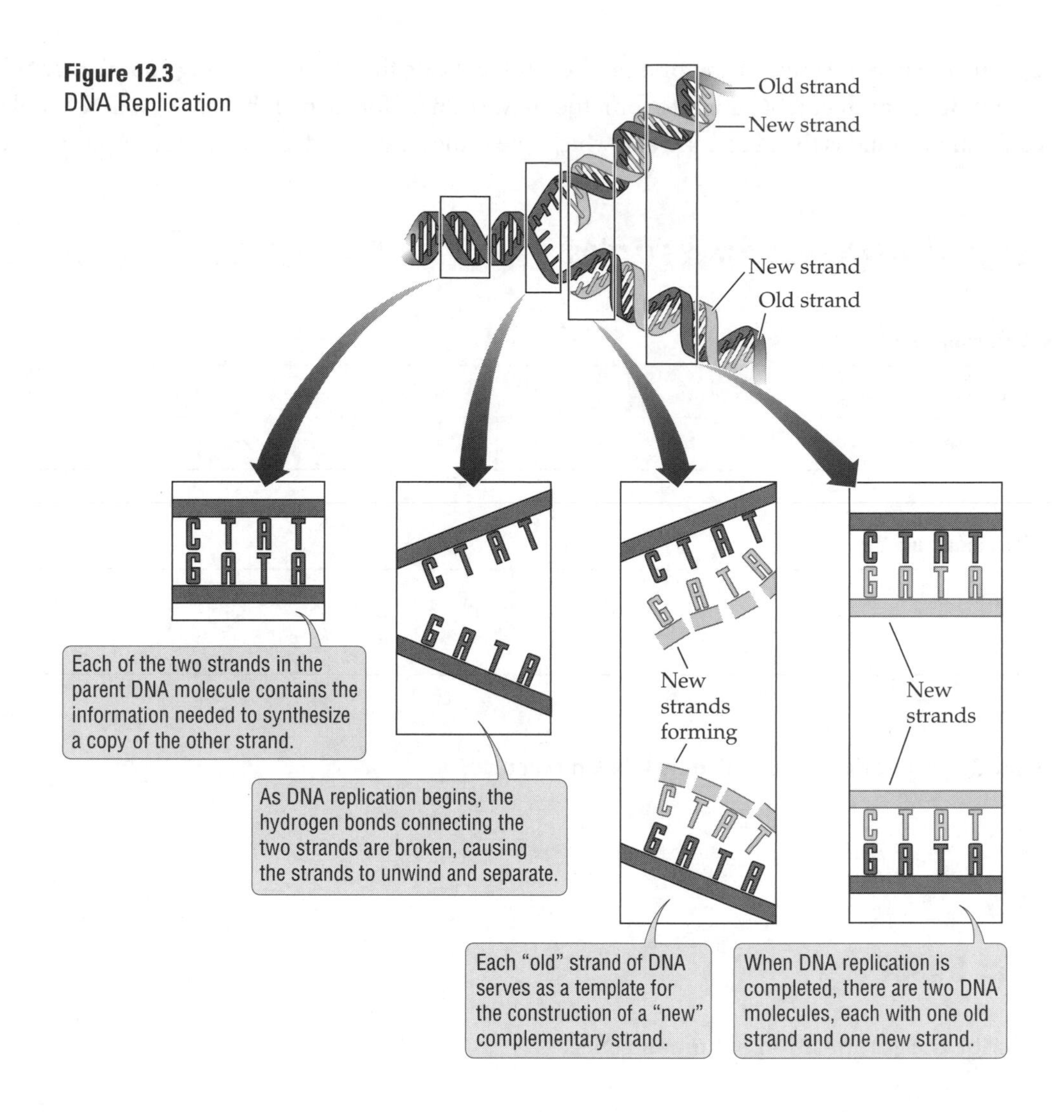

Figure 12.3
DNA Replication

1. Using DNA models provided by your instructor, create the first parent strand of DNA: GACGTAGACTTGTAC. Record the nucleotides in Table 12.2 as "Parent strand 1." Now, use the base-pairing rule to construct the second (complementary) parent strand of the DNA molecule, and record the complementary nucleotide sequence in the row labeled "Parent strand 2."

TABLE 12.2 COMPLEMENTARY PARENT STRANDS															
Parent strand 1															
Parent strand 2															

2. Separate the parent strands from one another and replicate the DNA by adding complementary nucleotides to each parent strand to form the "new strands" for each of the two new DNA molecules. Note which strand is the old strand and which one is the new strand. Record your results in Table 12.3.

TABLE 12.3 NEW COMPLEMENTARY STRANDS

New DNA molecule 1:

Parent strand 1															
New strand															

New DNA molecule 2:

Parent strand 2															
New strand															

3. What do you notice about the new DNA molecules?

Concept Check

1. Describe the composition of human DNA.

2. Which nucleotide bases are complementary to one another?

3. What is a base pair?

4. How can a scientist who has the sequence of one strand of DNA determine the sequence of the other strand?

5. Describe the steps of DNA replication.

6. What is the role of DNA polymerase?

ACTIVITY 2 Human DNA Isolation

DNA manipulation is valuable to many areas of **molecular biology**, the study of the molecules of life. Scientists can now isolate the DNA of anything from a virus to a human. Once the DNA has been isolated, scientists can perform a variety of tests, including determining the **sequence** of a DNA fragment or gene (the order of nucleotides), mapping the **genome** of a species (the sequence of nucleotides for all the DNA of an organism), determining whether an individual has a specific gene sequence, or comparing the DNA of two individuals of a species or even two different species of organisms. In all of these techniques, DNA isolation is the first step to perform these complex analyses. In this activity you will isolate your DNA in preparation for further analysis in Activity 3.

1. Obtain a paper cup, a pipette, microcentrifuge tubes, saline solution, a chelator solution, a water bath, a hot plate, and a buffer solution.
2. Start the water bath and add to the paper cup the amount of saline solution specified by your instructor. Swish the saline solution in your mouth and then spit the solution back into the cup. If you're trying to isolate your own DNA, what cells are you retrieving during this process?
3. Label one of the microcentrifuge tubes with your name, and pour your saline solution into the tube. Fill to the top of the tube, leaving just enough room to close the lid without losing any fluid.
4. Give your tube to your instructor to be centrifuged. Centrifugation spins the solution so that heavy material is deposited at the bottom of the tube. Your instructor will insert the tubes from the whole class in the microcentrifuge and spin them. Where in the tube should your cells be located after centrifugation?

5. When the machine is finished, your instructor will return your tube. Now you have a fluid and white material at the bottom of the tube. The white material consists of cells from your cheeks, which form a pellet at the bottom of the tube. Don't worry about the size of the pellet; sometimes it is very small. Simply pour off the fluid into the waste beaker, but be careful! You want to avoid dislodging the cells, so don't bang the tube.
6. After the fluid has been removed from the tube, you want to break open the cells to isolate your DNA. Add to the tube the amount of chelator solution specified by your instructor. This solution removes unwanted molecules such as proteins and RNA within the cell. Tap or flick the tube to mix the cells with the solution.
7. Heat the water bath to boiling, put the tube in the boiling water, and incubate it for 10 minutes. Boiling bursts open the cells to isolate DNA from the nucleus. After 10 minutes, remove the tube and tap/flick it for 15 to 20 seconds.
8. Give your tube to your instructor to be centrifuged. Centrifugation will separate any dissolved materials and those that precipitated during the addition of the chelator and the boiling.
9. When the machine is finished, your instructor will return your tube. Now you have a fluid and white material at the bottom of the tube again. This time you want to retrieve the fluid. Using the pipette, remove two drops of fluid and transfer it to a new microcentrifuge tube.
10. Label the tube and give it to your instructor for storage, unless you are moving on to Activity 3.

Concept Check

1. What does centrifugation do?

2. Describe the process of isolating DNA from human cells.

3. How is DNA isolation useful in a practical sense?

ACTIVITY 3 Polymerase Chain Reaction and Gel Electrophoresis

Through many great discoveries, humans now have the ability to copy portions of DNA in a process similar to that of DNA replication. The technique of copying sections of DNA is called the **polymerase chain reaction (PCR)**. It uses the enzyme DNA polymerase from a bacterium to make many copies of targeted DNA sequences. Scientists can use the PCR technique to make billions of copies of targeted sequences of DNA in just a few hours. This series of reactions occurs in a small test tube containing DNA with a target sequence, DNA primers, free nucleotides (A, T, C, and G), and DNA polymerase, along with other solutions to optimize the number of copies created during these reactions. The test tube is subjected to a variety of conditions to cause the following specific steps:

- *Step 1: DNA is broken apart.* DNA is heated to break the noncovalent bonds between the two complementary strands of DNA, separating them into single-stranded DNA. (Think of the two DNA strands as the two parts of a zipper that you're unzipping.)
- *Step 2: DNA primers bind to target DNA.* Two short segments of synthetic DNA called **DNA primers** are added, and they bind to either end of the target DNA through complementary base pairing. As the mixture cools slowly, one of two things occurs: either (1) the two original complementary DNA strands form noncovalent bonds again, allowing the two strands of DNA to come together, and PCR does not take place; or (2) the primer forms noncovalent bonds with short complementary regions along one DNA strand. The primers do not complete the nucleotide sequence of the complementary strand of original DNA; they are merely a starting point for the DNA polymerase to know what section of DNA to copy.
- *Step 3: The copy of the target sequence is completed.* Remember that PCR does not copy a full DNA molecule; rather, it targets a specific region of DNA, such as a gene. The DNA primers provide a starting point to copy the target DNA. In order to complete the sequence, DNA polymerase adds nucleotides (A, T, G, and C), all of which are in the test tube, to complete the target DNA sequence (filling in the missing complementary sequence).
- *Step 4: More copies are made.* The first three steps are repeated, with the primers binding either to the original DNA strands or to the new target DNA created during the first round of PCR (Figure 12.4). The process is done over and over again, producing many identical copies, or **clones**, of the targeted region of DNA, called **PCR products**. Each cycle of reactions copies the target region of DNA, producing more and more PCR products.

PCR is employed within DNA sequencing machines to determine the order of nucleotides within the genome of different species, including humans. When used to describe a species, the term "genome" refers to all of the genetic information held in the chromosomes of a species. **The Human Genome Project** is a collaborative effort (essentially complete) to determine the sequence of noncoding and coding regions of human DNA, giving scientists a much better idea of how our DNA is organized and the specific functions of our genes. This knowledge can be applied to the design of more specific medical treatments, drug design, and disease prevention. PCR is also valuable for scientists who want to replicate a specific gene and then study gene variation in a population and further determine the function of that gene. Also remember that the genes in your DNA are expressed in different cells at different times, and the regulation of DNA is very important for your body to produce the correct protein in the correct cell at the correct time. PCR enables scientists to understand this regulation by copying extremely small amounts of DNA, and it has expanded our ability to apply this scientific application in crime scene analysis.

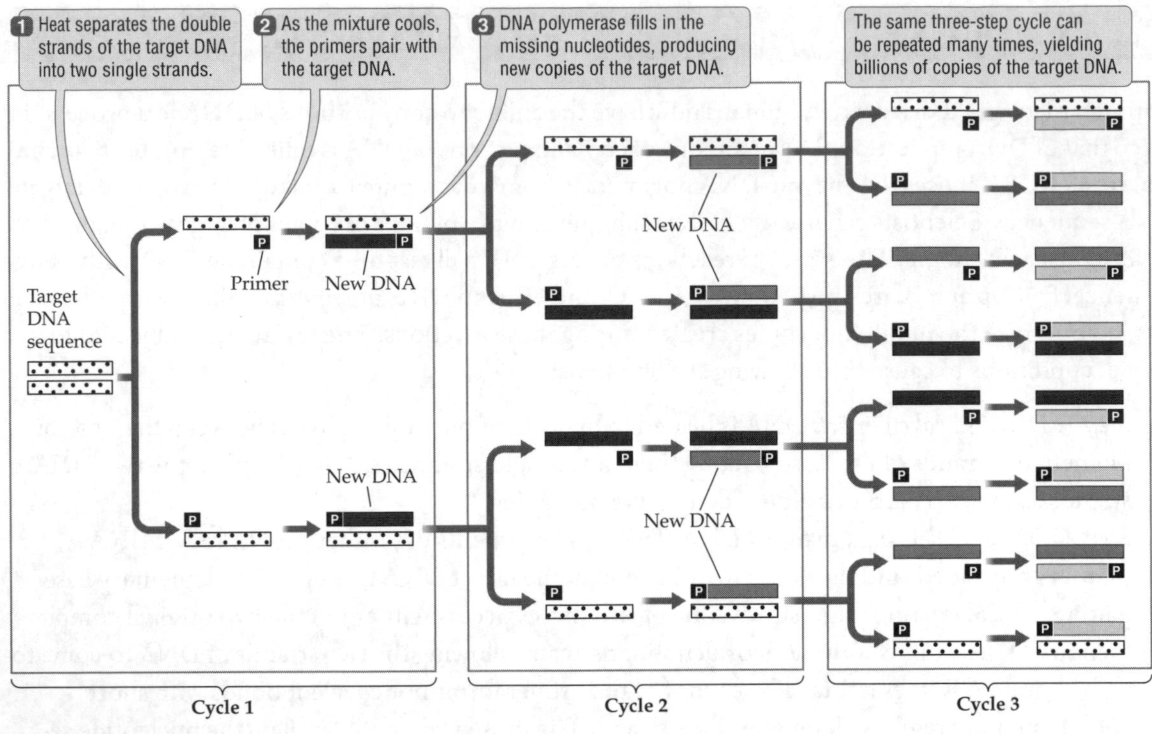

Figure 12.4
Polymerase Chain Reaction (PCR)

Every person's DNA is different. In any given population, each gene (and even each noncoding region) has many variations, so when PCR is performed on different people's DNA it can produce PCR products of different sizes. Every DNA fragment contains a specific number of nucleotides; for example, a PCR product from one individual may be 100 bases long, while another person may have a PCR product that is 200 bases long. PCR enables scientists to identify the type of gene variation that an individual has. A complementary technique called gel electrophoresis enables scientists to separate and identify the PCR product that they're studying.

Gel electrophoresis separates DNA fragments such as PCR products by their size. The gel acts like a sieve when an electrical charge is applied to it. Being negatively charged, DNA will migrate toward the positive side of the apparatus (Figure 12.5). After the electrical charge is applied to the gel containing a mixture of DNA fragments on one end, the shorter pieces of DNA will move faster through the gel.

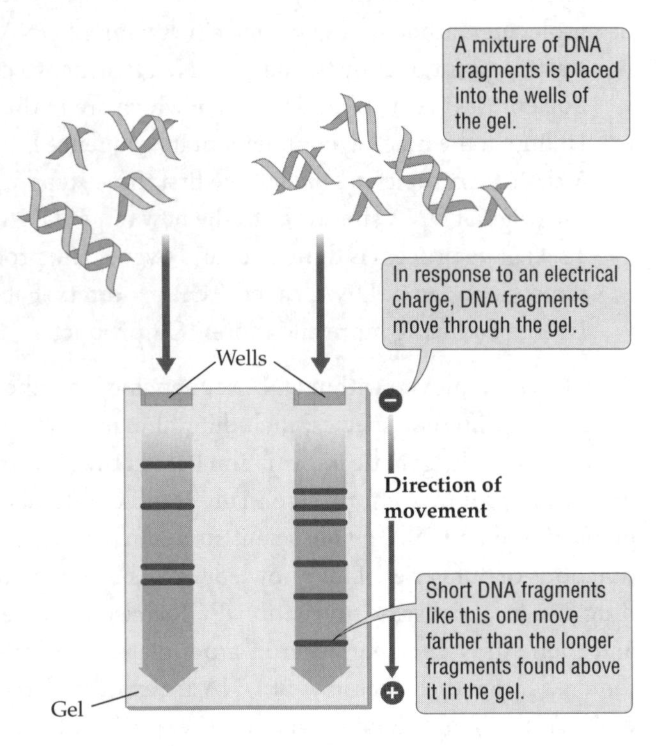

Figure 12.5
Separation of DNA Fragments by Gel Electrophoresis

In this activity you will use PCR to copy a specific portion of the DNA sample you isolated in Activity 2, and then examine it using gel electrophoresis.

Polymerase Chain Reaction

1. Obtain a PCR tube, gloves, the sample of DNA that you isolated in Activity 2 (or another sample if yours is not available), PCR Master Mix (which includes nucleotides A, T, G, and C), and the primers for the specific region of DNA that you'll be copying.

   BE SURE TO WEAR GLOVES SO THAT YOU DO NOT CONTAMINATE ANY OF THE DNA SAMPLES OR SOLUTIONS.

2. To the PCR tube, add the DNA sample and PCR Master Mix in the amounts specified by your instructor. Mix the contents by tapping or flicking the tube. Label the tube with your initials.
3. Give the tube to your instructor for insertion into a temperature-regulating machine called a thermocycler. The temperature needs to fluctuate to unwind the DNA (high temperature of about 90°C), then allow the primers to bind (at about 60°C), and finally allow DNA polymerase to add the nucleotides (A, T, G, and C) (at 70°C). These steps occur at very specific temperatures.
4. In the 30 or more times that this cycle is repeated in the thermocycler, many copies of the target region are made. Now we're ready for gel electrophoresis.

Gel Electrophoresis

Here you will use gel electrophoresis to visualize the PCR product that you just created.

1. Obtain tracking dye, a pipette, and an electrophoretic gel prepared by your instructor.
2. Using the pipette, add to your PCR tube 1 drop of tracking dye. This blue dye will travel through the gel at the same rate as small fragments of DNA.
3. Your instructor has prepared a gel. Following your instructor's directions, add your DNA sample to the gel for electrophoresis.
4. Once all students in the class have added their samples to the gel, your instructor will add a standard marker of DNA. A standard marker contains a mixture of DNA fragments of known sizes (for example, a 100 base-pair (bp) standard marker has pieces of DNA in 100-bp increments—100 bp, 200 bp, 300 bp, and so on). Because the speed at which DNA travels through the gel is based on its size, you can compare your PCR product to these standards to determine the size of your DNA.
5. After applying electrical current to the DNA apparatus, your instructor will remove the gel, apply a stain to visualize the DNA, and take a picture of your PCR products.
6. Everyone in your lab has unique DNA and possibly different sizes of DNA fragments (PCR products) from the PCR procedure. What is the size of your PCR product? The larger it is, the slower the DNA fragment will travel. How can you determine the approximate size of your DNA fragment?

7. On the gel representation in Figure 12.6, sketch what you see in the two lanes of your gel—the results for your PCR product and the standard marker.

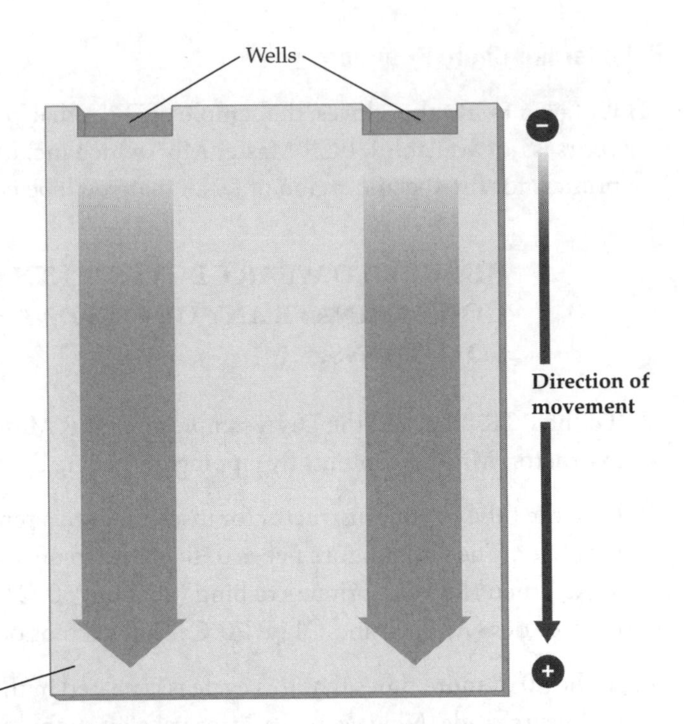

Figure 12.6
Gel Electrophoresis Results for PCR Products

Concept Check

1. Describe the polymerase chain reaction.

2. How can people have different PCR products?

3. Why does DNA travel through the gel during gel electrophoresis? How does size affect the way PCR products migrate during gel electrophoresis?

4. How can PCR be used to study a gene? What are some applications of PCR?

ACTIVITY 4 RNA and Protein Synthesis

Most genes contain instructions for manufacturing proteins, and the proteins encoded by genes are essential to life. Once a gene has been copied by means of PCR, it can be used to study the protein produced by the gene. A gene is a DNA sequence that directs the production of RNA molecules, which then direct the production of proteins in a two-phase process called **protein synthesis**. In this activity you will explore each of these two phases.

Transcription

The first phase of protein synthesis is **transcription**. Transcription transforms the information within the coding regions of genes into one of three types of RNA molecules (Table 12.4). The first type is the **messenger RNA (mRNA)** molecule, a copy of the code in the gene that contains the instructions for the protein sequence. Messenger RNA migrates to the cytoplasm to meet up with a ribosome. Ribosomes are composed of another type of RNA, called **ribosomal RNA (rRNA)**, and proteins. Ribosomes read the instructions in mRNA and position incoming **transfer RNA (tRNA)** molecules—which deliver to the ribosomes the appropriate amino acids specified by the mRNA(Figure 12.7). Ribosomes then join the amino acids to make the exact protein encoded by the gene.

TABLE 12.4 RNA MOLECULES AND THEIR FUNCTIONS

Type of RNA	Function	Shape
Messenger RNA (mRNA)	Specifies the order of amino acids in a protein.	
Ribosomal RNA (rRNA)	As a major component of ribosomes, assists in making the covalent bonds that link amino acids to make a protein.	
Transfer RNA (tRNA)	Transports the correct amino acid to the ribosome, according to the information encoded in the mRNA.	

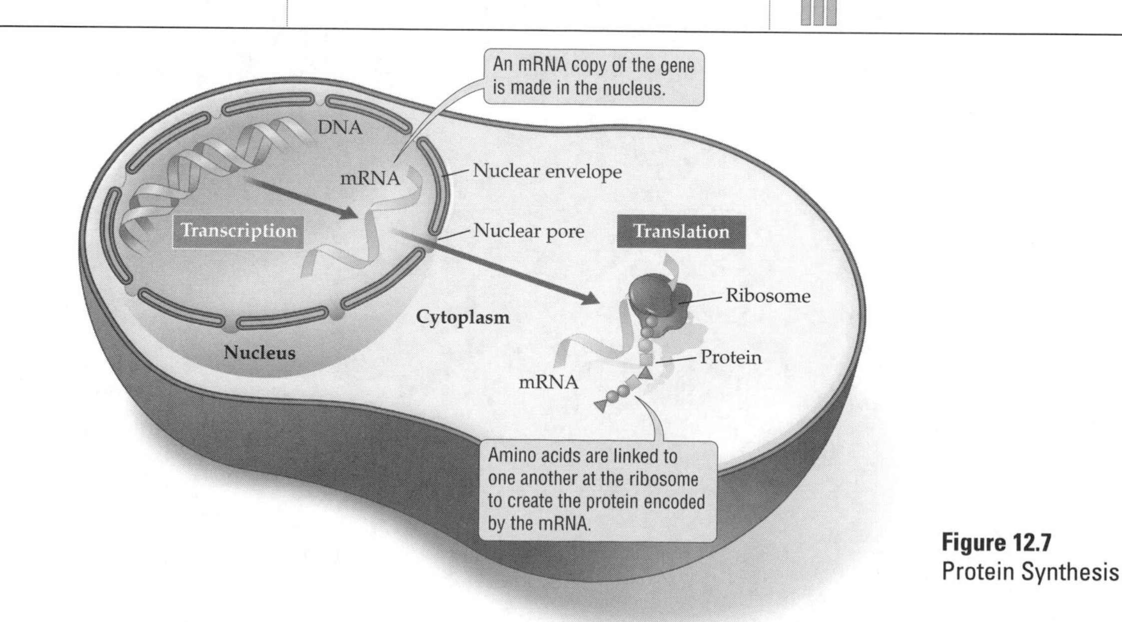

Figure 12.7
Protein Synthesis

RNA molecules are nucleic acids made of nucleotides covalently bonded to one another (Figure 12.8). Like DNA, the nucleotides of RNA molecules are composed of a sugar, a phosphate group, and one of four nitrogen-containing bases. But DNA and RNA molecules have three structural differences: RNA is single-stranded, its sugar is ribose instead of deoxyribose, and it has the base **uracil (U)** instead of thymine (uracil bonds with adenine in RNA).

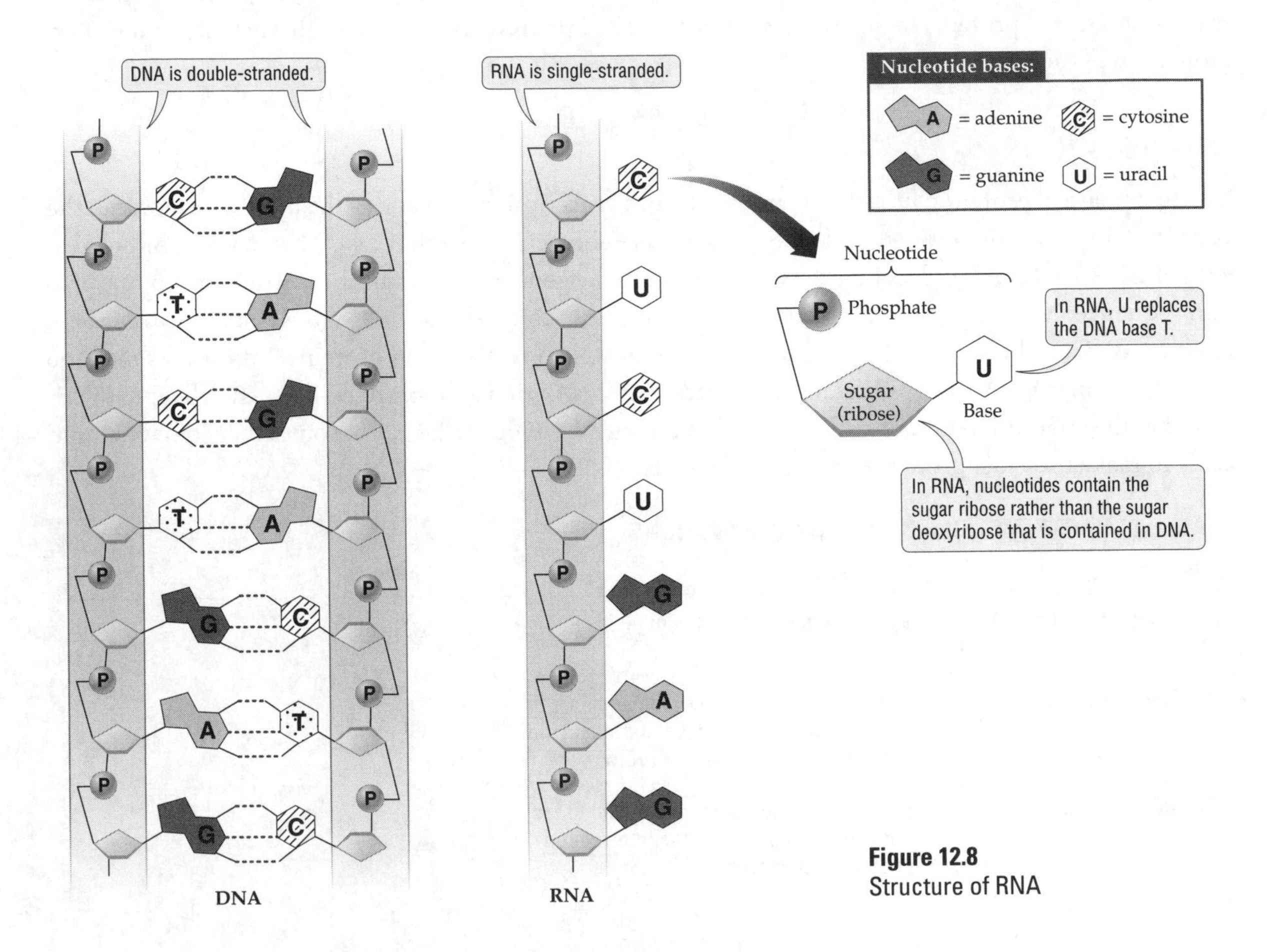

Figure 12.8
Structure of RNA

1. For each DNA strand depicted in Table 12.5, specify the bases for the mRNA strand that would be created during transcription. The base-pairing rule also applies to transcription.

   Example:

   DNA strand: GGT ACG AGT TTC ACT

   mRNA strand: CCA UGC UCA AAG UGA

TABLE 12.5 TRANSCRIPTION: PREDICTING mRNA STRANDS

Strand 1:

DNA strand	G	T	G	G	C	A	T	A	G	C	T	T	A	G	C
mRNA strand															

Strand 2:

DNA strand	T	C	C	G	A	T	A	C	C	T	T	G	C	G	A
mRNA strand															

Strand 3:

DNA strand	A	G	C	C	C	T	G	T	A	T	A	G	C	A	A
mRNA strand															

Strand 4:

DNA strand	C	G	A	A	A	T	C	G	T	T	G	T	C	A	T
mRNA strand															

Translation

Translation is the process of converting the code within a strand of mRNA into a protein sequence. Each sequence of three nucleotide bases in an mRNA molecule is called a **codon** and codes for one amino acid (Figure 12.9). The **genetic code** consists of all the possible codons of mRNA and the amino acids they specify (Figure 12.10). The genetic code is redundant, meaning that each amino acid, except methionine, is dictated by more than one codon.

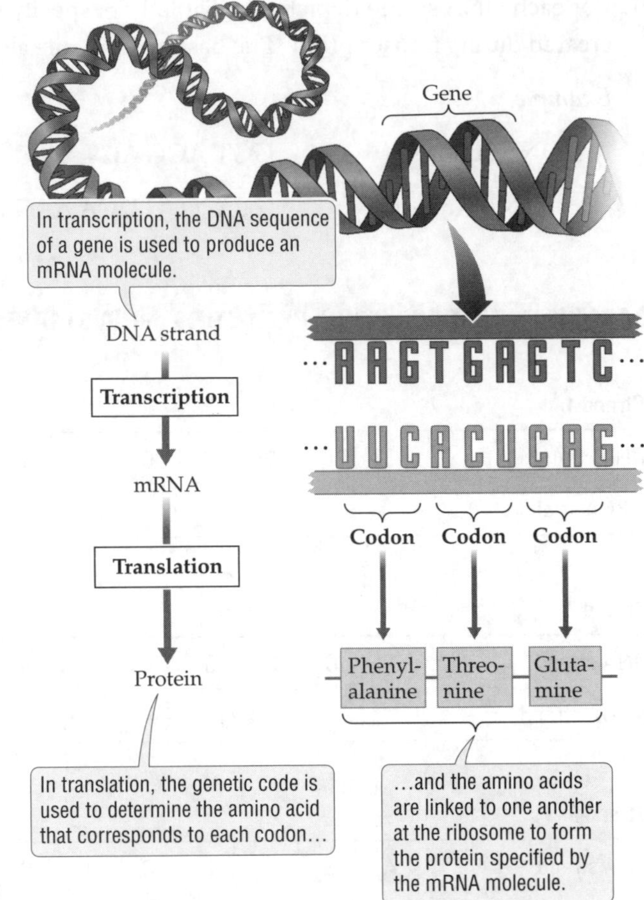

Figure 12.9
How Cells Use the Genetic Code

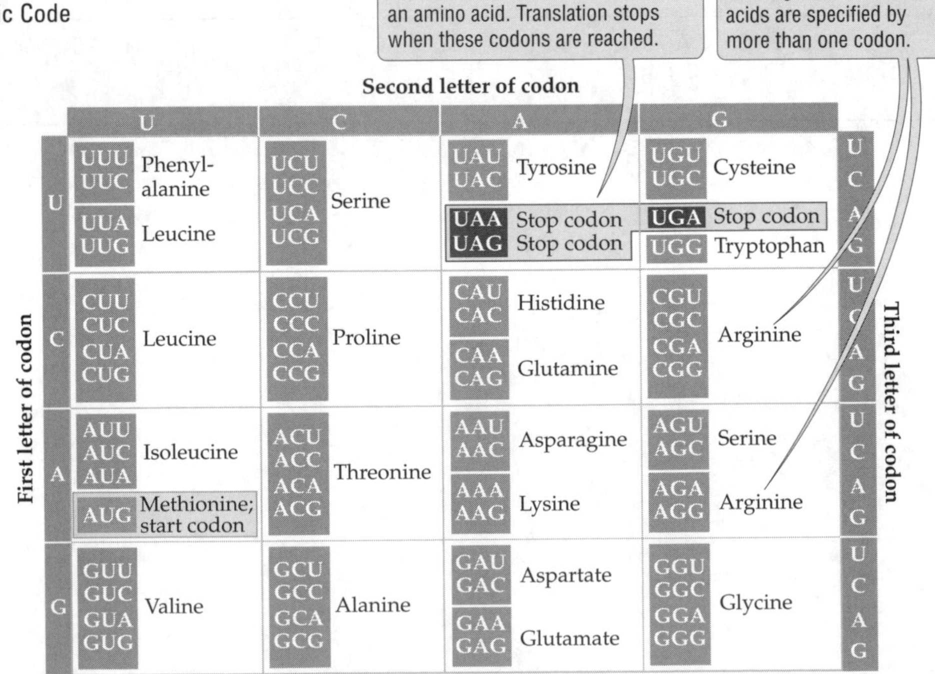

First letter	U	C	A	G	Third letter
U	UUU, UUC Phenyl-alanine; UUA, UUG Leucine	UCU, UCC, UCA, UCG Serine	UAU, UAC Tyrosine; UAA Stop codon; UAG Stop codon	UGU, UGC Cysteine; UGA Stop codon; UGG Tryptophan	U C A G
C	CUU, CUC, CUA, CUG Leucine	CCU, CCC, CCA, CCG Proline	CAU, CAC Histidine; CAA, CAG Glutamine	CGU, CGC, CGA, CGG Arginine	U C A G
A	AUU, AUC, AUA Isoleucine; AUG Methionine; start codon	ACU, ACC, ACA, ACG Threonine	AAU, AAC Asparagine; AAA, AAG Lysine	AGU, AGC Serine; AGA, AGG Arginine	U C A G
G	GUU, GUC, GUA, GUG Valine	GCU, GCC, GCA, GCG Alanine	GAU, GAC Aspartate; GAA, GAG Glutamate	GGU, GGC, GGA, GGG Glycine	U C A G

Figure 12.10
The Genetic Code

Translation begins when the ribosome reads the start codon within the mRNA. Next, the tRNA carrying the amino acid dictated by the codon binds to the mRNA. The tRNA binds to the codon because the tRNA contains the matching sequence (the same concept as the matching nucleotides of DNA), or **anticodon**. The code within the mRNA continues to be read by the ribosomes and the tRNAs continue to bind until one of three different stop codons indicates the end of the protein sequence (see Figure 12.10).

Here you will examine translation, focusing on the variety of codons that exist and the variety of combinations of amino acids that occur as a result.

1. Determine the amino acid that each strand of mRNA recorded in Table 12.5 would encode. Start by copying into Table 12.6 the mRNA strands that you predicted in Table 12.5.

2. Using the genetic code in Figure 12.10, determine the amino acid that each codon of the mRNA dictates. Record these amino acids in Table 12.6.

Example:

DNA strand:	G G T	A C G	A G T	T T C	A C T
RNA strand:	C C A	U G C	U C A	A A G	U G A
Amino acids:	proline	cysteine	serine	lysine	Stop

TABLE 12.6 TRANSLATION: PREDICTING AMINO ACIDS

Strand 1:

mRNA strand															
Amino acids															

Strand 2:

mRNA strand															
Amino acids															

Strand 3:

mRNA strand															
Amino acids															

Strand 4:

mRNA strand															
Amino acids															

3. Mutations within DNA occur all the time, triggered by a variety of causes, from environmental conditions (such as exposure to the sun) to mistakes during replication. Considering the likelihood of mutations, what is the advantage of having a redundant genetic code?

Concept Check

1. The code within DNA ultimately dictates the production of proteins. What two processes manufacture proteins from DNA and RNA during protein synthesis? Where in the cell do these processes occur?

2. Name the three types of RNA and describe their functions.

3. If an mRNA is 63 bases long, how many amino acids will it code for?

ACTIVITY 5 Restriction Enzymes and Gel Electrophoresis

After isolating DNA, scientists can mix the DNA from different organisms by cutting it with **restriction enzymes**, which are naturally occurring bacterial defense mechanisms. Restriction enzymes chop up the DNA of invading organisms, each type cutting at a specific sequence on the DNA. For instance, *Alu*I cuts DNA anywhere it contains the sequence AGCT, whereas *Not*I cuts at GCGGCCGC (Figure 12.11). This technique creates DNA fragments of manageable size. These fragments can then be separated by gel electrophoresis (described in Activity 3), which enables the scientist to separate pieces of cut-up DNA and to isolate this DNA by its size.

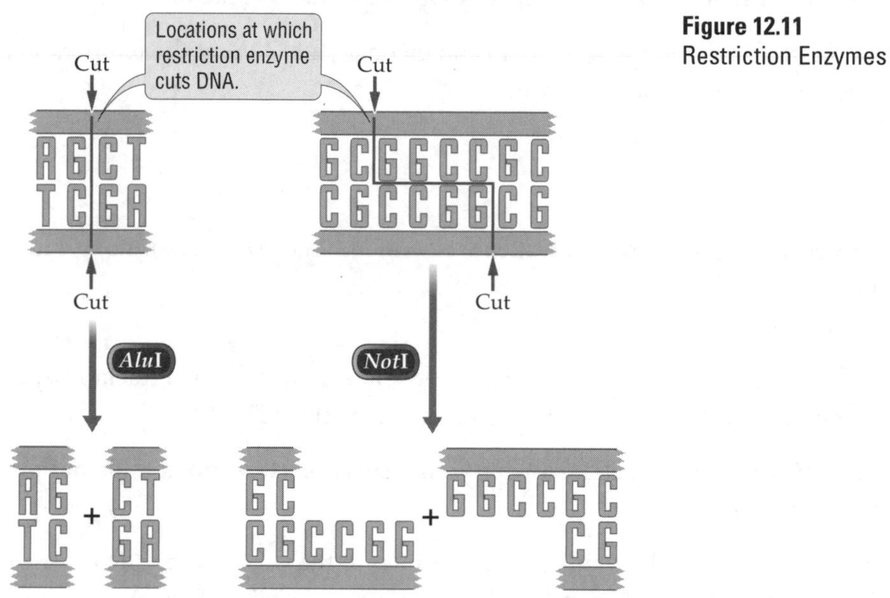

Figure 12.11
Restriction Enzymes

Restriction enzymes can be used to cut DNA from different organisms, isolate specific sections of DNA (genes), and combine fragments from two different organisms into recombinant DNA. **Recombinant DNA** is a synthetic combination of DNA fragments from different organisms. An enzyme called **ligase** acts like a glue to bind fragments of DNA together, creating recombinant DNA. Recombinant DNA can then be used to change or modify an existing organism, creating a genetically modified organism.

The application of gene manipulation through DNA isolation, cutting with restriction enzymes, PCR, and other techniques used to create recombinant DNA fall under the field of **genetic engineering**, in which a DNA sequence (often a gene) is isolated, modified, and then inserted back into an individual of the same or a different species. Genetic engineering is a broad field that uses these techniques to create new crops such as genetically modified soybeans, to produce pharmaceuticals like insulin for diabetic patients, or to treat patients with a genetic disease such as hemophilia by changing the patient's DNA through gene therapy. In **gene therapy**, scientists use many different techniques to repair genes, but all techniques clone (make an exact copy of) a good gene that is then inserted into the patient who has the genetic disorder.

In this activity you will examine how restriction enzymes cut a **plasmid**, a small circular segment of extrachromosomal DNA [a type of DNA that can replicate and function independent of the organism's chromosome(s)]. Plasmids typically carry one to a few genes that code for characteristics such as antibiotic resistance, and they are found naturally in bacteria and some eukaryotes. Then you will examine the fragments using gel electrophoresis.

Restriction Enzymatic Reactions

Here you will use a restriction enzyme to cut a plasmid.

1. Obtain a microcentrifuge tube, a pipette, a plasmid, a buffer solution, a set of restriction enzymes, and a water bath.
2. Label the tube containing the plasmid "Cut plasmid." Using the pipette, add one drop of the plasmid to a new tube and label this tube "Uncut plasmid" and set aside.
3. To the original ("Cut plasmid") tube, add one drop of buffer solution and mix by tapping or flicking the tube.
4. Add one drop of restriction enzyme and mix by tapping or flicking the tube.
5. Put the "Cut plasmid" tube in the water bath and incubate the DNA at the temperature and for the length of time specified by your instructor.

Gel Electrophoresis

Now you will use gel electrophoresis to examine the fragments you just isolated, along with the uncut plasmid you set aside in step 2 above.

1. Obtain tracking dye.
2. To both the "Cut plasmid" and the "Uncut plasmid" tubes, add 1 drop of tracking dye. This blue dye will travel through the gel at the same rate as small fragments of DNA.
3. Your instructor has prepared a gel. Following your instructor's directions, add your plasmid samples to the gel for electrophoresis.
4. Once all students in the class have added their samples to the gel, your instructor will add a standard marker of DNA so that you can determine the size of your plasmids. This solution contains a mixture of DNA fragments of known sizes.
5. After applying electrical current to the DNA apparatus, your instructor will apply a stain to visualize the DNA and take a picture of your plasmid DNA.
6. What size is your uncut plasmid DNA? How can you tell?

7. Into how many fragments did the restriction enzyme cut your plasmid? On the gel representation in Figure 12.12, sketch what you see in the three lanes of your gel—the results for your uncut plasmid, cut plasmid, and the standard marker. Label each lane.

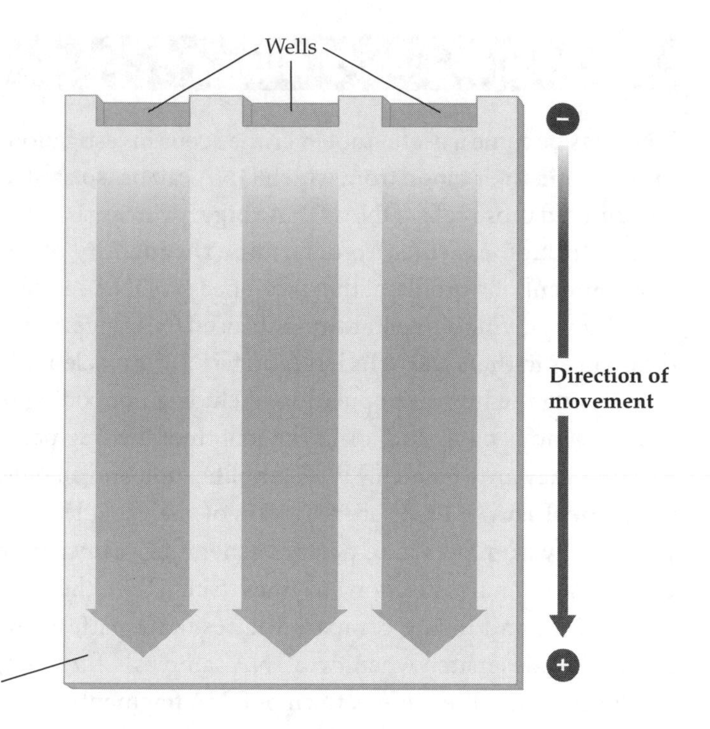

Figure 12.12
Gel Electrophoresis Results for Restriction Enzymatic Reactions

Concept Check

1. Why are restriction enzymes considered specific?

2. What is the natural function of restriction enzymes, and why are they useful in this capacity?

3. Why would a restriction enzyme produce more than one fragment of DNA?

ACTIVITY 6 DNA Fingerprinting

DNA has become a useful tool in crime scene investigation. People shed cells from their skin, within their mouth, or in their blood from which DNA can be isolated. It can then be tested to identify the person who contributed those cells/DNA. **DNA fingerprinting** is the use of a variety of DNA analyses, including PCR and restriction enzymes, to determine the identity of an individual or biological sample through the development of a profile of the differences in DNA sequences (a **DNA profile**).

Crime solving and paternity tests use DNA fingerprinting. All individuals except identical twins have differences in their DNA. In DNA fingerprinting, scientists focus on regions of DNA that are highly variable among the human population, including noncoding regions of DNA. DNA can vary in its sequence of nitrogenous bases and/or in the length of the fragment. Two profiles can be compared to determine whether they are a "match." PCR amplification and an older method called **restriction fragment length polymorphism (RFLP)** are two ways of analyzing DNA. PCR amplification copies specific regions that vary greatly from person to person, generating many patterns of amplified DNA fragments. In RFLP scientists use several restriction enzymes to cut DNA, then separate the resulting fragments through gel electrophoresis, and finally compare the resulting patterns from samples to determine whether there is a match. Profiles match when two DNA samples—for example, one from a crime scene and another from a suspect—have the same pattern of DNA fragments.

Figure 12.13 shows the DNA fingerprints related to a hypothetical crime. Each band represents a fragment of DNA created during RFLP analysis. The victim's DNA is in lane 1; evidence from the crime is in lanes 2, 3, and 4; and DNA from the suspects is in lanes 5 through 8. DNA at crime scenes is not always in the best condition—sometimes a criminalist will be able to isolate only a small amount of DNA. In such cases, the bands on the gel will be very light and may represent an incomplete fingerprint because the DNA is degraded. A darker or thicker band indicates a greater amount of DNA present. Sometimes, especially in sex-based crimes, the sample will contain a mixture of DNA from more than one individual. Mixtures display the profiles of more than one person. In this activity you will evaluate the results of profiles from RFLP testing and determine which suspect was at the crime scene.

1. Given the information provided in Figure 12.13, does the DNA fingerprint of any of the suspects resemble the DNA fingerprint found on evidence at the crime scene?

2. Does the victim's DNA fingerprint appear on evidence found at the crime scene?

3. Does any of the evidence contain degraded DNA or a sample with a low amount of DNA?

4. Examine the samples collected from the crime scene. Does any sample display a combination of DNA profiles from a suspect and the victim?

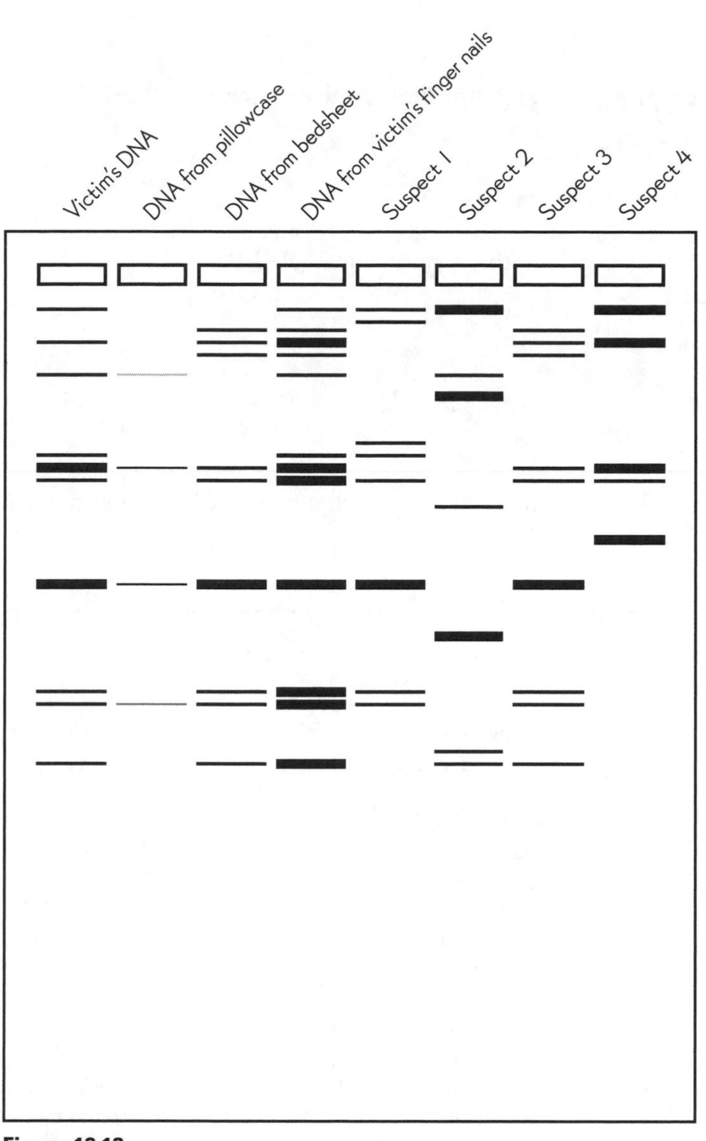

Figure 12.13
DNA Profile of Samples at a Crime Scene and Suspects

5. Which suspect is most likely the person who left DNA at the crime scene?

6. What other observations can you make about the crime scene?

Concept Check

1. How can DNA fingerprinting match one biological sample to another?

2. Describe restriction fragment length polymorphism (RFLP).

3. What is a DNA profile? When can a scientist say that two DNA profiles match?

Key Terms

adenine (A) (p. 12-3)
anticodon (p. 12-17)
base-pairing rules (p. 12-3)
chromosome (p. 12-1)
clone (p. 12-9)
coding region (p. 12-1)
codon (p. 12-16)
complementary strand (p. 12-3)
cytosine (C) (p. 12-3)
DNA (deoxyribonucleic acid) (p. 12-1)
DNA fingerprinting (p. 12-22)
DNA helicase (p. 12-4)
DNA polymerase (p. 12-4)
DNA primer (p. 12-9)
DNA profile (p. 12-22)
DNA replication (p. 12-4)
double helix (p. 12-3)
gel electrophoresis (p. 12-10)
gene (p. 12-1)
gene therapy (p. 12-19)
genetic code (p. 12-16)
genetic engineering (p. 12-19)
genome (p. 12-7)
guanine (G) (p. 12-3)
Human Genome Project (p. 12-9)
ligase (p. 12-19)
messenger RNA (mRNA) (p. 12-13)
molecular biology (p. 12-7)
noncoding region (p. 12-1)
nucleotide (p. 12-3)
PCR product (p. 12-9)
plasmid (p. 12-19)
polymerase chain reaction (PCR) (p. 12-9)
protein synthesis (p. 12-13)
recombinant DNA (p. 12-19)
restriction enzyme (p. 12-18)
restriction fragment length polymorphism (RFLP) (p. 12-22)
ribosomal RNA (rRNA) (p. 12-13)
RNA (ribonucleic acid) (p. 12-2)
sequence (p. 12-7)
thymine (T) (p. 12-3)
transcription (p. 12-13)
transfer RNA (tRNA) (p. 12-13)
translation (p. 12-16)
uracil (U) (p. 12-14)

Review Questions

1. Name the four nitrogenous bases found in DNA. Which ones are complementary to one another?

2. __________________, __________________, and __________________ are the three general components of a nucleotide.

3. The __________________ is the structure of DNA, in which two long strands of covalently bonded nucleotides are held together by noncovalent bonds twisted into a spiral coil.

4. __________________ is the use of DNA technology to determine the identity of an individual or biological sample through the development of a profile of sequence differences.

5. __________________, __________________, and __________________ are the three types of RNA molecules produced during transcription.

6. __________________ unwinds the DNA molecule by breaking the noncovalent bonds between the base pairs.

7. __________________ is the type of RNA that encodes the sequence of proteins.

8. Explain how scientists "cut" DNA.

9. What are two applications of DNA fingerprinting?

10. Describe the four steps of PCR.

11. How does DNA replication take place in a normally dividing human cell?

12. What is a chromosome? What are the different components of DNA?

13. What is the purpose of gel electrophoresis?

14. Why does DNA travel to the positively charged end of a gel apparatus?

15. What are some important uses of DNA isolation and analysis techniques?

16. Describe the processes involved in protein synthesis.

LAB **13**

Evolution of Resistance

Objectives

- Identify the mechanisms of microevolution.
- Demonstrate the impact of antibiotic resistance and the ease with which it can be conferred.
- Describe the methods by which prokaryotes can exchange genetic material.
- Perform transformation of a bacterial cell with a plasmid conferring antibiotic resistance.
- Test the effectiveness of antibacterial soaps.
- Explain how resistance evolves in organisms other than bacteria.
- Recognize the impact of human activity (both intentional and unintentional) on the evolution of resistance in populations.

Introduction

Evolution is the change in genetic characteristics in populations over time. **Microevolution** refers to small-scale changes in a population; macroevolution is large-scale change leading to the formation of new groups. There are four mechanisms that cause populations to evolve: mutation, gene flow, genetic drift, and natural selection. **Mutations** are changes to the sequence of DNA, potentially resulting in new characteristics within the populations. **Gene flow** is the change in the distribution of characteristics, or alleles, resulting from migration from one population to another. **Genetic drift** occurs when chance incidents dramatically reduce the size of a population, affecting the distribution of characteristics. Finally, **natural selection** is the result of unequal reproductive success and the subsequent survival of individuals with specific characteristics in a population. Each of these mechanisms contributes to microevolution within a population.

Macroevolution describes the formation and extinction of species over time. Macroevolution is difficult to observe because it occurs over such long periods of time, but microevolution occurs right before our eyes. Sometimes humans even affect the genetic makeup of populations, either intentionally or unintentionally. An unintentional outcome of human interaction with other populations is the evolution of resistance. In general, **resistance** is the ability to withstand an opposing force, such as antibiotics or pesticides. In this lab you will explore how easily resistance can evolve and how you can affect resistance in a population.

ACTIVITY 1 Sex in Bacteria

The concept of resistance is most often described in the context of bacteria. To evolve, a population has to change genetically. Prokaryotic cells undergo **asexual reproduction**, producing new genetically identical individuals through binary fission. How do the mechanisms of microevolution work to allow prokaryotic cells to evolve? Mutation can clearly change the DNA of an individual prokaryotic cell, but the other mechanisms of microevolution rely on genetic changes between individuals in a population. Bacteria do exchange genetic material—it just doesn't happen *during reproduction.* So don't be fooled—bacteria do have sex! Sex can be a separate process from reproduction. **Sex** is the process of genetic exchange between individuals in a population.

Prokaryotes have two types of DNA. Chromosomal double-stranded circular DNA contains the genes defining an individual bacterial cell while providing the information for a cell to function and reproduce. There may also be extrachromosomal DNA present, in the form of circular, single-stranded DNA called **plasmids.** Prokaryotes exchange genetic material by three different methods: conjugation, transformation, and transduction.

- **Conjugation** is the exchange of genetic material between bacterial cells through a tunnel connecting the two cells (Figure 13.1). This tunnel is an extension of the cell wall and plasma membrane called a **sex pilus** (plural "pili").
- **Transformation** is the means by which free plasmids left by dead cells can be taken up into a cell. Bacterial cells can import this DNA during specific periods when the cell wall and plasma membrane contain pores—a physiological condition called **competency** (Figure 13.2).
- **Transduction** is the process of genetic exchange facilitated through a viral infection by a **bacteriophage**, a special class of virus that infects prokaryotic cells (Figure 13.3). During infection by the bacteriophage, pieces of DNA are exchanged between viral DNA and the host cell DNA, and this virus then goes on to swap DNA with other bacterial cells.

These three methods of genetic exchange enable individual bacterial cells resistant to antibiotics to pass on their genes to other cells. Therefore, the process of sex enables bacteria to become genetically different and potentially more harmful to humans through resistance. In this activity you will examine each of these methods of sex in prokaryotes.

1. Study Figures 13.1 through 13.3 to understand the three prokaryotic methods of sex: conjugation, transformation, and transduction.

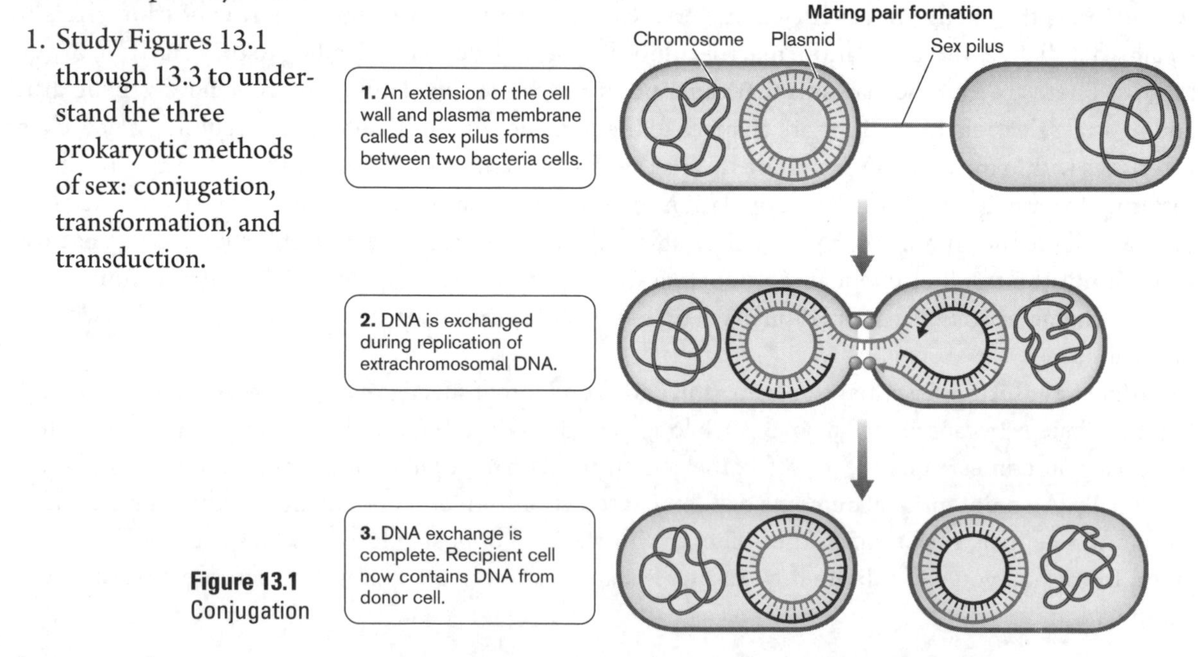

Figure 13.1 Conjugation

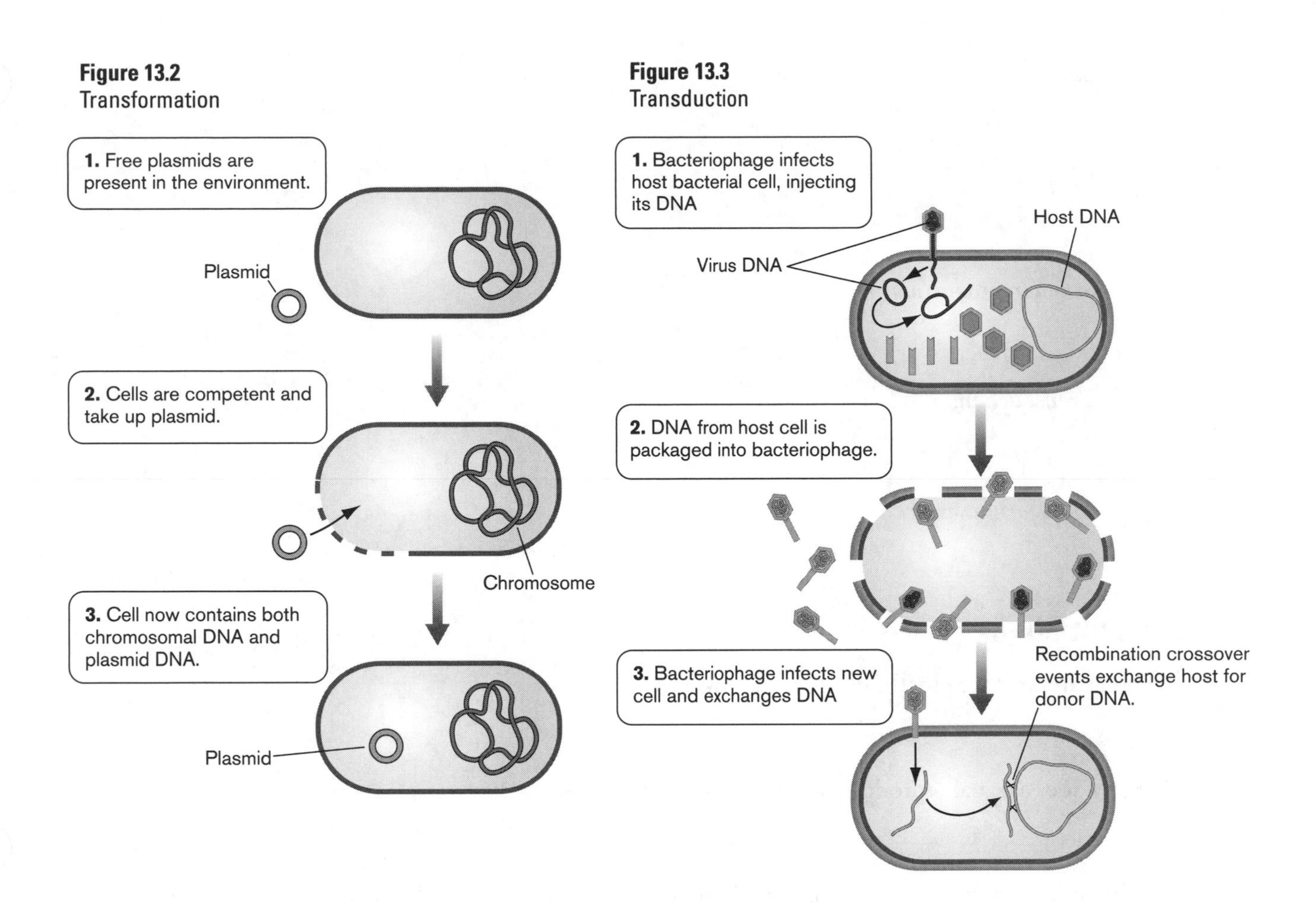

2. Define the following terms or processes using the figures above for reference:

Asexual reproduction:

Bacteriophage:

Chromosomal DNA:

Conjugation:

Plasmid:

Sex:

Sex pilus:

Sexual reproduction:

Transduction:

Transformation:

Concept Check

1. What are the three methods of sex in prokaryotes?

2. How does asexual reproduction allow an organism to evolve?

3. If an individual bacterial cell contained a penicillin-resistant gene and its colony was treated with penicillin, what would happen to the population?

ACTIVITY 2 Gram-Negative and Gram-Positive Bacteria

The advent of antibiotics and better hygiene added to the human population explosion by extending the survival rate of individuals. Like many other great discoveries, antibiotics were an accident. Alexander Fleming discovered that a contaminating fungus, *Penicillium notatum*, didn't die—and in fact flourished—in the presence of an infectious bacterium. This was the start of the antibiotic revolution. Many antibiotics inhibit formation of the cell wall of infectious bacteria.

There is a lot of variety in the composition of the **cell wall** and outer layers of prokaryotic cells. Most bacterial cell walls are either gram-positive or gram-negative, depending on the composition of their cell wall (Figure 13.4). The composition of the cell wall determines the cell's sensitivity to alcohol and its ability to retain a stain. In **gram-positive** bacteria, the cell wall contains a thick layer of a substance called **peptidoglycan.** This is a protein–sugar complex unique to bacteria. **Gram-negative** bacteria build a thinner, more complex cell wall containing a smaller amount of peptidoglycan, and have an extra outer membrane. Because many antibiotics attack the formation of the peptidoglycan layer, gram-positive bacteria are more vulnerable to antibiotics. The evolution of antibiotics compels a population of bacteria to evolve through natural selection. In this activity you will determine whether the bacteria provided to you are gram-positive or gram-negative.

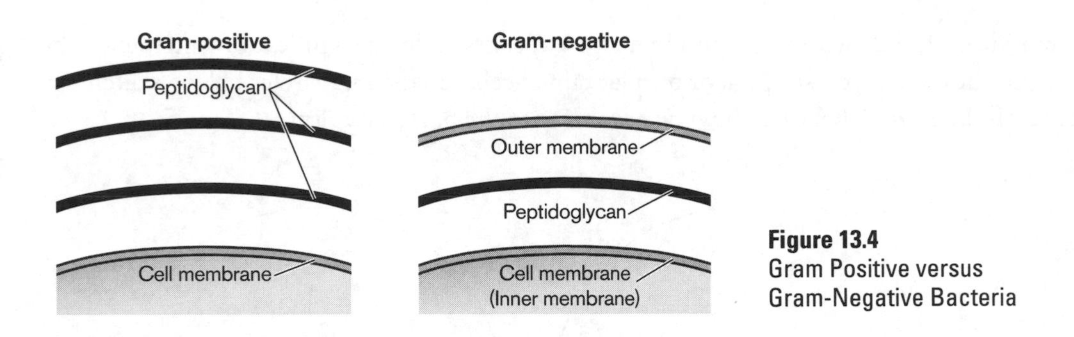

Figure 13.4
Gram Positive versus Gram-Negative Bacteria

1. Obtain a compound microscope, four bacterial cultures, four microscope slides, methanol, crystal violet, iodine, ethanol, safranin, and an inoculation loop or dropper.
2. Prepare each of the four cultures as follows:
   a. Label the slide with the name of the bacteria in the culture.
   b. Place a small sample of the culture on a microscope slide with either a loop or a small dropper.
   c. Spread the culture across the surface of the slide. Let the culture air-dry.
   d. Perform the following staining procedure, as illustrated in Figure 13.5:
      (1) Fix the cells to the slide by adding one drop of methanol. Again, let the cells air-dry.
      (2) Add one drop of crystal violet stain to the slide, and let the cell incubate for 1 minute.
      (3) Add one drop of iodine solution to bind the crystal violet to gram-positive cells, and incubate for 1 minute.
      (4) Wash with ethanol for 20 seconds.
      (5) Add one drop of safranin counterstain, and incubate for 1 minute.
   e. Wash with water.

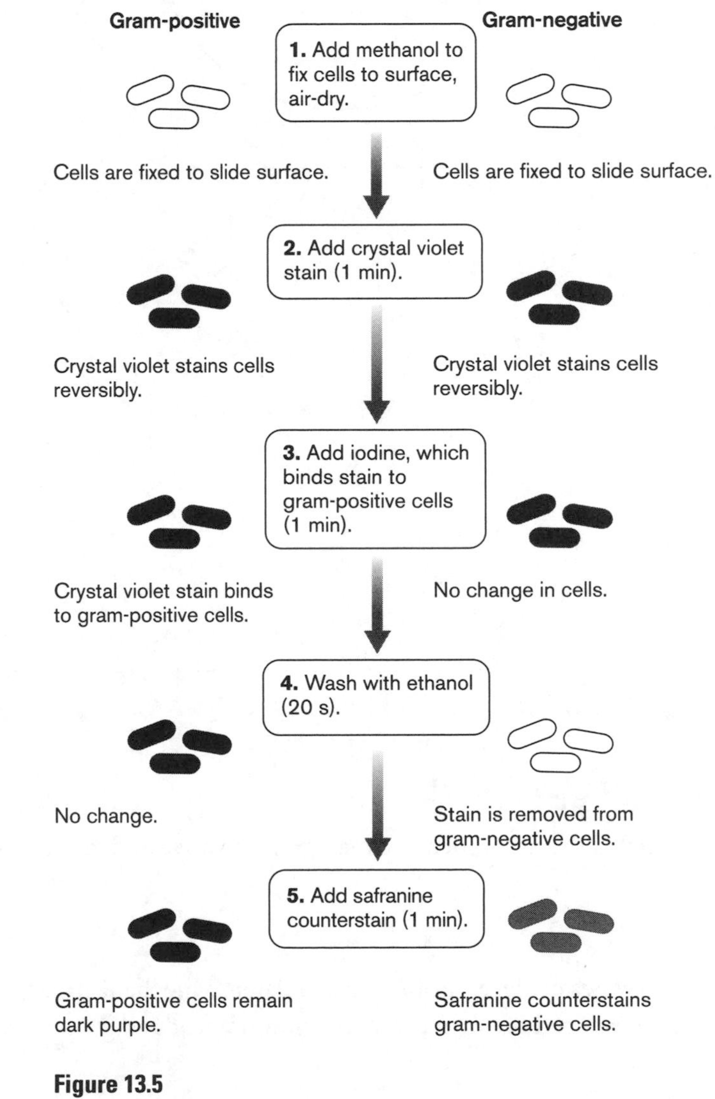

Figure 13.5
Gram Staining

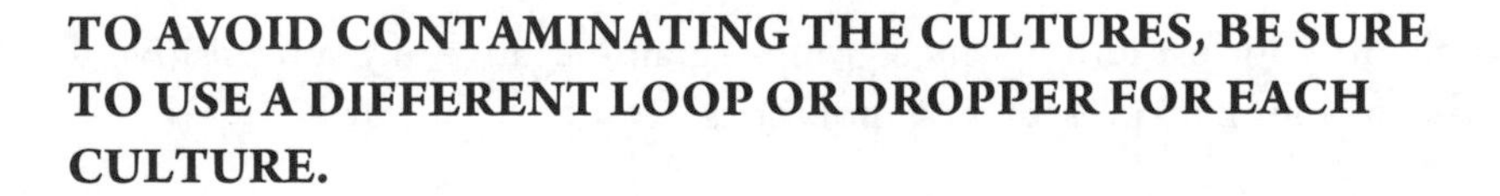
TO AVOID CONTAMINATING THE CULTURES, BE SURE TO USE A DIFFERENT LOOP OR DROPPER FOR EACH CULTURE.

3. Examine each slide with a compound microscope, first at low magnification and then at the highest magnification. Do you see pink or purple stained cells? In the four circles below, sketch what you see on each slide, and label each drawing according to the bacterial culture.

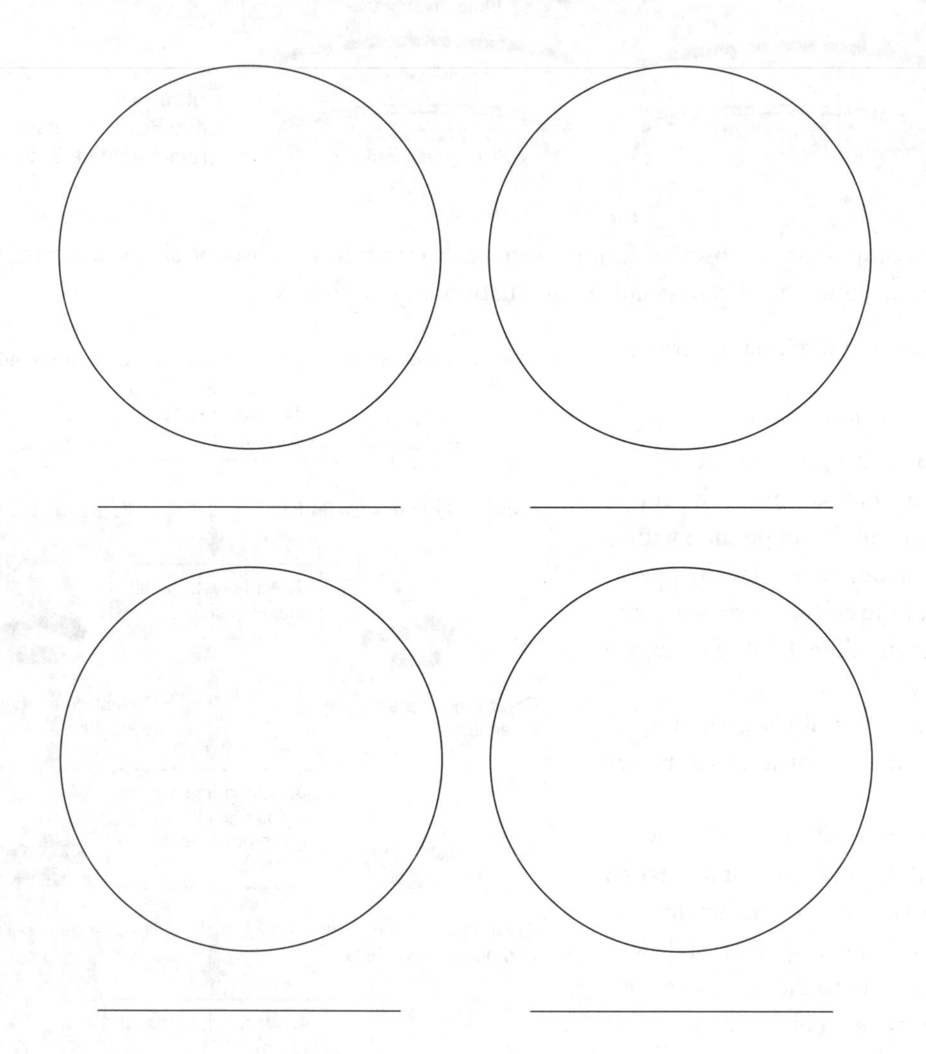

Concept Check

1. What is the difference between gram-positive and gram-negative bacteria?

2. Microbes evolved antibiotics as a defense mechanism. Explain the types of environmental pressures that induced the evolution of antibiotic-producing microbes.

3. In the context of antibiotic evolution, which group—gram-negative or gram-positive—would have evolved from the evolutionary pressures of antibiotics?

ACTIVITY 3 Bacterial Transformation

Like all other organisms, bacteria evolve. The mechanisms of microevolution change the distribution of alleles within a bacterial population over time. These changes are often the result of survival in the face of environmental factors such as antibiotics. Some individual bacterial cells, when exposed to an antibiotic, can render the antibiotic useless by destroying it. Remember that **antibiotics** are naturally occurring defense chemicals secreted by microbes to kill other microbes. They therefore attack a variety of structures or processes within microbial cells, including formation of the cell wall and membrane, protein synthesis, DNA replication, and enzymatic reactions.

Bacteria evolve in the face of many antibiotics. They do so through random mutations, which render an antibiotic useless and the bacteria resistant. This gene variation that causes resistance is often found within the extrachromosomal DNA called a plasmid. Remember, a plasmid is a self-replicating closed loop of DNA that is found naturally in bacteria. It contains several genes, all or none of which can confer resistance. Bacterial cells can import this DNA during specific periods when the cell wall and plasma membrane contain pores—a physiological condition called competency.

This activity will use a plasmid that confers ampicillin resistance with a marker gene called GFP. GFP is the abbreviation for green fluorescent protein, a protein that fluoresces under ultraviolet light. In this activity you will follow the process of transformation, which exemplifies how plasmids can confer antibiotic resistance.

1. Label two tubes as follows: "1: Experiment" (plasmid) and "2: Control" (no plasmid).
2. Add 20 drops of $CaCl_2$ to each tube. Be sure to use a clean pipette.
3. Obtain two colonies of bacteria for each tube by swiping a toothpick on the surface of the agar plate with the *E. coli*. Stir the $CaCl_2$ with the toothpick to dislodge the bacterial cells. You may use the same toothpick to retrieve both sets of the cells.
4. Your instructor will provide you with a plasmid. Add all of the plasmid to the experiment tube.
5. Close the tubes and incubate them on ice for 15 minutes.
6. Incubate the tubes in a water bath at 42°C for 90 seconds.
7. Incubate the tubes on ice again for 2 minutes.
8. Add 20 drops of recovery broth (LB broth) to the tubes.
9. Incubate the tubes at 37°C for 30 minutes.
10. Retrieve four agar plates (two with ampicillin and two without) and label them "1: Control," "2: Control with Ampicillin," "3: Experiment," and "4: Experiment with Ampicillin." Include your initials and the date on the bottom (agar-side) of all plates.
11. Add 10 drops from the control tube to plates 1 and 2, and 10 drops from the experiment tube to plates 3 and 4. Use a clean inoculating loop to spread the cells across each plate.
12. Give your instructor the plates that you have prepared.

NOTE: YOUR INSTRUCTOR MAY CHOOSE TO PROVIDE YOU WITH PREPARED PLATES, OR TO HAVE YOU EXAMINE THE PLATES YOU PREPARED DURING THE NEXT PERIOD.

13. In Table 13.1 below, record the number of green bacterial colonies that appear on each agar plate.

TABLE 13.1 ACTIVITY 3: NUMBER OF BACTERIAL COLONIES ON AGAR PLATES

	Control (no plasmid)	Experiment (plasmid)
Agar without ampicillin		
Agar with ampicillin		

14. Which plate(s) contain(s) no bacterial growth? Why?

Concept Check

1. How does competency affect the ability of a cell to take up a plasmid?
2. Do all plasmids have antibiotic resistance?
3. How does transformation relate to the evolution of antibiotic resistance in bacterial populations?
4. Are all bacteria that take up plasmids with antibiotic resistance detrimental to the human population? Explain your reasoning.

ACTIVITY 4 Testing Popular Antibacterial Substances

Companies use gimmicks because gimmicks sell products. People demand foods labeled as natural because they believe "natural" foods are better for them. The same is true for antibacterial soaps, lotions, dishwashing detergents, cleaning products, and now even toys. Killing bacteria seems like a logical approach to preventing infection; but the major lesson lost here is that bacteria are also beneficial, and when antibacterial products are used they kill both the vulnerable beneficial and the infectious bacteria, contributing directly to the proliferation of antibacterial-resistant bacteria. Before people used antibiotics and disinfectants with such frequency and on such a wide scale, these substances could kill most bacteria. But as antibiotics and antibacterial substances became more popular, the few individual resistant bacteria survived and continued reproducing, thriving in the absence of weaker bacteria killed by widespread antibiotic and disinfectant use. Over time natural selection continued and the frequency of resistant individuals increased.

The sieves in Figure 13.6 represent the evolutionary pressure exerted by a disinfectant and three different antibiotics. Often genes that confer resistance to antibacterial products can also lead to antibiotic resistance, and mutations and gene transfer within and among bacterial species allow some bacteria to develop resistance to multiple antibiotics. Antibacterial soap may be trendy, but good old soap and water are the most effective in fighting infection. In this activity you will test the effectiveness of different antibacterial products at completely killing microbes.

1. Obtain two sterile agar plates. Your instructor will tell you which treatment to perform.
2. Draw a line down the middle of the bottom of a sterile agar plate—the side that contains the agar. Labeling the lid is useless because the lid can turn.
3. Label one side of the line "Before" and the other "After."
4. Follow the instructions for the treatment assigned to you by your instructor:
   a. **Treatments A–C**
      (1) Touch the "Before" side of the plate with your thumb and index finger, and wipe slowly and softly across the agar.
      (2) Wash your hands for 30 seconds using one of the following treatments, as specified by your instructor, and then rinse and dry your hands:
         Treatment A: plain soap
         Treatment B: alcohol solution
         Treatment C: antibacterial soap
      (3) On the "After" side of the plate, wipe the same two fingers across the plate.
   b. **Treatments D–F**
      (1) Take a labeled agar plate to the bathroom. Use the restroom and, before washing your hands, touch the "Before" side of the plate with your thumb and index finger; wipe slowly and softly across the agar.
      (2) Wash your hands for 30 seconds using one of the following treatments, as specified by your instructor, and then rinse and dry your hands:
         Treatment D: plain soap
         Treatment E: alcohol solution
         Treatment F: antibacterial soap
      (3) On the "After" side of the plate, wipe the same two fingers across the plate.
5. Give the plates to your instructor for incubation.

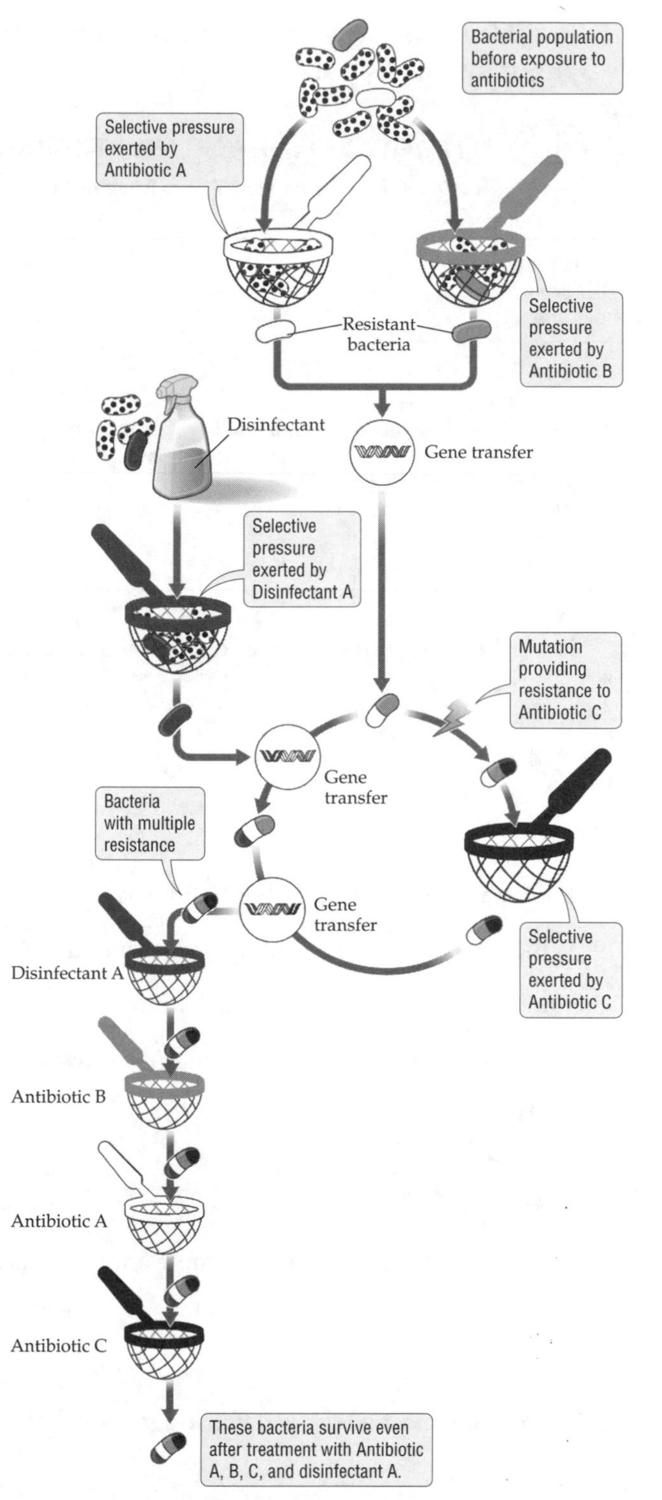

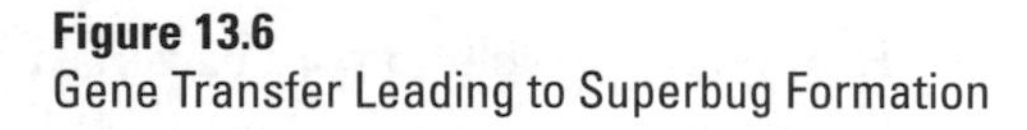
Figure 13.6
Gene Transfer Leading to Superbug Formation

NOTE: YOUR INSTRUCTOR MAY CHOOSE TO PROVIDE YOU WITH PREPARED PLATES, OR TO HAVE YOU EXAMINE THE PLATES YOU PREPARED DURING THE NEXT PERIOD.

6. In Table 13.2, identify how many unique types of bacterial or fungal colonies appear on the agar plates that you treated.

TABLE 13.2 ACTIVITY 4: NUMBER OF DIFFERENT COLONIES PRESENT ON CULTURE PLATES

	Treatment A, before	Treatment A, after	Treatment B, before	Treatment B, after	Treatment C, before	Treatment C, after
Number of Different Colonies						

	Treatment D, before	Treatment D, after	Treatment E, before	Treatment E, after	Treatment F, before	Treatment F, after
Number of Different Colonies						

7. Describe these different types of colonies, differentiating them by color, shape, and size of colony.

8. Fill in the remainder of Table 13.2 by using the data collected by your classmates.
9. Which treatment produced the most types of colonies?
10. Which treatment appears to be the most effective at killing the microbes?
11. Does it appear that any type of microbe is resistant to antibacterial soap?

Concept Check

1. How do antibacterial soaps promote antibacterial resistance in bacterial populations?

2. What is the most effective defense against infectious bacteria?

3. How does the evolution of a bacterial population after exposure to antibacterial substances demonstrate natural selection?

ACTIVITY 5 Resistance in Other Organisms

An **adaptation** is any advantageous characteristic that increases an organism's ability to survive and reproduce. Both genetically engineered organisms and naturally occurring organisms can have resistance to a variety of conditions on the basis of their genetic makeup.

A whole field of scientific research called **genetic engineering** is dedicated to genetically modifying organisms using naturally occurring processes such as transformation. This technology enables scientists to produce "new" organisms with a specific favorable trait that makes them more adapted to their environment. New technologies also enable humans to mix DNA from different organisms. This foreign DNA may allow resistance, but whether naturally occurring or produced by humans, all resistance is genetically based.

The development of resistance is very advantageous for the microbe that is becoming resistant. It is important to acknowledge how this process can easily create a **superbug**, a microbe that has evolved resistance to most commonly known antibiotics (see Figure 13.6). Recently, many cases of methicillin-resistant *Staphylococcus aureus* (MRSA) have cropped up in schools and health care facilities across the country. The danger of uncontrolled antibiotic misuse is the evolution of this type of resistant bacterium.

Humans have encountered a similar problem with pesticide resistance in insect populations. Bt toxin was a commonly used pesticide, isolated from the bacterium *Bacillus thuringiensis*, which naturally produces the Bt toxin that kills insects. In conjunction with the widespread use of this pesticide, individual insects resistant to the toxin survived and reproduced while those that were not resistant died. When scientists noticed that insect resistance was evolving, the agricultural industry began to incorporate the Bt toxin into crops such as corn and cotton, so that only insects that fed on the plant would be exposed to the toxin. Isolating the Bt toxin in the plant as opposed to spraying it as a pesticide helps limit exposure to the toxin for beneficial insects that do not eat the plant. By contrast, spraying plants with this pesticide causes many more populations of insects to evolve resistance, because the toxin is more widely exposed.

Finally, plants have also been genetically engineered to be resistant to the herbicide (weed killer) Roundup®. These crops are called Roundup Ready® crops. Over 80% of American soybean crops are Roundup Ready®. When Roundup® is applied to a field of these resistant crops, any other plant in the area will be killed. Not unlike the case of bacteria and antibiotics, or of insects and pesticides, certain plants have natural resistance to Roundup®. The artificial treatment of populations by humans produces populations of genetically resistant organisms.

Imagine a scenario in which a plasmid is created to contain naturally existing genes conferring the following characteristics (Figure 13.7):

- *Ampicillin resistance.* This gene enables an organism to grow in the presence of ampicillin.
- *Bt toxin.* This gene enables an organism to produce the toxin that kills insects.
- *Roundup® resistance.* This gene enables an organism to survive when exposed to the herbicide Roundup®.

In this activity you will evaluate what would happen if each of three targeted organisms—a bacterium, a plant, and an insect—took up this plasmid and then was introduced into a natural field.

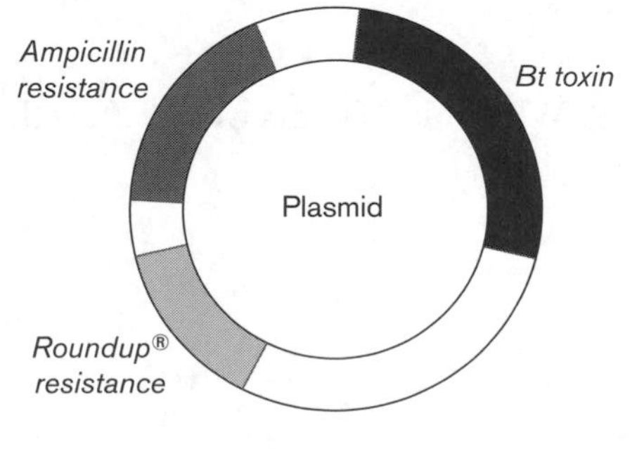

Figure 13.7
Hypothetical Plasmid

Bacterium

1. How would the genes shown in Figure 13.7 affect the bacterium?

2. Would the bacterium become a superbug?

3. What other factors would affect the bacterium's success?

4. How would the gene products (proteins) affect other organisms?

5. What could you change in the plasmid to affect the bacterium's survival?

Plant

6. How would the genes shown in Figure 13.7 affect the plant?

7. Would the plant survive in greater or fewer numbers than other plants?

8. How could the genes expressed affect the survival rate of other plants?

9. What other factors would affect the plant's success?

10. How would the gene products (proteins) affect other organisms?

11. What could you change in the plasmid to affect the plant's survival?

Insect

12. How would the genes shown in Figure 13.7 affect the insect?

13. Would the insect survive in greater or fewer numbers than other insects?

14. How could the genes expressed affect the survival rate of other insects?

15. What other factors would affect the insect's success?

16. How would the gene products (proteins) affect other organisms?

17. What could you change in the plasmid to affect the insect's survival?

Concept Check

1. Why is resistance an adaptation?

2. Will all organisms evolve resistance to human interventions?

3. What are the evolutionary dangers of genetically modified organisms?

Key Terms

adaptation (p. 13-11)
antibiotic (p. 13-7)
asexual reproduction (p. 13-2)
bacteriophage (p. 13-2)
cell wall (p. 13-4)
competency (p. 13-2)
conjugation (p. 13-2)
evolution (p. 13-1)
gene flow (p. 13-1)
genetic drift (p. 13-1)
genetic engineering (p. 13-11)
gram-negative (p. 13-4)
gram-positive (p. 13-4)
macroevolution (p. 13-1)
microevolution (p. 13-1)
mutation (p. 13-1)
natural selection (p. 13-1)
peptidoglycan (p. 13-4)
plasmid (p. 13-2)
resistance (p. 13-1)
sex (p. 13-2)
sex pilus (p. 13-2)
superbug (p. 13-11)
transduction (p. 13-2)
transformation (p. 13-2)

Review Questions

1. What is an antibiotic?

2. Describe the difference between gram-negative and gram-positive bacteria.

3. How do prokaryotes have sex?

4. How can overuse of antibiotics lead to a superbug?

5. What is the difference between microevolution and macroevolution?

6. How can a plasmid be used to create a new organism?

7. What are bacteriophages?

8. How have humans accelerated the evolution of resistance?

9. How does genetic engineering produce organisms different from those that exist naturally?

10. Why is using antibiotics to treat a disease caused by a virus detrimental?

11. ________________ are changes to the sequence of DNA potentially resulting in new characteristics within populations.

12. ________________ is the process of producing new genetically identical individuals in a population.

13. ________________ occurs when chance incidents dramatically reduce the size of a population, affecting the distribution of characteristics.

14. ________________ is a tunnel extending from the cell wall and plasma membrane to connect two prokaryotic cells.

15. ________________ describes a change in the distribution of characteristics, or alleles, resulting from migration from one population to another.

16. ________________ is the result of unequal reproductive success and the subsequent survival of individuals with specific characteristics within a population.

17. ________________ is a protein–sugar complex unique to bacteria, forming the cell wall.

LAB 14

General Mammalian Anatomy (Fetal Pig Dissection)

Objectives

- Identify the general anatomical structures of mammals.
- Determine the gender of a fetal pig.
- Locate major structures of the oral cavity, neck region, thoracic cavity, and abdominal cavity in a fetal pig, and describe their functions.
- Recognize similarities and differences between human and fetal pig anatomy.
- Describe how the major mammalian organs function in relation to one another.

Introduction

To comprehend why pigs are useful in learning about human anatomy, it is advantageous to trace the common lineage of humans and pigs. Humans and pigs belong to the same phylum (chordates) and both are vertebrates, animals that have a backbone. Pigs and humans also belong to the same class (mammalians) and both are classified as eutherians, animals that provide milk to their young and surround the developing fetus with a placenta. Pigs and humans evolved very similar anatomical structures.

Pigs contain the same tissue types as humans: epithelial, connective, muscle, and nervous. Pigs and humans share the same general anatomy, including the same or similar organs in their digestive, respiratory, circulatory, immune, and endocrine systems. Current research is producing genetically engineered pigs raised to serve as an organ supply for the human population. Medical doctors and research scientists also use pigs and their organs for transplants and medical treatments. The pig circulatory system follows paths similar to those in humans—transporting blood to the heart, to the lungs, back to the heart, and then to the rest of the body. Pigs and humans also have similar skin. Forensic scientists use pig carcasses to simulate the effects of gunshot wounds on humans. One difference between pigs and humans is that the gestation period (development of a full-term offspring) is much shorter in pigs—only 16 to 17 weeks, compared to 36 weeks in humans.

In this lab you will dissect a fetal pig and examine its anatomical structure to gain a deeper understanding of mammalian anatomy and how organs function together.

ACTIVITY 1 Preparing for Dissection

To understand different anatomical structures, along with their functions and locations throughout this lab, it is important to become familiar with some commonly used terminology. In this activity you will use diagrams of the fetal pig to learn basic anatomical terms.

1. Examine Figure 14.1 to understand the following terms:
   - The **dorsal** side of the pig is the "back" side of the animal.
   - The **ventral** side is the "belly" side—the opposite of dorsal.
   - The **anterior** part is the "head" end.
   - The **posterior** part is the "tail" end—the opposite of anterior.

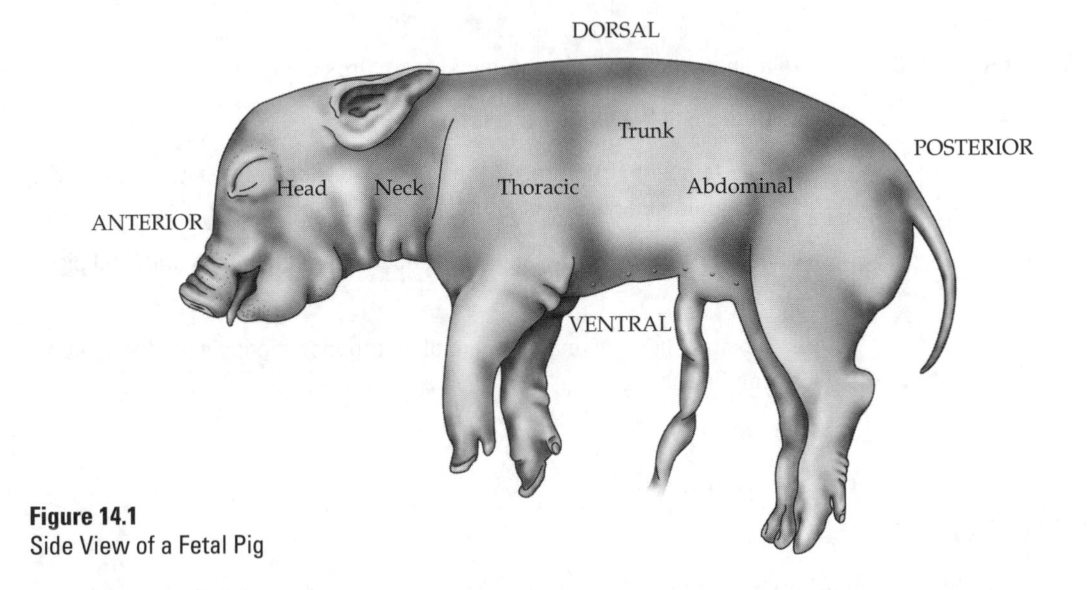

Figure 14.1
Side View of a Fetal Pig

2. Two other useful terms are "cross section" and "longitudinal section." Incisions are made either across the body (**cross section**) or along the length of the body (**longitudinal section**). In Figure 14.2, draw a letter "t," with the vertical line reaching from the anus to the mouth, and the horizontal line crossing the vertical line from shoulder to shoulder to complete the "t" shape. A longitudinal section is any section parallel to the line from the anus to the mouth (the vertical line of the "t"). A cross section is any section parallel to the line from shoulder to shoulder (the horizontal line of the "t").

3. Anatomical structures are designated according to the perspective of the animal. In Figure 14.2, you might be tempted to refer to the anterior foot on the right as a "right foot," but think of your own body. When you tell someone that your right foot hurts, you don't refer to it as your left foot just because that person is looking at you, do you? For a pig oriented as in Figure 14.2, the structures on your right are on the pig's left. When the terms "right" or "left" are used throughout the dissection in this lab, remember that they refer to the pig's left or right and not yours. With this distinction in mind, label the right anterior foot in Figure 14.2.

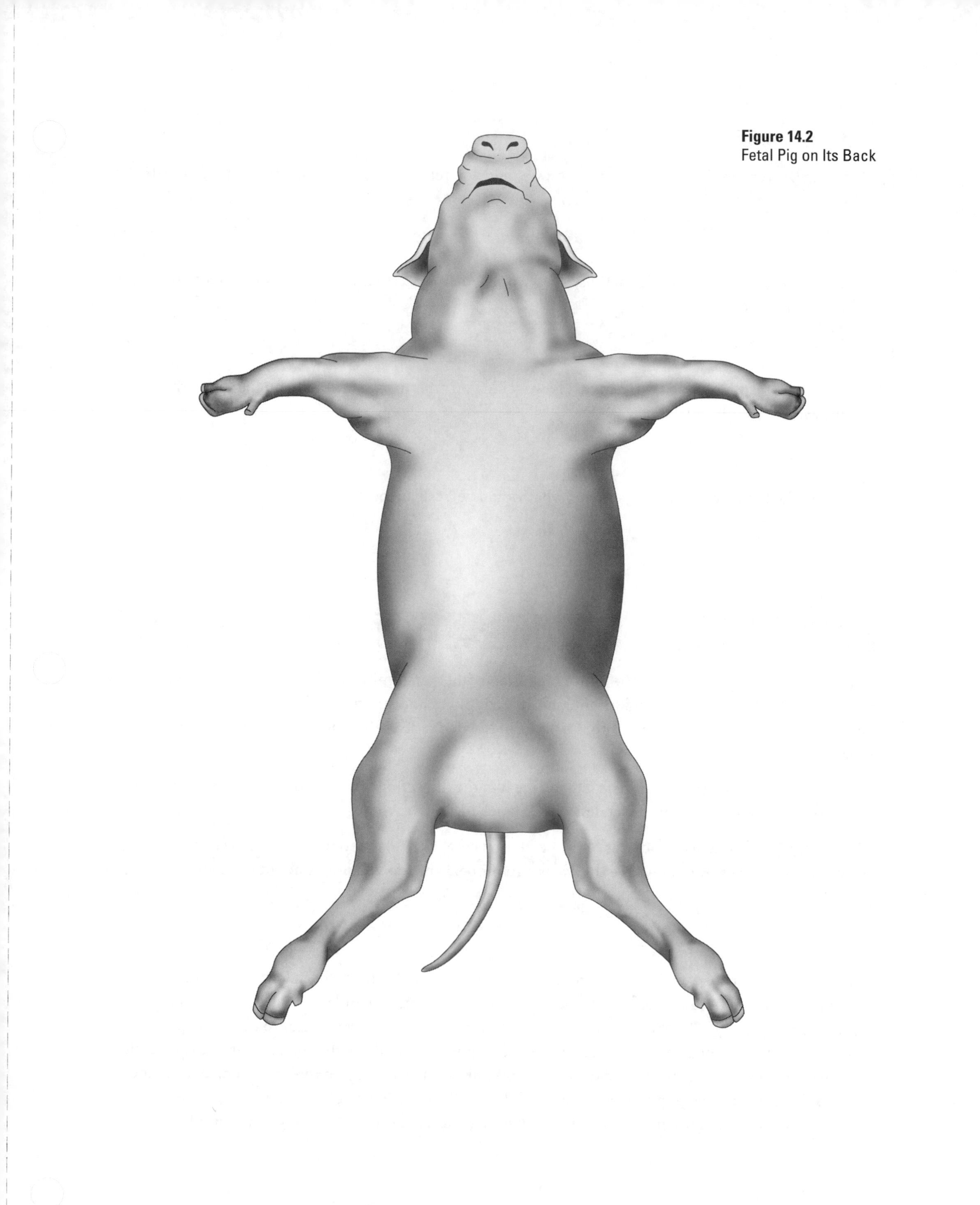

Figure 14.2
Fetal Pig on Its Back

Concept Check

1. Using the terminology of this activity, how would you refer to a structure located near your feet?

2. Using the terminology of this activity, how would you refer to the side of the human body that shows the face?

3. Using the terminology of this activity, how would you refer to the side of the human body that shows the spine?

4. Using the terminology of this activity, what would you call an incision made from your right hip to your left hip?

ACTIVITY 2 External Anatomy

In this activity you will examine the major organs and other anatomical structures of the fetal pig. As you proceed, note the color, size, and orientation of each organ and structure. As you identify these organs and structures, remember that they do not perform isolated tasks—that they are components of a greater organ system.

1. Obtain a fetal pig from your instructor. Place it in the dissecting tray. The head, neck, and trunk are the three main regions of the fetal pig (see Figure 14.1).
2. Identify the following structures of the head: ears (pinnae), nose and nasal passages (nares), and eyes. Open an eyelid to observe the eye with a probe. What do you notice about the eye?

3. There are no main external structures along the neck. Posterior to the neck is the trunk, composed of the **thoracic region**, and then the **abdominal region** with appendages attached to each region (see Figure 14.1). The thoracic region extends from the neck to just anterior to the belly. Directly posterior to the neck you will notice the injection point where the veins and arteries of the fetal pig have been injected with latex. Between the forelimbs, from the chest to the abdomen, identify the **nipples** (**mammary papillae**; singular "papilla") of the pig (Figure 14.3). Compare your pig with one of your classmates' pigs of the opposite sex. Do both male and female pigs have nipples?

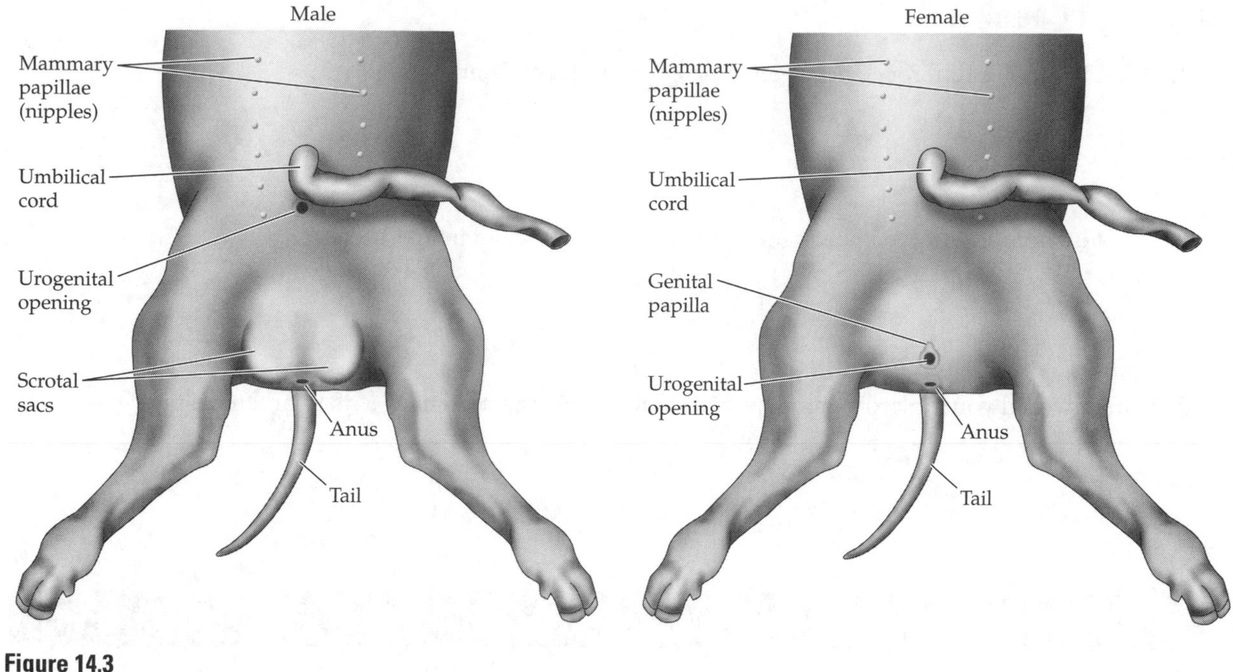

Figure 14.3
External Anatomy and Sex Determination

4. In the abdominal region are several external structures to identify. The most obvious is the one protruding from the belly of the fetal pig: the **umbilical cord**. The umbilical cord provides nutrient- and oxygen-rich blood through the umbilical vein (labeled with blue latex) while removing waste- and carbon-laden blood through the umbilical arteries (labeled with red latex).

5. The next structures to identify depend on whether your pig is a male or female. Does your pig have an opening posterior to the umbilical cord?

   a. If the answer is yes, you have a male and this is the **urogenital opening** of your male pig (see Figure 14.3). "Urogenital" is an appropriate name because this is the opening of the **penis**, which has both urinary and reproductive functions. From the opening, move longitudinally toward the posterior of the pig and find the **anus.** Do you see two protuberances along the thighs of the male fetal pig? These are the **scrotal sacs** containing the testes.
   b. If your pig does not have an opening posterior to the umbilical cord, move in a longitudinal direction toward the posterior. You should find a flap of skin, which is part of the **genital papilla,** housing the **urogenital opening** that leads to the vagina and the bladder of the female pig. Note that both male and female pigs have a urogenital opening. Finally, locate the **anus**, the opening directly posterior to the genital papilla.
   c. Be sure that you have the opportunity to examine one of your classmates' pigs of the opposite sex.

Concept Check

1. What is a major difference between female pigs and female humans?

2. What is a major difference between male pigs and male humans?

3. Where would you look to immediately determine whether you have a male or female pig?

ACTIVITY 3 The Head: Oral Cavity

Within the mouth are structures providing the first means by which the pig starts digestion. The **teeth** macerate (chop) food, and the enzyme amylase in the saliva starts digesting the sugars in food. The **tongue** pushes food into the throat (**pharynx**)—where the nasal cavities (**nasopharynx**) and the back of the mouth meet (Figure 14.4). The pharynx contains the flap of skin called the **epiglottis**, which either allows air into the **larynx** and the rest of the respiratory system, or covers the larynx while food enters the **esophagus** of the digestive system.

In this activity you will examine the anatomical structures of the fetal pig's oral cavity.

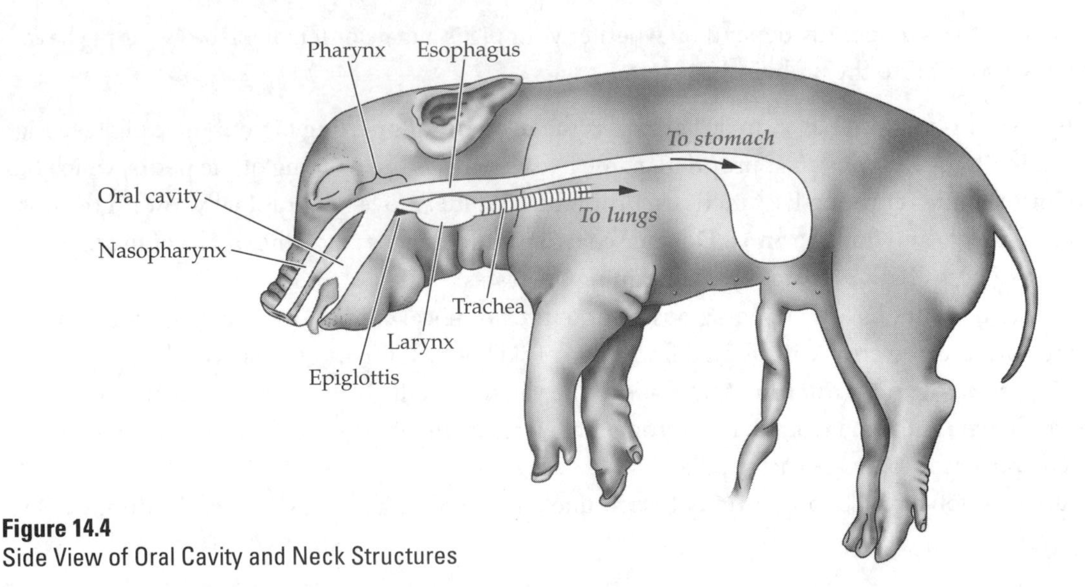

Figure 14.4
Side View of Oral Cavity and Neck Structures

1. Obtain the tools that you will need during your dissection: a pair of scissors with a blunt (curved) end and a pointed end, a blunt probe, a sharp probe, and a pair of straight forceps. Using Figure 14.5 for reference, make a cross-sectional incision from the corner of each side of the mouth toward each ear and the dorsal side of the head.

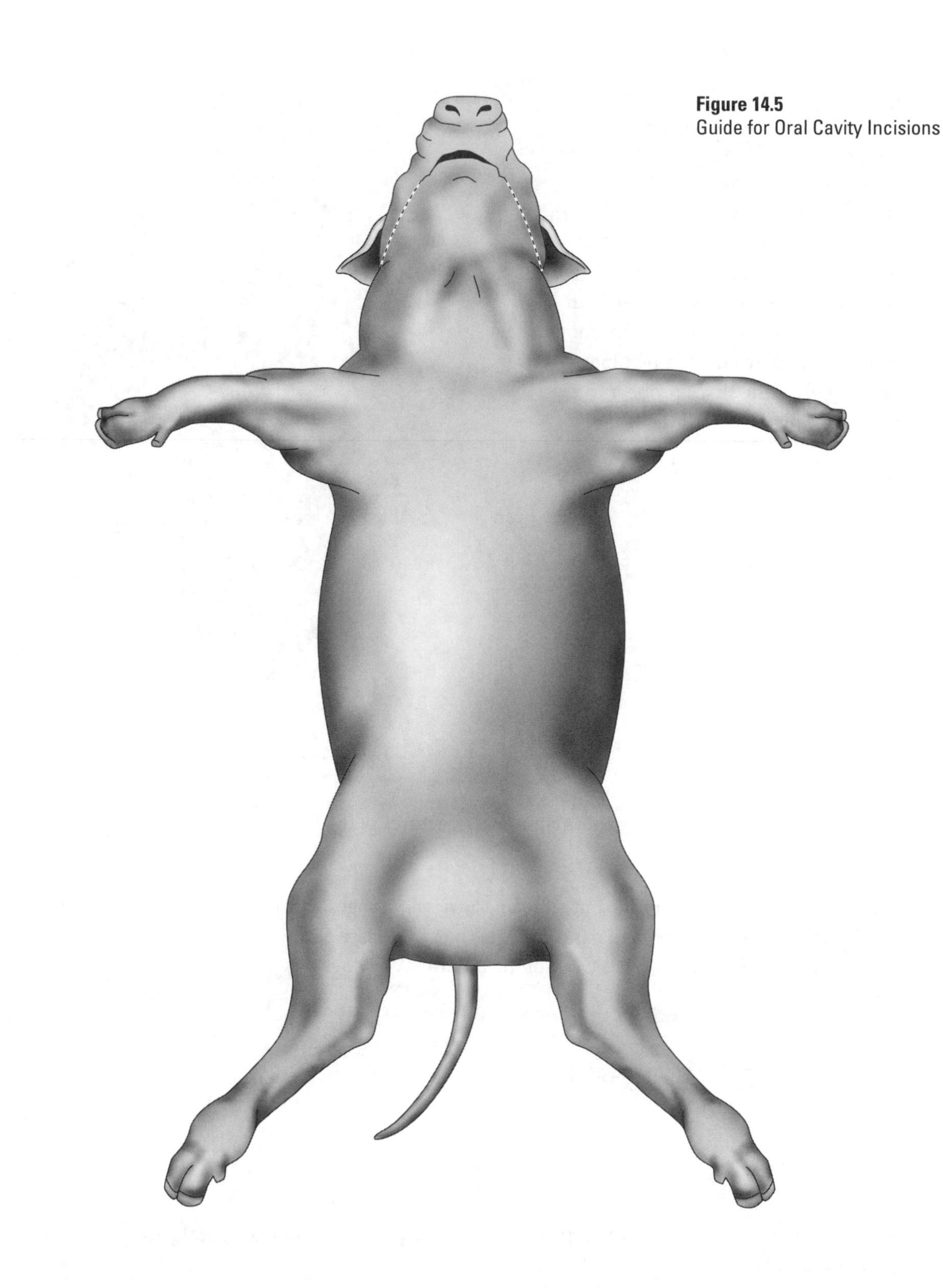

Figure 14.5
Guide for Oral Cavity Incisions

2. Open the oral cavity. The tongue is easily identified, as it typically sticks out. What does the roof of the mouth look like? Using the blunt probe, feel the ridges and note where the tissue on the roof becomes soft. The hard-ridged area is the **hard palate**; the soft area is the **soft palate** (see Figure 14.6).

3. Identify the three openings of the oral cavity. The anterior opening is the nasopharynx, where air from the nares flows into the pharynx (see Figure 14.6). The dorsal opening is the esophagus, delivering food from the mouth to the stomach. The epiglottis covers the larynx, or voice box, which closes over the trachea to prevent food from passing down the wrong pipe. Carefuly place the sharp probe into these openings to determine where they lead.

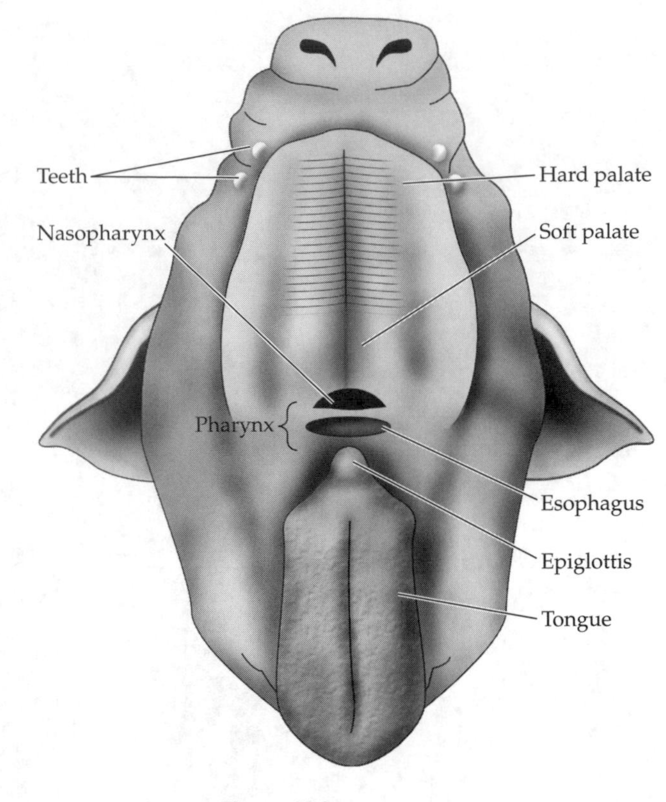

Figure 14.6
Oral Cavity of a Fetal Pig

Concept Check

1. The palates of the oral cavity separate it from the nasal passages. During chewing, the soft palate permits breathing through the nasal passage while the hard palate of the oral cavity aids in food maceration. Which palate is closer to the front of the mouth? Why does this placement make sense?

2. How is food passed down to the esophagus?

3. Have you ever experienced food or drink moving down the trachea instead of the esophagus? Why does this happen?

ACTIVITY 4 The Neck Region

Dorsally, the neck contains the esophagus and then ventral to the esophagus is the larynx, which leads to the **trachea** (see Figure 14.4). The esophagus is a flexible muscular tube connecting the pharynx to the stomach. Meanwhile, air moves from the larynx to the trachea to the lungs when you inhale, and then from the lungs to the trachea to the larynx when you exhale. The larynx is the voice box composed of vocal cords made of cartilage. When the air moves past the vocal cords, they vibrate to create sound. The trachea is a ribbed tube, much more rigid than the esophagus. The trachea connects the larynx to the lungs.

On either side of the trachea are the tissues of the **thymus gland** (Figure 14.7). The thymus gland is involved primarily in the immune system, where T (thymus) cells mature and play a central role in fighting disease. On the ventral side toward the posterior end of the trachea is the **thyroid gland**. This gland is part of the endocrine (hormone-producing) system, producing thyroxine and regulating metabolism.

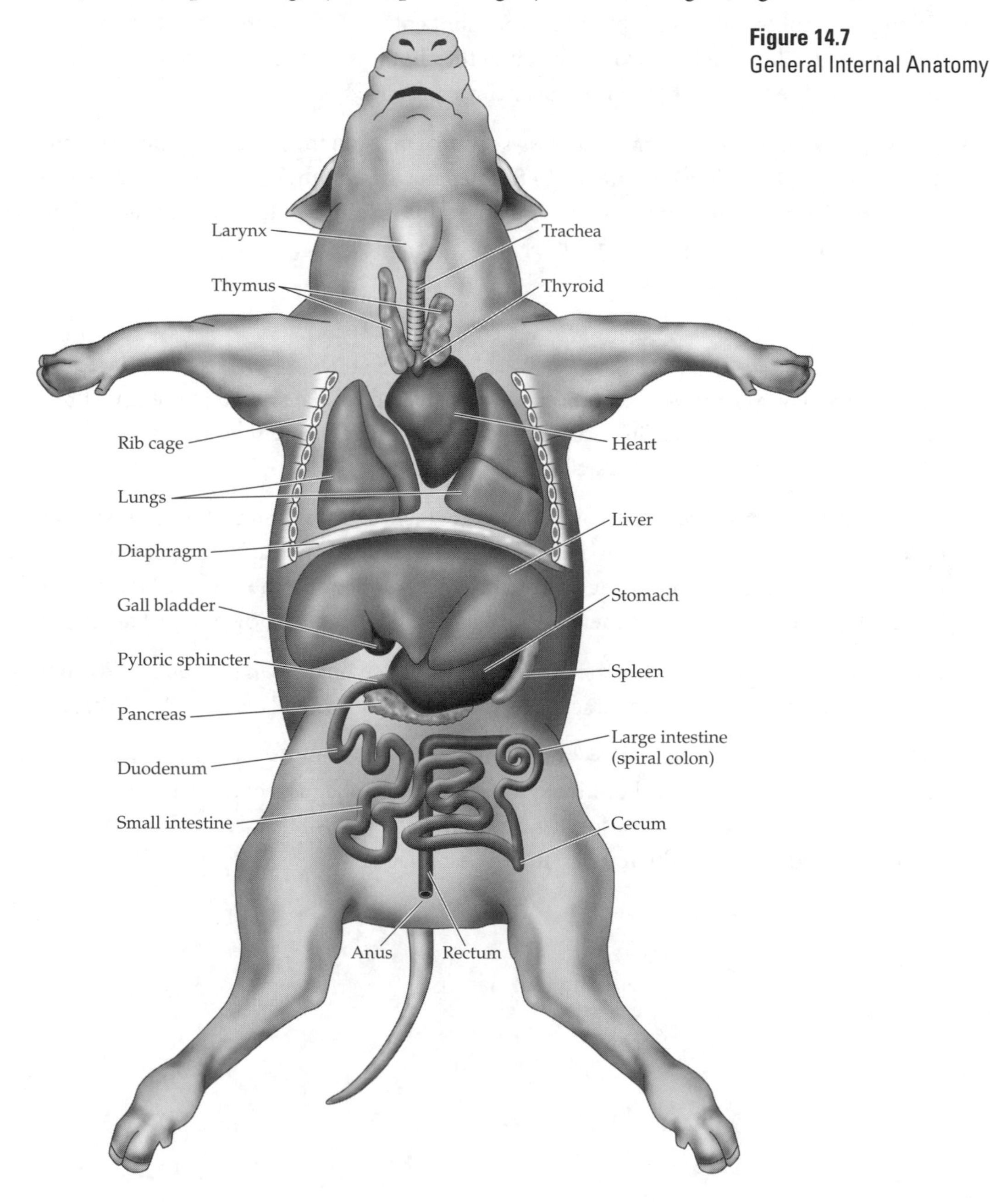

Figure 14.7
General Internal Anatomy

In this activity you will investigate the anatomical structures of the fetal pig's neck region.

1. Secure your pig to the dissecting tray. Use string to tie one anterior leg, and then wrap the string around the dissecting tray and tie the other leg down taut. Do the same to the posterior legs. Are both sets of legs fully spread apart? If not, you can wrap the string around either leg to tighten up the set of legs.

TO AVOID DAMAGING THE PIG'S INTERNAL STRUCTURES, USE ONLY THE BLUNT END OF THE SCISSORS, INSERTING THEM ONLY DEEP ENOUGH FOR THE BLUNT END TO BE SUBMERGED BELOW THE SKIN, AND PULL THE SCISSORS UP EVERY TIME YOU CUT.

2. Taking care not to cut too deeply, place the blunt end of the scissors against the pig's body, and make a longitudinal incision from the hair of the chin toward the posterior end of the pig until the incision is at the forelimbs (the vertical line in Figure 14.8).
3. Along each side of the chin, make a cross-sectional incision from the longitudinal incision that you made in step 2, as depicted in Figure 14.8. Then make two other cross-sectional incisions, at the shoulders. Now use the blunt probe to pull the flaps on either side of the throat back to expose the interior. Can you identify any structures?
4. Connective tissue covers some of the structures that you will be asked to identify. For example, the larynx is at the base of the chin. Use the forceps to remove the connective tissue covering the larynx, which is a hard, oblong bulb of cartilage.
5. Follow the larynx posteriorly; it connects to the trachea. Some connective tissue will cover the trachea as well; remove this tissue to gain a better view of the trachea. Refer to Figures 14.4 and 14.7 for assistance in identifying the neck structures. Can you locate the thymus? It is a grayish white, lobular tissue on either side of the trachea. The left lobe extends down to just anterior to the heart.
6. Posterior to the larynx, identify the thyroid gland, which is a brown to reddish brown triangular structure that sits ventral to the trachea toward the posterior end close to the rib cage.
7. The pharynx splits into two tubes—one leading to the respiratory system via the larynx and trachea, and the other leading to the digestive system via the esophagus, which is dorsal to the trachea. Move the trachea to the side and the esophagus should become visible. What color is the esophagus? What is the difference between the trachea and the esophagus?

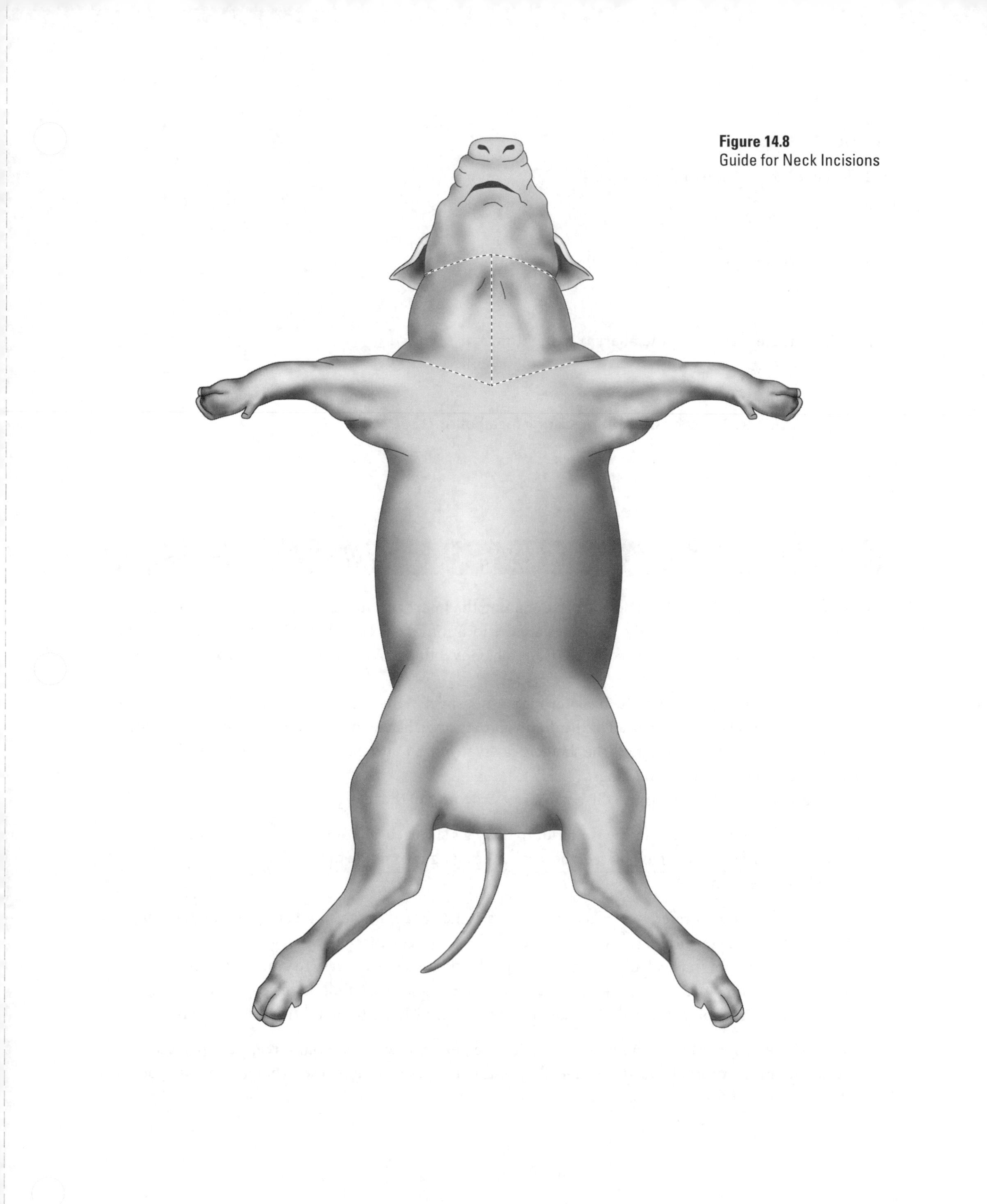

Figure 14.8
Guide for Neck Incisions

Concept Check

1. What is the function of the thymus gland?

2. What structures connect the pharynx to the lungs?

3. What structure connects the pharynx to the stomach?

4. Which structure is responsible for regulating metabolism?

ACTIVITY 5 The Thorax

Posterior to the neck is the thoracic cavity, containing the **heart** and **lungs** (see Figure 14.7). The heart moves the blood around the body; the lungs are the location of gas exchange. Oxygen diffuses into the bloodstream while carbon dioxide collected from the body diffuses out into the air. The **rib cage** protects these vital organs.

The esophagus passes down the back through a thin sheet of muscle that forms a cross section at the base of the thorax. Here a muscle called the **diaphragm** forms a barrier between the thoracic and abdominal cavities. The diaphragm expands the thoracic cavity to pull air into the lungs, and when the diaphragm is relaxed, air is pushed out of the lungs.

In this activity you will examine the anatomical structures of the fetal pig's thoracic cavity.

1. Use your finger to feel the rib cage anterior to the pig's stomach and posterior to the neck. Do the same on your own body. Do you feel the hard bone along the middle of the rib cage? This bone is called the **sternum**.

2. Using Figure 14.9 for reference, place the scissors at the base of the sternum and cut longitudinally through the rib cage, moving anterior toward the cut you made for the neck region. The bones of the fetal pig are not fully calcified, so the rib cage should be relatively easy to cut. Make a cross-sectional cut along the base of the rib cage, as shown in Figure 14.9. Pull the rib cage toward each side to open the thoracic cavity. Can you identify the heart? Refer to Figure 14.7 for assistance.

3. The heart is covered by a thin membrane called the **pericardium**. Use your forceps to open the pericardium to expose the heart. Is there fluid within the pericardium? If so, what purpose do you think fluid in this sac would serve?

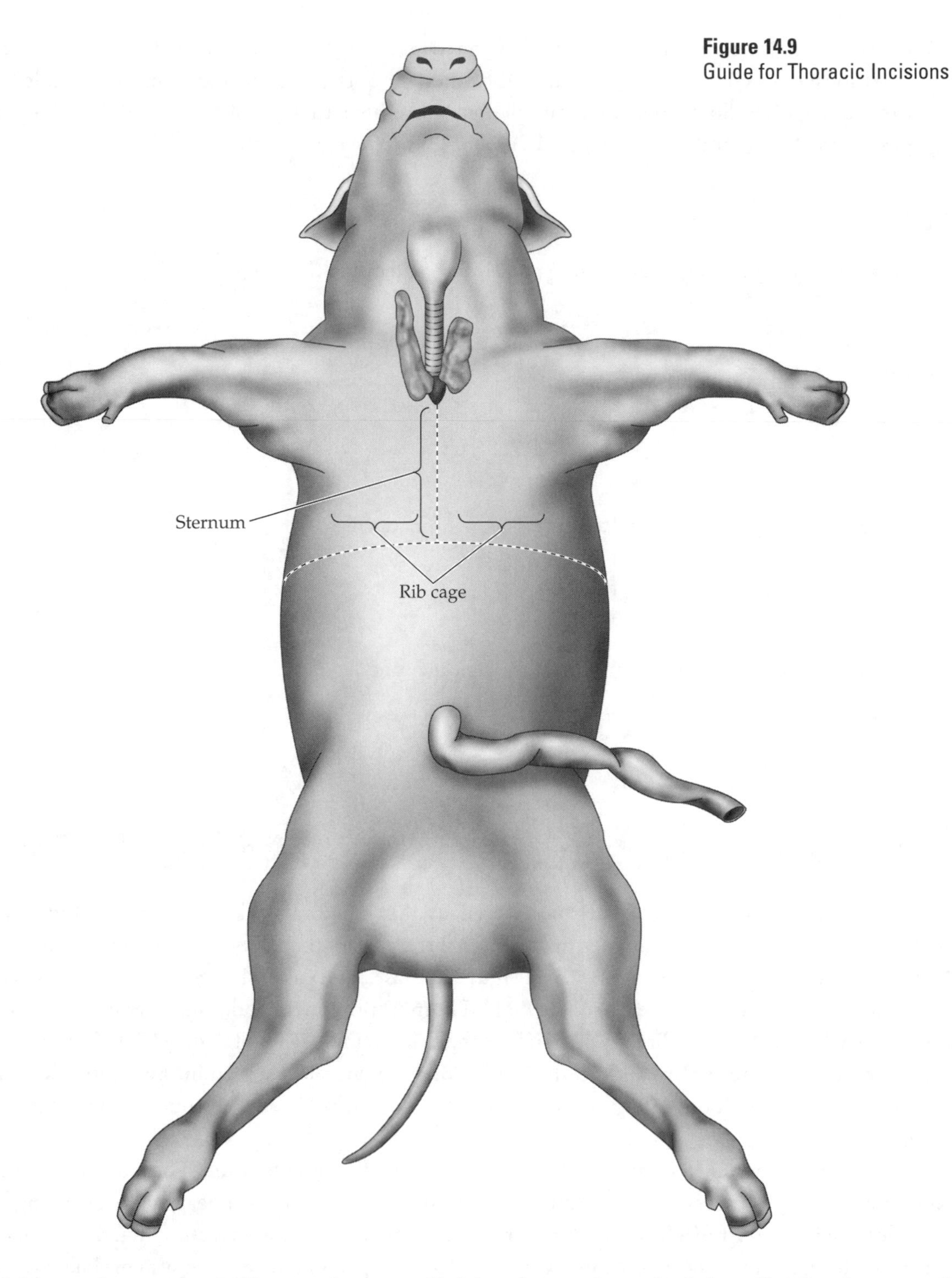

Figure 14.9
Guide for Thoracic Incisions

4. Each lung is also surrounded by membranes, called the **pleurae** (singular "pleura"). The pleurae and pericardium have similar functions: they protect the organ and supply oxygen and nutrients to it. Use your forceps to remove this sac as well. Have you ever heard of pleurisy? It is an infection of the pleurae. How many lobes/sections do the lungs have on each side?

5. Move the lungs and heart to one side. Can you identify any tubes running dorsally? There should be two tubes: one, a flexible muscular tube; the other, the aorta, is tough and firm. The flexible tube is the esophagus. Follow it posteriorly until it reaches a thin structure that forms a cross section at the base of the thorax and the top of the abdomen. This is the diaphragm.

Concept Check

1. How does the diaphragm function to pull air into and out of the lungs?

2. What is the function of the rib cage?

3. Where are the pleurae located and what is their function?

4. What is the hard bone in the middle of the rib cage called?

ACTIVITY 6 The Abdomen

The abdomen contains a variety of organs covered in a thin membrane, like the pleurae of the lungs or the pericardium of the heart. Membranes that protect the organs of the abdominal region are called **mesenteries**. They contain blood vessels to provide nutrient, waste, and gas exchange for abdominal organs, and nerves to communicate with the rest of the body. The most prominent abdominal organ is the **liver**, a reddish brown, lobed structure directly posterior to the diaphragm (see Figure 14.7). The liver detoxifies the bloodstream, regulates sugar, and produces bile. Bile is a chemical stored in the **gallbladder**, a green sac located underneath the right dorsal lobe of the liver, and released into the small intestine to break down fats.

Posterior to the liver is the **stomach**, the second site of digestion for food. The stomach does not perform the majority of digestion, but it starts the breakdown of proteins by releasing pepsin along with hydrochloric acid. Along the left side of the stomach, is another organ, the **spleen**, a fingerlike structure that skirts the stomach. The spleen recycles red blood cells and is part of the immune system. The stomach connects to the **small intestine**. This organ performs the majority (about 80%) of digestion and absorption in the digestive system. The small intestine receives bile from the liver or gallbladder, and many enzymes from the **pancreas**. The pancreas also produces hormones that it releases into the bloodstream. Between the small intestine and the **large intestine** (or colon) is a protruding structure: the appendix of the pig, called the **cecum**. The cecum is equivalent to the human **appendix**, which has a minor role in the immune system. From the small intestine, food then moves on to the large intestine, which conserves water, absorbs vitamins, and produces feces.

In this activity you will examine the anatomical structures of the fetal pig's abdominal cavity.

1. Using Figure 14.10 as an incision guide, open the abdominal cavity to expose the major organs without damaging them. The liver should almost jump out at you. It is a reddish brown organ with many lobes. Why is the liver so large?

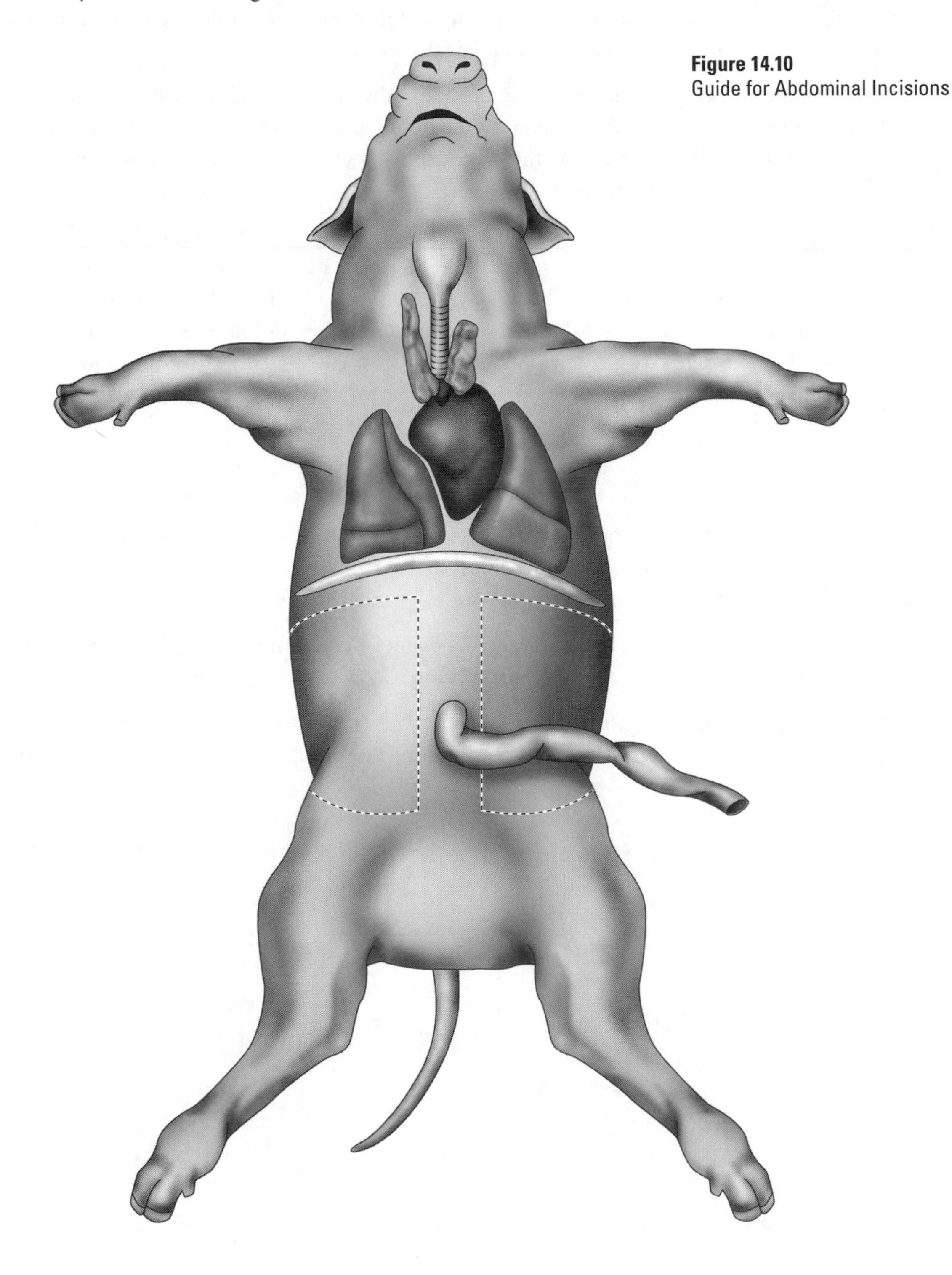

Figure 14.10
Guide for Abdominal Incisions

2. Lift the right lobe of the liver to find the green-gray sac embedded in the liver; this is the gallbladder. Can you see the duct leading from the gallbladder where bile flows to the small intestine?
3. The stomach, an off-white sac, is located dorsal to the liver and to the left of the gallbladder. The anterior portion of the stomach is connected to the esophagus. The spleen is a reddish brown organ protruding like a finger around the stomach. The spleen is thin and flimsy, so be careful as you poke around. Refer to Figure 14.7 for assistance in identifying these structures.
4. The stomach then swings around from the left side of the pig to the right, where you will find the **pyloric sphincter** at the base of the stomach. This structure is a thick, round muscle that allows food to enter the small intestine. Find the pyloric sphincter by using your fingers to feel for a thick, hard structure at the base of the stomach.
5. Posterior to the sphincter is the first part of the small intestine, called the **duodenum**. This is the portion that receives the bile and enzymes to start digestion. Move all the intestines to the right side of the pig (remember this is opposite of your right!). At the same time, lift up the stomach to expose the pancreas. The pancreas lies posterior to the stomach and on the left side of the duodenum because it secretes the digestive enzymes into this portion of the small intestine. What does the pancreas look like? What color is it?
6. The small intestine is a long, convoluted tube. Follow the tube from the pyloric sphincter down. At the end of the small intestine is a terminal point sticking out; this is the cecum. Can you find the endpoint between the small and large intestines?
7. From the cecum, the intestine becomes larger in diameter and forms a spiral, this is the majority of the large intestine and is also called the **spiral colon**. Can you locate the spiral colon?
8. From the spiral colon, the intestine moves posteriorly (**rectum**) toward the anus. Locate each of these sections of the large intestine. Does the intestine become larger (thicker in diameter) as you move along?

Concept Check

1. What abdominal organs are not involved in digestion?

2. What are the functions of the liver?

3. Where is the appendix of the pig? What is it called?

4. Which abdominal structure secretes enzymes and hormones?

ACTIVITY 7 Applying Knowledge of the Fetal Pig to Human Anatomy

Some important external and internal structural differences exist between pigs and humans. First, whereas pigs are tetrapods (that is, they move on four appendages), humans are bipeds (moving on two appendages). Humans have more intricate and dexterous limbs. Whereas the female pig has only one common opening for its urinary and reproductive systems, human females have two openings. Several organs have different shapes or sizes. For example, the thymus gland in humans is much smaller and resides closer to the heart and not around the trachea as in the fetal pig. The spleen is significantly smaller in humans. The large intestine of pigs is structurally different from its human counterpart, and longer than in humans. The bladder is also a different shape and in a different location. The general structures of the thoracic cavity between pigs and humans differ only in relative size. Although these differences exist, the functions of the basic anatomical structures in pigs and humans are the same, and analyzing how organs work together in the pig helps elucidate the role of the structures and organs of your body.

In this activity you will assemble all the information that you collected in Activities 1 through 6 to understand the organ systems of your own body. The digestive system removes nutrients from food you consume and then gets rid of the solid wastes. The respiratory system allows for gas exchange, with oxygen diffusing into your body while carbon dioxide diffuses out. The circulatory system is the main internal transport system. The immune system defends your body against infection and injury. The endocrine system is a collection of glands and other tissues in animals that produce signaling molecules called hormones and release them directly into your circulatory system. For each structure in Table 14.1, check off all the systems in which that structure functions in some capacity.

TABLE 14.1 COMPARISON OF BODY STRUCTURES AND THEIR ROLES IN ORGAN SYSTEMS

Structure	Digestive system	Respiratory system	Circulatory system	Immune system	Endocrine system
Appendix					
Diaphragm					
Esophagus					
Gallbladder					
Heart					
Large intestine					
Larynx					
Liver					
Lungs					
Pancreas					
Pharynx					
Small intestine					
Spleen					
Stomach					
Teeth					
Thymus					
Thyroid					
Tongue					
Trachea					

Key Terms

abdominal region (p. 14-4)
anterior (p. 14-2)
anus (p. 14-5)
appendix (p. 14-14)
cecum (p. 14-14)
cross section (p. 14-2)
diaphragm (p. 14-12)
dorsal (p. 14-2)
duodenum (p. 14-16)
epiglottis (p. 14-6)
esophagus (p. 14-6)
gallbladder (p. 14-14)
genital papilla (p. 14-5)
hard palate (p. 14-7)
heart (p. 14-12)
large intestine (p. 14-14)
larynx (p. 14-6)
liver (p. 14-14)
longitudinal section (p. 14-2)
lung (p. 14-12)
mammary papilla (p. 14-4)
mesentery (p. 14-14)
nasopharynx (p. 14-6)
nipple (p. 14-4)
pancreas (p. 14-14)
penis (p. 14-5)
pericardium (p. 14-12)
pharynx (p. 14-6)
pleura (p. 14-13)
posterior (p. 14-2)
pyloric sphincter (p. 14-16)
rectum (p. 14-16)
rib cage (p. 14-12)
scrotal sac (p. 14-5)
small intestine (p. 14-14)
soft palate (p. 14-7)
spiral colon (p. 14-16)
spleen (p. 14-14)
sternum (p. 14-12)
stomach (p. 14-14)
teeth (p. 14-6)
thoracic region (p. 14-4)
thymus gland (p. 14-9)
thyroid gland (p. 14-9)
tongue (p. 14-6)
trachea (p. 14-9)
umbilical cord (p. 14-5)
urogenital opening (p. 14-5)
ventral (p. 14-2)

Review Questions

1. Which structures and organs are involved in the digestive system?

2. Which external and internal structures are involved in the respiratory system?

3. The location of which external structure identifies whether the pig is a male or female?

4. When the diaphragm contracts, what happens to the lungs? When the diaphragm relaxes, what happens to the lungs?

5. The ________________ is where the majority of digestion and absorption occurs.

6. Vitamin absorption, water reabsorption, and feces production occur in the ________________.
7. What enzyme breaks down sugars? Where is it produced?

8. The ________________ and the ________________ connect the pharynx to the lungs.
9. The ________________ is a flexible tube located dorsal to the heart and lungs.
10. The ________________ provides a barrier between the thoracic and abdominal cavities.
11. The ________________ produces bile that is stored in the ________________.
12. What is the membrane covering of the abdominal organs called?

13. What is the membrane covering of the lungs called?

14. What is the membrane covering of the heart called?

LAB 15 Blood and Circulation (with Fetal Pig Dissection)

Objectives

- Observe the components of human blood and discuss their functions.
- Identify the major structures of the heart on a human heart model.
- Explain how the four chambers of the heart, veins, and arteries function within the pulmonary and systemic circuits of the mammalian cardiovascular system.
- Identify the major differences between the fetal and adult mammalian cardiovascular system.
- Trace the major arteries and veins of the mammalian cardiovascular system through fetal pig dissection.

Introduction

The **circulatory system** is the network of tubes transporting fluids carrying nutrients and wastes to and from cells in an animal. The vertebrate circulatory system, called a **cardiovascular system**, is closed; that is, it has no intentional openings and is a complex network of blood vessels completing a full circuit back to their origin in the heart.

Within the cardiovascular system, blood and cells move to distribute necessary molecules and remove wastes within two main pathways: the pulmonary circuit and the systemic circuit. The **pulmonary circuit** refers to blood streaming from the heart to the lungs, and then returning from the lungs to the heart. In the pulmonary circuit, the heart pumps the oxygen-poor blood that it receives from the body to the lungs, and the lungs then return oxygenated blood that the heart will pump throughout the body. The **systemic circuit** begins when the heart pushes oxygen-rich blood to the body—this circuit delivers oxygen to cells and picks up their carbon dioxide. The systemic cycle is complete when the oxygen-poor blood returns to the heart from the body, to be delivered to the lungs via the pulmonary circuit.

In this lab you will explore each component of the cardiovascular system: the blood, the heart, and the vessels that enable this system to meet the needs of the organism.

ACTIVITY 1 Blood

Blood is the fluid communication and delivery highway of the body. The cardiovascular system is a series of interconnected tubular vessels through which blood is pumped. The liquid portion of blood, called **plasma**, is mostly water and contains many dissolved molecules, such as proteins; the solid components are red blood cells, white blood cells, and platelets.

- Red blood cells, or **erythrocytes**, transport gases via a protein called **hemoglobin**. Hemoglobin carries oxygen and carbon dioxide. Erythrocytes last only a few months, and are constantly generated and recycled.
- White blood cells, or **leukocytes**, come in many forms and function in the immune system. White blood cells attack specific invaders or release specific molecules to combat infection or injury. Since there is such a variety of leukocytes, they have a variety of life spans—some lasting only a few hours for immediate action against invaders, and some lasting many years to remember and attack the same invader.
- Platelets, or **thrombocytes**, are fragments of bone marrow cells that clog up unintended openings such as puncture wounds in the cardiovascular system. Along with proteins, platelets form the clots that stop bleeding. **Hemophilia** is a genetic disorder in which proteins involved in clotting do not form properly, so the individual never forms the correct meshwork of proteins and platelets to stop bleeding. Platelets last only about 7 days within the bloodstream.

Bone marrow contains stem cells that produce red blood cells, white blood cells, and platelets. All of the cellular components of blood form within bone marrow, and these components constantly need to be regenerated.

In this activity you will examine blood, observing both healthy and diseased blood cells.

Healthy Blood Cells

1. Obtain a human-blood slide and a compound microscope.
2. Using the compound microscope, examine the slide of human blood. The liquid part of the blood, the plasma, does not stain.
3. Look at the sample of human blood. What color are the majority of cells on your slide?

4. By far the majority of cells on your slide are red blood cells (erythrocytes). Can you identify any cellular structures within the red blood cells? The erythrocyte has no nucleus, but it is filled with the protein hemoglobin. In the circle below, sketch red blood cells as they appear in your field of view under the highest magnification.

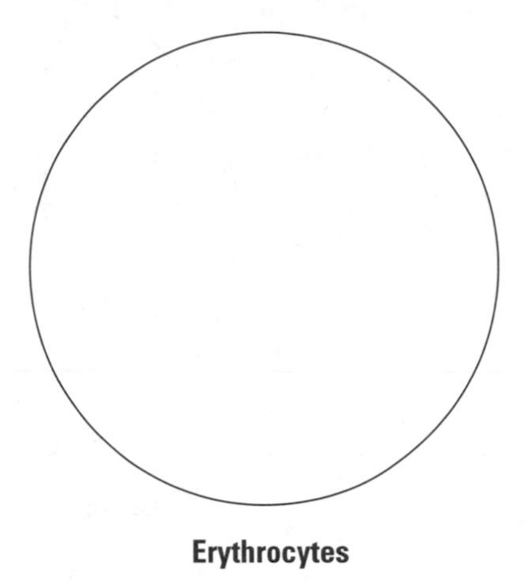

Erythrocytes

5. Can you think of a reason why it might be advantageous for red blood cells not to have nuclei?

6. Aside from erythrocytes, the other types of blood cells on your slide are the various white blood cells (leukocytes). Move around the slide until you locate a cell that is not a red blood cell. What difference do you immediately notice?

7. All leukocytes contain nuclei and function mainly in the defense system of the body. The two most common white blood cells are the **lymphocyte** and the **neutrophil**. The lymphocytes that you'll see in this slide fight infections. Lymphocytes have a large, round nucleus that nearly fills the cell. Neutrophils eat infectious bacteria. The nucleus of a neutrophil is shaped almost like an hourglass. In the first circle below, sketch the cell that you're examining, and label it either "lymphocyte" or "neutrophil."

8. Which type of leukocyte did you draw in step 7? If you drew a neutrophil, locate a lymphocyte. If you drew a lymphocyte, locate a neutrophil. In the second circle below, sketch this second type of white blood cell, and label it appropriately.

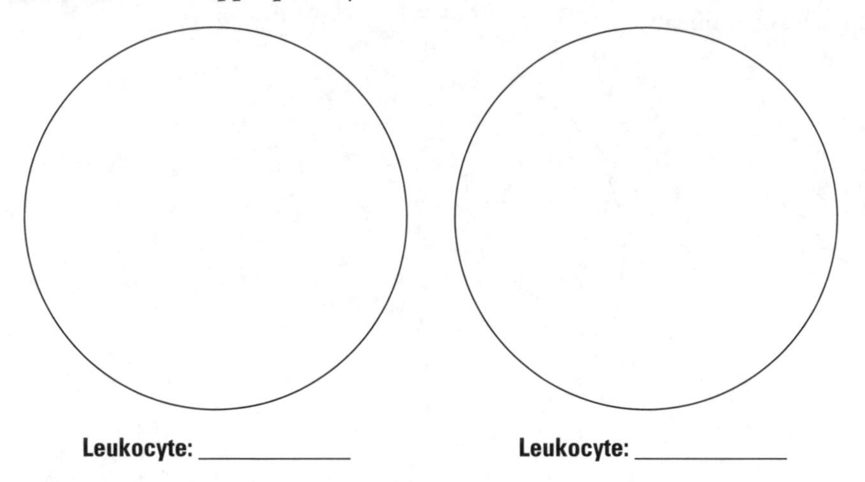

Leukocyte: ____________ **Leukocyte:** ____________

Diseased Blood Cells

Oxygen attaches to the iron in hemoglobin. Up to four oxygen molecules bind to the iron within a hemoglobin molecule. Iron is an essential element that the human body requires at low concentrations, and an insufficient level of iron leads to a condition called **anemia**. Any condition reducing the amount of oxygen carried by blood is referred to as anemia.

A genetic disease called **sickle-cell anemia** is named after the resulting shape of red blood cells in affected individuals. The hemoglobin molecule takes on an odd shape and causes the red blood cell to assume a strange curved, sharp shape as well. Sickle-cell anemia is protective against the plasmodium that causes malaria and therefore combats malaria, but sickle cells are larger than normal red blood cells, making it difficult for them to move through the bloodstream. The sickle cells puncture **capillary beds**—the extremely small vessels where gas, nutrient and waste exchange occurs—leading to organ failure at about 40 years of age.

1. Obtain and examine the sickle-cell anemia slide provided by your instructor. What is the first thing you notice about the red blood cells?

   In the circle below, sketch these cells.

Sickle-cell anemia

2. Obtain and examine the leukemia slide provided by your instructor. Leukemia is a cancer that affects white blood cells. **Cancer** is the uncontrolled division of a specific type of cell, in this case a type of leukocyte. Remember how many white blood cells were present on the human blood slide you examined at the beginning of this activity? Do you notice anything different about the number and type of white blood cells present in the slide you're examining now?

   In the circle below, sketch both the white blood cells and the red blood cells as they appear in your field of view under the highest magnification.

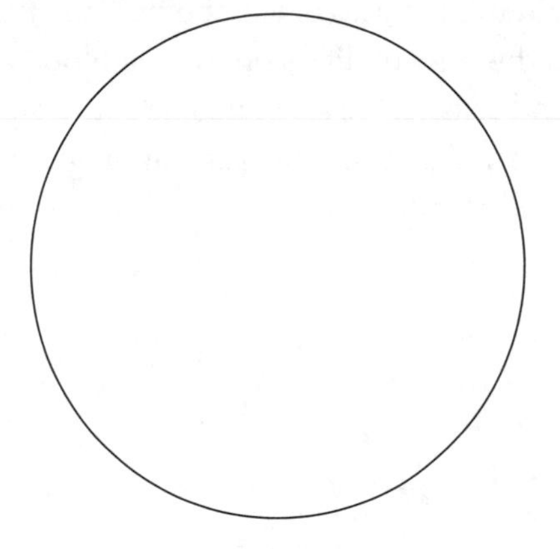

Leukemia

Concept Check

1. What are the cellular components of blood and their general functions?

2. What type of blood cell has the longest life span? Why?

3. What is hemophilia?

4. How does insufficient dietary iron affect your body? What causes the shape change of red blood cells in sickle-cell anemia?

ACTIVITY 2 Human Heart Model

The heart and the vessels that deliver blood to the body make up the human cardiovascular system. Arteries carry blood from the heart, and veins carry blood to the heart. In the pulmonary circuit, **arteries** carry oxygen-poor blood from the heart to the lungs, and **veins** carry oxygenated blood from the lungs to the heart. In the systemic circuit, arteries carry oxygenated blood from the heart to the body, and veins carry oxygen-poor blood from the body to the heart.

In both circuits, arteries become thinner **arterioles** that lead to capillary beds. Capillary beds are extremely thin, small enough for one red blood cell to move through, and composed of a thin wall of epithelial cells. Capillary beds penetrate every area of your body. Their thinness allows water and other molecules to move freely according to the rules of diffusion, so that blood can exchange gases, nutrients, and wastes with the surrounding fluid and cells. At the arteriole side of the capillary bed (near the arteries), heart pressure pushes water out from the capillaries into the surrounding **interstitial fluid**. Nutrients and oxygen diffuse into the fluid and into cells. In the opposite direction, wastes and carbon dioxide diffuse from

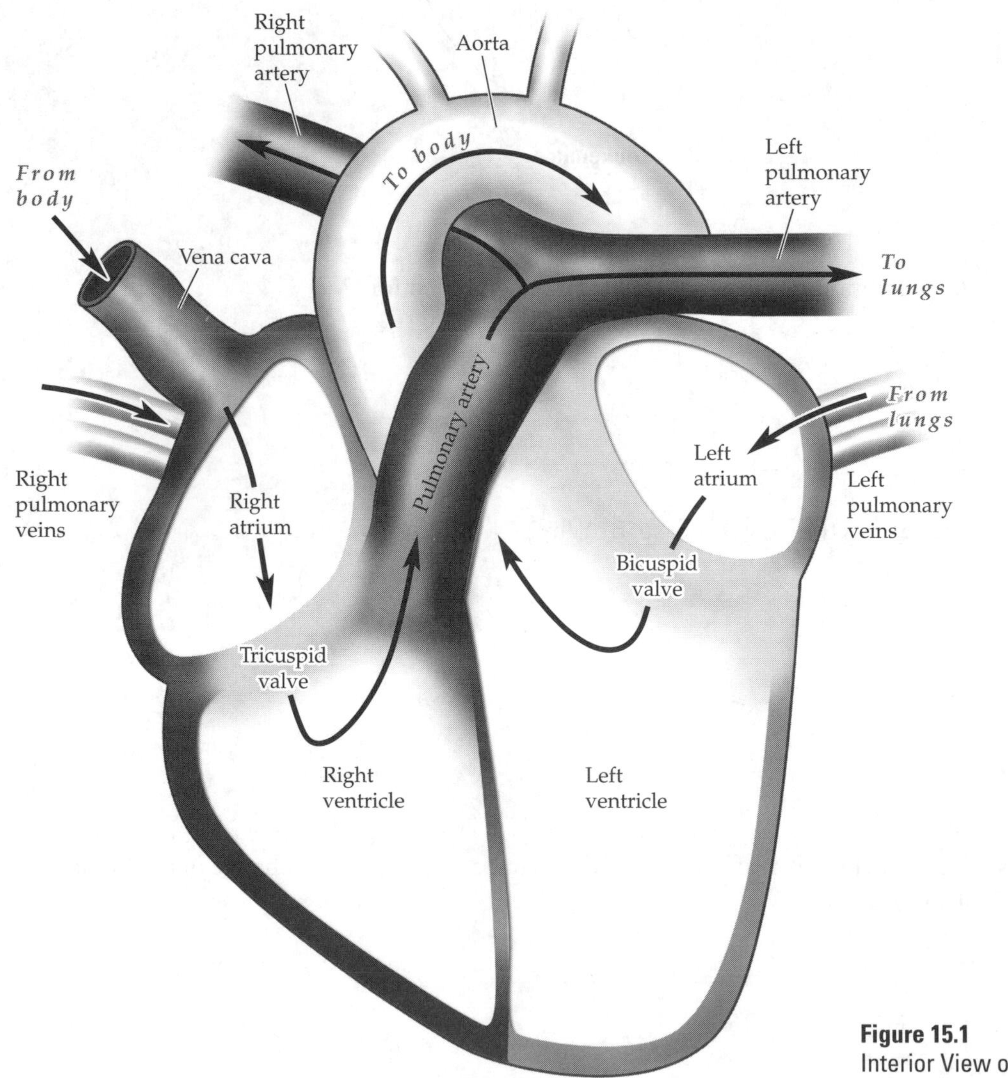

Figure 15.1
Interior View of the Human Heart

cells into the interstitial fluid and then into the venule side of the capillary bed (near the veins). From the capillary bed, blood moves back to the heart via small veins called **venules**, which lead to larger veins and to the **vena cava** leading to the right atrium of the heart.

The heart has four chambers: right atrium, right ventricle, left atrium, and left ventricle (Figure 15.1). The two chambers on the right side of the heart receive blood that is low in oxygen and pump it through the pulmonary circuit. The chambers on the left receive oxygenated blood and pump it to respiring cells throughout the body via the systemic circuit. The **right atrium** collects blood that is low in oxygen from the vena cava; the blood then travels through the **tricuspid valve** to the **right ventricle**. From the right ventricle, the blood is pumped into the **pulmonary artery**, which branches into the left and right pulmonary arteries leading to their respective lung, where gas exchange takes place. Two sets of **pulmonary veins** bring oxygenated blood from the lungs and dump it into the **left atrium**. Blood then travels through the **bicuspid valve** and into the **left ventricle**. From the left ventricle the heart pushes the blood into the **aorta**, the main artery of the body.

In this activity you will trace the path of blood in the human heart.

1. Obtain a human heart model from your instructor. In the steps that follow, use the human heart model and Figure 15.1 to help you trace the path of blood through the heart. Remember that every structure is named according to the perspective of the individual, so what is on the right from your perspective is actually named "left."

2. Observe the external structures of the heart model and locate the four chambers: right atrium, right ventricle, left atrium, and left ventricle. The atria collect blood, and the ventricles pump blood to the two circuits. Do you notice a difference in size between the atria and the ventricles? The ventricles need to push blood into each circuit and are therefore larger than the atria.

3. Locate the blood vessel leading to the right atrium: this is the vena cava. The vena cava coming from above the right atrium is referred to as **anterior**. The **anterior vena cava** collects oxygen-poor blood from the body above the heart. The vena cava coming from below the right atrium is referred to as **posterior**. The **posterior vena cava** collects oxygen-poor blood from the body below the heart. Both the anterior and posterior portions of the vena cava dump oxygen-poor blood into the right atrium.

4. Open the heart model to observe the internal structures, and then trace the path of blood from the right atrium, through the tricuspid valve. The tricuspid and bicuspid valves are named for the number of flaps (cusps) that make up the valve. Blood enters the right ventricle through the tricuspid valve and then travels to the lungs via the pulmonary artery. How many pulmonary arteries are there?

5. Now trace the blood traveling back to the heart from the lungs—through the pulmonary veins, to the left atrium, through the bicuspid valve, to the left ventricle, and then to the aorta. An easy way to remember which valve is located on which side of the heart is that "tricuspid" has an "r" in it and the tricuspid valve is on the right side of the heart.

6. Close your heart model. Now, leaving the heart and the pulmonary circuit, trace the aorta as it loops toward the dorsal side of the model. The first branch of the aorta is the **coronary artery**. Whereas the pulmonary artery and veins transport blood from and to the heart for the purpose of gas exchange, the coronary artery is the blood vessel providing nutrient- and oxygen-rich blood to

muscles of the heart. Locate the coronary artery. This artery is the most common place for a clot to form leading to a heart attack. High blood pressure can have serious health consequences, including heart attack, stroke, and kidney failure. The most common cause of high blood pressure is the hardening and constriction of blood vessels because of fatty deposits in vessel walls, especially the coronary artery.

7. Oxygen-poor blood is drained from the heart tissue via the **cardiac veins** depicted along with the coronary arteries in the external view of the heart in Figure 15.2. The cardiac veins drain their oxygen-poor blood directly into the right atrium, completing the systemic circuit of the heart. Find the cardiac veins on the exterior of the heart model.

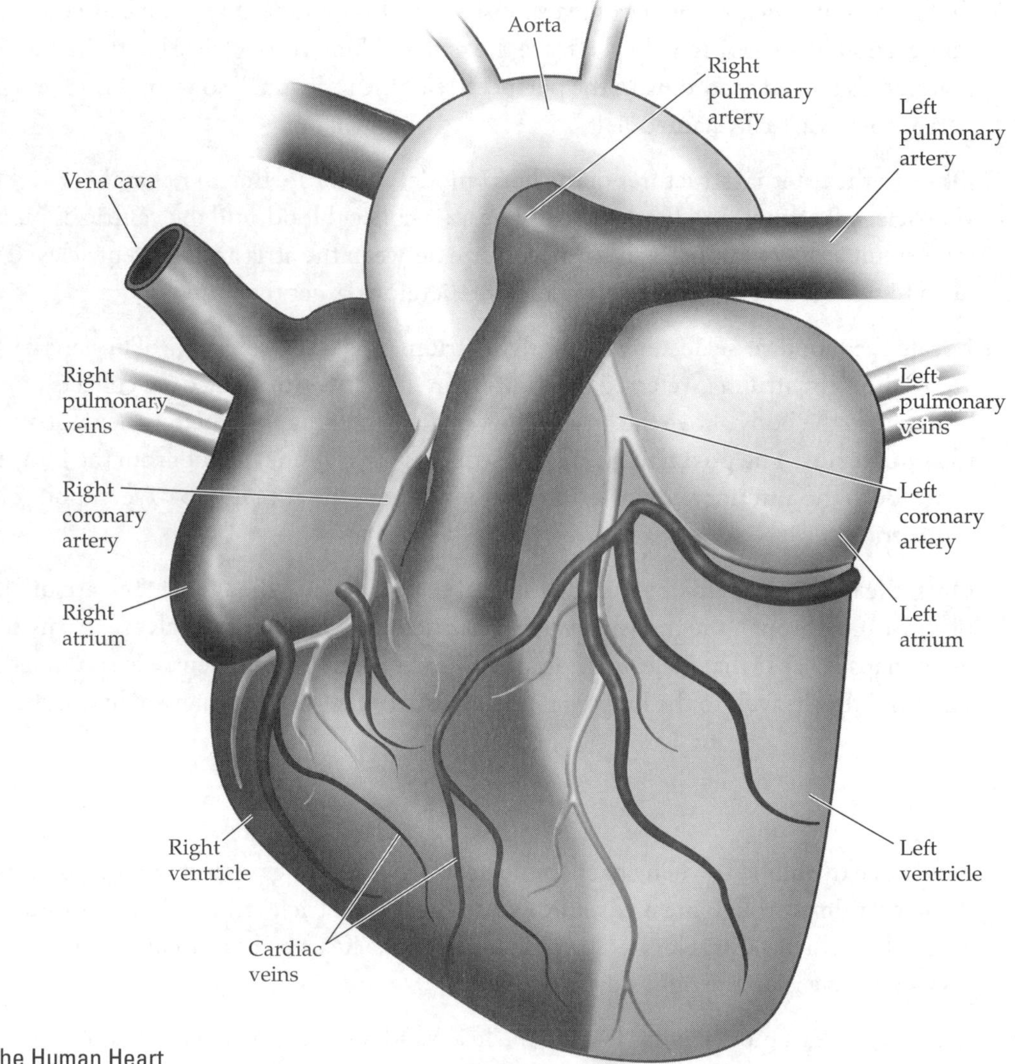

Figure 15.2
Exterior View of the Human Heart

Concept Check

1. What is the purpose of the pulmonary circuit? What is the purpose of the systemic circuit? Explain how the two function together in the cardiovascular system.

2. What is the general path of blood flow from the vena cava to the pulmonary artery in the pulmonary circuit? What is the general path of blood flow from the left and right pulmonary veins to the aorta in the systemic circuit?

3. Heart attacks usually result from blockage of which blood vessel? Why does blocking this blood vessel affect the heart so dramatically?

4. How do capillary beds assist in the function of the cardiovascular system?

ACTIVITY 3 Cardiovascular System of the Fetal Pig

Although there is some variation of structure in different species of mammals, the fetal pig's cardiovascular system illustrates the general movement of blood and the function of the cardiovascular system in all mammals. Like humans, the heart of the fetal pig has four chambers: two large ventricles and two small atria. The atria of the fetal pig are dark and cap the ventricles. The blood leaves the heart via the aorta, which extends down the back to the abdomen. In the abdomen, the aorta branches to the digestive system, the kidneys, and the genitals, and then on to the umbilical cord and legs. The blood flowing through the digestive system collects nutrients, and this blood then flows to the liver, where nutrients are distributed as needed. The kidney is the cleansing organ for the bloodstream. The blood then travels to the vena cava, which leads back to the heart.

The cardiovascular system in the fetal pig differs from the adult pig's cardiovascular system in only three major ways:

- The umbilical vein brings oxygen- and nutrient-rich blood from the mother directly to the liver of the fetus, while the umbilical arteries remove wastes to be disposed of by the mother.
- The fetal pig heart has a structure called the **ductus arteriosus** connecting the pulmonary artery and aorta. The ductus arteriosus bypasses the pulmonary circuit and delivers the blood directly to the systemic circuit, since the blood already has the oxygen it needs from the umbilical vein (and the fetus's lungs cannot perform gas exchange yet).
- A hole between the right and left atria, called the **foramen ovale**, also shunts blood to the systemic circuit.

These are minor differences in the fetus and adult in both pigs and humans, and the fetal pig provides a very good example of the cardiovascular system found in all mammals.

In this activity you will examine major arteries and veins of the fetal pig's cardiovascular system.

TO AVOID DAMAGING THE PIG'S INTERNAL STRUCTURES, USE ONLY THE BLUNT END OF THE SCISSORS, INSERTING THEM ONLY DEEP ENOUGH FOR THE BLUNT END TO BE SUBMERGED BELOW THE SKIN, AND PULL THE SCISSORS UP EVERY TIME YOU CUT.

Veins

1. After placing the fetal pig in the dissecting pan and acquiring your dissecting instruments, start at the heart and, using Figures 15.1 and 15.2, identify the four chambers: right atrium, left atrium, right ventricle, and left ventricle. Push the heart down softly, and find the three tubes attached it. Remember that the arteries are labeled with red latex and the veins with blue.

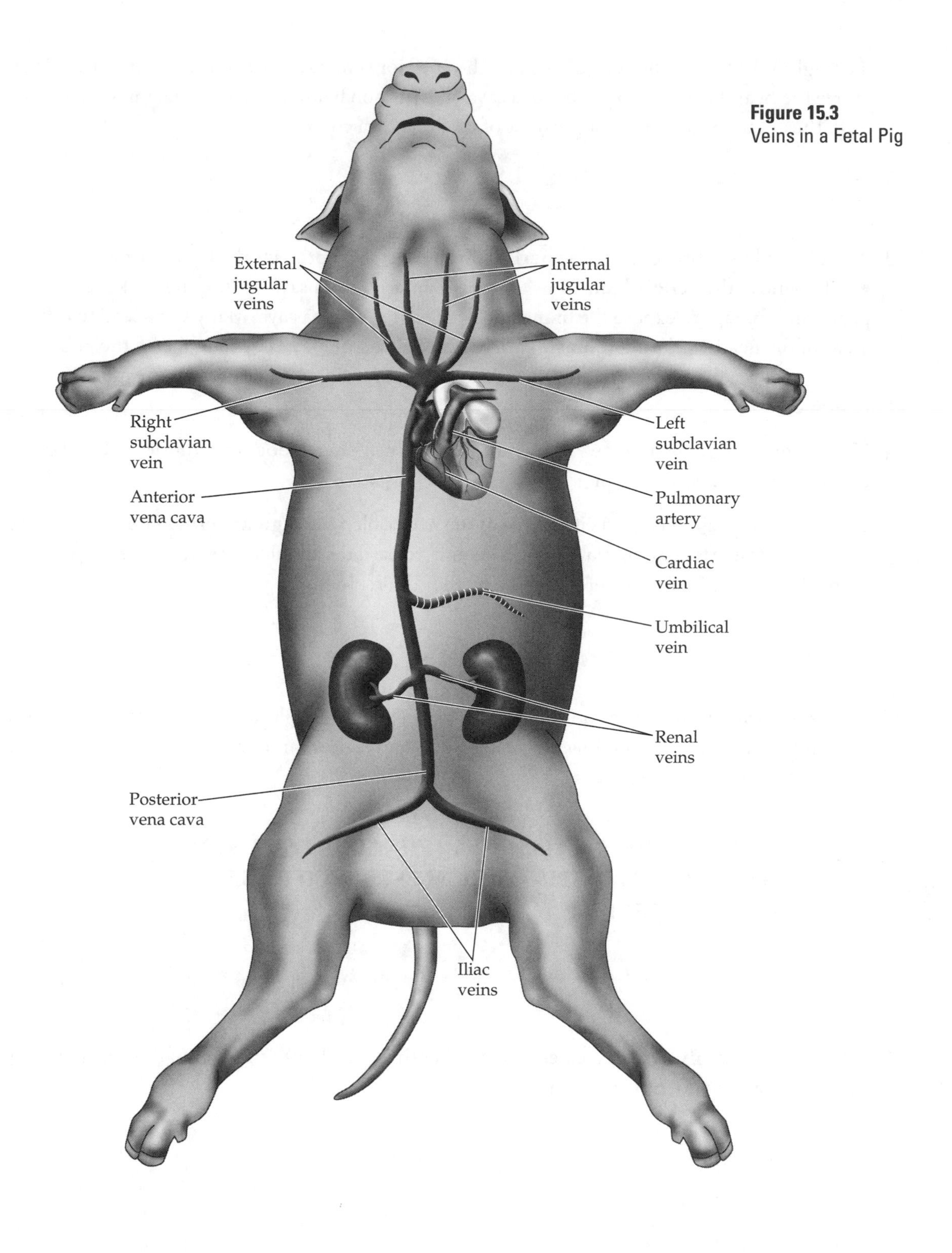

Figure 15.3
Veins in a Fetal Pig

2. The rightmost tube is the vena cava. Note the anterior vena cava (the portion of the vessel located above the heart) and the posterior vena cava (the portion below). To which chamber of the heart do both the anterior and posterior portions of the vena cava connect?

3. A fluid-filled membrane called the **pericardium** covers and protects the heart. Using your forceps, gently remove the pericardium. Now follow the anterior vena cava up into the neck area. Use your probe and forceps to expose the branching of the anterior vena cava. Many veins lead to different areas of the head, neck, and thoracic cavity. Your instructor may ask you to locate the pulmonary veins; these are the most difficult veins to identify in the fetal pig, so use the heart model as a reference.

4. Using Figure 15.3, identify the **subclavian veins**, which drain blood from the arms. Can you locate the subclavian veins in your pig?

5. There are two major veins on either side of the neck called the **jugular veins**. The veins along the trachea are the **internal jugular veins**. The ones toward the shoulder are the **external jugular veins**. Find these four veins. From where do they drain blood?

6. The arteries and veins of the abdomen typically run parallel to one another, leading, respectively, into and out of the organ. Follow the posterior vena cava down into the abdomen. Find the **renal veins**, which lead to the kidneys.

YOU WILL REUSE THE FETAL PIG, SO DO NOT REMOVE ANY ORGANS OR TISSUES UNLESS SPECIFICALLY INSTRUCTED.

7. Where is the **umbilical vein** located? Remember that it leads into the liver of the fetal pig. Why does the umbilical vein go into the liver?

Arteries

Now you will trace the major arteries of the fetal pig.

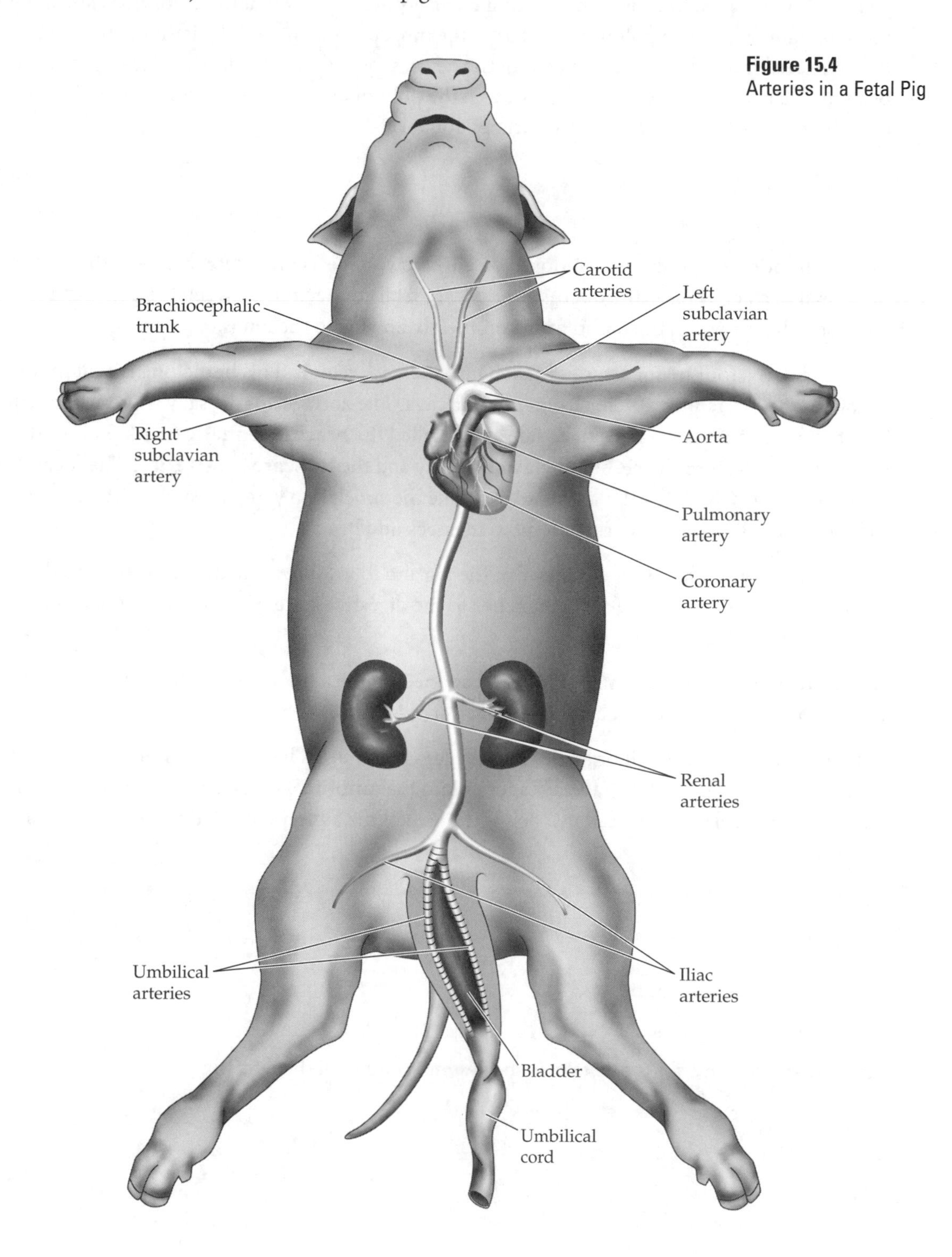

Figure 15.4
Arteries in a Fetal Pig

1. Return to the vena cava, the rightmost tube projecting from the top of the heart on the fetal pig. The other two tubes are the aorta and the pulmonary artery (Figure 15.4). Although the arteries are injected with red latex, the aorta and pulmonary artery are so thick that the color does not show, so these remain white. The pulmonary artery is the most prominent and the leftmost tube. The aorta is the central tube and tucked in the back of the heart, so you may have difficulty seeing it unless you push on the heart softly. Locate the aorta and the pulmonary artery on your fetal pig. To which chambers of the heart do each of these tubes connect?

2. Follow the aorta on your pig as it loops dorsally. Do you see a connection between the aorta and the pulmonary artery? This is the ductus arteriosus, a fetal connection between the pulmonary artery and aorta that prevents the blood from completely flowing to the lungs.

3. From the aorta, four major arteries branch into the neck region. This branching is a bit more complex than the branching of veins from the vena cava. The aorta loops dorsally with two branches. The first branch is on the pig's right side and is called the **brachiocephalic trunk**. Follow the trunk up where it splits into the **right subclavian artery** and the two **carotid arteries**. The second branch from the aorta is the **left subclavian artery**. Like the subclavian vein, the subclavian artery leads to the forelimbs. The carotids bring blood to the neck and head.

4. Move the lungs and heart to either side of the pig and follow the aorta down the back to the abdomen. In the abdomen, the aorta branches to the digestive system, the kidneys, the genitals, the legs, and the umbilical cord.

5. The aorta branches into **renal arteries** that connect to the kidneys, which filter blood to remove wastes and toxins. Can you locate the renal arteries?

6. Follow the aorta to where it branches into the legs of the pig. These are the **iliac arteries**, supplying the legs with blood. Next, the aorta branches into the **umbilical arteries**. These surround the bladder, a long muscular sac located just below the skin of the abdomen. What is the function of the umbilical arteries?

Concept Check

1. How do arteries and veins function in the systemic circuit of the fetal pig?

2. How do the ductus arteriosus and foramen ovale affect the pulmonary circuit in fetal mammals?

3. How is the fetal pig a good example of the mammalian cardiovascular system?

4. How do the kidneys and liver affect blood in the cardiovascular system?

5. Arteries have thick walls that can expand with blood pressure. Veins have weaker walls, and valves that ensure the proper direction of blood flow. Given what you know about the function of arteries and veins in the systemic circuit, why do their different structures make sense?

Key Terms

anemia (p. 15-4)
anterior (p. 15-7)
anterior vena cava (p. 15-7)
aorta (p. 15-7)
arteriole (p. 15-6)
artery (p. 15-6)
bicuspid valve (p. 15-7)
blood (p. 15-2)
brachiocephalic trunk (p. 15-14)
cancer (p. 15-5)
capillary bed (p. 15-4)
cardiac vein (p. 15-7)
cardiovascular system (p. 15-1)
carotid artery (p. 15-14)
circulatory system (p. 15-1)
coronary artery (p. 15-7)
ductus arteriosus (p. 15-10)
erythrocyte (p. 15-2)
external jugular vein (p. 15-12)
foramen ovale (p. 15-10)
hemoglobin (p. 15-2)
hemophilia (p. 15-2)
iliac artery (p. 15-14)
internal jugular vein (p. 15-12)
interstitial fluid (p. 15-6)
jugular vein (p. 15-12)
left atrium (p. 15-7)
left subclavian artery (p. 15-14)
left ventricle (p. 15-7)
leukocyte (p. 15-2)
lymphocyte (p. 15-3)
neutrophil (p. 15-3)
pericardium (p. 15-12)
plasma (p. 15-2)
posterior (p. 15-7)
posterior vena cava (p. 15-7)
pulmonary artery (p. 15-7)
pulmonary circuit (p. 15-1)
pulmonary vein (p. 15-7)
renal artery (p. 15-14)
renal vein (p. 15-12)
right atrium (p. 15-7)
right subclavian artery (p. 15-14)
right ventricle (p. 15-7)
sickle-cell anemia (p. 15-4)
subclavian vein (p. 15-12)
systemic circuit (p. 15-1)
thrombocyte (p. 15-2)
tricuspid valve (p. 15-7)
umbilical artery (p. 15-14)
umbilical vein (p. 15-12)
vein (p. 15-6)
vena cava (p. 15-7)
venule (p. 15-7)

Review Questions

1. Why are the ventricles larger than the atria?

2. Explain the flow of blood through the heart.

3. The ________________ are the veins of the neck; the ________________ are the arteries of the neck.

4. The ________________ collects blood from the veins to bring it to the right atrium, while the ________________ distributes blood from the left ventricle to the arteries.

5. The ________________ are the tiny vessels of the cardiovascular system where nutrients, wastes, and gases are exchanged throughout the body.

6. ________________ is the fluid surrounding the cells of your body where nutrients, wastes, and gases are exchanged into and out of the cardiovascular system.

7. The ________________ covers the heart, in order to protect it from the movement of the body.

8. ________________ is the name for the veins and arteries supplying blood to the kidneys.

9. Describe anemia. How does the genetic disease sickle-cell anemia cause anemia?

10. Describe the differences between the fetal and adult mammalian cardiovascular systems.

11. What are the functions of the umbilical vein and the umbilical arteries?

12. How do veins and arteries function in the pulmonary circuit? In the systemic circuit?

13. What is the first branch of the systemic circuit?

14. Name the different solid components of blood and their functions.

LAB 16 The Cardiovascular System

Objectives

- Trace the flow of blood through the cardiovascular system.
- Recognize the consequences of clots in the vessels of the cardiovascular system.
- Detect heartbeats and identify heart sounds.
- Measure pulse rate and the effects of different conditions on the pulse.
- Evaluate various conditions affecting blood pressure.
- Describe how oxygen and carbon dioxide levels in the blood affect respiration.

Introduction

The human organism comprises a collection of complex, interacting **organ systems** that work in maintaining the health of the individual person. The **cardiovascular system** is a complex superhighway connecting all the systems of the body. Blood flows from the heart through blood vessels, carrying nutrients and oxygen to cells. Blood also carries wastes and carbon dioxide away from cells to the appropriate systems that remove them from the body. Blood vessels come in three types: **arteries**, **veins**, and **capillary beds**. The veins and arteries provide the transportation system connecting different organ systems of the body; the capillary beds allow for the exchange of gases, nutrients, and wastes with surrounding fluids and cells.

The blood of the human circulatory system flows in one direction and remains within the structures of the system. Hence, it is a closed circulatory system, and it is called the "cardiovascular system" because it contains a heart ("cardio-") and vessels ("vascular"). The cardiovascular system provides a means for the body to communicate and balance the needs of different organs and tissues. In this lab you will examine the vital role of the cardiovascular system in the human body.

ACTIVITY 1 Blood Flow

Blood travels from the heart through the body and then back to heart along two pathways: the **pulmonary circuit** and the **systemic circuit**. The pulmonary circuit is responsible for delivering oxygen-poor blood to the lungs, exchanging carbon dioxide for oxygen in the blood, and returning the oxygenated blood to the heart. The systemic circuit is responsible for moving the oxygenated blood all around the body and returning the deoxygenated blood to the heart to be pumped back to the lungs in the pulmonary circuit.

Arteries carry blood away from the internal chambers of the heart, and veins bring blood back to the internal chambers. In general, veins are low in oxygen content and carry blood back to the heart, from which it travels to the lungs for gas exchange; arteries generally carry oxygen-rich blood from the heart to the body. Both the arteries and the veins branch into smaller vessels, called **arterioles** and **venules**, respectively. The arterioles deliver nutrients and oxygenated blood to the capillary beds, where they diffuse into the surrounding fluid and cells. Wastes and carbon dioxide diffuse into the venule side of the capillary beds (see the bottom of Figure 16.1 to help you visualize a capillary bed).

In this activity you will examine the structures of the human cardiovascular system and trace the path of blood as it flows through this system.

1. The heart is composed of four chambers. The two on top, called the **atria** (singular "atrium"), collect the blood. The two on the bottom, the **ventricles**, pump the blood into the pulmonary and systemic circuits. In Figure 16.1, label the right and left atria and the right and left ventricles.

REMEMBER THAT THE STRUCTURES ON THE RIGHT ARE ACTUALLY NAMED "LEFT" AND STRUCTURES ON THE LEFT ARE NAMED "RIGHT" BECAUSE THESE DESIGNATIONS ARE ASSIGNED FROM THE PERSPECTIVE OF THE INDIVIDUAL.

2. The main vein, the **vena cava**, brings blood that is low in oxygen from the body to the first chamber of the heart, called the right atrium. The right atrium collects the blood and passes it to the right ventricle. The right ventricle then pumps blood to the lungs, where gas exchange occurs. The blood is pumped from the right ventricle to the lungs via the **pulmonary artery**. (This is the only artery that does not carry oxygenated blood.) In Figure 16.1, label the vena cava and the pulmonary artery.

3. After gas exchange in the lungs, the oxygenated blood is pumped into which circuit? Why?

4. In Figure 16.1, use a blue pen or pencil to trace the flow of blood with low oxygen content in both the full figure and the inset diagram of a capillary bed. Remember that oxygen-poor blood will appear in both the pulmonary and systemic circuits. At what point in the systemic circuit will oxygen-poor blood enter the bloodstream?

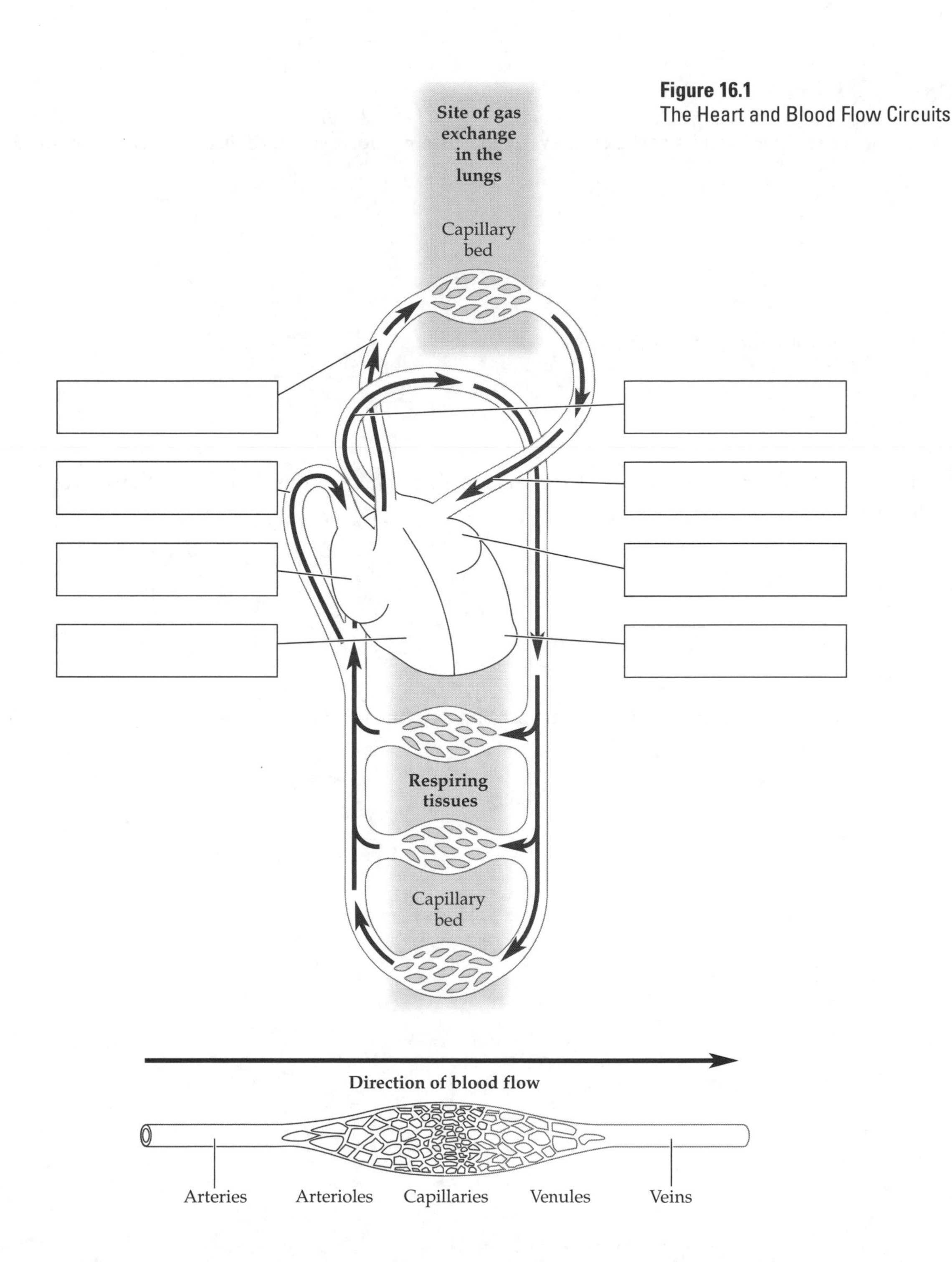

Figure 16.1
The Heart and Blood Flow Circuits

5. The **pulmonary veins** bring oxygen-rich blood from the lungs to the left atrium. (These are the only veins that contain oxygenated blood.) Blood collected in the left atrium is passed on to the left ventricle, where it is pumped into the systemic circuit via the **aorta**. In Figure 16.1, label the pulmonary veins and the aorta.

6. In Figure 16.1, use a red pen or pencil to trace the flow of oxygen-rich blood in both the full figure and the inset diagram of a capillary bed. Remember that oxygen-rich blood will appear in both the pulmonary and systemic circuits.

Concept Check

1. What are the two circuits of the cardiovascular system, and what are their respective functions?

2. Name the four chambers of the heart.

3. Describe the functions of veins and arteries, specifying whether the blood being transported by the respective vessels is oxygen-rich or oxygen-poor.

4. What are the two exceptions to arteries carrying oxygen-rich blood and veins carrying oxygen-poor blood?

ACTIVITY 2 Blockage of Blood Vessels

The vessels of the cardiovascular system are composed of three tissue types. The veins and arteries differ in composition according to the roles and stresses of each. (Arteries must endure the pressure from the heart and are therefore composed of thicker layers of tissues.) The exterior layer contains an elastic but sturdy **connective tissue** (Figure 16.2). The next layer, **smooth muscle**, contracts and relaxes to move blood along the vessels. The interior layer, consisting of epithelial cells, is called the **endothelium**.

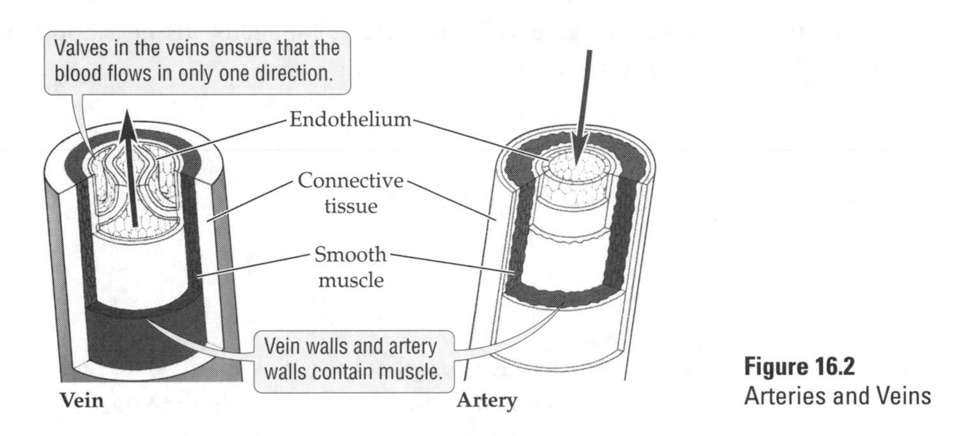

Figure 16.2
Arteries and Veins

Veins contain the same tissues but are not surrounded by as much smooth muscle as arteries are. Arteries need this smooth muscle to bounce back from the pressure exerted on them by the heart. Veins also have **valves**; these are flaps within the vessel that prevent backflow of the blood. The flow of blood through the veins is also controlled by the contraction of **skeletal muscle**, a type of voluntary muscle that moves the blood back to the heart.

Arterioles and venules contain the same types of tissues that the larger arteries and veins have, but they are smaller in diameter. Capillary beds have such a small diameter that only one red blood cell can pass through them at a time. The capillary beds have only one layer of tissue (epithelial), allowing for easy diffusion between the blood and the surrounding fluid, which is called the **interstitial fluid**.

In this activity you will examine the effects of blockage of both arteries and veins, and you will investigate how the diameter of a blood vessel affects rate of blood flow.

Arteries and Cholesterol

In addition to gases, nutrients, and wastes, the cardiovascular system moves raw materials that are needed to build and maintain cell structures. Synthesized from saturated fats that are derived from animal products like meat, cheese, and eggs, **cholesterol** is one such raw material and is a necessary component of every cell in your body. Cholesterol is a fatty compound and does not dissolve well in the watery medium of the blood. For this reason, specific proteins act as carriers of cholesterol, conveying it to cells and returning surplus cholesterol to the liver for storage. These cholesterol-carrying proteins are of two forms: **low-density lipoproteins (LDLs)** transport cholesterol to cells, and **high-density lipoproteins (HDLs)** transport cholesterol back to the liver for storage.

A healthy person has a higher percentage of HDLs compared to LDLs, implying that surplus cholesterol is being efficiently removed from the bloodstream. A shift in the balance causing the percentage of

LDLs to be higher, however, may lead to significant health problems because too much cholesterol is moving around in the blood, is not being used in cell maintenance, and is not being returned to the liver for storage.

If this is the case, the excess cholesterol in the blood may work its way underneath the endothelium of the blood vessel, forming a **plaque**. Plaques may ultimately form a constriction at a point in the blood vessel that reduces the blood flow and may cause the condition known as **atherosclerosis**. Plaques within the coronary artery cause over 80 percent of heart attacks, in which the supply of oxygenated blood to a region of heart muscle is prevented. Symptoms may include a lack of responsiveness and abnormal breathing in the individual.

1. In Figure 16.2, note the three layers of tissue within the artery: connective tissue, smooth muscle, and endothelium. Which layer is affected when plaques form?

2. Why is it so dangerous if a plaque forms in the coronary artery?

3. Have you ever tried to wash your car with a water hose that has a kink in it? Water pressure at the faucet tends to build up in response to the kink, sometimes causing the hose to become detached. Using this analogy, what may happen to blood flow if numerous "kinks" (plaques) occur in the cardiovascular system?

Veins and Clots

A clot develops when blood coagulates into a solid form, creating a blockage in a vessel, including veins. The valves within veins catch cells or other material, often after an injury that elicits an immune response to repair any break in the blood vessel. This buildup, or clot, is called a **thrombosis** (plural "thromboses"). An **embolism** is a clot that becomes detached and moves via the bloodstream to another part of the body, forming a new clot in a different part of the body. These two types of restriction in the bloodstream—thromboses and embolisms—reduce blood flow.

1. What types of cells or materials might catch on the valves of veins?

2. Oxygen-poor blood is moved back to the heart through the interaction of one-way valves in veins and skeletal muscle (Figure 16.3). Add the labels "Thrombosis" and "Potential embolism-forming clot" where appropriate on the figure.

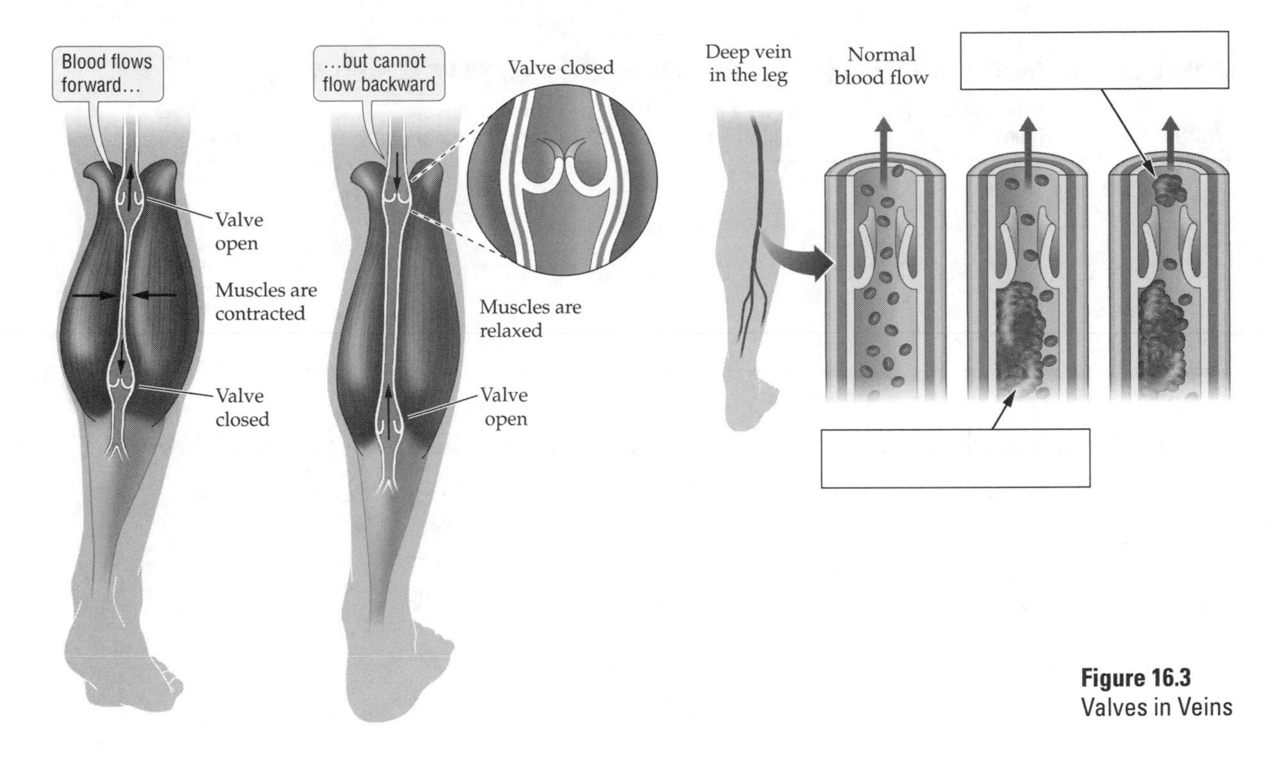

Figure 16.3
Valves in Veins

3. After a while, the thrombosis may loosen and the clot may travel through the bloodstream. Where might this clot go?

Simulating Rate of Flow in Blood Vessels

1. Obtain a stopwatch, ruler, beaker, and bottles. Your instructor will provide you with four bottles that have openings of different sizes, to simulate the range of diameters a blood vessel may have. Record the diameter of each bottle in Table 16.1.

2. Add the same amount of water (as specified by your instructor) to each bottle. Record this amount in Table 16.1, under "Amount of liquid in bottle."

3. For each bottle, determine the rate of flow by turning the bottle upside down and pouring the water into the beaker. Start the timer as soon as the bottle is turned over and stop it when the last drop of liquid exits the bottle. In Table 16.1, record the amount of time required for the water to exit the bottle.

4. For each bottle, convert the time required for the water to exit the bottle into the rate of flow by dividing the number of milliliters of water by the number of minutes. Record the results in the last column of Table 16.1.

TABLE 16.1 ACTIVITY 2: MEASURING RATE OF FLOW IN RELATION TO VESSEL DIAMETER

Bottle	Bottle diameter (mm)	Amount of liquid in bottle (ml)	Time required for water to exit (s)	Rate of flow (ml/min)
1				
2				
3				
4				

5. What can you conclude about decreased blood vessel diameter and its effect on blood flow?

Concept Check

1. Why are veins more prone to clots than arteries are?

2. Why are clots dangerous?

3. What life-threatening condition can excess LDL buildup in the bloodstream cause?

ACTIVITY 3 Heartbeat and Pulse

Two groupings of nervous tissue called nodes regulate the contraction of the heart. The atria start the contraction cycle through an electrical impulse from the **pacemaker**, also known as the **sinoatrial (SA) node**. As a result, both atria contract simultaneously. The message is then relayed to the **atrioventricular (AV) node**, and the ventricles contract simultaneously (Figure 16.4).

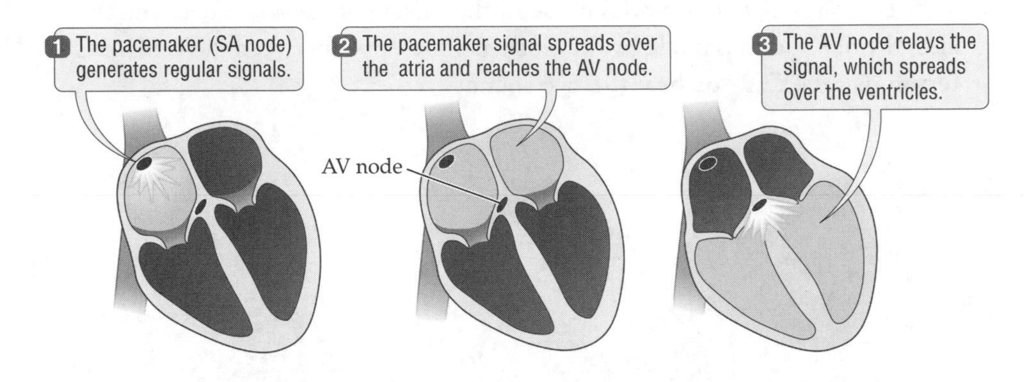

Figure 16.4
The Pacemaker

Heartbeat is a measure of the cardiac cycle of blood flowing into the relaxed heart, followed by the contraction of the atria moving the blood into the ventricles and the subsequent contraction of the ventricles moving the blood into the pulmonary and systemic circuits. The ventricles contract to push approximately 80 ml of blood out of the heart 60 to 75 times per minute on average. The cardiac cycle is typically divided into two parts: ventricular contraction is known as **systole**; ventricular relaxation, as **diastole** (Figure 16.5). During contraction, the ventricles push blood out of the heart; relaxation enables the ventricles to refill.

Pulse is a measure of the cycle in which arteries increase in diameter and then relax as a result of the blood pressure wave originating from the left ventricle during the systolic and diastolic cycles. The number of pulses per minute, or **heart rate**, indicates how fast your heart is beating, reflecting the expansion and contraction of the arteries.

In this activity you will listen to your own heartbeat and measure your own pulse.

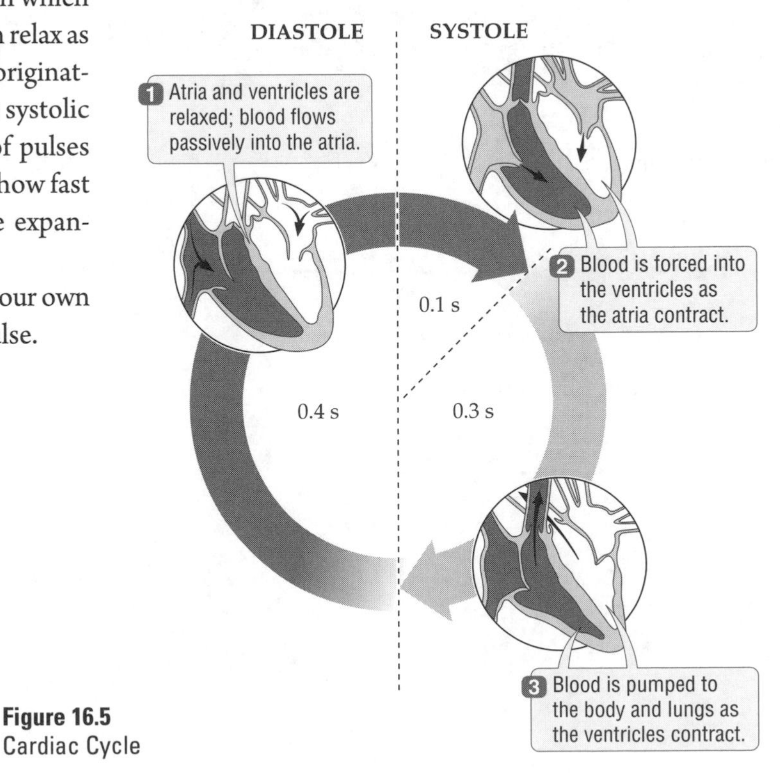

Figure 16.5
Cardiac Cycle

Heartbeat

1. Obtain a stethoscope. Wipe the earpieces with an alcohol wipe to make sure they are clean.
2. Put the earpieces in your ears. Can you hear anything?
3. Place either side of the stethoscope on your chest. Do you hear anything? If not, try the other side.
4. Listen for your heart sounds. These are often described as *lub-dub, lub-dub,* and so on. As Figure 16.6 illustrates, the *lub* sound is made by the closing of the valves between the atria and the ventricles; the *dub* sound is made by the closing of the valves between the ventricles and the pulmonary artery and aorta. Can you hear the two sounds?

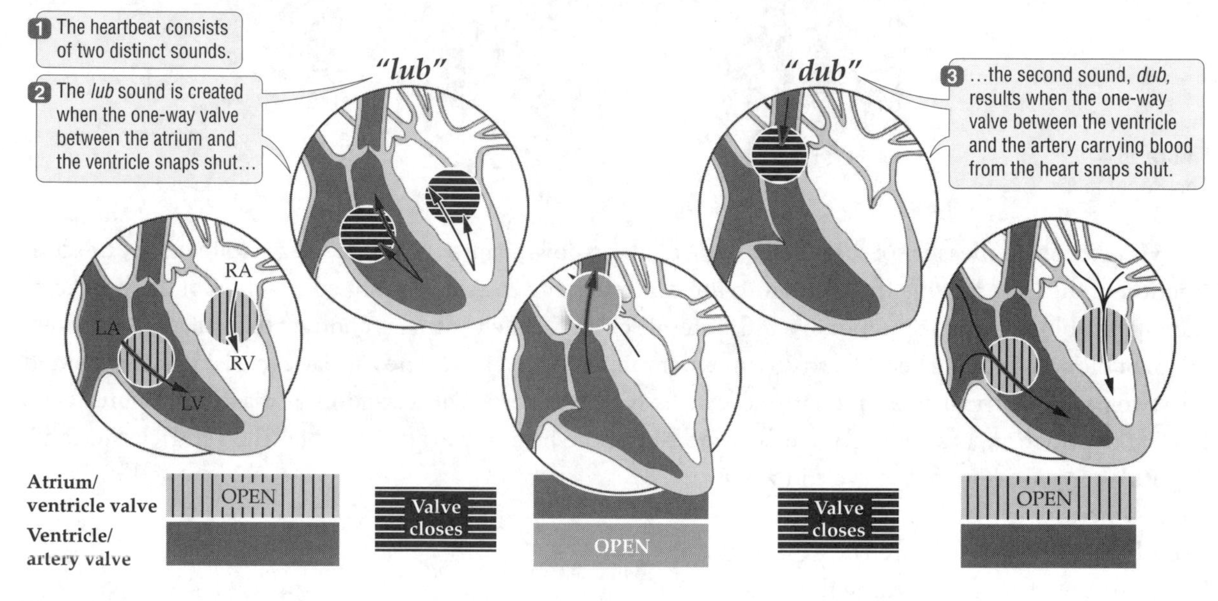

Figure 16.6
Human Heartbeat

Pulse

1. Obtain a timer or stopwatch.
2. Flip one of your hands so that it is palm side up. Slowly move the tips of your index and middle fingers on the inner part of the opposite wrist until you locate your pulse. If you have trouble locating the pulse in your wrist, try your neck. Place your index and middle fingers under your chin directly below your eye. Can you feel your pulse?
3. Measure your heart rate by counting the beats of your heart for 1 minute under the conditions listed in Table 16.2 and record your heart rate for each condition.

TABLE 16.2 ACTIVITY 3: MEASURING HEART RATE

Condition	Heart rate (beats/min)
After sitting still for 5 minutes	
Immediately after standing up	
After walking up and down one flight of stairs	
After running in place for 30 seconds	

4. How does your heart rate vary from activity to activity? Why does the heart rate change?

5. Describe what is happening in your heart and in your blood vessels when your heart rate increases.

6. Why do you think it is necessary for your heart rate to increase when you increase activity?

Concept Check

1. What are systole and diastole?

2. From the ventricles, where does the blood flow next?

3. A heart murmur can be detected as an irregular heart sound. What parts of the heart do you think a heart murmur affects?

4. What is pulse?

5. When you take your pulse, why aren't you monitoring the contraction and relaxation of the veins?

ACTIVITY 4 Blood Pressure

Blood pressure measurement detects the change in pressure in the arteries that is exerted by the heart. The systolic phase pushes blood from the heart into the arteries, increasing pressure on the arteries; the diastolic phase is the resting of the arteries while blood refills the heart (see Figure 16.5). Therefore the pressure is higher during the systolic phase and lower during the diastolic phase. The blood pressure in the systemic circuit is much higher than in the pulmonary circuit because the blood has to travel a much greater distance in the systemic circuit.

A blood pressure cuff is used to measure the changes in systolic and diastolic pressures. The average blood pressure is 120 mm (measured in mercury, which is abbreviated "Hg") and 80 mm. Blood pressure is normally expressed as systole/diastole—in this example, 120/80. Since systole exerts more pressure, it is the higher number. Figure 16.7 illustrates the changes in pressure as the blood flows through the vessels of the cardiovascular system.

In this activity you will measure blood pressure and design an experiment to test how a chosen factor affects blood pressure.

DO NOT PARTICIPATE IN THIS ACTIVITY IF YOU HAVE OR THINK YOU HAVE ANY HEART OR OTHER PHYSICAL CONDITION THAT MIGHT BE ADVERSELY AFFECTED.

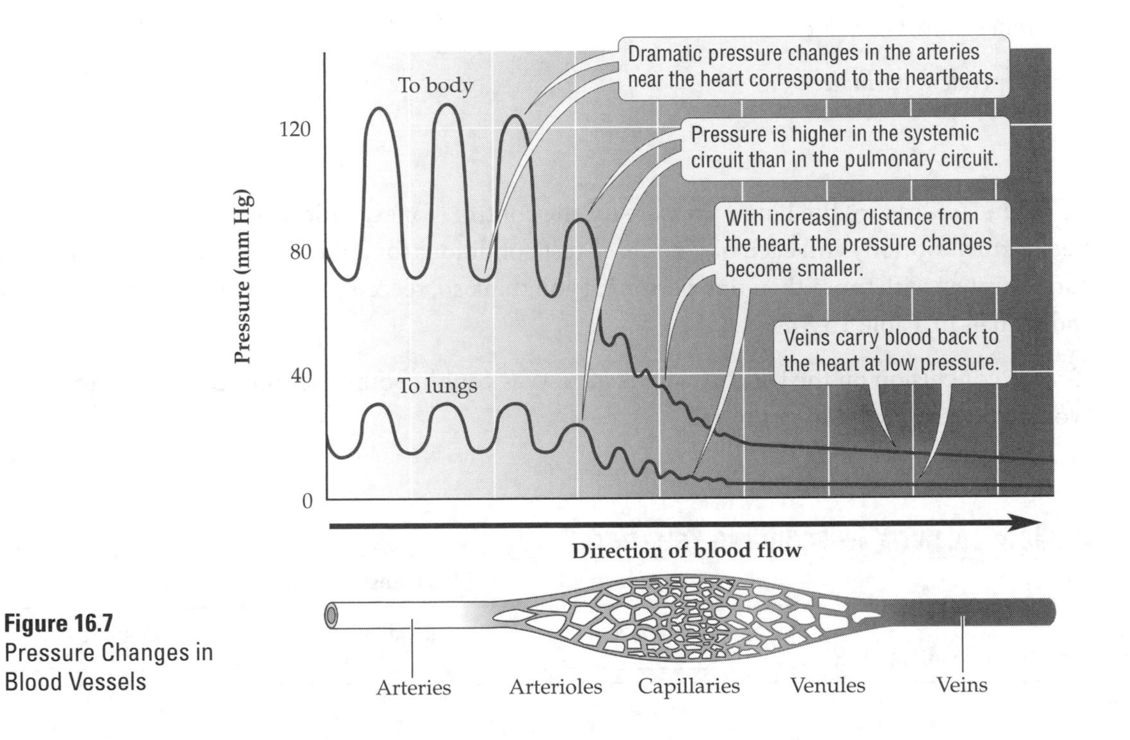

Figure 16.7
Pressure Changes in Blood Vessels

1. Obtain a manual or automatic blood pressure cuff from your instructor.
2. Design an experiment to test differences in blood pressure.

3. Determine the question you would like to answer. Here are some suggestions:
   a. Does blood pressure change if a person is lying down versus sitting?
   b. Does height affect blood pressure?
   c. Does blood pressure change if a person is standing versus sitting?
   d. Does blood pressure change after a person walks versus runs?

   In the space provided here, write the question you're going to test.

4. Formulate a hypothesis based on your question, and record it here.

5. To test your hypothesis, choose the conditions that you will measure. Will blood pressure increase, decrease, or stay the same under your given conditions? Record these conditions (A, B, and C) in Table 16.3, and write your predictions here. (For example, if you are testing the effect of body position on blood pressure, your conditions might be "A: standing," "B: lying down," and "C: bending over.")

6. Test your hypothesis, following the directions provided by your instructor on how to measure blood pressure. What is your independent variable (the variable controlled by the experimenter)? What is your dependent variable (the variable measured by the experimenter)?

   Independent variable:

   Dependent variable:

7. In Table 16.3, record the blood pressure for the control (for example, at rest if you select the "walks versus runs" option) and each of your three test conditions for a single individual. Should you perform this test with more than one person? If you think so, record your measurements for additional individuals in Table 16.3 as well.

8. Draw a conclusion on the basis of your results. Was your hypothesis supported or rejected? How would you revise your hypothesis?

TABLE 16.3 ACTIVITY 4: RECORDING YOUR DATA

	Blood pressure (mm Hg)			
Person	**Control**	**Condition A:**	**Condition B:**	**Condition C:**
1				
2				
3				

Concept Check

1. What is blood pressure?

2. What blood pressure is considered normal?

3. What are some physical characteristics that may affect blood pressure?

ACTIVITY 5 Regulation of Gaseous Exchange

The level of carbon dioxide, and hence oxygen, in our blood is monitored by different tissues to increase or decrease respiration. When tissues in the human body detect high levels of carbon dioxide, they send messages to the brain to increase respiration. When carbon dioxide levels are low (high oxygen levels), the brain receives messages to slow breathing. This process of monitoring and adjusting regulates the levels of oxygen in the blood.

The lungs are cavities that open to pull air in during **inhalation** and then collapse to push air out during **exhalation**. Lung tissue does not control this movement. The **diaphragm** is a thin muscle separating the thoracic and abdominal cavities. This muscle contracts along with muscles between the ribs, the **intercostal muscles**, to expand the thoracic cavity and the lungs. These muscles then relax, collapsing the lungs to push air back out. Other muscles aid this activity when an increased rate of breathing is needed, such as during exercise.

In this activity you will measure how long you're able to hold your breath under a variety of conditions, and you will observe the effect of low oxygen levels on respiration rate.

DO NOT PARTICIPATE IN THIS ACTIVITY IF YOU HAVE OR THINK YOU HAVE ANY HEART OR OTHER PHYSICAL CONDITION THAT MIGHT BE ADVERSELY AFFECTED.

Holding the Breath

1. Obtain a stopwatch. You and your lab partner should take turns performing steps 2 and 3, so that each have an opportunity to have your breathing tested.
2. Observe the movements of your lab partner during quiet breathing. Note your partner's physical changes during inhalation and exhalation.

3. While sitting down, measure how many seconds your lab partner can hold his or her breath (without marked discomfort!) under the following conditions:
   a. After a normal exhalation
   b. After a normal inhalation
   c. After a very deep exhalation
   d. After a very deep inhalation
   e. After several deep inhalations

   Record your results in Table 16.4.

TABLE 16.4 ACTIVITY 5: MEASURING BREATH-HOLDING CAPACITY

Condition	Amount of time my lab partner's breath was held (s)	Amount of time my breath was held (s)
After a normal exhalation		
After a normal inhalation		
After a very deep exhalation		
After a very deep inhalation		
After several deep inhalations		

4. Now switch roles with your partner and repeat steps 2 and 3. Make sure you record your own measurements as well in the appropriate column in Table 16.4.
5. Using the information provided in the introduction to this activity, how would you explain your results?

6. How soon after taking a very deep breath while sitting does your partner breathe again?

7. Cessation of breathing, called **apnea**, often occurs immediately after deep breathing. Why?

Effect of Low Oxygen Levels on Respiration Rate

BEFORE BEGINNING THIS PART OF THE ACTIVITY, WAIT 5 MINUTES UNTIL THE BREATH HAS RETURNED TO NORMAL.

1. Obtain a brown paper bag. You and your lab partner should take turns performing steps 2 and 3, so that you each have an opportunity to have your breathing tested.
2. Count how many breaths your partner takes in a 2-minute interval while breathing normally.
3. Now have your partner cover both nose and mouth with the bag so that there are no air leaks. Instruct your partner to breathe deeply into the bag for 2 or 3 minutes and count how many breaths are taken in a 2-minute interval. How is the rate of respiration affected?
4. Now switch roles with your partner and note the number of breaths you take in a 2-minute interval while breathing normally and while breathing into the bag.
5. Why are the respiration rates while breathing into the paper bag different? Can you think of examples when this occurs naturally?
6. What causes the variation in respiration between you and your partner?

Concept Check

1. Pregnancy puts a lot of pressure on a woman's abdominal and thoracic cavities. How does pregnancy affect breathing in the female?
2. Is it possible to hold your breath indefinitely? If not, why not?
3. Apnea occurs because of an increase in what gas?
4. Why does climbing a flight of stairs require a higher breathing rate?

Key Terms

aorta (p. 16-3)
apnea (p. 16-16)
arteriole (p. 16-2)
artery (p. 16-1)
atherosclerosis (p. 16-6)
atrioventricular (AV) node (p. 16-9)
atrium (p. 16-2)
blood pressure (p. 16-13)
capillary bed (p. 16-1)
cardiovascular system (p. 16-1)
cholesterol (p. 16-5)
connective tissue (p. 16-5)
diaphragm (p. 16-15)
diastole (p. 16-9)
embolism (p. 16-6)
endothelium (p. 16-5)
exhalation (p. 16-15)
heart rate (p. 16-9)
heartbeat (p. 16-9)
high-density lipoprotein (HDL) (p. 16-5)
inhalation (p. 16-15)
intercostal muscles (p. 16-15)
interstitial fluid (p. 16-5)
low-density lipoprotein (LDL) (p. 16-5)
organ system (p. 16-1)
pacemaker (p. 16-9)
plaque (p. 16-6)
pulmonary artery (p. 16-2)
pulmonary circuit (p. 16-2)
pulmonary vein (p. 16-3)
pulse (p. 16-9)
sinoatrial (SA) node (p. 16-9)
skeletal muscle (p. 16-5)
smooth muscle (p. 16-5)
systemic circuit (p. 16-2)
systole (p. 16-9)
thrombosis (p. 16-6)
valve (p. 16-5)
vein (p. 16-1)
vena cava (p. 16-2)
ventricle (p. 16-2)
venule (p. 16-2)

Review Questions

1. What chambers of the heart are responsible for pulmonary circulation?

2. What chambers of the heart are responsible for systemic circulation?

3. What are the structural differences between veins and arteries?

4. The ________________ are the heart chambers that collect blood; the ________________ are the heart chambers that pump blood.

5. The arteries branch into smaller-diameter vessels called ________________.

6. The veins branch into smaller-diameter vessels called ________________.

7. ________________ carry cholesterol to cells; ________________ deliver cholesterol to the liver.

8. Explain how a heart attack occurs.

9. Describe a thrombosis and the danger it may cause if the clot breaks off and moves through the bloodstream.

10. __________________ is the contraction of the ventricles; __________________ is the relaxation of the ventricles.

11. What causes heart sounds?

12. Describe the differences between pulse, heartbeat, and blood pressure.

13. What structure(s) are responsible for blood pressure and pulse?

14. How do oxygen and carbon dioxide levels regulate respiration?

15. How does increased respiration affect circulation?

16. __________________ is the cessation of breathing due to overventilation.

17. Capillary beds exchange nutrients and oxygen for wastes and carbon dioxide with the __________________ surrounding cells.

18. Which ventricle can you deduce to be larger? Why?

19. The __________________ triggers contraction of the atria; the __________________ triggers contraction of the ventricles.

20. Within the lungs, on which end of the capillary bed would carbon dioxide diffuse out of the capillaries? On which end of the capillary bed would oxygen diffuse into the capillaries?

LAB 17

Senses and Reflexes

Objectives

- Examine the structure and function of the organs that comprise the five human senses and appreciate their combined role in the nervous system.
- Identify the different types of sensory receptors and explore the processes by which sensory receptors relay information to the brain about what the body is perceiving.
- Experience the blind spot that exists in human vision because of the presence of the optic nerve in the photoreceptor-rich retina.
- Describe the pathways of a reflex arc and the distinction between the central nervous system and peripheral nervous system.

Introduction

The specialized cells of the nervous system, **sensory receptors** that include sensory neurons, enable humans to detect a stimulus. A **stimulus** is any change to the body or environment generating a signal by a cell. The outcome of these signals are your senses of **taste**, **touch**, **smell**, **hearing**, and **vision**. The senses first detect a change through a sensory receptor cell, sending this signal to an intermediate neuron, called an **interneuron**. The interneuron relays the signal either to the brain, which interprets the message; or directly to a motor neuron, causing a reflex arc; or to both the brain and a motor neuron. A sensory receptor is typically a type of **sensory neuron**, but it can also be a specialized cell, parts of a cell, or a group of cells working closely with a sensory neuron.

Sensory receptors detecting these messages are **chemoreceptors**, **mechanoreceptors**, **photoreceptors**, **pain receptors**, or **thermoreceptors**. Each sense uses one or more of these receptor types, as described in Table 17.1.

TABLE 17.1 Different Ways to Sense the World

Receptor type	Stimulus	Sense
Chemoreceptors	Chemicals	Taste, smell
Mechanoreceptors	Physical changes	Touch, hearing, proprioception (body position), balance
Thermoreceptors	Moderate heat and cold	Thermoreception (gradations of heat and cold)
Pain receptors	Injury, noxious chemicals, chemical and physical irritants	Pain, itch
Photoreceptors	Light	Vision
Electroreceptors*	Electrical fields (especially those generated by muscle contractions of other animals)	Electrical sense
Magnetoreceptors*	Magnetic fields (mainly of Earth)	Magnetic sense

*The sensory receptors listed here are found in many animals, including most vertebrates. Those marked with an asterisk, however, are not found in humans.

The nervous system is composed of interacting neurons that form the **peripheral nervous system (PNS)** and the **central nervous system (CNS)**. The PNS contains the sensory neurons and the motor neurons; the CNS consists of the brain and the spinal cord (Figure 17.1). The sensory neurons of the PNS detect stimuli and pass these on to the spinal cord or brain, and the motor neurons produce reactions directed either automatically via the spinal cord (as in the reflex arc) or consciously by the brain. The CNS integrates and processes the messages sent by the PNS.

In this lab you will examine, through personal experience, how external stimuli are received and acted upon by the PNS and CNS of your nervous system.

FIGURE 17.1
The Human Nervous System Comprises the PNS and the CNS

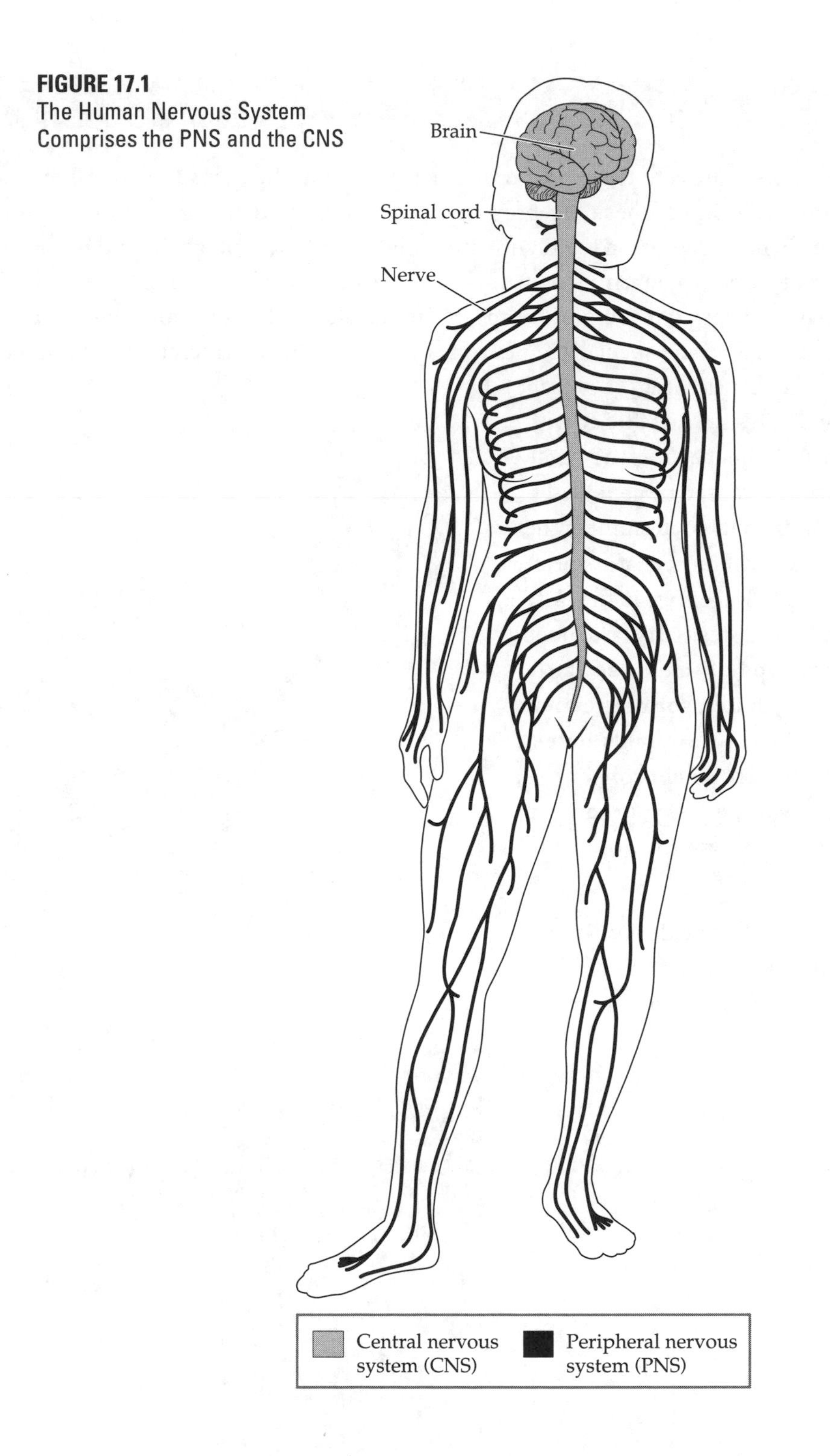

ACTIVITY 1 Taste

Humans use chemoreceptors to detect the molecules present in the items they consume. Scientists hypothesize that the evolution of these receptors enabled animals to detect good or bad/poisonous food. Therefore, these chemoreceptors in the mouth are evolutionarily advantageous and persisted in the population. Chemoreceptors are concentrated in pockets of the human tongue called **taste buds** (Figure 17.2). When a substance touches your tongue, the specific molecules of that substance bind according to their shape or characteristics to the specialized neurons in the taste buds, which then send a message to the brain telling it what chemical type is present.

Not everyone tastes the same thing with the same intensity when eating the same food. This variation is caused by genetic differences among individuals determining the types and amounts of chemoreceptors present in their taste buds. Receptors in the taste buds detect four basic tastes: **salty** taste buds clearly detect any type of salt, **sweet** taste buds detect carbohydrates, **sour** taste buds detect acids, and **bitter** taste buds detect alkaloids such as those found in coffee.

In this activity you will use your sense of taste to identify the flavors in four substances.

1. Your instructor will provide you with four different substances, none of which are harmful.

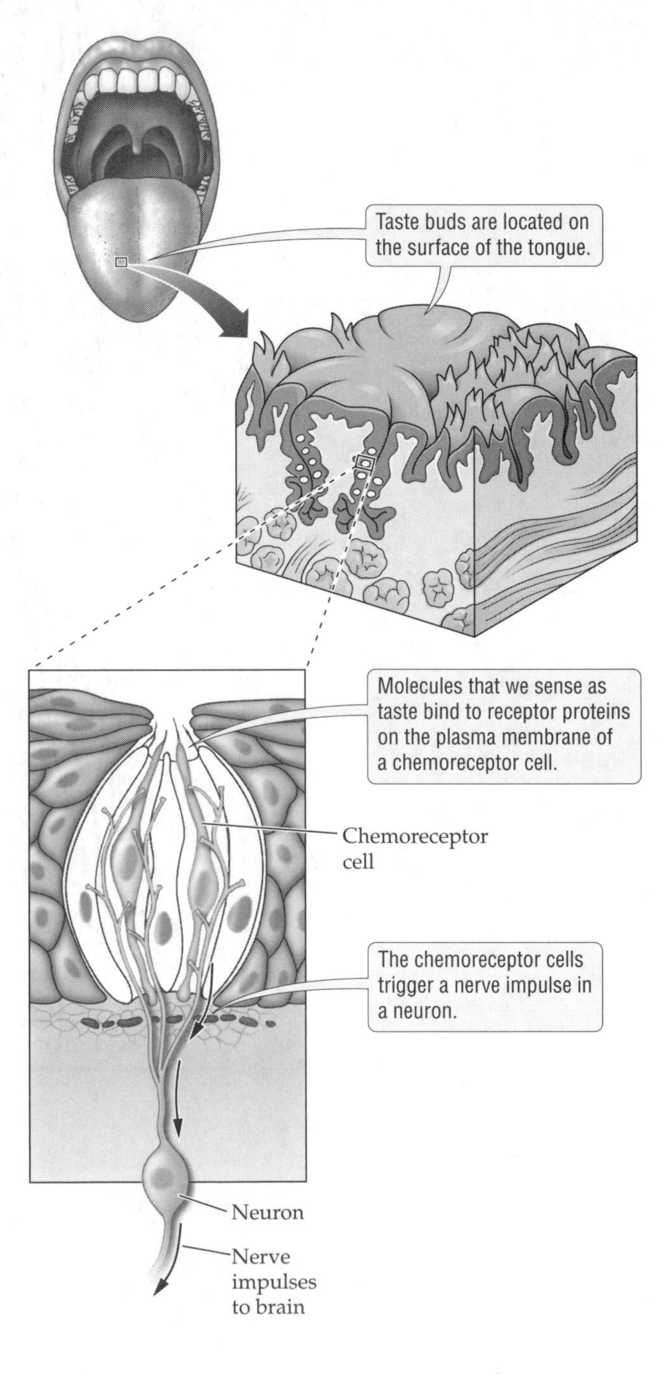

FIGURE 17.2
Taste Buds

2. Consume a sample of each substance and record in Table 17.2 the taste that you detect for each. Did you dislike the taste of any of the substances?

TABLE 17.2 Activity 1: Tasting Substances

Substance	Salt	Sweet	Sour	Bitter
1				
2				
3				
4				

3. After everyone has completed the activity, compare your results with the rest of the class. Was everyone able to identify the basic tastes of all the substances?

Concept Check

1. What two flavor(s) do you think might indicate a poisonous or dangerous food?

2. What two flavor(s) do you think would indicate a good food?

3. How do chemoreceptors distinguish between a carbohydrate and an acid?

ACTIVITY 2 Smell

Smell helps animals detect chemical stimuli that are helpful in finding food, mates, predators, or prey. Molecules present in the air reach us through the process of diffusion. They bind to the chemoreceptors present in hairs projecting from cells in the nasal passage (Figure 17.3). The receptors are located on hair to increase the likelihood of exposure to the molecule. As with taste, the genetic makeup of the individual determines the types and the density of chemoreceptors present in the nose, and therefore the sensitivity of the individual. The message is relayed to the brain and interpreted as a specific scent.

In this activity you will use your sense of smell to detect and identify an odor.

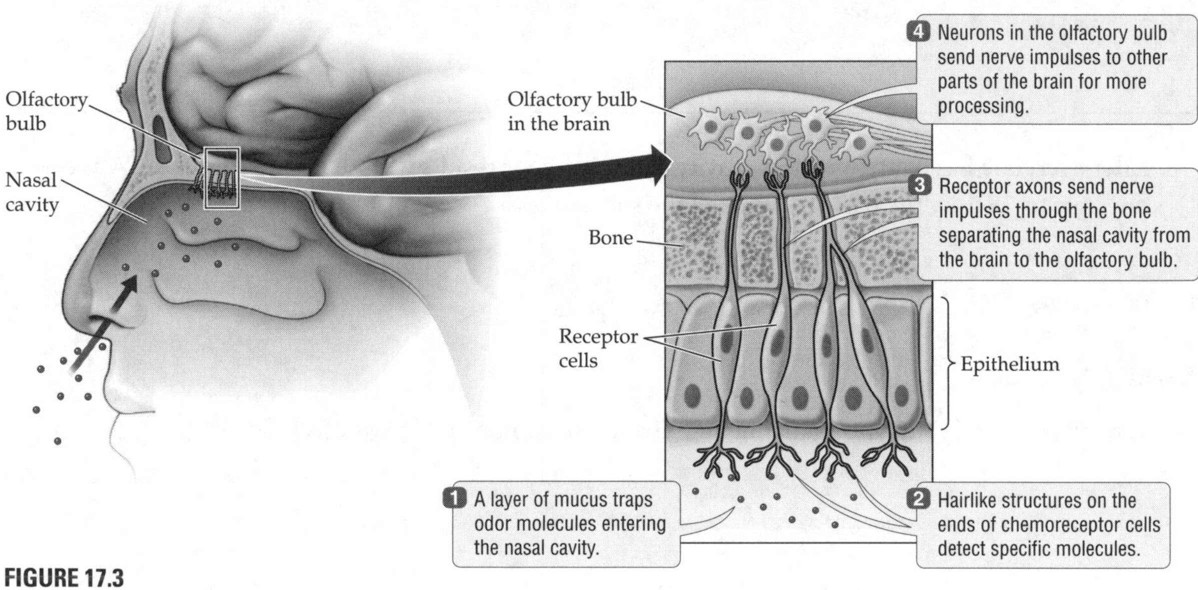

FIGURE 17.3
Chemoreceptors of the Nose

1. Your instructor will provide you with four different substances, none of which are harmful. All of these are common substances that you have probably smelled in the past.
2. Obtain a ruler and a timer. Place the bottle of substance A 1 foot from the edge of your lab bench and position your chair so that you're sitting at the edge of the lab bench.
3. Open the bottle and start the timer. When you are able to smell the odor from the bottle, note in Table 17.3 how much time has elapsed. Do not stop the timer.
4. Can you identify the odor? Keep the timer running until you are able to identify the odor and record in Table 17.3 how much time has elapsed.

TABLE 17.3 Activity 2: Smelling Substances

Substance	Time at detection	Time at identification
A		
B		
C		
D		

5. Repeat steps 2 through 4 for each of the remaining three bottles. Did you dislike the smell of any of the substances? Were some stronger than others?

6. Compare your results with those of the rest of the class. Was everyone able to identify the substances? Who was the most sensitive?

Concept Check

1. Why do some people have a better sense of smell than others?

2. Why does having a cold affect an individual's sense of smell?

3. Give an example of information about your surroundings that is provided by smell alone.

ACTIVITY 3 Touch

Nerve impulses generated by touch mechanoreceptors located all over the body enable an animal to detect physical change. Humans perceive a range of pressures, from a slight breeze to a punch on the arm, through the presence of different types and densities of mechanoreceptors. The upper portion of the skin is the epidermis, where hairs protrude; sensitive touch receptors are just below this layer in the dermis (Figure 17.4). Deeper down in the hypodermis lie less sensitive touch and pain receptors, along with the base of the hair, called the follicle. Some mechanoreceptors surround the hair follicle, detecting gentle changes in pressure, such as a light breeze. Other mechanoreceptors are found at the boundary between the dermis and hypodermis, providing a sense of more intense pressure on the body. Like hair, mechanoreceptors are present at different densities throughout the body.

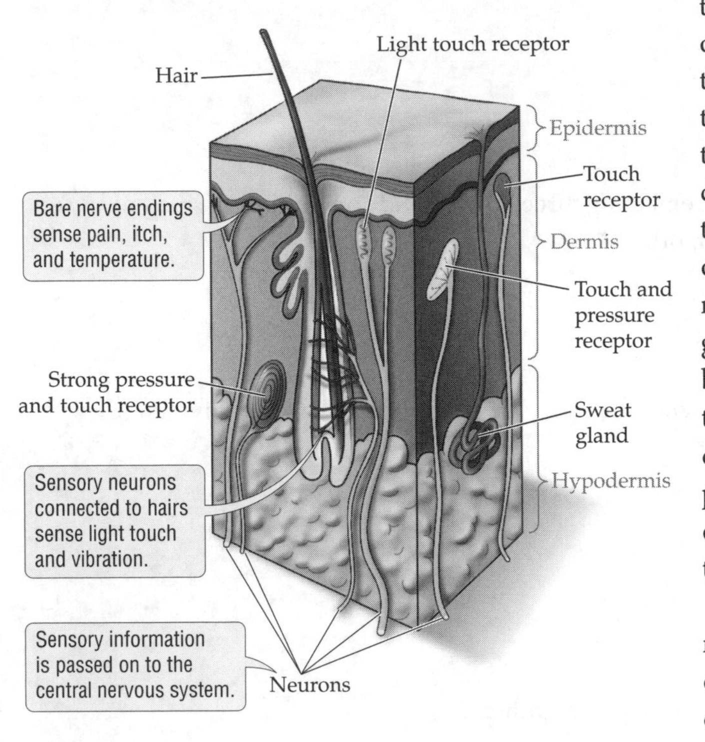

FIGURE 17.4
Receptors of the Skin

The two-point discrimination test is one means of observing the density of mechanoreceptors. In this activity you will use this test to determine the distance between mechanoreceptors in the skin.

1. Obtain a pair of scissors and a ruler. You and your lab partner should take turns performing steps 2 through 4, so that you each have an opportunity to experience the touch sensations.

2. Instruct your lab partner to close his/her eyes. For each area of skin listed in Table 17.4, gently hold the points of the scissors on the given area, with the scissors closed. Ask your partner to tell you whether he/she experiences one or two touch sensations.

3. Open the distance between the scissor points slightly and again ask your partner to tell you whether he/she experiences one or two touch sensations.

4. Repeat step 3 until your partner reports feeling two sensations instead of one. As soon as your partner detects two points, measure the distance between the tips of the scissors and record this value in Table 17.4.

5. Now switch roles with your partner. Make sure you record your own measurements as well in the appropriate column in Table 17.4.

6. When you detect one point, only one touch receptor receives the message. When you detect two points, more than one touch receptor receives the message. This test determines the density of nerve endings (touch receptors) in an area of your body, which also identifies the sensitivity of the area. Do the areas in Table 17.4 have different densities of touch receptors?

7. Compare your values with those of other students in class. Are they the same?

TABLE 17.4 Activity 3: Two-Point Touch Discrimination

Area of the body	My two-point discrimination distance (mm)	My partner's two-point discrimination distance (mm)
Inside of forearm		
Back of neck		
Back of palm		
Surface of palm		
Palmar surface of index finger		
Back side of index finger		

Concept Check

1. Why would different areas of the skin have different densities of mechanoreceptors?

2. Why are there different types of mechanoreceptors?

3. Which areas of the body would you expect to have particularly high or low densities of mechanoreceptors?

ACTIVITY 4 Hearing

The ability to hear is based on a complex system of specialized structures formed to concentrate sound. This system includes structures such as membranes, hairs, and mechanoreceptors that convert sound to a chemical signal. A **sound wave** is the change in air pressure resulting from the collision of molecules. The human ear detects the frequency of a sound wave (**pitch**) and the intensity of the wave (**loudness**). The **basilar membrane** of the inner ear vibrates more or less in relation to the loudness. By contrast, pitch is determined by a specific area of the membrane vibrating to a specific frequency of a sound wave.

In this activity you will identify the structures of the ear to understand their function and perform some tests on hearing acuity.

Structures of the Ear

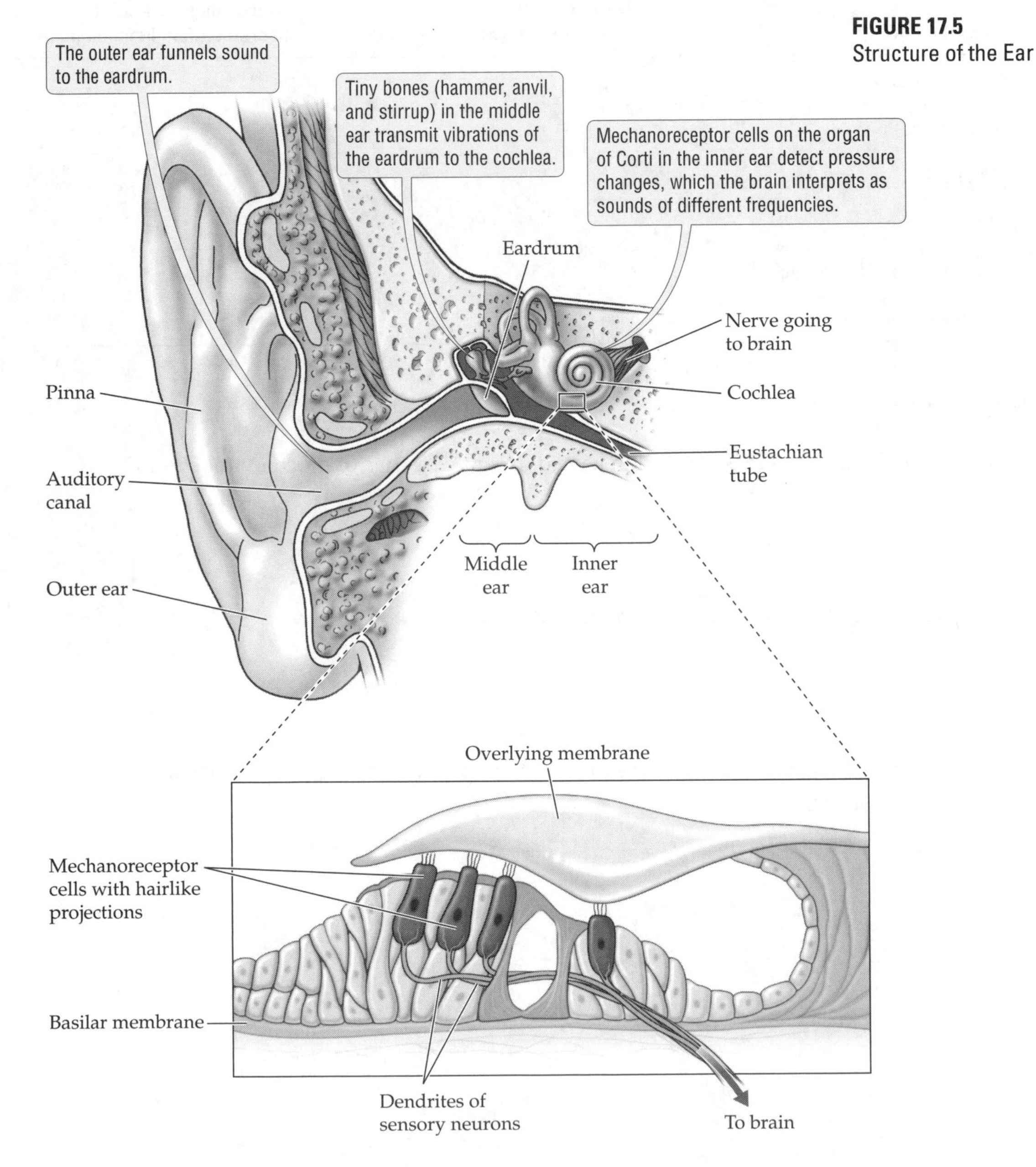

FIGURE 17.5
Structure of the Ear

1. The **outer ear** is composed of the **auditory canal** and the **pinna**. The pinna funnels the sound waves into the ear through the auditory canal. Using Figure 17.5 and a model of the ear if your instructor has one, locate the end of the auditory canal.
2. A membrane at the inner end of the auditory canal, the **eardrum**, converts the sound waves into a physical movement. This movement is passed on to the structures of the **middle ear**.
3. Locate the tube in the middle ear that connects the ear to the throat. This tube, the **eustachian tube**, allows the pressure on either side of the eardrum to be equal. Have you ever experienced your ears "popping"? This phenomenon is caused by an imbalance of pressure on either side of the eardrum due to altitude changes.
4. The vibration from the sound waves pass into the middle ear, which contains three small bones—**hammer**, **anvil**, and **stirrup**—which transmit the signal to another membrane that relays the physical movement to the **cochlea**. Locate the structure that looks like a snail shell. This is the cochlea, within the **inner ear**.
5. Within the cochlea, the **organ of Corti** converts the mechanical vibrations to a chemical signal. The organ of Corti is a complex structure containing mechanoreceptors with hairlike projections and the basilar membrane. The vibrations from the initial sound wave that entered the auditory canal force the basilar membrane against the overlying membrane, in turn forcing the hairlike projections to initiate action potentials in sensory neurons that ultimately send the signal to your brain.
6. Now that you have traced the process of hearing, describe the function of each structure in Table 17.5.

TABLE 17.5 Activity 4: Structures of the Ear

Structure	Function
Auditory canal	
Cochlea	
Eardrum	
Eustachian tube	
Middle ear	
Organ of Corti	
Pinna	

Auditory Acuity

You have probably experienced a reduction in your ability to hear a number of times during your life. This may have been while coming in to land after an airplane flight, where the rapid descent has not allowed your eustachian tubes to equalize the air pressure on either side of your eardrums. Your hearing may have been impinged when you suffered your last cold or if you had too much wax in your auditory canal. These are, thankfully, temporary causes of hearing impairment; other, more long-term causes may be structural damage caused by injury or disease.

Recall that your hearing depends on the detection of sound waves that are directed down the auditory canal by the pinna, converted into a vibration at the eardrum that passes the stimulus through the middle-ear bones, and finally to hairlike mechanoreceptors within the organ of Corti. These mechanoreceptors then convert the stimulus into a chemical impulse that is passed on to your brain.

Here you will conduct a basic auditory acuity test for both of your ears.

1. Obtain balls of cotton and silver coins (dimes, nickels, or quarters) from your instructor.
   You and your lab partner should take turns performing steps 2 through 5, so that you each have an opportunity to test your auditory acuity.
2. Have your partner place a cotton ball in his or her left ear.
3. *Lightly* tap two silver coins together right next to your partner's right ear. Slowly move away from your partner's right ear while continually tapping the two coins. Stop when your partner says the tapping is no longer audible.
4. Using a meter stick, measure the distance from your partner's right ear to where you are now standing. Record this distance in Table 17.6.
5. Repeat steps 2 through 4 for your lab partner's left ear, switching the cotton ball to the right ear. Record the distance in Table 17.6.
6. Now switch roles with your partner and repeat steps 2 through 5. Make sure you record your own measurements as well in the appropriate column in Table 17.6.
7. Do both of your ears have the same auditory acuity? Do both of your partner's ears have the same auditory acuity?

TABLE 17.6 Activity 4: Measuring Auditory Acuity

	Left-ear distance (cm)	Right-ear distance (cm)
My ears		
My lab partner's ears		

Concept Check

1. How do mechanoreceptors detect sound?

2. Why is the eustachian tube important?

3. What is the difference between pitch and loudness?

4. Name two factors that can reduce auditory acuity.

ACTIVITY 5 Vision

The human eyes are complex structures specialized for focusing and detecting light. Their design includes a **lens**, which bends light, projecting the image onto the back surface of the eyeball containing photoreceptors. Photoreceptors are the sensory receptors of the eye, converting light energy into a chemical signal. The **optic nerve** carries the signal to the brain, where it is processed.

In this activity you will identify the structures described in the text and perform a test of visual acuity to understand the visual ability of each eye.

Structures of the Eye

1. Referring to Figure 17.6, along with a model of the eye if provided by your instructor, identify the structures of the eye. Starting from the outside, a thin membrane called the **cornea** covers the eye.

2. The most obvious part of the eye, the **iris**, is often a characteristic used to describe a person's appearance. The iris is the colored portion of the eye, which is determined by the presence of different pigments.

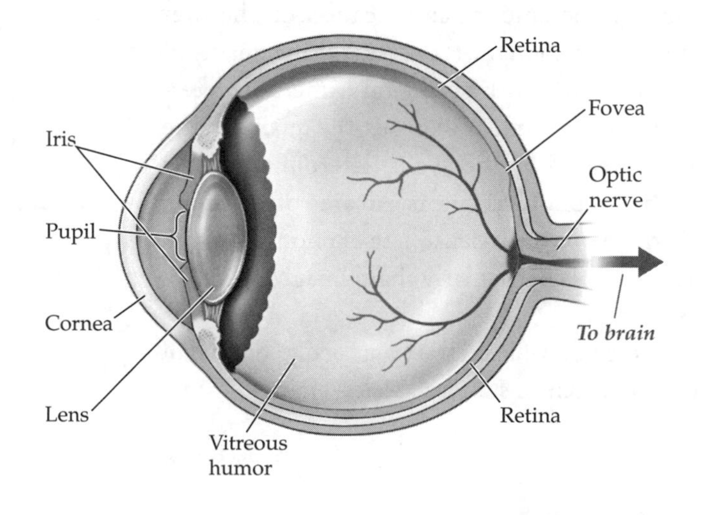

FIGURE 17.6
Structure of the Eye

3. Can you locate the opening of the iris? This is the **pupil**. Pupils change size depending on the lighting. The iris adjusts to increase or decrease the diameter of the pupil, allowing more light into the eye in dark conditions and less light into the eye in bright conditions.
4. The light reflected from an image is collected onto the lens, where the image appears inverted because of the bending of the light. Can you locate the lens? The lens is adjusted to focus the image onto the tissue of the **retina** in the back of the eyeball.
5. A fluid, the **vitreous humor**, fills the inside of the eyeball. The light image passes through this fluid to the back of the eyeball onto the retina. This is where the image is translated from light energy into chemical energy through the photoreceptors.
6. The chemical signal created by the photoreceptors is sent to the brain through the neurons of the optic nerve. Where is the optic nerve located?

Visual Acuity

Many structures are integral to providing the sharpest image possible. **Visual acuity** is a measure of the eye's ability to focus. As illustrated in Figure 17.7, muscles in the eye change the shape of the lens to focus the image directly onto the retina, which contains the photoreceptors. If the lens bends the light too much or not enough, then images may appear blurry at certain distances. External lenses, such as glasses or contacts, compensate for such lens problems.

Other factors affecting the quality of the image that is received, processed, and sent to the brain are the amount and location of photoreceptors present in the retina. Just as the quality of an image created by a digital camera is determined by the number of pixels, the quality of the image created by the eye is determined by the number of photoreceptors. An area of the retina called the **fovea** is a dense area of photoreceptors that provides particularly sharp images (see Figure 17.6).

Here you will test the visual acuity of each of your eyes using a standard vision chart.

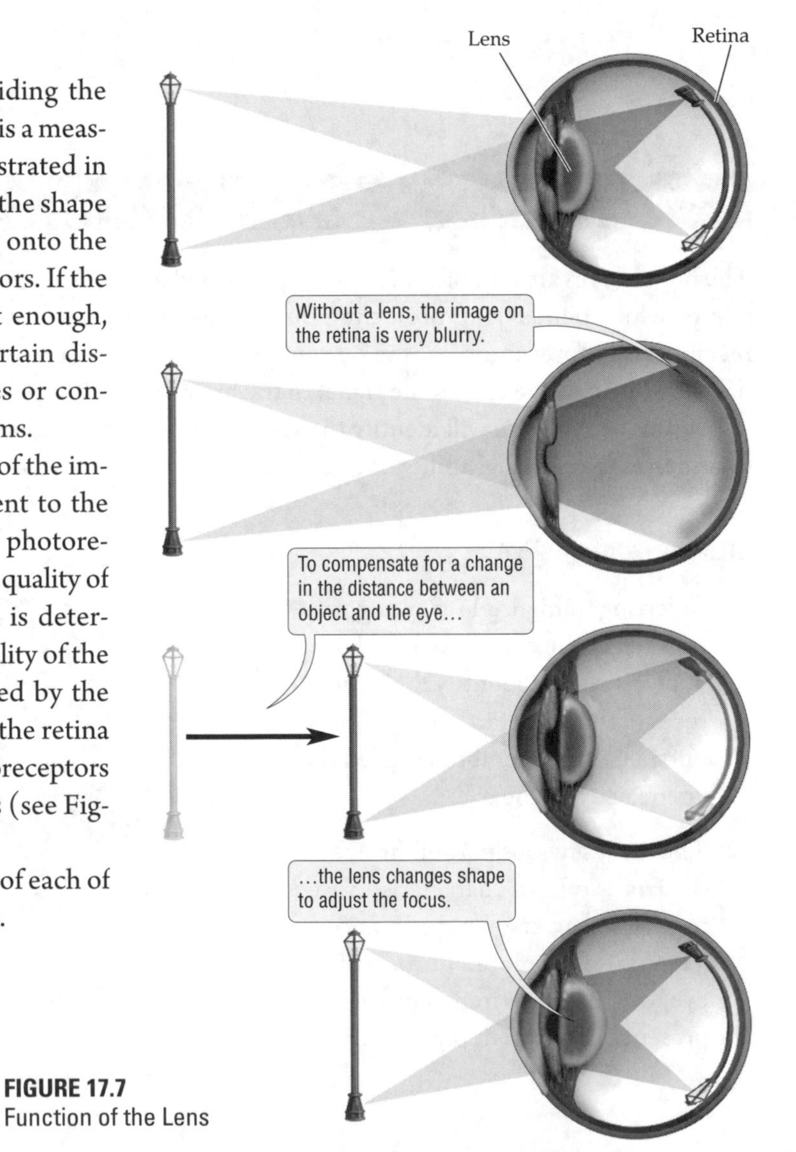

FIGURE 17.7
Function of the Lens

1. You and your lab partner should take turns performing steps 2 and 3, so that you each have an opportunity to test your vision.
2. If your partner has glasses on, instruct him/her to take them off. Have your partner stand 20 feet from the eye chart and cover his/her right eye. Starting at the top of the chart, have your partner read the letters on each line from left to right. What is the last line your partner can read without making a mistake? In Table 17.7, record the ratio that is listed to the right of this line on the eye chart. This is the **visual acuity score**. The top number (numerator) is the distance you are standing from the eye chart: 20 for 20 feet. The bottom number (denominator) is the distance at which a person with average visual acuity could read that line. For example, if line 5 is the last line your partner can read, then your partner's visual acuity score is 20/40. A person with 20/20 vision would be able to read line 5 from a distance of 40 feet.
3. Repeat step 2 with your lab partner, covering the left eye this time. In Table 17.7, record the acuity results for your partner's right eye.
4. Now switch roles with your partner. Make sure you record your own measurements as well in the appropriate column in Table 17.7.
5. Do your two eyes have the same visual acuity? Do your partner's two eyes have the same visual acuity?
6. The average human has 20/20 vision. Can a person have greater than 20/20 vision—for example, 20/10?

TABLE 17.7 Activity 5: Measuring Visual Acuity

	Visual acuity (ratio from eye chart)	
	Left eye	**Right eye**
My visual acuity		
My partner's visual acuity		

Concept Check

1. What structures of the eye affect visual acuity? Describe the functions of each.

2. What type of cells convert light energy into chemical energy in the eye?

3. Which structure are we referring to when we talk about eye color?

4. What factors account for differences in the visual acuity scores of different people?

ACTIVITY 6 Blind Spot

A photoreceptor is a cell that responds to light. The human retina has two types of photoreceptors: rods and cones. **Rods** detect changes in the level of light and are more sensitive to different shades of gray. **Cones** detect colors. Both types of photoreceptors relay the image from the eye to the brain by way of the optic nerve. The optic nerve exits the retina and contains no photoreceptors (see Figure 17.6). Because the lack of photoreceptors prevents light detection where the optic nerve exits, this area is referred to as the **blind spot**.

In this activity you will determine the location of the blind spot in each of your eyes.

1. Your instructor will provide you with a piece of paper and a ruler. The paper will have a dot and an "X" a few inches apart. You and your partner should take turns performing steps 2 and 3, so that you each have an opportunity to determine the distance of your blind spot for each eye. The distance represents the location of your blind spot in your field of view.

2. If you wear glasses, keep them on. Position the paper so that the "X" is on the right and the dot is on the left. Hold the paper 1 foot away from your eyes. Close your *right eye only*. Stare at *only* the "X" with your *left* eye. Slowly move the paper toward you until the dot disappears, at which point have your partner measure the distance from your eye to the piece of paper. Record this distance in Table 17.8.
3. Repeat step 2 for your right eye and record the results in Table 17.8.
4. Now switch roles with your partner. Be sure to record the data for both of your lab partner's eyes as well as your own in the appropriate columns in Table 17.8.
5. Compare your results with those of the rest of the class. Does everyone have the same distance to the blind spot? Why or why not?

TABLE 17.8 Activity 6: Finding the Blind Spot

	Distance from eye to paper at blind spot (cm)	
	Left eye	**Right eye**
My blind spot distance		
My partner's blind spot distance		

Concept Check

1. Why does a blind spot exist?

2. Why does the blind spot occur only when you hold the paper a certain distance from your face?

3. How would the loss of an eye affect the impact of your blind spot on your overall vision?

ACTIVITY 7 Reflex Arc

The body does not require that sensory signals be processed in the brain in order for a reaction to occur. Sometimes it is necessary for your body to respond to its environment without even communicating with the brain. A **reflex** is an effortless or unconscious reaction elicited directly by sensory stimulation. Reflexes are also known as **spinal reflexes** or **reflex arcs**. In a reflex reaction (Figure 17.8), sensory neurons relay information to interneurons, which then send the message directly to motor neurons (instead of to the brain).

In this activity you will test your body's ability to react unconsciously by testing your patella and foot reflexes.

1. Obtain a rubber mallet. You and your partner should take turns performing steps 2 and 3 so that you each have an opportunity to experience the reflex arc.

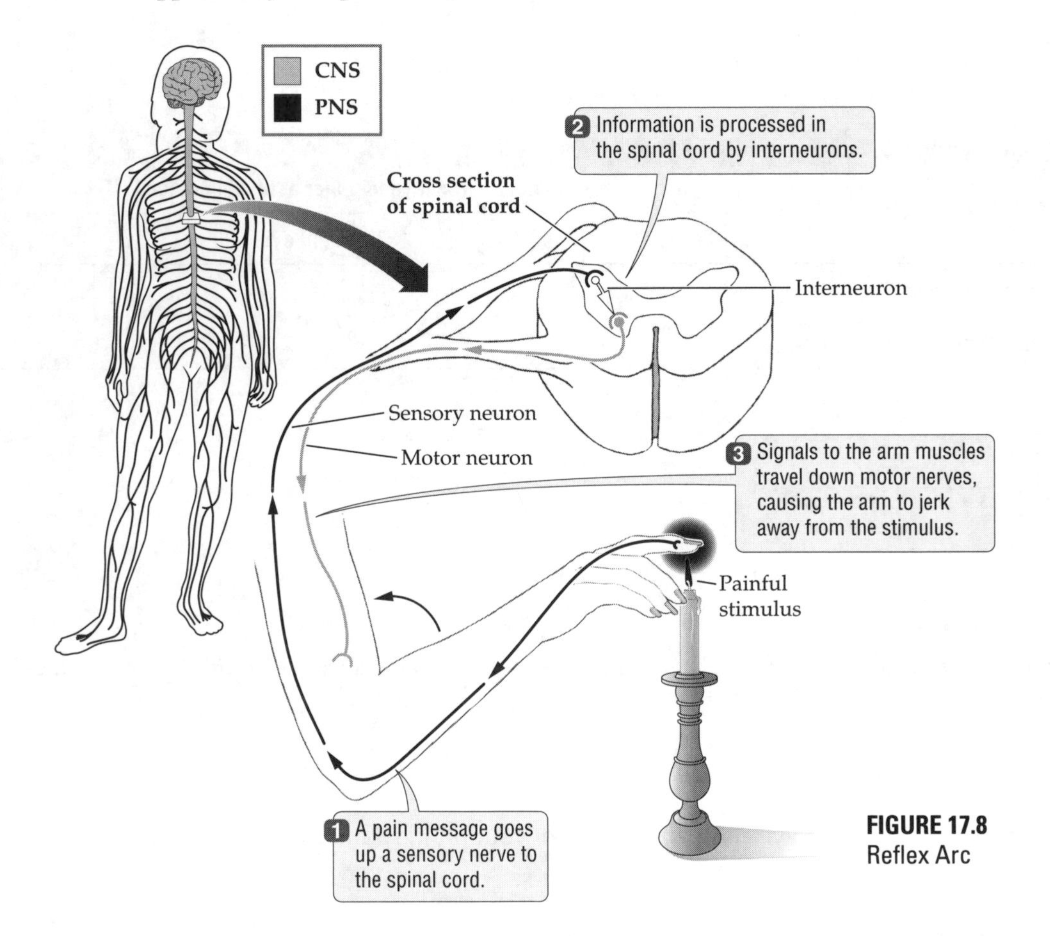

FIGURE 17.8
Reflex Arc

2. Have your partner sit on a chair with his/her knee at the edge dangling freely. Use your fingers to feel the bottom edge of your partner's kneecap. Gently tap the area directly below the kneecap with the rubber mallet. Did your partner's leg move? Try again until it does. You are tapping the patellar ("patella" = kneecap) tendons. Tendons attach muscles to bone.

3. Have your partner remove a shoe. Use your fingers to feel the back of your partner's foot; find the top edge (toward the head) of the bone along the back of the heel. Using the rubber mallet, gently tap the area directly above the bone, which contains the soft Achilles tendon. Did your partner's foot move? Try again until it does. It may take a few attempts to locate the correct area that will elicit the reflex.
4. Now switch roles with your partner and repeat steps 2 and 3.

Concept Check

1. What is the role of motor neurons in a reflex arc?

2. What is the advantage of reflex arcs?

3. Other than during today's activity or at a doctor's office, have you ever experienced a reflex? If so, describe what happened.

Key Terms

anvil (p. 17-11)
auditory canal (p. 17-11)
basilar membrane (p. 17-10)
bitter (p. 17-4)
blind spot (p. 17-16)
central nervous system (CNS) (p. 17-2)
chemoreceptor (p. 17-1)
cochlea (p. 17-11)
cone (p. 17-16)
cornea (p. 17-13)
eardrum (p. 17-11)
eustachian tube (p. 17-11)
fovea (p. 17-14)
hammer (p. 17-11)
hearing (p. 17-1)
inner ear (p. 17-11)
interneuron (p. 17-1)
iris (p. 17-13)
lens (p. 17-13)
loudness (p. 17-10)
mechanoreceptor (p. 17-1)
middle ear (p. 17-11)
optic nerve (p. 17-13)
organ of Corti (p. 17-11)
outer ear (p. 17-11)
pain receptor (p. 17-1)
peripheral nervous system (PNS) (p. 17-2)
photoreceptor (p. 17-1)
pinna (p. 17-11)
pitch (p. 17-10)
pupil (p. 17-14)
reflex (p. 17-18)
reflex arc (p. 17-18)
retina (p. 17-14)
rod (p. 17-16)
salty (p. 17-4)
sensory neuron (p. 17-1)
sensory receptor (p. 17-1)
smell (p. 17-1)
sound wave (p. 17-10)
sour (p. 17-4)
spinal reflex (p. 17-18)
stimulus (p. 17-1)
stirrup (p. 17-11)
sweet (p. 17-4)
taste (p. 17-1)
taste bud (p. 17-4)
thermoreceptor (p. 17-1)
touch (p. 17-1)
vision (p. 17-1)
visual acuity (p. 17-14)
visual acuity score (p. 17-15)
vitreous humor (p. 17-14)

Review Questions

1. The five major senses are ________________, ________________, ________________, ________________, and ________________.
2. Which sense(s) use mechanoreceptors?
3. Which sense(s) use chemoreceptors?
4. Which sense(s) use photoreceptors?
5. The ________________ contain the chemoreceptors that detect molecules in food.
6. Which four tastes can a human detect?
7. Where are the chemoreceptors located in the nasal passage?
8. Why are there different types of mechanoreceptors to detect pressure on the human body?
9. How does hair detect wind?
10. The ________________ funnels sound into the ________________ and to the eardrum.
11. The ________________ contains the organ of Corti, where the mechanoreceptors of the ear are located.
12. In a reflex arc, a stimulus is received by a(n) ________________, which relays information to a(n) ________________, which relays information to a(n) ________________.
13. The ________________ contains the photoreceptors of the eye.

14. Where does an image enter the eye?

15. Which structures of the eye aid in producing a sharp image?

16. Can you name an instance in which two or more senses act together to cause a reaction?

17. What are the two types of photoreceptors in the eyes, and how do they function?

LAB 18 Human Urinary and Reproductive Systems

Objectives

- Identify the structures of the urinary and reproductive systems in humans.
- Compare the functions of male and female urinary and reproductive systems.
- Recognize how hormones regulate the menstrual cycle, and describe how birth control affects this cycle.
- Simulate the transmission of disease through the exchange of body fluids and observe firsthand how easy it is for diseases to spread in this way.

Introduction

The urinary and reproductive systems are closely connected. In human males, these two systems use some of the same structures; in human females, these two systems are separate. The **urinary system** removes wastes in the bloodstream created during metabolic activity. The wastes are then stored until enough urine has accumulated to trigger its release. The **reproductive system** produces gametes: either **sperm** or **eggs**. If a sperm fuses with an egg, fertilization occurs and the reproductive structures of the female provide a development chamber for the growing fetus. A pregnant female also provides all of the necessary nutrients and oxygen to the fetus while removing all of the wastes generated by the fetus.

In this lab you will examine the structures of the urinary and reproductive systems, learn about the menstrual cycle and birth control, and examine how the exchange of body fluids can facilitate the transmission of disease.

ACTIVITY 1 Urinary System

The human body functions on a daily basis through the coordinated effort of trillions of cells working together to perform millions of chemical reactions every day. These reactions require components supplied by the digestive and respiratory systems. In the end, these reactions produce necessary products and also wastes. The urinary system functions to rid the body of wastes deposited into the bloodstream by cells throughout the body.

In this activity you will examine human models to learn about the structures of the urinary system and their specific functions. The models you will use display the normal structures of the human urinary system.

1. Obtain female and male human models showing the urinary system.

2. Note whether the model you're examining is a male or female. Locate the two large **kidneys** on the model. The kidneys are composed of thousands of tubules, which filter blood, concentrating waste in the form of **urine** to be excreted. The female and male both filter blood in the kidneys.

3. Do you see the glands "capping" the kidney? These are the **adrenal glands**. They release hormones such as epinephrine (adrenaline), norepinephrine (noradrenaline), and cortisone. These hormones regulate the body's response to stress by signaling the liver to release high levels of glucose into the bloodstream, providing the body with energy to initiate the fight-or-flight response.

4. Find the tubes leading out of the kidneys. These are the **ureters**. The smooth muscle surrounding these tubes contracts in waves (peristalsis) to move urine from the kidney directly to the top of the bladder, a muscular bag shaped like a triangle.

5. The **bladder** stores urine until the accumulation of urine signals the brain to tell the bladder that it is ready to release its urine. A human physically controls the release of urine from the bladder through a tube called the **urethra**. At the base of the bladder and the start of the urethra lies the **internal sphincter**. A sphincter is a circular muscle like a doughnut that regulates the passage of a substance. The internal sphincter is composed of smooth muscle and is under involuntary control. Locate the internal sphincter on your model.

6. Farther down the urethra is another sphincter, called the **external sphincter**; it is composed of skeletal muscle and is therefore under voluntary control. After consciously deciding to release urine, a person relaxes the external sphincter, triggering the automatic relaxation of the internal sphincter. The internal sphincter can also act on its own. If the bladder has become too full, the internal sphincter will relax automatically, releasing the external sphincter. How do you think the size of an individual's bladder affects how often he or she will urinate?

7. Be sure to examine both the male and the female models. On the male model, look for the urethra as it exits the bladder and enters the penis. The many structures that surround the urethra on its way to the penis secrete fluids into the urethra to create semen. You may examine these in Activity 2. On the female model, look for the urethra as it exits the bladder. In humans, the female urethra remains separate from the reproductive system and exits the body on its own. Are the male and female urethras the same length? Why or why not?

Concept Check

1. How does a human release urine voluntarily?

2. What is the function of the kidneys?

3. What is the difference between the urethra and the ureters?

4. What is the general function of the urinary system?

ACTIVITY 2 Reproductive System

In this activity you will examine the structures and functions of the human female and male reproductive systems.

Female

The female reproductive system (Figure 18.1) remains a separate entity from the urinary system. The **ovaries** are located on either side of the spinal cord. The ovaries produce the female gamete: the **ovum** (plural "ova"), or egg. The two ovaries take turns releasing an egg each month into the **oviduct** (also known as the **fallopian tube**), which carries the egg or developing embryo to the **uterus**. If the egg is fertilized in the oviduct, the developing embryo will implant into the uterus, where it will grow into a fetus. The fetus will then pass through the **cervix** and into the **vagina** during birth.

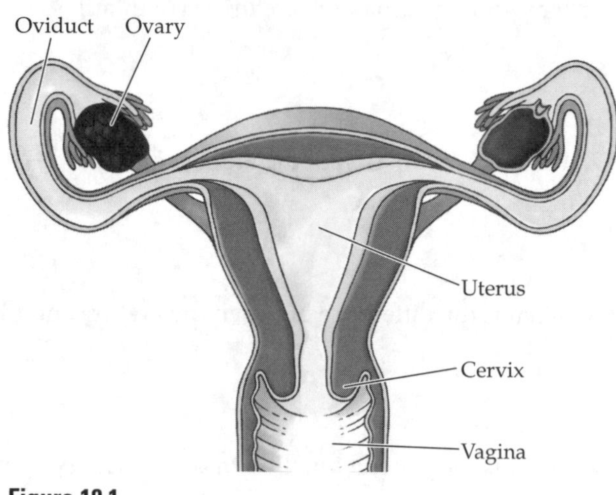

Figure 18.1
Female Reproductive Anatomy

1. Obtain a model of the human female that shows the reproductive system.
2. On the model, locate the ovaries. From the ovaries, follow the oviducts. Do these tubes resemble any other structures in this region?
3. The oviducts deliver the egg or developing embryo to the uterus. What do you notice about the structure of the uterus?
4. From the uterus, follow the chamber to the exterior by tracing the path a fetus would take during childbirth. Within this passage, identify the cervix and the vagina. A fetus ready for birth is clearly larger than the cervix and vagina. What qualities do you think allow these structures to recover from childbirth?

Male

The male reproductive system (Figure 18.2) is a bit more complex, with many more structures. Each male **testis** (plural "testes") resides in a **scrotal sac** exterior to the abdominal cavity. Producing sperm is a temperature-sensitive process, so the scrotal sac regulates temperature by moving the testis closer to or farther away from the male's body. Within the testes, sperm are produced. The sperm move into the **epididymis**, where they mature and are stored until ejaculation. During ejaculation, sperm move through the tube called the **vas deferens** to the urethra.

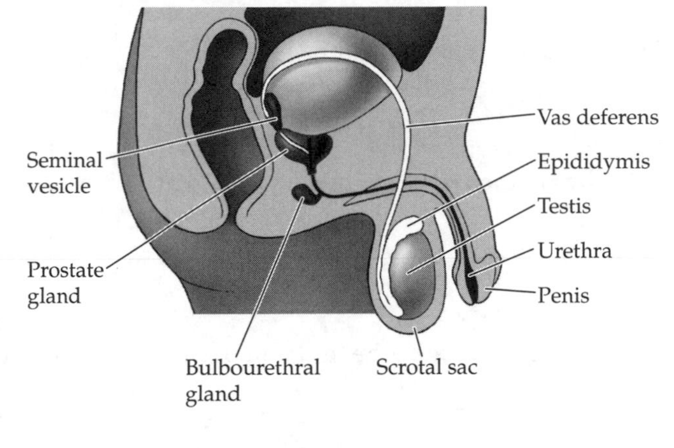

Figure 18.2
Male Reproductive Anatomy

Along the way to the **penis**, the material for semen is provided by three main glands: the **seminal vesicles**, the **prostate gland**, and the **bulbourethral gland**. Semen is a mixture of sperm, nutrients, an alkaline fluid, and a lubricant. The seminal vesicles add nutrients to help sperm survive the long journey searching for an egg within the female reproductive tract. The prostate gland produces an alkaline (basic) fluid to help sperm survive in the environment of the female reproductive tract, which is acidic to prevent infection. Lastly, the bulbourethral gland secretes a lubricant to decrease friction within the urethra.

1. Obtain a model of the human male that shows the reproductive system.

2. On the model, trace the process of sperm and semen production. Locate the scrotal sacs containing the testes. The testes contain many tubules where sperm are generated.

3. Do you see the cap surrounding the testes? This is the epididymis. What is the function of the epididymis?

4. Find the vas deferens as it exits the epididymis toward the penis. Just as the male testis is analogous to the female ovary, the male vas deferens is analogous to the oviduct of the female reproductive system. What does the vas deferens connect to next?

5. Do you see the glands along the way to the penis? First are the small seminal vesicles; next, the large reddish-colored gland surrounding the urethra is the prostate; and finally, the bulbourethral gland is situated just before the urethra enters the penis.

Concept Check

1. Where are eggs produced? Where is sperm produced?

2. Where does an egg become fertilized?

3. Why are the testes located in the scrotal sacs outside of the abdomen?

4. Name and describe the function of each component of semen.

ACTIVITY 3 Menstrual Cycle and Birth Control

The production of female gametes (eggs) is tied to the cycle that prepares a woman's womb for the development of a baby. This closely regulated cycle is called the **menstrual cycle**. During the menstrual cycle, the tissue of the uterus thickens and then is shed each month. At the same time, an egg matures in the ovaries each month and is subsequently released for fertilization. How does the body know when to build tissue or release an egg?

Hormones regulate this cycle. A hormone is a signaling molecule produced by specialized tissues or glands and distributed throughout the body. Two different sets of hormones control this cycle: the **gonadotropins** (luteinizing hormone and follicle-stimulating hormone) and the **ovarian hormones** (estrogens and progesterone). *These hormones are present throughout the menstrual cycle, but their specific levels correspond to distinct physical changes during the course of the cycle.* The average woman's menstrual cycle is 28 days long, but a menstrual cycle can deviate from the norm for a variety of reasons, including genetics and stress.

A major medical innovation was the advent of birth control medications. They prevent unwanted pregnancies and provide more stable menstrual cycles. In this activity you will analyze the changes in hormone levels over the course of the menstrual cycle and consider how birth control medications regulate the menstrual cycle.

Natural Menstrual Cycle (without Birth Control)

Follow the steps below to trace the activity of hormones throughout the menstrual cycle.

Days 0–5

During the first five days, the lining of the uterus is shed. Within the brain, the pituitary gland produces two gonadotropins: follicle-stimulating hormone (FSH) and luteinizing hormone (LH). These hormones are produced at the beginning of the menstrual cycle. FSH and LH stimulate the start of egg maturation in the ovary. The developing egg and the surrounding cells are called the **follicle**. The follicle produces an **estrogen**, and as more follicle cells are produced, more of the estrogen is produced.

1. In Figure 18.3a, measure the height of LH and FSH (in millimeters) at day 0 and day 5. Record these measurements in Table 18.1 (p. 18-10). In Figure 18.3b, do the same for the estrogen and progesterone levels. Which hormone levels are changing?

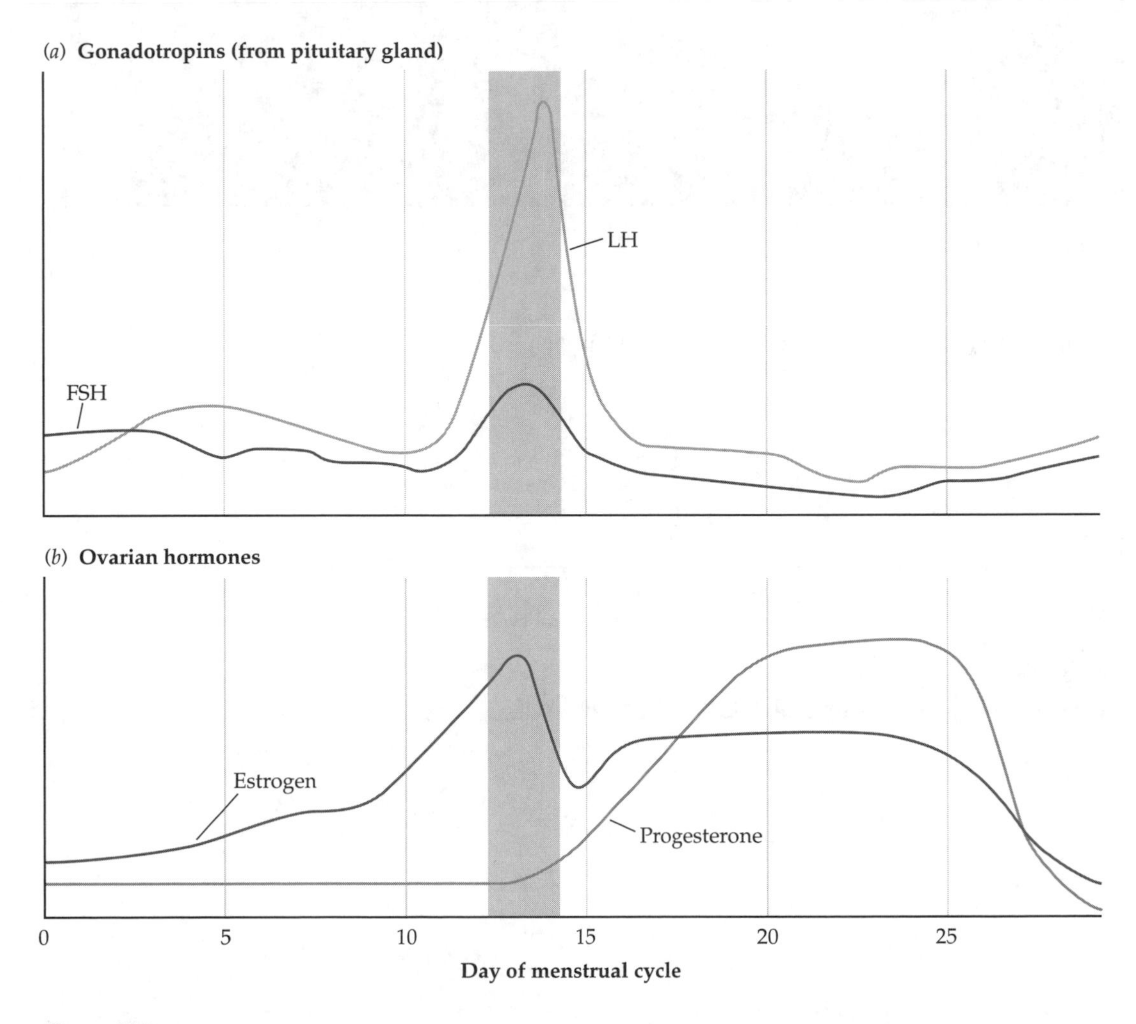

Figure 18.3
Changes in Gonadotropins and Ovarian Hormones during the Menstrual Cycle

2. Examine the events in the lining of the uterus and in the ovary depicted in Figure 18.4. Describe what is happening in the lining of the uterus and in the ovary from day 0 to day 5.

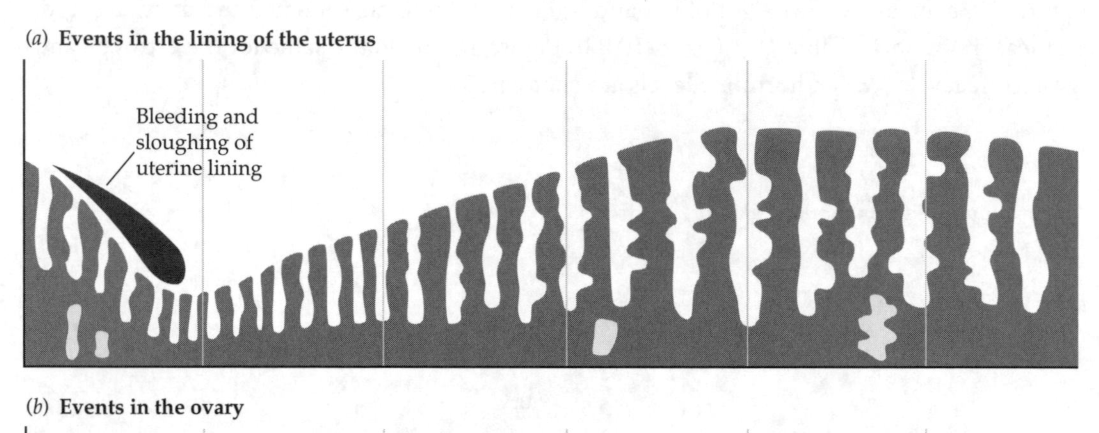

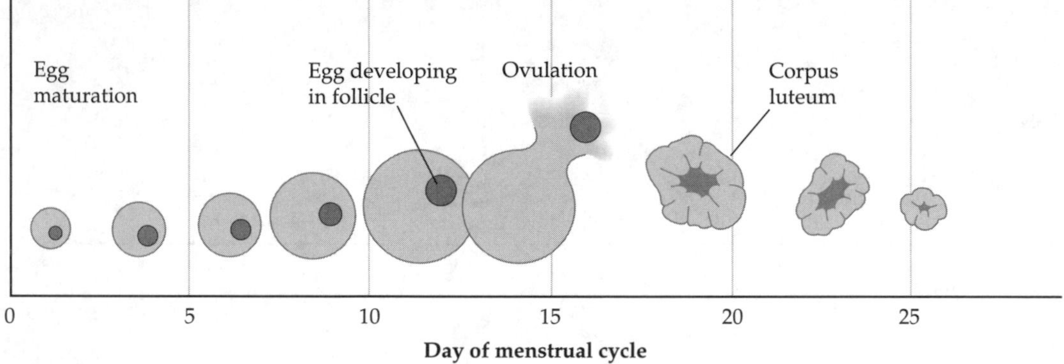

Figure 18.4
Changes in the Uterus and Ovary during the Menstrual Cycle

Days 6–15

During days 6–15, the lining of the uterus rebuilds. Until we approach day 15, estrogen production by the cells surrounding the developing egg suppresses the production of FSH and LH to lower background levels. At 15 days, the lining of the uterus is ready to accept a developing embryo and the egg is ready to be fertilized by a sperm. When the estrogen level reaches a threshold concentration, it causes a spike in FSH and LH, thereby stimulating the ovary to release the egg into the oviduct (**ovulation**).

3. In Figure 18.3a, measure the height of LH and FSH (in millimeters) at day 10 and then at the maximum height of each hormone during the highlighted spike time (at about day 12 or 13). Record these numbers in Table 18.1. In Figure 18.3b, do the same for the estrogen and progesterone levels. Which hormone levels are changing?

4. Examine the events in the lining of the uterus and in the ovary depicted in Figure 18.4. Describe what is happening in the lining of the uterus and in the ovary from day 6 to day 15.

Days 16–20

During days 16–20, FSH and LH return to low background levels while estrogen remains at an intermediate level. After the ovary releases the egg into the oviduct, the follicle cells that surrounded the egg until ovulation remain in the ovary. These cells develop into the **corpus luteum**. The corpus luteum produces **progesterone**, which, along with estrogens, causes the uterus to secrete mucus to trap any embryo that may have formed.

5. In Figure 18.3a, measure the height of LH and FSH (in millimeters) at day 20. Record these numbers in Table 18.1. In Figure 18.3b, do the same for the estrogen and progesterone levels. Which hormone levels are changing?

6. Examine the events in the lining of the uterus and in the ovary depicted in Figure 18.4. Describe what is happening in the lining of the uterus and in the ovary from day 16 to day 20.

Days 21–28

During days 21–28, if the egg is not fertilized during its journey down the oviduct and therefore no embryo embeds itself in the uterus, progesterone and estrogen levels start to diminish as the corpus luteum degrades. The secretion of substances by the uterus starts to slow. At the end of 28 days, the corpus luteum is completely degraded and progesterone and estrogen levels are now back to what they were at the beginning of the cycle.

7. In Figure 18.3a, measure the height of LH and FSH (in millimeters) at day 25 and the last day of the cycle at 28 days. Record these numbers in Table 18.1. In Figure 18.3b, do the same for the estrogen and progesterone levels. Which hormone levels are changing?

8. Examine the events in the lining of the uterus and in the ovary depicted in Figure 18.4. Describe what is happening in the lining of the uterus and in the ovary from day 21 to day 28.

TABLE 18.1 HORMONE REGULATION OF THE HUMAN MENSTRUAL CYCLE

	Measurement of hormone level (mm)			
Day	**LH**	**FSH**	**Estrogen**	**Progesterone**
0				
5				
10				
Spike (about day 12 or 13)				
20				
25				
28				

Birth Control

Birth control is a regimen of ovarian hormones during the first 3 weeks of the menstrual cycle. Here you will consider the composition and function of birth control.

1. At the end of the natural menstrual cycle, both progesterone and estrogen levels drop, initiating the increase in LH and FSH to start a new cycle. The primary function of LH and FSH is to initiate development of the egg and control release of the egg from the ovary during the spike around day 12 (see Figure 18.3a). From the measurements you entered in Table 18.1, when in the menstrual cycle is the FSH level at its lowest (*hint:* the lowest height measured)? What is the level of LH at this time?

2. At this time the levels of LH and FSH are not regulating the menstrual cycle. What are the estrogen and progesterone levels? How are they affecting the uterus and egg at this time?

Birth Control Weeks 1–3

3. A variety of birth control medications are available, but ultimately they all have the same objective: to suppress LH and FSH levels, as in the stage of the menstrual cycle that you identified in the previous part of this activity. The first 3 weeks of birth control provide a means to prevent egg development and to prevent the release of an egg from the ovary. From your answers above, which hormone(s) do you think birth control medications contain and why?

4. Which stage of the natural menstrual cycle do the first 3 weeks of birth control mimic? What do you think is happening to the uterine lining of a woman on birth control during these weeks?

5. How do you think birth control affects what is happening to the egg in the ovary during the first 3 weeks of the menstrual cycle?

Birth Control Week 4

During the fourth week of birth control, a **placebo** (a pill containing no drug), is administered instead of a hormone. As a result, ovarian hormone levels drop significantly.

6. What effect does this drop in ovarian hormones have on LH and FSH? What happens to the uterine lining?

7. Reexamine the events in the ovary shown in Figure 18.4b. How long does it take for an egg to mature fully and be ready for ovulation?

8. Considering this information, will an egg develop by the end of the week in which the birth control placebo is taken?

9. How does the start of the ovarian hormones again prevent ovulation even if an egg has developed?

Concept Check

1. What is a hormone?

2. Which hormones does the pituitary gland produce?

3. What are the physiological effects of high LH and FSH levels?

4. What are the physiological effects of very high estrogen levels?

5. What are the physiological effects of high progesterone levels?

ACTIVITY 4 Sexually Transmitted Disease

During sexual intercourse, body fluids are exchanged through secretions in the form of semen and fluids within the female reproductive tract. This exchange allows the potential not only for a sperm to fertilize an egg, but also for the transmission of infectious diseases known as **sexually transmitted diseases (STDs)**. STDs typically require the exchange of body fluids, which can occur during sexual intercourse or other sexual acts, during the development of a fetus in a mother's uterus, or during the accidental exchange of body fluids such as through the reuse of contaminated needles. Oral contraceptives, along with other birth control treatments, do not prevent body fluids from being exchanged. Condoms significantly reduce the risk of spreading an STD, but they are not 100 percent effective.

The effects of widespread unprotected sex and the reuse of infected needles continue to plague countries around the world as they battle the epidemics of STDs such as herpes, chlamydia, syphilis, and HIV (human immunodeficiency virus) and HPV (human papilloma virus) infections. In the United States, 50 percent of people have contracted or will contract an STD at some point in their lives. Although many STDs are treatable, medications can treat only the symptoms of herpes and infections caused by HPV or HIV but not cure the disease. These STDs are particularly dangerous because they can lead to the development of other infectious diseases or cancer, eventually causing death. Current rates of HIV in Washington, DC, are as high as 3 percent (affecting about 18,000 people). Diseases caused by viruses such as HIV are very hard to treat because the virus continually mutates, preventing medications from working.

One way the human body attacks infections is to produce **antibodies** (specialized proteins) to detect and eliminate a disease from the body. Antibodies do this by binding specifically to the infectious disease at an area called an **antigen**. The word "antigen" is a contraction of "antibody-generating." Many STD tests detect the presence of an antigen in blood. In this activity you will demonstrate the ease with which diseases spread when people share body fluids, as happens with STDs when people engage in unprotected sex. The activity uses a harmless chemical during the simulation (not a real STD) to ensure your safety. You will use an antigen test to identify infected students and determine the person who started the spread—mimicking the huge undertaking that doctors faced during the 1980s when identifying HIV/AIDS and trying to determine the identity of **patient zero** (an infected individual who starts an outbreak).

1. Obtain a body fluid sample, six clean transfer pipettes, a testing plate, rinse solution, detection solution, color solution, paper towels, and positive and negative controls. In Table 18.2, record your name next to the body fluid sample number that has been assigned to you.

2. Choose a partner from your class and swap body fluids by using your transfer pipette to suck up all the fluid from your tube and squirt it into your partner's tube. Mix the two fluids by drawing them up into the transfer pipette several times. Throw away the used transfer pipette after the exchange. In Table 18.2, record in the "Partner 1" column the number of the person with whom you exchanged fluids.

3. After everyone in the class has exchanged body fluids with one partner, retrieve a clean transfer pipette. Now perform another transfer, choosing someone in a different part of the room. In Table 18.2, record in the "Partner 2" column the number of the person with whom you exchanged fluids.

4. After everyone in the class has exchanged fluids with a second person, retrieve a clean transfer pipette. Now perform another transfer, choosing someone in a different part of the room. Throw away the used transfer pipette after the exchange. In Table 18.2, record in the "Partner 3" column the number of the person with whom you exchanged fluids.

TABLE 18.2 STD DATA FOR ACTIVITY 4

Body fluid number	Name	Partner 1	Partner 2	Partner 3	Result (+ or –)
1					
2					
3					
4					
5					
6					
7					
8					
9					
10					
11					
12					
13					
14					
15					
16					
17					
18					
19					
20					
21					
22					
23					
24					

5. Test your body fluid for the presence of an antigen. Perform this test at a lab bench. The testing plate provided has rounded wells in which you will place your sample. Each testing plate requires a positive and negative control, provided by your instructor. Label one well "Positive control" and another well "Negative control." Next, label one well with your initials; this is the well into which you will place your sample.

6. With a clean transfer pipette for each control and your body fluid sample, add four drops of each solution to its appropriate well. Incubate the testing plate for 15 minutes. This incubation allows antigens within any STD-containing fluid to bind to the plastic of the well.

7. Turn the plate over onto a paper towel, emptying out the fluid from all the wells. Tap the plate a few times on the paper towel to make sure all the fluid is removed. Next, fill the wells with the rinse solution. Flip the plate again, emptying all the rinse solution from the wells onto the paper towels. Repeat the rinsing procedure twice, each time emptying the liquid onto the paper towels. After the third rinse, make sure there is no liquid remaining in the wells.

8. The wells with STD-containing samples now have the antigens bound to the bottom of the well. Add four drops of the detection solution to each well. Incubate the detection solution for 5 minutes. This incubation allows the detection solution to bind with any wells that are coated with the STD, forming a detection–STD complex.

9. Turn the plate over onto a paper towel, emptying out the fluid from all the wells. Tap the plate a few times on the paper towel to make sure all the fluid is removed. Next, fill the wells with the rinse solution. Flip the plate again, emptying all the rinse solution from the wells onto the paper towels. Repeat the rinsing procedure twice, each time emptying the liquid onto the paper towels. After the third rinse, make sure there is no liquid remaining in the wells.

10. Add four drops of the color solution. Incubate for 5 minutes. The color solution will allow any wells that contain the detection–STD complex to turn blue. In the last column of Table 18.2, record your result: positive (+) or negative (–).

11. Once everyone has recorded results, gather the class data for each fluid sample, including the name of the original possessor, exchanges, and results. Record all of this information in Table 18.2.

12. In Table 18.2, highlight the students infected by the end of the exchange process. Examine the results of the partners of all these students. Do any of these students have a partner who was *not* infected in the end? Eliminate any student who has an uninfected partner at the end as the possible patient zero. (A positive person who exchanged with someone who is negative at the end cannot be patient zero.)

13. Among the remaining students, is there more than one person in your class who was initially infected? Can you distinguish between the originally infected person (patient zero) and the first person with whom patient zero shared body fluid?

14. If only one student was originally infected, then no more than eight students can possibly be infected. Figure 18.5 shows one possible route of transmission. Patient zero (6) infects the first partner (15), who then goes on to infect two new people during exchanges 2 (24) and 3 (3). Patient 24 will in turn infect patient 16 during exchange 3. Patient zero (6) then infects 4, who in turn infects 12 during exchange 3. Finally, patient zero (6) infects 20 during both of their last exchanges. Note that 4 was not infected during the first exchange with a different person; likewise, 16, 3, 12, and 20 were not infected until their last exchanges. All partners with whom patient zero swaps fluid are considered primary contacts (15, 4, and 20) because they were directly infected. Any partner infected by a primary contact is a secondary contact (24, 3, and 12). Finally, 16 is considered a tertiary contact because it took three exchanges for that person to be infected. Using Figure 18.5 as a model, depict in the box below how the infection spread from patient zero in your class to his or her primary and secondary contacts.

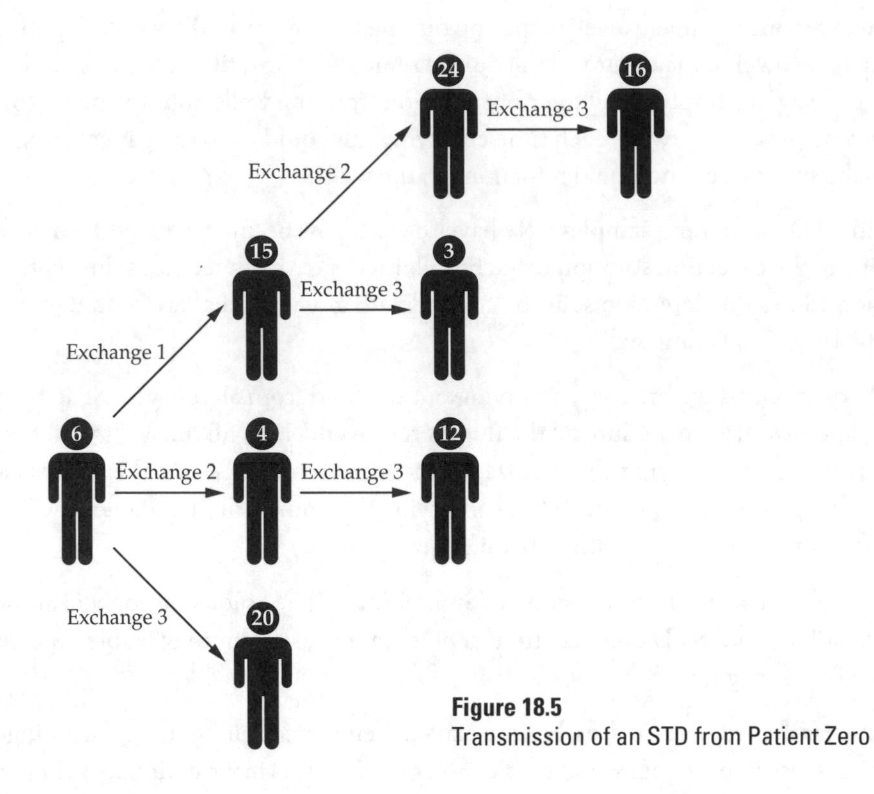

Figure 18.5
Transmission of an STD from Patient Zero

Transmission from patient zero in your lab simulation

Concept Check

1. How is an STD spread from one sexual partner to another?

2. What is a patient zero?

3. Why is contraction of an STD such as herpes or infection by HPV or HIV so dangerous?

Key Terms

adrenal gland (p. 18-2)
antibody (p. 18-13)
antigen (p. 18-13)
birth control (p. 18-11)
bladder (p. 18-2)
bulbourethral gland (p. 18-5)
cervix (p. 18-4)
corpus luteum (p. 18-9)
egg (p. 18-1)
epididymis (p. 18-5)
estrogen (p. 18-7)
external sphincter (p. 18-2)
fallopian tube (p. 18-4)
follicle (p. 18-7)
gonadotropins (p. 18-6)
hormone (p. 18-6)
internal sphincter (p. 18-2)
kidney (p. 18-2)
menstrual cycle (p. 18-6)
ovarian hormones (p. 18-6)
ovary (p. 18-4)
oviduct (p. 18-4)
ovulation (p. 18-9)
ovum (p. 18-4)
patient zero (p. 18-13)
penis (p. 18-5)
placebo (p. 18-12)
progesterone (p. 18-9)
prostate gland (p. 18-5)
reproductive system (p. 18-1)
scrotal sac (p. 18-5)
seminal vesicle (p. 18-5)
sexually transmitted disease (STD) (p. 18-13)
sperm (p. 18-1)
testis (p. 18-5)
ureter (p. 18-2)
urethra (p. 18-2)
urinary system (p. 18-1)
urine (p. 18-2)
uterus (p. 18-4)
vagina (p. 18-4)
vas deferens (p. 18-5)

Review Questions

1. Describe how urine is produced and eliminated in the human male.

2. Describe how sperm and semen are produced.

3. How does a male regulate the temperature of the testes?

4. Judging by the male anatomy, what structure is cut during a vasectomy?

5. The ________________ stores urine until it is released into the ________________.

6. The ________________ filter the blood, removing all the wastes to be excreted.

7. How does birth control work to prevent pregnancy?

8. ________________ and ________________ are the gonadotropin hormones produced by the pituitary gland that initiate the maturation of an egg during the menstrual cycle.

9. ________________ and ________________ are the ovarian hormones produced during the menstrual cycle. These hormones are responsible for preparing the lining of the uterus for pregnancy and stimulating the secretion of mucus to trap any embryo that may have formed.

10. A(n) ________________ is any disease spread through sexual intercourse.

11. What urinary-system structures do males and females have in common?

12. What is the most effective way to prevent the spread of STDs?

13. What is the most effective form of preventing an STD during sexual intercourse?

14. Do oral contraceptives protect against STD infections?

15. Name the female anatomical structures involved in the production of an egg and those involved in fetal development and birth.

LAB **19**

Animal Development and Histology

Objectives

- Observe the early developmental stages of a deuterostome.
- Connect the early embryonic stages of development to later stages of differentiation.
- Describe the process by which tissues form during early development.
- Study tissue layers and discuss the role and function of tissues in a developed organ.

Introduction

Some of the evolutionary innovations that set animals apart from other organisms are the presence of specialized tissues, organs, and organ systems. **Tissues** are specialized, coordinated collections of cells. They evolved in the phylum Cnidaria. Later, during the evolution of flatworms, organs and organ systems provided a means for tissues to function collectively. **Organs** have a defined boundary composed of different tissues organized to carry out a specialized function such as digestion or respiration. Four discrete tissue types exist in humans: epithelial, connective, muscular, and nervous tissues. Most organs are composed of all four tissue types, exemplifying the integral role that each plays in the function of an organism.

Developmental biology explores the process of development from a **zygote**, the first cell formed after a sperm fertilizes an egg, to an adult organism. An important part of development is the formation and differentiation of tissues. The more complex an organism is, the more intricate is this process. **Histology** is the study of these tissues and their functions. In this lab you will first explore the process of development in animals and then use histological techniques to observe the tissues and organs that result from development.

ACTIVITY 1 Early Embryonic Stages

The sea star, also known as the starfish, is an appropriate animal to help us understand the development of a human being. Although the developmental processes of sea stars and humans are not exactly the same, both groups are **deuterostomes**, meaning that, during formation of the gut, the anus is the first opening to form and the mouth forms later. In this activity you will examine the main stages of development of a sea star: cleavage, gastrulation, and differentiation.

Cleavage

Human gametes—the egg and sperm—contain 23 chromosomes each, and the zygote is formed upon fertilization of an egg by a sperm. After production of the zygote, the **embryo** develops through rapid cell division, forming a hollow sphere of cells called the **blastula** (Figure 19.1). The blastula is the same size as the original egg. This stage exhibits no cell growth but merely cell division, also called **cleavage**.

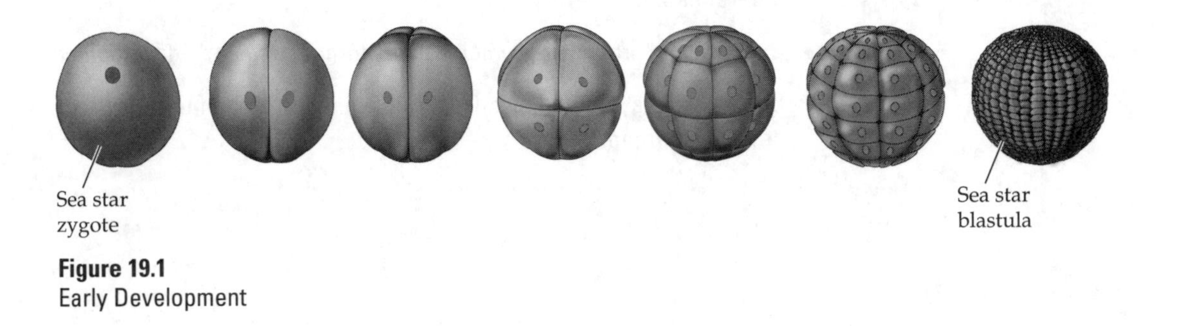

Figure 19.1
Early Development

1. Using a compound microscope, examine the slide labeled "sea star development."
2. There will be many sea star embryos in your field of view. Do you notice any differences between them?
3. Locate a specimen in which the circle is a consistent color throughout the circle. Does your specimen have a halo around it? Focus up and down. (Remember to use the fine-focus knob to do this!)
   a. If no halo is present, this is the unfertilized egg of a sea star.
   b. If your specimen has a halo, it is a fertilization membrane, meaning this is the zygote of a sea star.
   c. Locate both an unfertilized egg and a zygote, and sketch them in the first two circles below.
4. Next, you should notice several other stages of cell division—for example, two- and four-cell stages. What is the difference between these stages? What are the similarities? In the third circle on the next page, sketch a four-cell stage.

5. The blastula is often mistaken for the unfertilized egg or the zygote. Do you see any embryos with a dark ring on the outside and a lighter-colored middle?

6. Think of a blastula as a balloon, the cells composing the rubbery outer part. But where air occupies the space inside the balloon, the blastula has a fluid-filled cavity called the **blastocoel**. What size is the blastula compared to the zygote? In the fourth circle below, sketch the blastula.

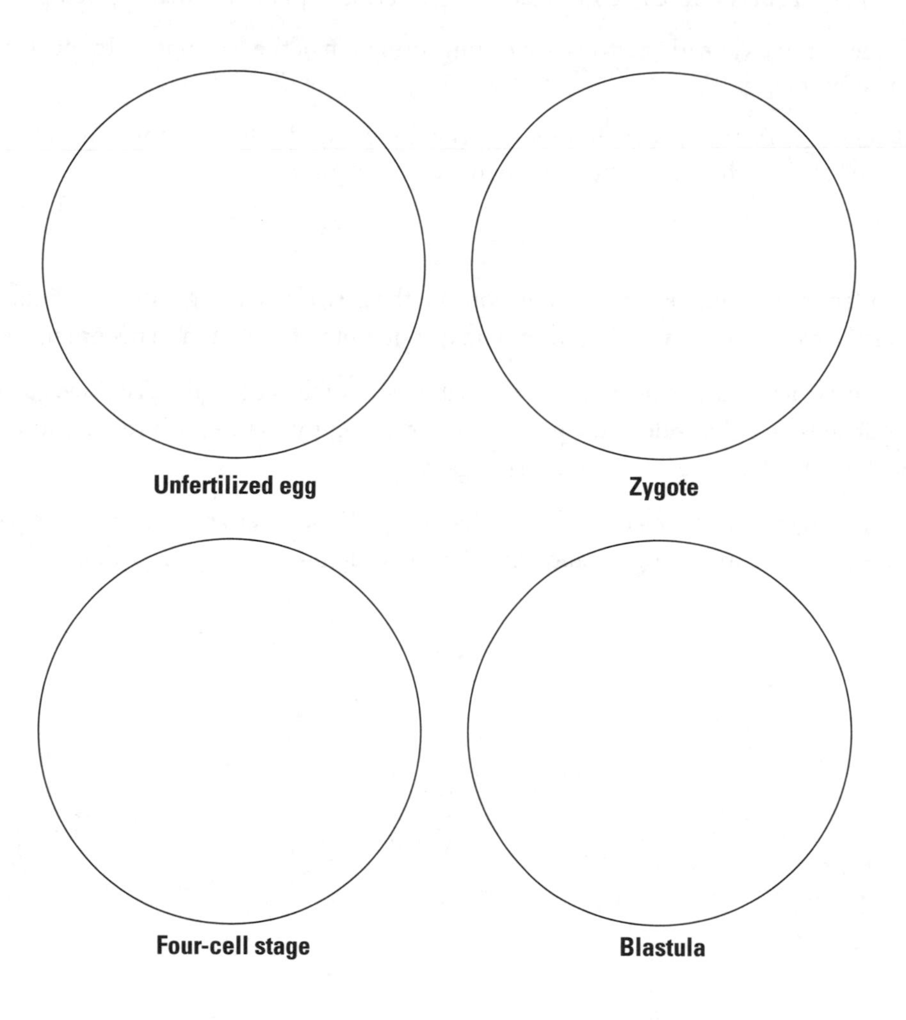

Gastrulation

Once the blastula has formed, cells start to move during **gastrulation**. The ball of cells starts forming a pore, the **blastopore**, in one area. The cells migrate inside into the blastocoel. The **gastrula** is the stage during pore formation until a complete gut forms. To visualize this process, imagine pushing your finger into a balloon. The pore formed by your finger pushing inward is equivalent to the blastopore of the gastrula. Imagine poking your finger all the way through the balloon to the other side. The cavity formed is like one continuous tube and corresponds to the **digestive tube**, or **gut**, formed during development in the embryo. Remember that in deuterostomes, the blastopore becomes the anus (whereas in **protostomes** it develops into the mouth).

1. Identify the embryos at early gastrulation during formation of the blastopore. In the first circle below, sketch an example.
2. Find a later stage of the gastrula. In the second circle below, sketch an example. Is there now a difference in size between the late gastrula compared to the zygote?
3. Can you identify the outer layer of cells making up the gastrula? During differentiation, these cells have more specific roles, such as forming the outer layer of protective tissue and becoming nervous tissue.
4. In the late gastrula, pouches begin to form on either side of the end of the primitive gut. Can you identify the pouches on either side of the gut? These are a primitive tissue layer representing cells that will become the bulk of the organism's internal organs.
5. Can you identify the inner layer of cells forming the gut? These cells become more specific in their role to line the outside of major organs and form specific structures and organs within the adult.

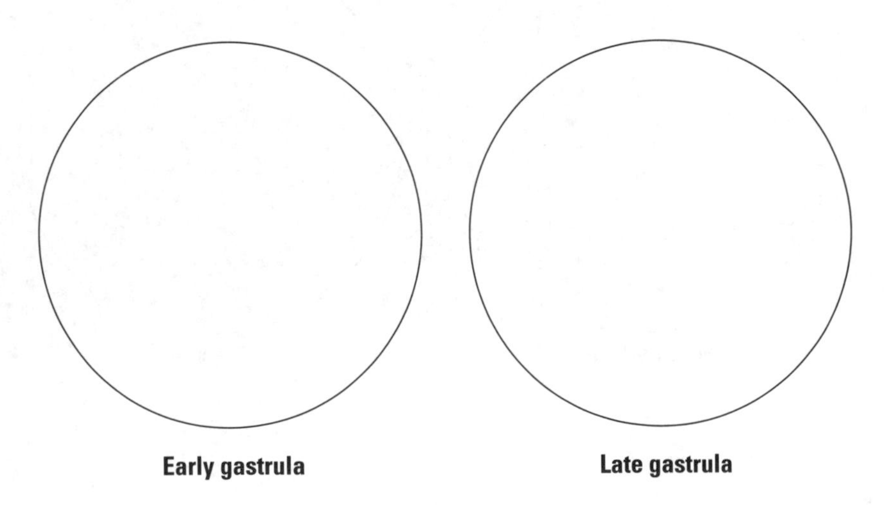

Differentiation

As deuterostomes, both sea stars and humans exhibit three layers of early embryonic tissues. These tissues lead to the development of tissues of specific types and functions, such as muscle tissue, in the resulting adult. Gastrulation sets the scene by positioning the cells in layers that later undergo **differentiation**, the process of becoming more specific in function. The tissue on the outside of the gastrula is the **ectoderm**. Aptly named, the middle layer of tissue is the **mesoderm** (those small side pouches at the end of the gut). On the inside is the **endoderm**. Figure 19.2 illustrates these layers, and Table 19.1 identifies the tissues that each embryonic layer becomes in an adult human.

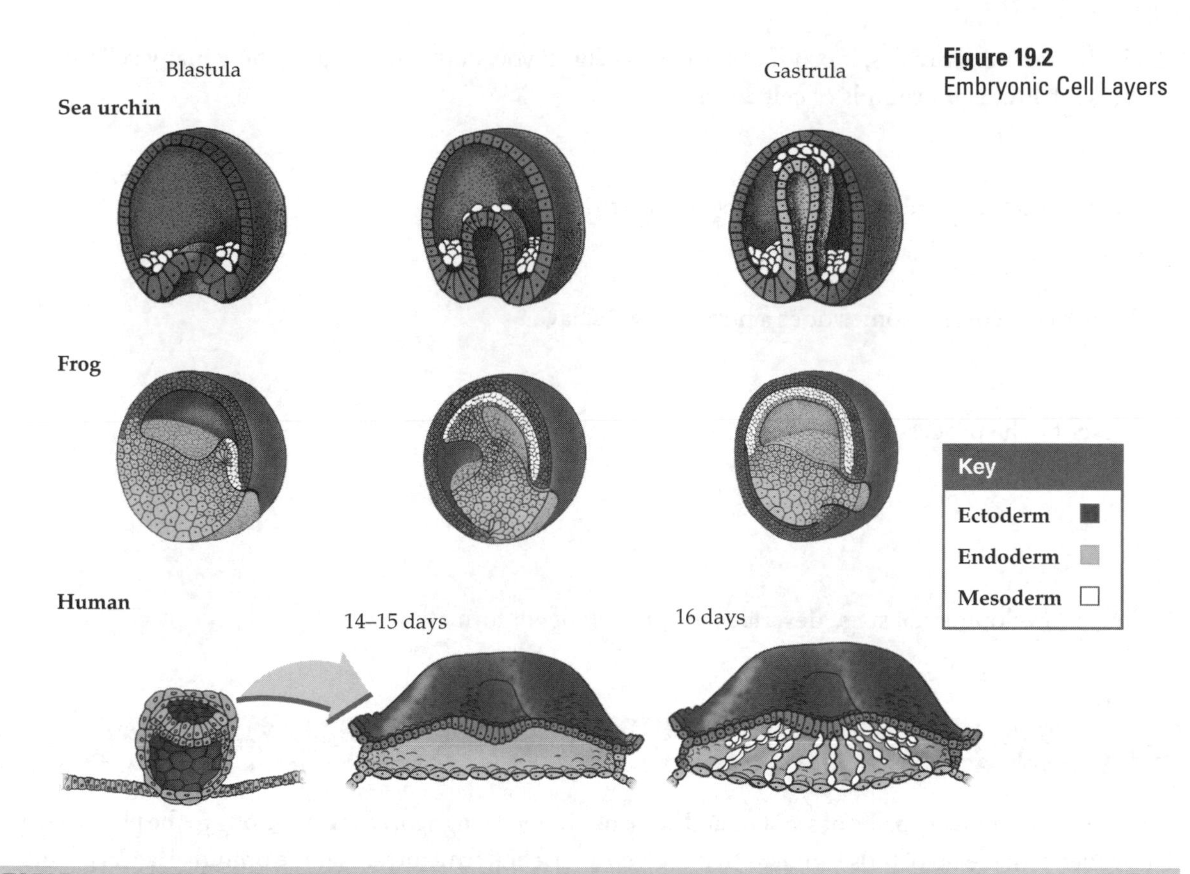

Figure 19.2
Embryonic Cell Layers

TABLE 19.1 FATES OF THE THREE CELL LAYERS IN THE DEVELOPING HUMAN EMBRYO

Cell layer	Corresponding adult structures
Endoderm	Liver, pancreas, thyroid, and linings of the gut and lungs
Mesoderm	Skeleton, muscles, reproductive structures, kidneys, circulatory and lymphatic systems, blood, and the inner layer of skin
Ectoderm	Skull, nerves and brain, the outer layer of skin, and teeth

Here's an easy way to remember these tissue names: The root *derm* means "skin"; just think of "dermatology." For the prefixes indicating where the tissue is located, *ecto*, like "outside," has a "t" as the third letter; *endo*, like "inside" has an "n" as the second letter; and *meso* and "middle" start with the same letter. These early embryonic tissue types develop further into more specific roles in the adult organism during differentiation.

1. In the sketch you made of the late gastrula above, label where each tissue layer would be located.

Concept Check

1. Each time a cell divides, it produces two new cells. If you start with a zygote, how many cells are present after eight rounds of cell division?

2. What structures does the endoderm give rise to?

3. How many chromosomes does a human zygote have?

4. Describe the process of differentiation.

5. What developmental stage describes the process of gut formation?

ACTIVITY 2 Early Vertebrate Development

Clearly, the evolutionary paths of sea stars and humans diverged long ago. Humans belong to the phylum Chordata, which is made up of those animals that evolved a vertebral structure. After the primitive layers of embryonic development, the embryo's cells specialize further. Each layer produces the specific tissues listed in Table 19.1. Differentiation is the process by which cells become increasingly limited or restricted in their range of potential fates. Cell fate is mostly fixed by the time the animal is born.

In the chicken embryo, familiar structures such as the eye, the backbone (neural tube), the head, and the limbs are easily discernible. These and other structures normally arise from cells that migrate in groups and form tissues. A specific group of cells called **somites** are the small repeating blocks lining the length of the neural tube. They differentiate from the mesoderm. The presence of these repetitive structures suggests evolution from a segmented ancestor. The somites develop into skeletal muscle, a layer of skin called the dermis, and cartilage that surrounds the spinal cord in an adult. Ultimately, all the structures you see in this chicken embryo will differentiate into the four tissue types seen in adult animals. In this activity you will examine the more defined structures of a chicken embryo that form during the developmental process of differentiation. Use Figure 19.3 to locate the structures present in the embryo.

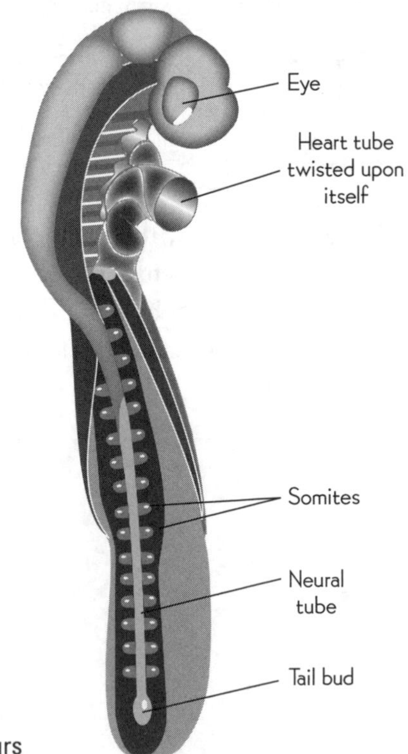

Figure 19.3
Chicken Embryo at 48 Hours

1. Using a compound microscope with the 10× objective in place, examine the embryo. Can you discern any familiar structures?
2. In the circle below, sketch the embryo. Do you see the somites? Why do you think the somites are located where they are?
3. What is the function of the blood vessels that surround the embryo?

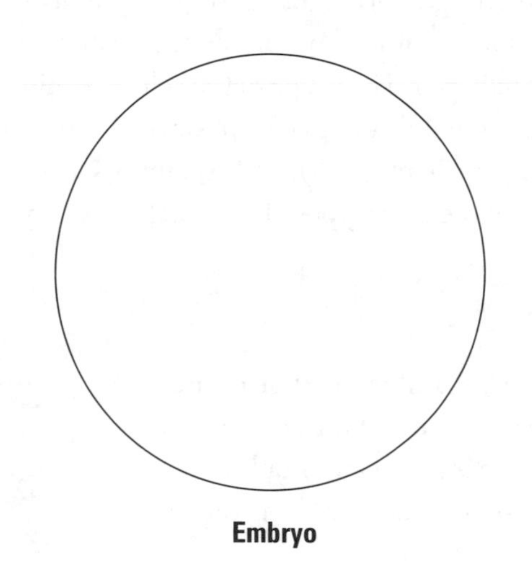

Embryo

Concept Check

1. What does it mean that cells become specialized?
2. What are somites? Why are these types of structures important to differentiation during development?
3. Are somites completely differentiated (is this their final fate within the organism)?

ACTIVITY 3 Epithelial Tissue

The human body is composed of over 300 types of cells, all arising from the same zygote that undergoes cleavage, gastrulation, and differentiation. The layers of cells within a tissue work with other tissue types to completely form the organs of all of your organ systems. No organ functions independently; rather, all organs require the mosaic of cells within the human body to smoothly perform all of the tasks needed for survival. In this activity and the three that follow, you will explore the composition and function of each tissue in an adult animal.

Working from the outside, the **epithelial tissue**, or **epithelium** (plural "epithelia"), is any tissue covering the body and lining the spaces of the gut, all organs, and organ systems. Cells of the epithelium are the protective cells of all your organs and body. Epithelium consists of either one layer of cells ("simple epithelium") or multiple layers ("stratified epithelium"). The shape of the cell determines its name. The cells of epithelial tissue come in many shapes and sizes, but the three main types are **cuboidal**, like a cube or square (Figure 19.4); **columnar**, like a column or rectangle (Figure 19.5); and **squamous**, like a squished circle (Figure 19.6). In this activity you will examine each of the three types of cells that can compose a tissue layer of epithelium.

Simple Cuboidal Epithelium

1. Using a compound microscope, examine the slide of the thyroid gland. Remember that you're looking at a cross section of a human organ. If this is an organ, should more than one type of tissue be present? Where should you look for the epithelium?

2. The thyroid gland is composed of thousands of follicles (balls of cells) where hormones that the thyroid produces are stored. A single layer of epithelium surrounds the central storage area of each follicle. Scan the whole slide with the scanning lens in place and then increase the magnification. Can you locate the follicles?

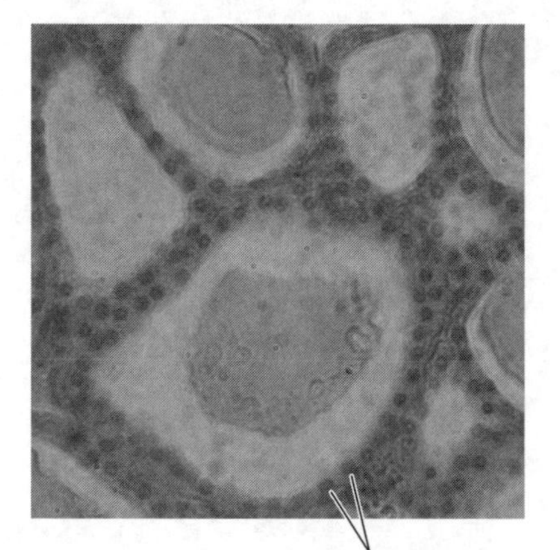

Figure 19.4
Simple Cuboidal Epithelium

3. Once you have the highest-power objective in place, focus in on the epithelium. Note that it is not always easy to see the plasma membranes of adjacent cells. Use the nuclei to help you determine the cell's boundaries. In the circle below, sketch the tissue.

Simple cuboidal epithelium

Simple Columnar Epithelium

1. Using a compound microscope, examine the slide of the gallbladder. How does the gallbladder differ from the thyroid gland?

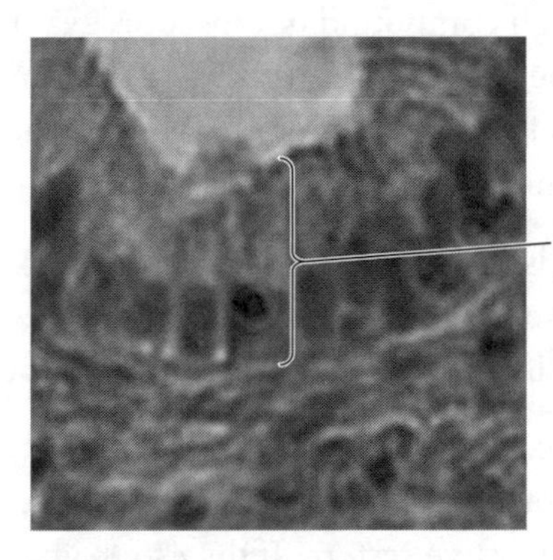

Figure 19.5
Simple Columnar Epithelium

2. The gallbladder stores bile, a digestive material produced by the liver. This is not a complete cross section of the gallbladder. If this is just a portion of the larger organ, how do you determine where the epithelium is located?

3. Once you've found the simple columnar epithelium, what structures do you see below it?

4. Notice that the cells making up this epithelium are tall and thin (column-shaped in two dimensions). Can you identify any cellular compartments or organelles? In the circle below, sketch an individual columnar cell and label any of the specific compartments or organelles that you can identify.

Columnar cell

Stratified Squamous Epithelium

1. Using a compound microscope, examine the trachea and esophagus slide. Only the esophagus contains squamous epithelium, so you will be examining only this structure.
2. This is a longitudinal section through the esophagus, so the lumen is long and narrow. Can you identify the stratified squamous epithelium?
3. Notice that there are many layers of cells. The esophagus connects the mouth to the stomach in the digestive system. If the esophagus has such a thick layer of epithelial cells, do you think nutrients are absorbed while food passes through it to the stomach? In the circle below, sketch the layered cells.

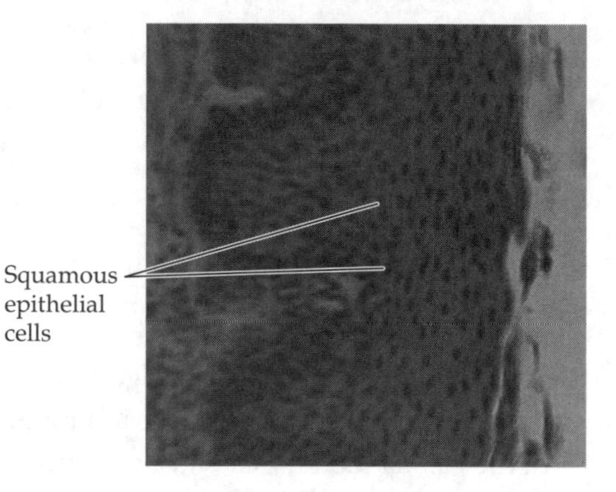

Figure 19.6
Stratified Squamous Epithelium

Stratified squamous epithelium

Concept Check

1. How are epithelial cells named?

2. What is the function of epithelial cells?

3. What organs have epithelial cells?

ACTIVITY 4 Connective Tissue

Of the four different tissue groups, **connective tissue** contains the most variety in cell shape, but its function is its defining characteristic. Derived from the mesoderm, connective tissue does just what its name suggests: connects one tissue to another, serving as a passage for blood vessels and nerves to travel from one area in the body to another. Connective tissue forms the "ropes" that attach muscle to bone (tendons) and one bone to another bone (ligaments). Fatty (**adipose**) tissue (Figure 19.7), cartilage, bone, and blood are four other examples of connective tissue. The epithelium slides in Activity 3 contained connective tissue, but it is hard to see in those examples. In this activity you will examine three types of special connective tissue that are very easy to identify: adipose tissue, bone tissue, and blood.

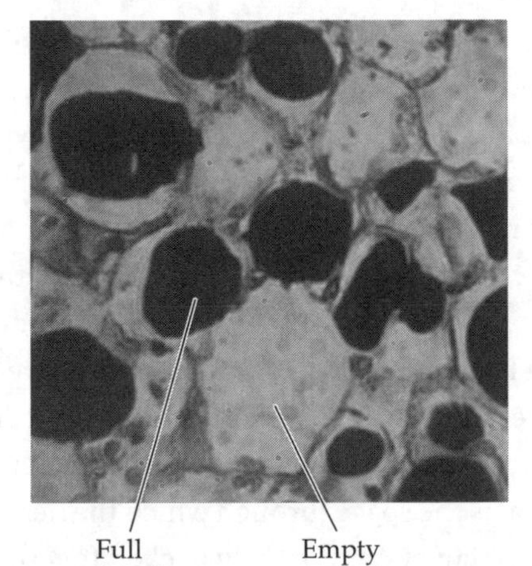

Figure 19.7
Adipose Tissue

Adipose Tissue

Adipocytes are fat storage cells (Figure 19.7). They are larger than most other cells in the human body. Humans store fat because it provides more energy than the equivalent weight of sugar or proteins.

1. Using a compound microscope, examine the adipose tissue slide. The slide contains cells stained black with a dye that adheres to the triglycerides stored in them. Do you see any cells that are not stained black?
2. In the circle below, sketch what you see. Describe the appearance of the tissue. How many of the cells on your slide contain fat?

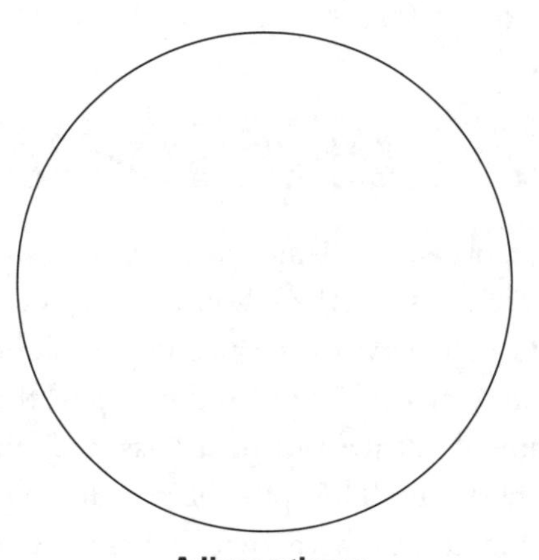

Adipose tissue

Bone Tissue

Your bones are constantly adding new deposits of minerals throughout every day of your life. This constant mineral replenishment combats the loss of bone over time. Bones are typically composed of a thick outer layer called **compact bone**, which you will be examining along with an inner compartment called **spongy bone**. Spongy bone houses the bone marrow where stem cells produce another connective tissue type called blood (red blood cells, white blood cells, and platelets). Bones are structural and protective, providing the ability to move, a source of calcium, and a home for the stem cells of blood cells.

Compact bone, the dense outer layer of bones, is composed of concentric layers called **haversian systems** or **osteons** (Figure 19.8). Haversian systems contain **osteocytes**, cells that deposit calcium and other minerals, forming the matrix of concentric layers. In the center of the haversian system is a central canal called the **haversian canal**. This canal is the space through which the nerves and blood vessels travel. Compact bone is composed of many haversian systems tightly packed together.

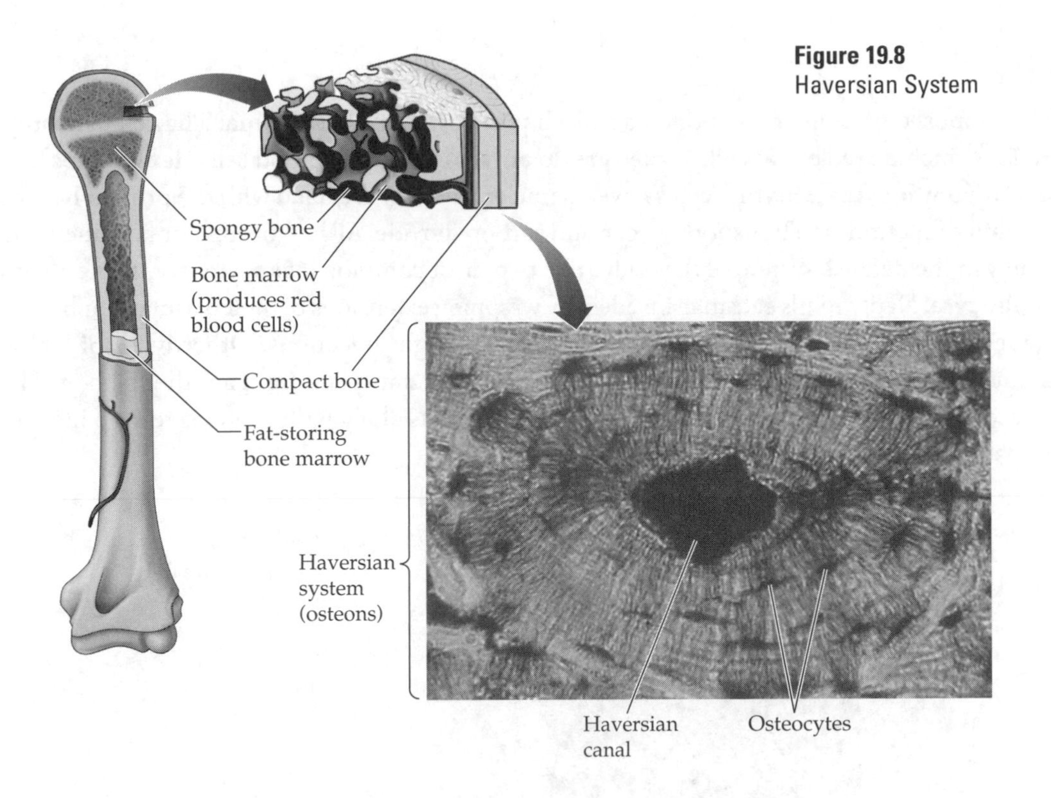

Figure 19.8
Haversian System

1. Using a compound microscope, look at the bone under the scanning objective. Note the overall density of the compact tissue on the slide. Why would compact bone differ in composition from spongy bone?

2. Increase the magnification. Can you identify the haversian canal? Do you see the concentric circles around the haversian canal?

3. Along the circles, do you see black spots? In living bone tissue, the area within the black spots is occupied by the osteocytes. In the circle below, on the basis of what you see in the microscope, sketch an example of the haversian system, including the haversian canal and osteocytes of bone.

Haversian system

Blood

Blood is composed of cellular components and a liquid portion called **plasma**. The cellular components (Figure 19.9) include red blood cells (called **erythrocytes**) and white blood cells (**leukocytes**), as well as platelets (**thrombocytes**). Erythrocytes have no nuclei, but they are filled with the protein **hemoglobin**. Hemoglobin's function is to transport oxygen and carbon dioxide. All leukocytes contain nuclei and function mainly in the defense system of the body. The two most common leukocytes are the **neutrophil** and the **lymphocyte**. Neutrophils eat small invaders by wrapping extensions of their plasma membrane around the invaders. B lymphocytes produce the antibodies that fight infections. Other types of lymphocytes are responsible for tissue rejection when someone receives a transplant from another person. Thrombocytes, pieces of cells produced in the bone marrow, form clots along with proteins to stop bleeding from the bloodstream.

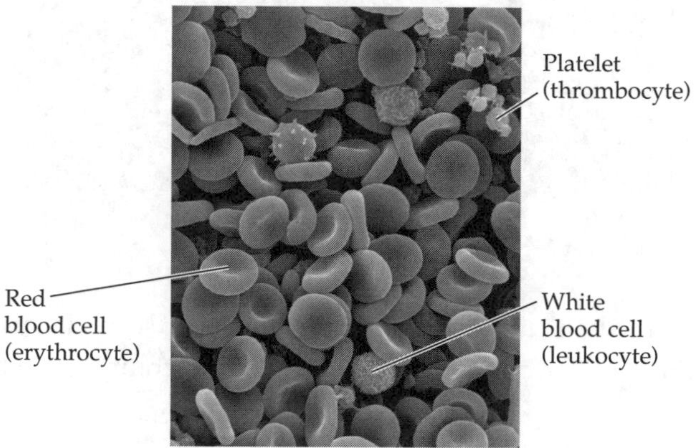

Figure 19.9
Human Blood Cells

1. Under the compound microscope, examine the slide of human blood. Note that the plasma does not stain. What color are most of the cells on your slide?

2. By far the majority of the cells on your slide are red blood cells. Can you identify any compartments within the red blood cells? In the first circle below, sketch the cell.

3. Move around the slide until you locate a cell that is not a red blood cell. What difference do you immediately notice?

4. In the second circle below, sketch the cell that you're examining. If this cell has an unusually shaped nucleus (almost an hourglass), it's a neutrophil. If the cell has a large, round nucleus that encompasses almost the entire cell, it's a lymphocyte. In the label below the second circle, identify the type of cell you sketched.

5. Locate the other type of white blood cell and sketch it in the third circle below. In the label below the third circle, identify the type of cell you sketched.

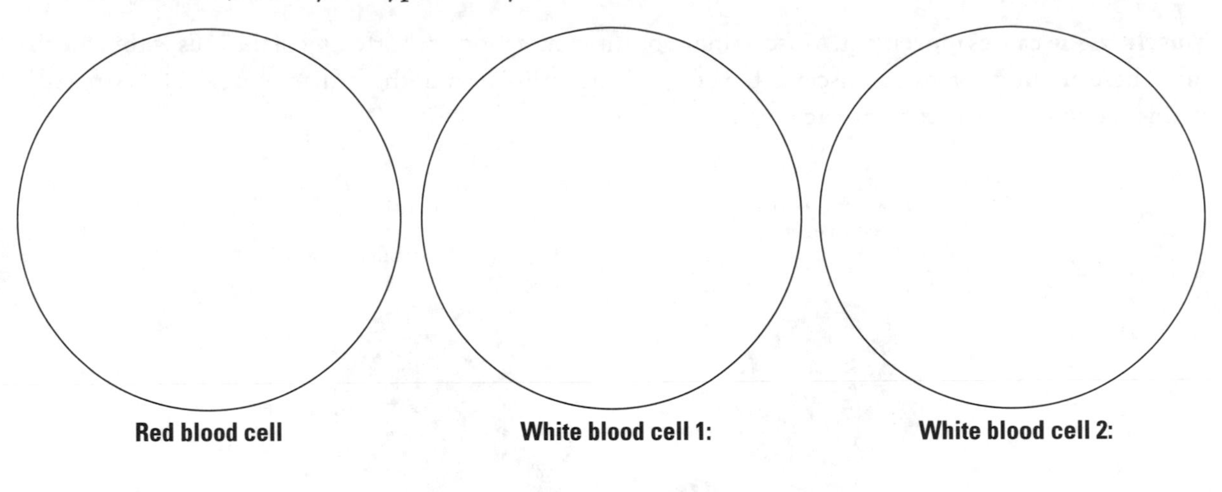

Red blood cell | **White blood cell 1:** ______________ | **White blood cell 2:** ______________

Concept Check

1. What is the function of connective tissue?

2. Describe how osteocytes lay down the minerals forming bone tissue.

3. Why are some cells of adipose tissue stained black while others appear clear?

4. Why is blood tissue important? Describe a general function for each type of cell composing blood tissue.

ACTIVITY 5 Muscle Tissue

Muscle tissue causes movement. Muscles move by the contraction or shortening of the cells within the tissue. There are three types of muscle in the body (Figure 19.10): smooth, cardiac, and skeletal (striated). In this activity you will examine each type.

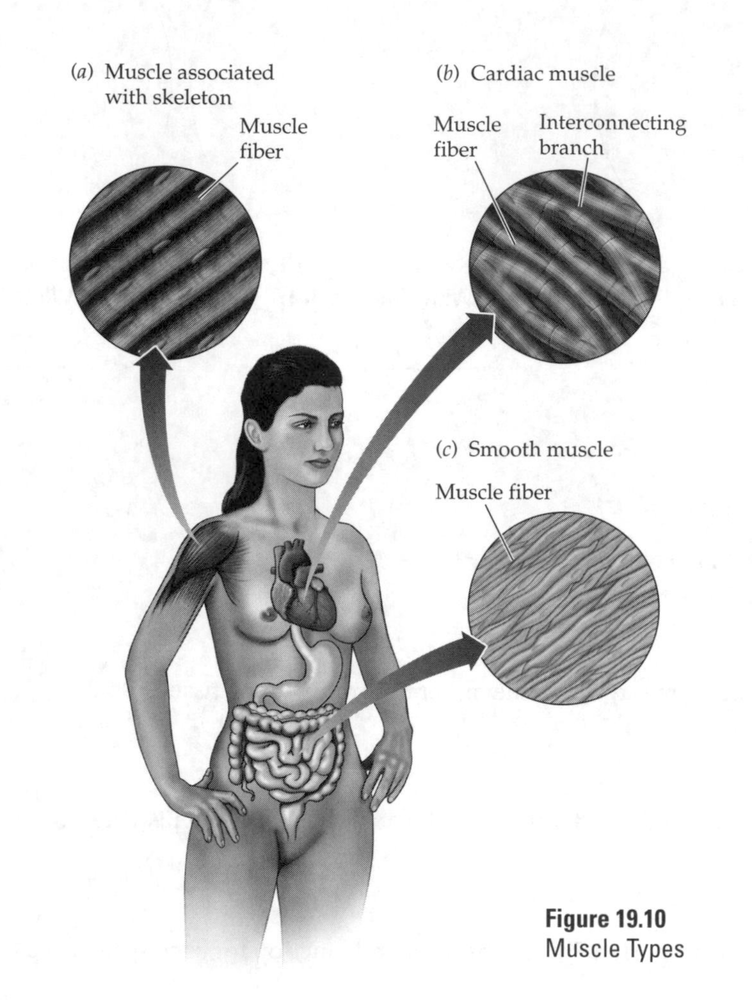

Figure 19.10
Muscle Types

Smooth Muscle

Smooth muscle is the muscle that surrounds many tubelike organs in your body: blood vessels, the digestive tube, vas deferens in the male, the urethra, and so on.

1. Examine the slide of smooth muscle under the microscope.
2. In the circle below, sketch this tissue at the highest magnification. Can you see different cells?

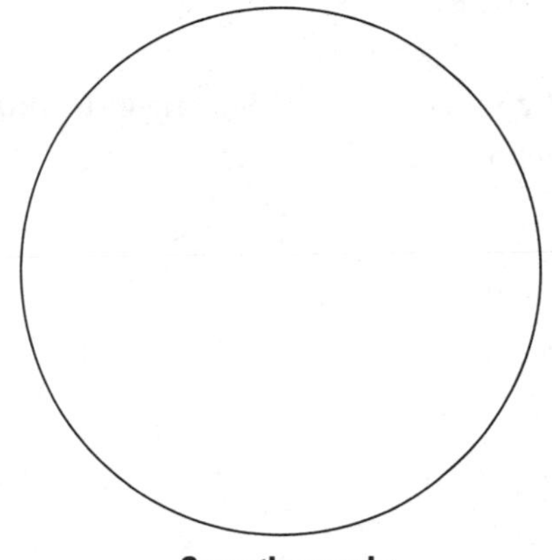

Smooth muscle

3. In the esophagus, stomach, and intestines, smooth muscle contracts around these tubes to move food along the passage, in the process known as **peristalsis**—the alternating contraction and relaxation of smooth muscle to move substances around the body. The contraction of smooth muscle around your arteries is responsible for raising your blood pressure. Is smooth muscle under your voluntary control? Do you consciously think to contract your smooth muscles?

Cardiac Muscle

Cardiac muscle is the specialized muscle of the heart. The functioning of the heart is a coordinated effort of multiple cardiac muscles forming the chambers of the heart. These chambers push blood into and out of the lungs and around the body.

1. Examine the slide of cardiac muscle under the microscope.
2. In the circle below, sketch this tissue at the highest magnification. Do you notice more detail than in the smooth muscle? Is cardiac muscle under your voluntary control?

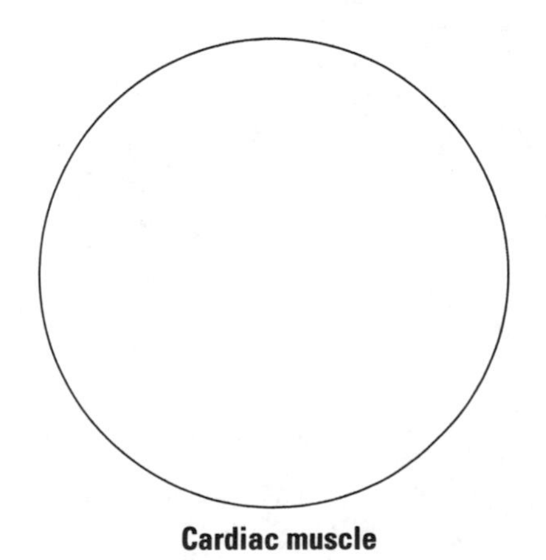

Cardiac muscle

Skeletal Muscle

Skeletal muscle is sometimes called striated muscle. Skeletal muscles enable us to move our arms and legs, as well as the diaphragm for breathing.

1. Examine the slide of human skeletal muscle under the microscope. You will easily see why it is called "striated." The proteins (actin and myosin) responsible for contraction form very distinct bands, or striations, in this muscle type, making it one of the easiest tissues to identify. Each skeletal muscle cell contains many nuclei.
2. In the circle below, sketch what you see. Identify the long individual cells with striations and label the nuclei and plasma membrane of each cell.

Skeletal muscle

Concept Check

1. Which muscles are involuntary? Which are voluntary?
2. How do muscles cause the body to move?
3. What muscles control the movement of food through the digestive system?

ACTIVITY 6 Nervous Tissue

Nervous tissue varies quite a bit throughout the body. The cells of the nervous system are **neurons**. The neuron has three main structures: the cell body (where the nucleus resides); dendrites that project out from the cell body like little hairs; and an axon, which is a long projection that leads to a terminal. The **dendrites** receive information, which is then processed in the cell body; the message is then sent on to the **axon**, causing a chemical reaction to occur at the axon terminal. This terminal can end at another neuron, a muscle cell, an endocrine cell, or any other cell, depending on the function of the neuron. Typically, people think the nerves are the cells of the nervous tissue, but **nerves** are bundles of the axons of many neurons along with connective tissue and a blood supply. In this activity you will examine nervous tissue.

1. Using a compound microscope, examine the slide of nervous tissue. You will not be able to distinguish an axon from a dendrite. Using Figure 19.11 to help you focus on the types of cells you should examine, locate some neurons. What is the density of neurons in your field of view? Why would an organ have varying concentrations of neurons?

2. In the circle below, sketch the tissue. Can you identify the nucleus in any of the cells?

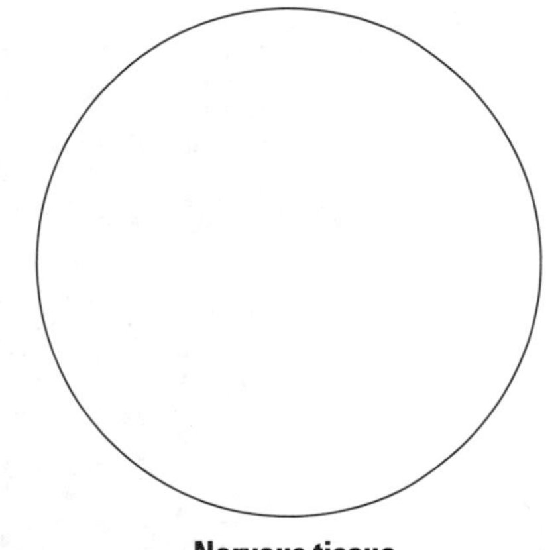

Nervous tissue

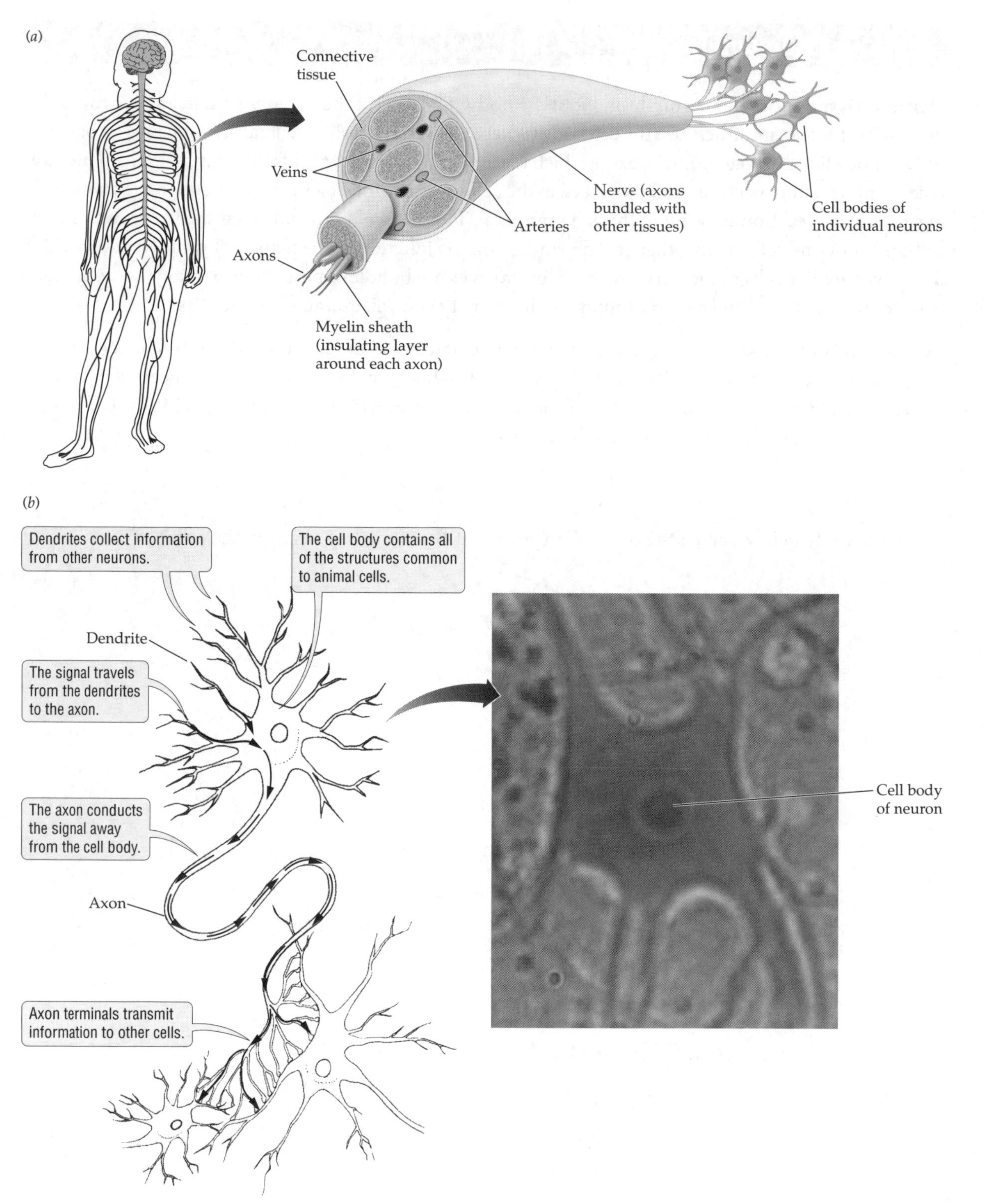

Figure 19.11
Neurons

Concept Check

1. How does a neuron receive, process, and send information?

2. Why do you think neurons have so many dendrites?

3. In what parts of the body would you expect to find nervous tissue?

ACTIVITY 7 Human Organs

As you have seen throughout Activities 1 through 6, no slide displaying human tissue types contained only one specific tissue. Even bone contains the haversian canal, where neurons and blood vessels are to communicate and provide nutrients to osteocytes. In this next activity, you will examine an illustration of human skin. This organ protects your body from the outside world, by forming thick layers of cells. The skin also contains glands that secrete antimicrobial substances. Hair and neurons provide sensory information to the body about the world around you. The arrector muscle pulls the hair follicle closer to the skin surface in reaction to different sensory information. Thermoreceptors in the skin detect whether an environment is hot or cold. Touch and pain receptors provide additional sensory input. Blood vessels deliver the required oxygen and nutrients to all of these cells while whisking away the wastes.

In this activity you will observe a cross section of skin on the slide provided by your instructor and use Figure 19.12 to identify which parts of the skin contain which types of tissues.

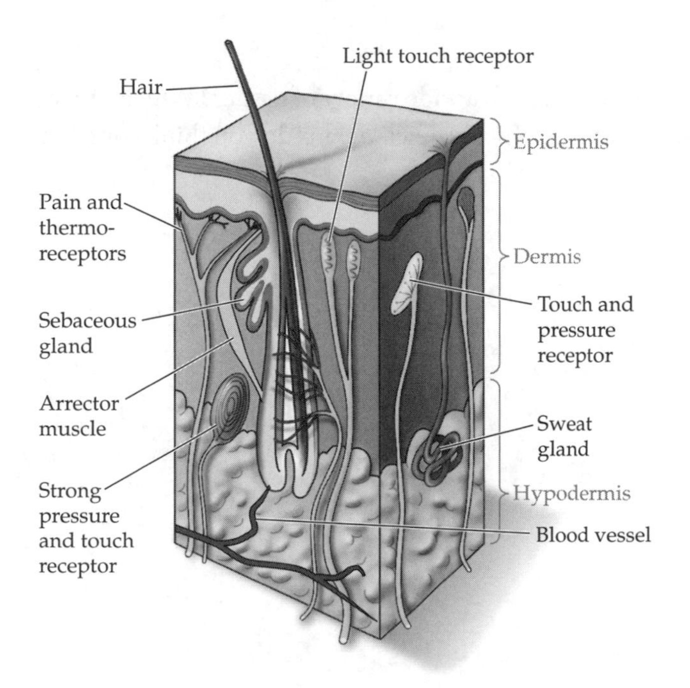

Figure 19.12
Cross Section of Skin

1. Using a compound microscope, examine the slide of human skin and identify the outer edge, epidermis. In Table 19.2 check off the tissue types that you find in the epidermis.
2. Now, using Figure 19.12 for guidance, locate on your slide the other structures of skin listed in Table 19.2. Check off the tissue types that you find present in each specific structure.

TABLE 19.2 TISSUE TYPES FOUND IN SKIN STRUCTURES

Structure	Epithelial tissue	Connective tissue	Muscle tissue	Nervous tissue
Epidermis				
Arrector muscle				
Dermis				
Touch and pain receptors				
Blood vessels				
Hypodermis				
Sebaceous gland				

Concept Check

1. What part of the skin contains all four tissue types?
2. What part of the skin contains only one type of tissue?
3. Would you expect other organs to contain such a wide variety of tissues? Which organs do you think would contain the lowest number of tissue types? Which organs do you think contain the highest number of tissue types?

Key Terms

adipocyte (p. 19-12)
adipose (p. 19-11)
axon (p. 19-19)
blastocoel (p. 19-3)
blastopore (p. 19-4)
blastula (p. 19-2)
cardiac muscle (p. 19-17)
cleavage (p. 19-2)
columnar epithelium (p. 19-8)
compact bone (p. 19-12)
connective tissue (p. 19-11)
cuboidal epithelium (p. 19-8)
dendrite (p. 19-19)
deuterostome (p. 19-2)
developmental biology (p. 19-1)
differentiation (p. 19-4)
digestive tube (p. 19-4)
ectoderm (p. 19-4)
embryo (p. 19-2)
endoderm (p. 19-4)
epithelial tissue (p. 19-8)
epithelium (p. 19-8)
erythrocyte (p. 19-14)
gastrula (p. 19-4)
gastrulation (p. 19-4)
gut (p. 19-4)
haversian canal (p. 19-12)
haversian system (p. 19-12)
hemoglobin (p. 19-14)
histology (p. 19-1)
leukocyte (p. 19-14)
lymphocyte (p. 19-14)
mesoderm (p. 19-4)
muscle tissue (p. 19-16)
nerve (p. 19-19)
nervous tissue (p. 10-19)
neuron (p. 19-19)
neutrophil (p. 19-14)
organ (p. 19-1)
osteocyte (p. 19-12)
osteoporosis (p. 19-24)
osteon (p. 19-12)
peristalsis (p. 19-17)
plasma (p. 19-14)
protostome (p. 19-4)
skeletal muscle (p. 19-18)
smooth muscle (p. 19-17)
somite (p. 19-6)
spongy bone (p. 19-12)
squamous epithelium (p. 19-8)
thrombocyte (p. 19-14)
tissue (p. 19-1)
zygote (p. 19-1)

Review Questions

1. What are the three embryonic tissue layers? Name some structures that they produce in later development.

2. A __________________ is a hollow, fluid-filled ball of cells formed during animal development.

3. __________________ produce the anus from the blastopore, whereas __________________ produce the mouth from this same structure.

4. What is cleavage?

5. Why do biologists think somites indicate that vertebrates evolved from segmented animals?

6. Describe the relationship between cells, tissues, and organs.

7. Square, blocklike cells of the epithelium are called ________________.

8. ________________ are small, biconcave cells in blood that transport oxygen.

9. What is the function of bone?

10. ________________ are cells that store fats such as triglycerides.

11. **Osteoporosis** is when bone loss is greater than bone replacement. If bone is constantly replenishing itself, how does osteoporosis occur?

12. Describe peristalsis. What muscles control peristalsis?

13. Why are skeletal muscles striated?

14. What are the cells of the nervous system?

15. Describe how a neuron receives and transmits a signal.

16. List the four types of tissue found in the human body and describe a function for each one.

LAB 20

Seed Plant Anatomy and Life Cycle

Objectives

- Identify the major anatomical structures of seed plants and the important variations among them.
- Identify differences between primary and secondary growth.
- Describe the composition and function of the flower.
- Explain the alternation of generations and how this differs from reproduction in animals.
- Examine the parts of a seed and describe their purposes.
- Explore the development and function of fruits, and distinguish the difference between types of fruits.
- Identify differences between monocots and dicots.

Introduction

Plants literally feed the world. At the base of each food chain are the **producers**, any organisms that can make their own food (**autotrophs**). Consumers, including humans, depend on the survival and proliferation of producers for their own survival. The seeds of grasses such as rice, corn, and wheat feed the human population. Beyond this ecological significance and being aesthetically beautiful, plants provide humans with essential materials such as wood and cotton. In addition, both traditional uses and modern-day applications display the value of plants in treating human disease. For example, one of the most prescribed medications to improve muscle contractions in patients suffering from heart disease is digitalis, an extract from the plant foxglove.

The evolution of land plants resulted in terrestrial domination by two types of seed plant groups: the angiosperms and the gymnosperms. **Gymnosperm** means "naked seed," and gymnosperms are often referred to as conifers or cone-bearing plants, because conifers are the dominant group. **Angiosperm** means "hidden seed," and angiosperms are flower-producing plants. Cones contain the reproductive organs in gymnosperms, and flowers contain the reproductive organs in angiosperms. Otherwise, the basic structure of angiosperms and gymnosperms is very similar.

The angiosperms are further subdivided into two groups: monocots and dicots. **Monocots** derive their name from having a single (*mono*) embryonic leaf in the seed; **dicots** have two (*di*) embryonic leaves. Monocots include the grasses and palms; dicots include most familiar trees (oaks and elms) and many nonwoody plants (roses and rhododendrons). Differences between monocots and dicots include parallel venation in monocots versus netlike venation in dicots. You will discover other differences during the course of this lab.

In this lab you will learn about the general nonreproductive structures of the seed plant and then the vital role of the flower in reproduction and the life cycle in angiosperms. The final two activities will reveal the developmental stages in the seed and the structure and function of fruits.

ACTIVITY 1 Seed Plant Anatomy

The basic structure of a seed plant consists of the root, the stem and the leaf. In this activity you will examine each of these structures (Figure 20.1).

The Root

The **root** is the structure typically located belowground, providing the plant with an anchor. It also functions to absorb water and nutrients from the soil. Projections called root hairs increase the surface area of the root system. At the tip of each root is an area of active growth and cell division located behind the root cap (a protective "cap" of cells at the root tip). Seed plants display one of two root systems: a **taproot system,** with one dominant root; or a **fibrous root system,** forming a dense mat of roots. Here you will compare these two root systems.

1. Obtain samples of two angiosperms—one monocot (a grass) and one dicot (a radish).
2. Examine the root systems of the two samples. How do the roots differ?

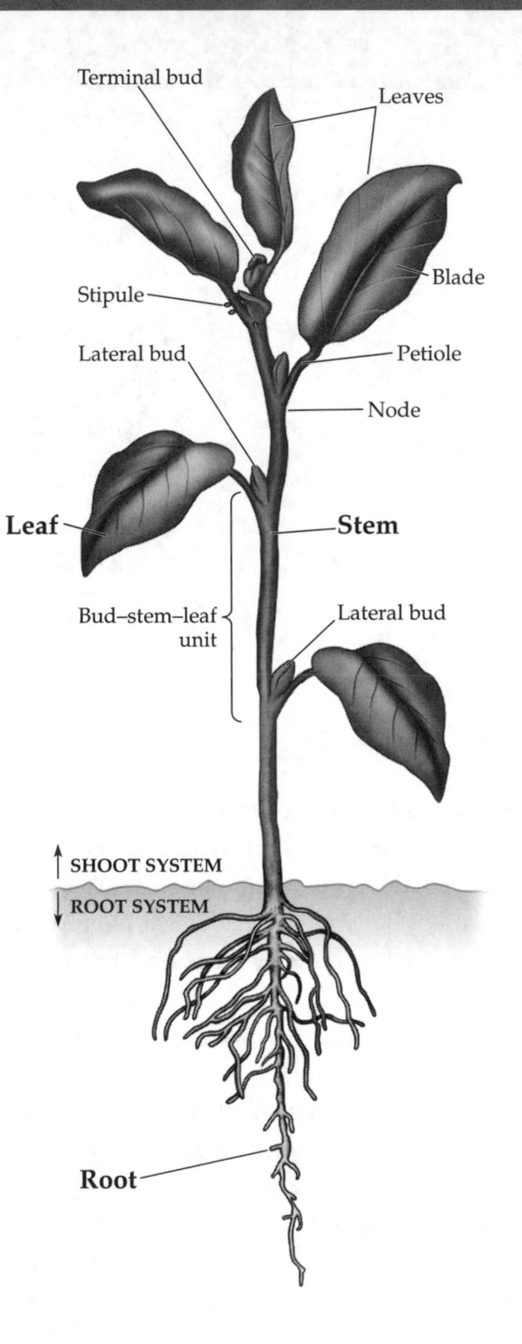

Figure 20.1
General Anatomy of a Plant

3. Study Figure 20.2. Part (a) shows a taproot system; part (b), a fibrous root system. For the plants that you're observing, identify whether each root has a taproot or a fibrous roots.

Figure 20.2
Two Kinds of Root Systems

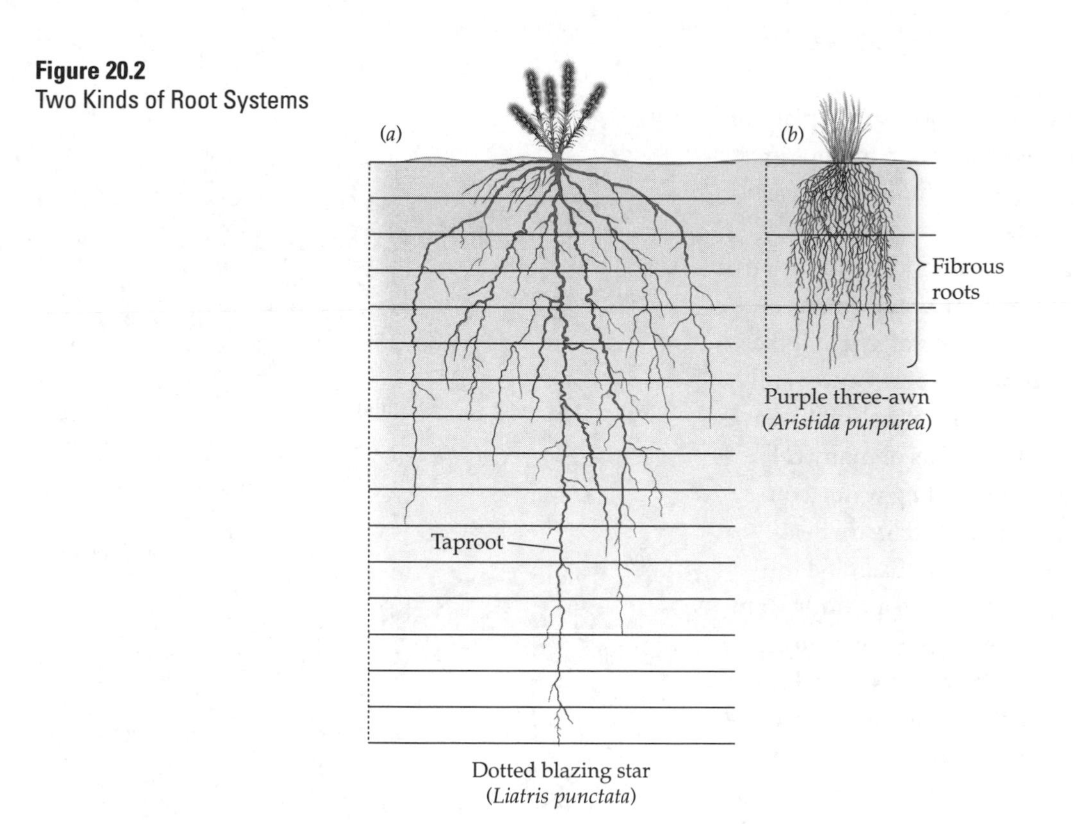

4. What are the advantages of a taproot system?

The Stem

The **stem** is normally the dominant structure of a seed plant, and for deciduous plants (those that lose their leaves) it is the only structure visible during the winter months. The stem functions to support the leaves and to transport water and sugars between the belowground and aboveground structures. The base of the stem is usually thicker than the top of the plant. Along the stem are two areas of growth leading to the formation of leaves: The uppermost part of the stem, called the **terminal bud**, develops into both leaves and new stem tissue (see Figure 20.1). On the sides (laterally) of the stem are areas of potential growth called **lateral buds**. These are located above the **nodes**, where leaves attach to the stem. The nodes are also where leaf scars form if the plant loses its leaves. Here you will examine a typical woody plant stem.

1. Obtain a twig of a sample gymnosperm.

2. Compare the stem of the monocot (grass) and dicot (radish) whose roots you examined in the previous section to the gymnosperm twig. Are there noticeable differences between the angiosperm and gymnosperm stems?

3. In **primary growth**, a plant increases in length but not in thickness. Woody plants are characterized by secondary growth, an increase in the girth of the stem as additional, secondary layers of xylem form. **Xylem** transports water and minerals from the soil to leaves, and it consists of many cell types, including water-conducting cells that are dead when mature. Examine the cross section of a woody stem provided by your instructor. For each year of **secondary growth**, a new ring of secondary xylem forms, as Figure 20.3 illustrates. How many rings of secondary xylem do you see?

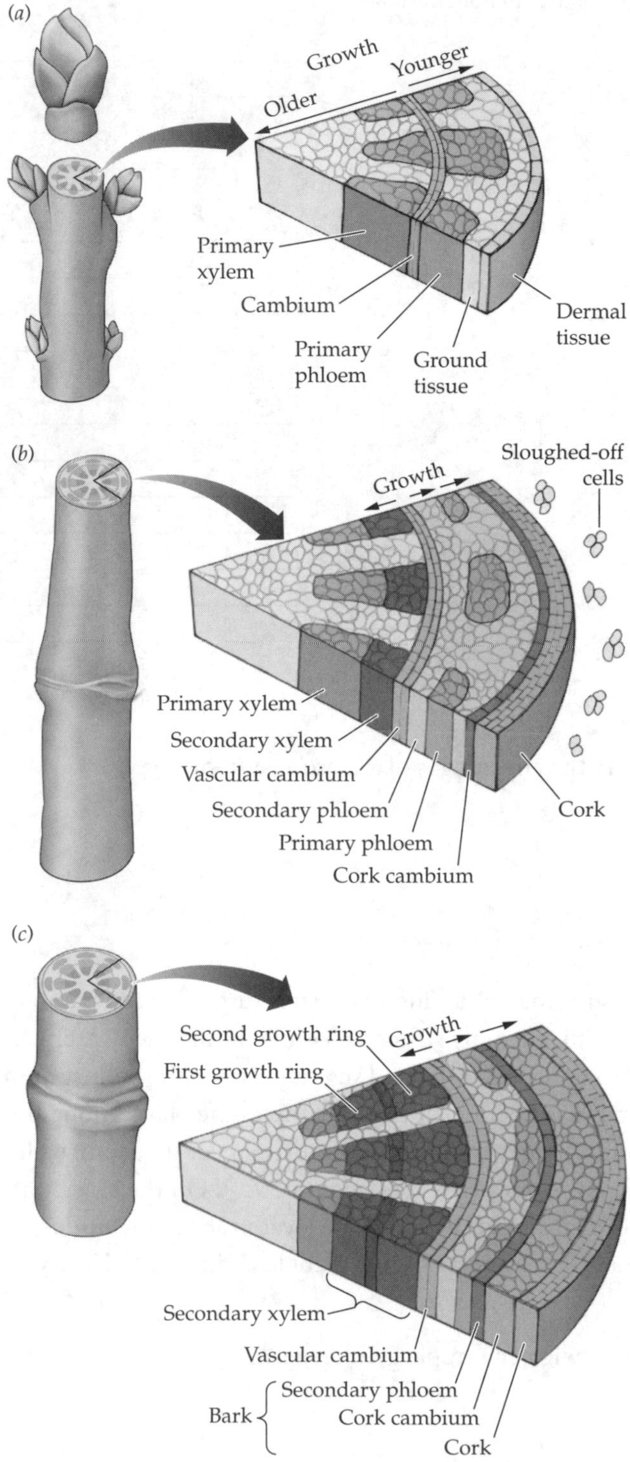

Figure 20.3
Secondary Growth in a Woody Plant Stem

4. In the box below, sketch the cross section of the woody plant stem. Label the oldest and youngest rings of secondary growth.

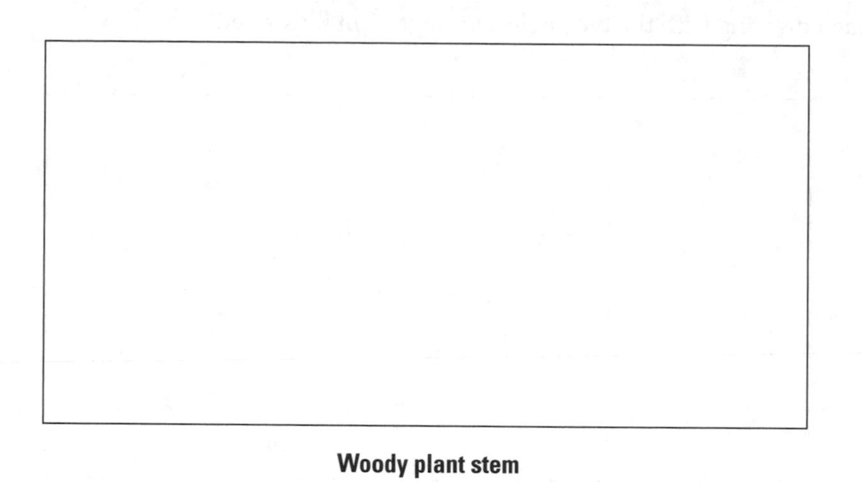

Woody plant stem

The Leaf

Leaves provide the plant with most of its food, through the process of photosynthesis. They originate from either the terminal or lateral buds (see Figure 20.1). Leaf structure maximizes surface area for increased gas exchange and sun exposure, and leaves are adapted to suit their local environments. The blade, petiole, and stipule are the three main structures of a leaf, although some plants have modified or lost these structures. The **blade**, the main photosynthetic structure, can be flat or pointy, with various shapes and sizes. It attaches to the stem at the node, which is normally connected by a structure called the **petiole**, or arm, of the leaf. Like an arm moves a hand, the petiole moves the leaves of maple trees to maximize sun exposure. Corn leaves lack a petiole and are therefore described as "sessile." **Stipules** are small, often leaflike structures at the base of a leaf in angiosperms. They are often adapted for specific functions, like protection in roses (thorns) and support in prickly ivy (climbing tendrils).

Leaf arrangement, venation, and size vary greatly across species as plants have become adapted to different environmental conditions.

Leaf Arrangement

Leaf arrangement depends on the location of lateral buds (Figure 20.4). Leaves positioned directly across from one another on the stem are said to have an **opposite** arrangement. The arrangement in which more than two leaves are positioned around the stem at the same node is called **whorled**. When node positions alternate between left and right sides of the stem, the leaves are described as having an **alternate** arrangement. Here you will compare the different types of leaf arrangement.

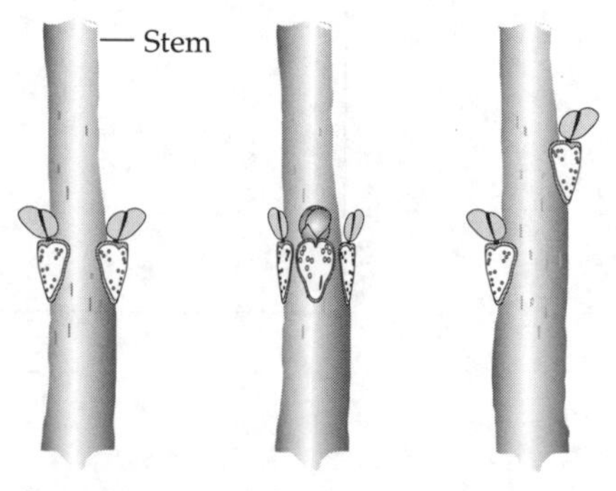

Figure 20.4
Leaf Arrangement

1. Obtain three different leaf samples—one of each type of arrangement.
2. Determine which sample represents each arrangement. In the three boxes below, sketch the samples and label each drawing with the type of leaf arrangement illustrated.

Leaf Venation

In some plant species, venation of the leaf is also highly variable, depending on adaptation to a specific environment. **Venation** is the arrangement of veins in the leaves of the plant (Figure 20.5). Much like our circulatory system, these channels transport nutrients and water to cells throughout the body of the plant. **Palmate** veins stretch out like the bones of our hands, lacking a main vein and having many smaller branches covering the surface of the leaf. **Parallel** veins do not cross each other, but run in the same directions down the length of the leaf. **Pinnate** venation displays one main vein with evenly spaced smaller veins running perpendicularly down the side of the leaf. Here you will compare the different types of leaf venation.

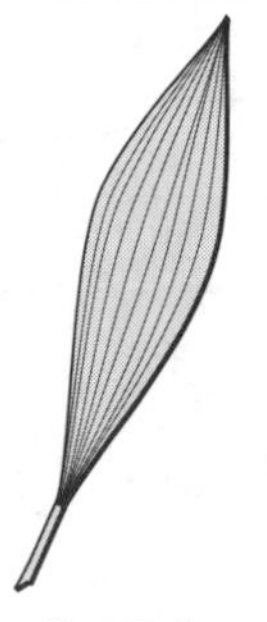

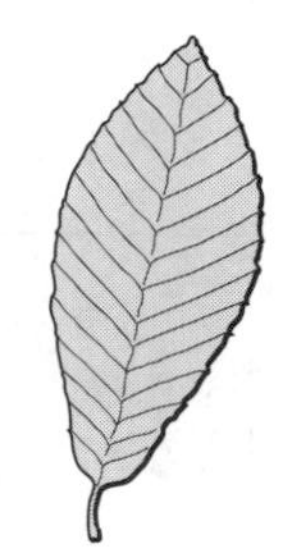

Figure 20.5
Leaf Venation

1. Obtain three different leaf samples—one of each type of venation.

2. Determine which sample represents each venation. In the three boxes below, sketch the samples and label each drawing with the type of venation illustrated.

3. Examine your dicot and monocot specimens. Which type of venation do they have? Do you think venation is an inherited characteristic?

Concept Check

1. Describe the three main structures of seed plants. What functions do they serve?

2. Why have plants evolved such a variety of leaf types, arrangements, and venation?

3. What is the function of the actively dividing cells behind the root cap?

4. How is the terminal bud different from a lateral bud?

5. Are there consistent differences between the physical characteristics of dicots and monocots?

ACTIVITY 2 Flower Anatomy

Like leaves, flowers come in all sizes and shapes, but most of the parts are common among all flowers. Think of a flower like an onion with four layers. Genes determine the ultimate structure, creating highly modified leaves: the sepals, petals, stamen, and carpel (Figure 20.6). The outer layer, consisting of **sepals**, protects the developing flower. The next layer consists of the **petals**, usually modified to attract a pollinator or spread the seeds of the flower. The last two layers consist of the reproductive structures: the **stamen** (male sex organ) and the **carpel** (female sex organ).

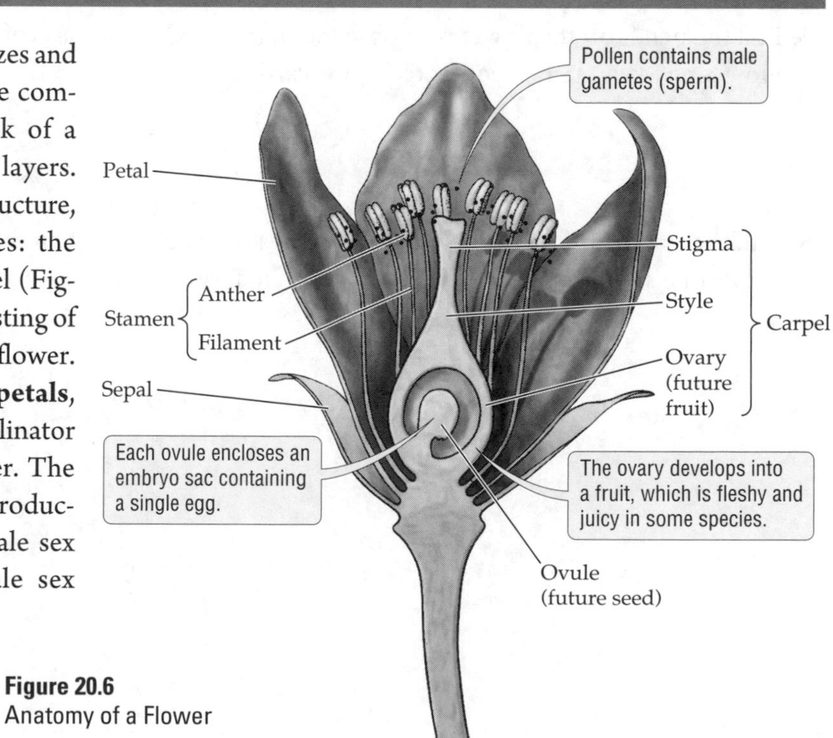

Figure 20.6
Anatomy of a Flower

The flower is usually a **hermaphroditic** reproductive structure: it contains both male and female sex organs (stamen and carpel). The stamen is composed of the **filament** and the **anther**. The filament is a long structure that permits greater exposure of the anther. The anther produces pollen grains where sperm are produced. The carpel includes the **stigma**, the **style**, and the **ovary**. The stigma secretes a sticky substance to trap pollen grains. The style connects the stigma and the ovary. Pollen delivers sperm to the carpel, and sperm fertilizes the egg in the ovule where it resides.

In this activity you will examine a flower to identify these structures.

1. Obtain a fresh flower and a razor blade.
2. Starting from the outside of the flower, identify the concentric rings, or whorls, of modified leaves. A typical flower has four whorls: sepals, petals, stamens, and carpels. Does your flower have sepals?
3. In the box below, sketch each whorl.

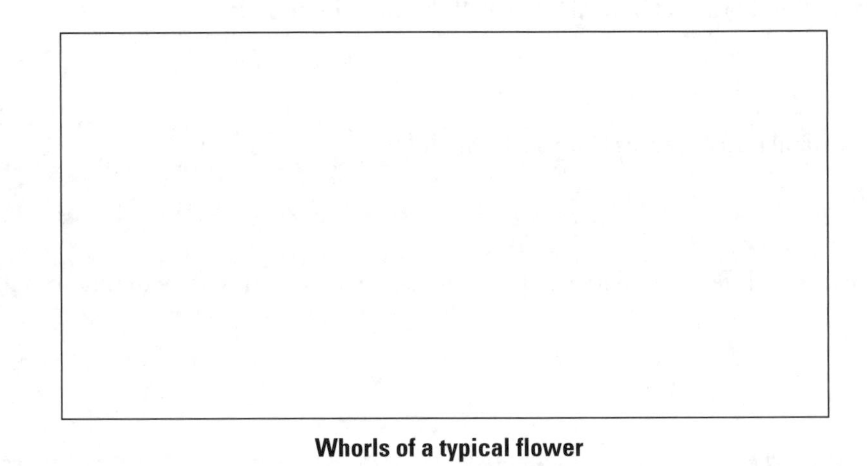

Whorls of a typical flower

4. Pull the petals off the flower to expose the stamen. Label the parts of the stamen on your drawing. How long is the stamen compared to the carpel?
5. Pull the stamen off to expose the carpel. Does your carpel contain one ovary? Is the base of the carpel one smooth, round structure, or does it have sections?

6. Using a razor blade, cut a cross section of the ovary. Examine the cross section under a dissecting microscope. Now do you think the ovary has more than one ovule? In the box below, sketch the ovary.

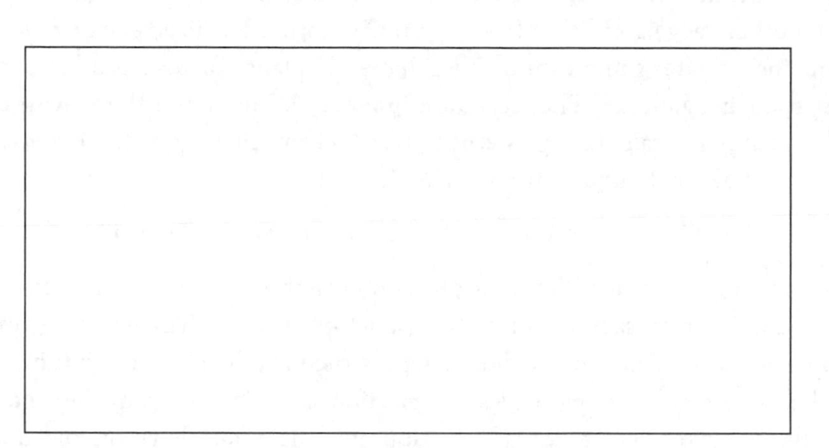

Ovary

Concept Check

1. Explain why a flowering plant is usually a hermaphrodite.

2. What is the function of the filament?

3. Describe the functions of the four whorls of a typical flower.

ACTIVITY 3 Plant Reproduction and Life Cycle

Almost all eukaryotes undergo sexual reproduction, receiving DNA from two sources, thereby promoting genetic diversity. Humans receive their DNA in the form of two sets of chromosomes from the mother and father. Organisms that have two sets of chromosomes are known as **diploid** (**2*n***). Organisms that have only one set of chromosomes are known as **haploid** (***n***). In humans, meiosis creates gametes that are haploid. **Gametes** from a mother (egg) and father (sperm) unite to form a fertilized egg or **zygote** that is diploid.

In animals, meiosis creates gametes and only gametes. In plants, meiosis produces intermediate haploid cells before producing gametes. These cells are **spores** (*n*). The spores then divide during mitosis to produce a multicellular tissue called the **gametophyte**, which produces gametes. Here's a condensed version of the sequence of events that produces gametes in plants:

Meiosis → mitosis → differentiation of egg or sperm

In both the anther and the ovule of flowering plants, mitosis produces multicellular tissue from the spore. The pollen grain and the embryo sac are the resulting gametophytes, which then undergo mitosis to produce sperm and egg. The pollen grain attaches to the stigma. A structure called the **pollen tube** forms to allow the sperm access to the ovule, where the egg resides. Fertilization takes place, forming the zygote. The zygote divides to produce the embryo, which eventually develops into the multicellular diploid adult that is called a **sporophyte** because it produces the spores.

During its life cycle every plant produces two distinct multicellular "generations": (a) haploid, multicellular gametophytes, which then give rise to (b) diploid, multicellular sporophytes that comprise the next generation. For this reason, scientists describe the life cycle of a plant as an **alternation of generations**. In this activity you will examine the life cycle of a seed plant and compare it to that of a seedless plant, moss.

1. Obtain two moss samples—one of the gametophyte generation and one of the sporophyte generation.
2. Study Figure 20.7, illustrating the alternation of generations in a flowering seed plant. Sexual reproduction ensures genetic diversity through the combination of DNA from two sources. With this in mind, is the diploid stage more genetically diverse than the haploid stage? Why?

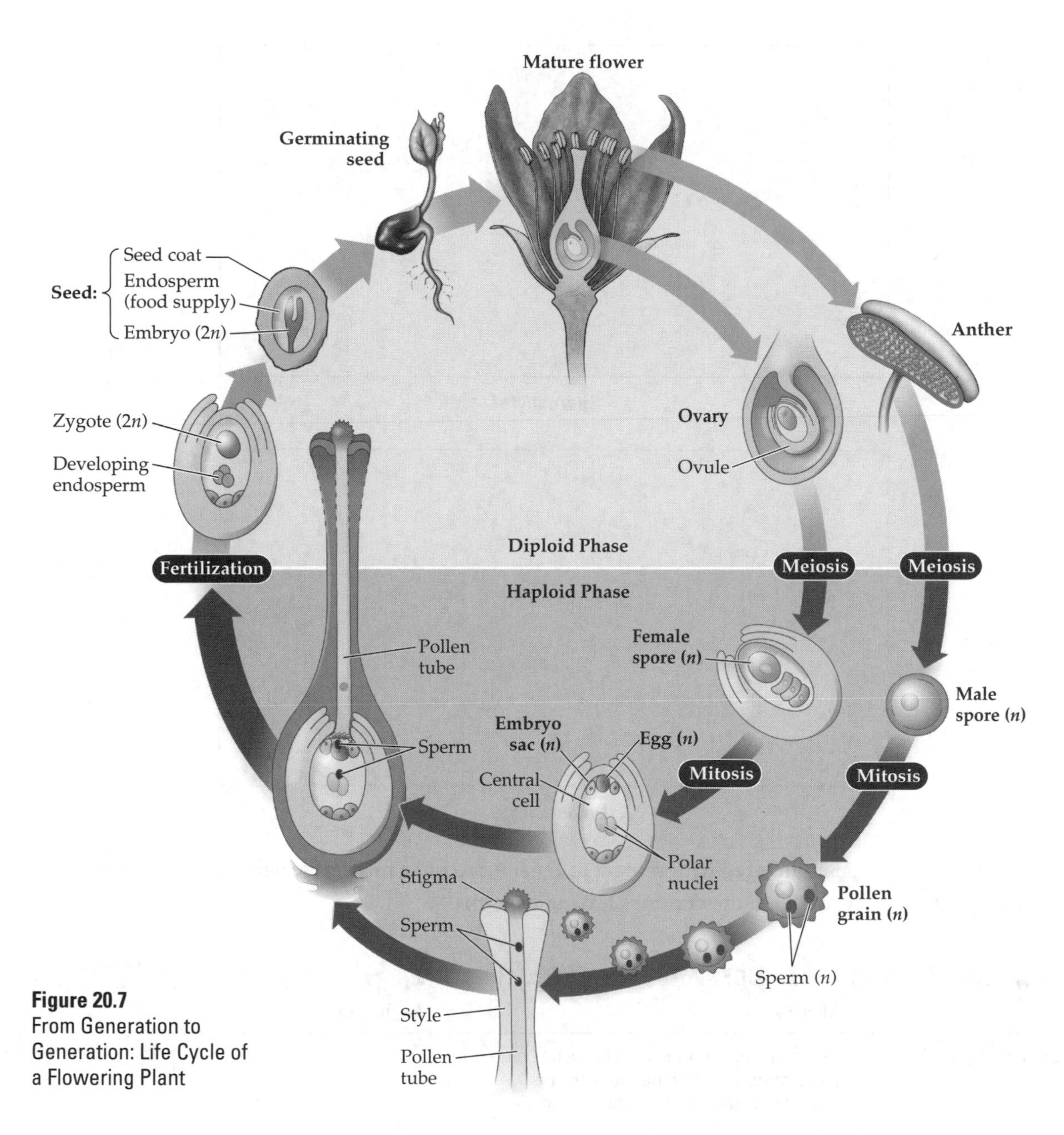

Figure 20.7
From Generation to Generation: Life Cycle of a Flowering Plant

3. Are pollen grains single-celled or multicellular? Are embryo sacs single-celled or multicellular?

4. Unlike seed plants, mosses have conspicuous gametophytes. The spores released from the mother plant (sporophyte) give rise to free-living gametophytes that are often photosynthetic. Examine the moss samples. The sporophyte is smaller than the gametophyte. Sketch the two samples in the two boxes on the next page.

Sporophyte

Gametophyte

5. Table 20.1 has been completed with respect to the life cycle of a moss. Use what you know about seed plants to help you fill in the columns for angiosperms.

TABLE 20.1 COMPARISON OF MOSS AND ANGIOSPERM LIFE CYCLES

	Mosses	**Angiosperms**
Gametophyte (*n*)	Dominant structure in the life cycle. Long-living structure, with no roots or true vascular system, that produces gametes.	
Sporophyte (2*n*)	Dependent on the gametophyte. Small, chlorophyll-containing structure that produces spores.	
Reproduction	Water is necessary for sperm to reach eggs. Spores hibernate until conditions favorable for fertilization exist.	

Concept Check

1. How is a gametophyte different from a gamete?

2. Describe alternation of generations in seed plants.

ACTIVITY 4 Development of the Seed

In seed plants, the female gametophyte is the **embryo sac.** One cell in each embryo sac differentiates into an egg cell, but the embryo sac also contains a large cell with a pair of nuclei lying free in the cytoplasm. When pollen lands on a stigma, it creates the pollen tube, which grows through the style and enters the embryo sac inside the ovule. One sperm fertilizes the egg, and the other fertilizes the two nuclei—which is why sexually reproducing flowering plants are said to have **double fertilization**. The fertilized egg becomes the diploid zygote, and the two fertilized nuclei form the **endosperm**, which provides food for the developing embryo. As the embryo and endosperm mature, the outer layers of the ovule differentiate into the **seed coat**, the hardened covering of a seed.

In this activity you will examine the embryos inside corn and bean seeds.

1. Obtain bean seeds, a razor blade, a dissecting probe, bean embryo slides, corn embryo slides, and a dissecting microscope.
2. Carefully remove the seed coats from the bean seeds.
3. Using Figure 20.8 (on the following pages) for reference, pull apart the two large **cotyledons**, where food is stored and sometimes processed. Plants having two cotyledons are dicots. Do you recall the differences between monocots and dicots that you listed in Activity 1?
4. Using a dissecting microscope, examine the embryo located between the cotyledons.
5. In the box below, sketch the embryo of a bean seed. Can you distinguish between the future root and leaves?

Bean embryo

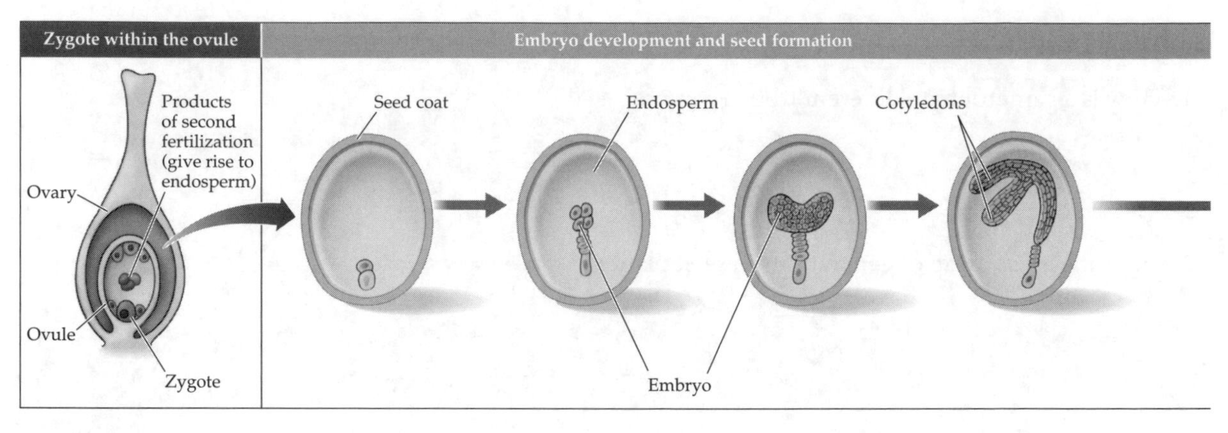

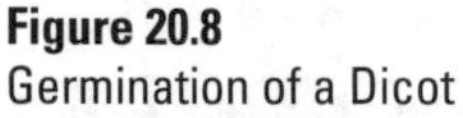
Figure 20.8
Germination of a Dicot

6. Using the slides provided by your instructor, examine the various stages of bean germination. Refer to Figure 20.8. Which stages of embryo development can you see on your bean embryo slides?

7. A plant having one cotyledon, such as corn, is called a monocot. Using the slides provided by your instructor, examine the stages of embryo development in a corn seed.

8. Seeds develop in the ovary, which is composed of one or many ovules. How is this different from the development of gymnosperm ("naked") seeds, and what benefit does seed development within an ovary confer?

9. Inside each ovule is one egg. Therefore, inside an ovary with one ovule, one seed can develop; and inside an ovary with many ovules, many seeds will develop. Can you think of any plants that have flowers containing an ovary with many ovules? Think of the food you eat.

Concept Check

1. What are the parts of a seed?

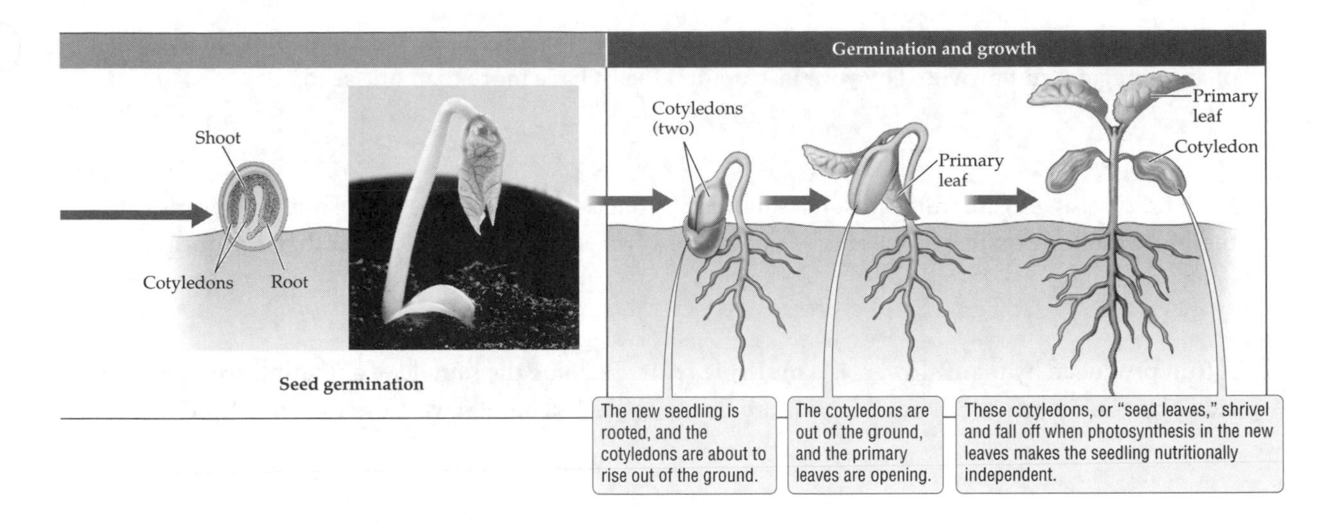

2. Describe the difference between monocot and dicot seeds.

3. What is the purpose of the seed coat?

ACTIVITY 5 Fruits

Throughout evolution, ovaries evolved into many forms to attract animals to aid in seed dispersal. The seeds are swallowed by animals and, being indigestible, generally deposited with the feces miles away from the plant. Some ovaries are dry (such as acorns), attracting bears and squirrels; others are fleshy (such as grapes), attracting birds. But all modifications of the ovary are fruits. Hence, the fruit serves as a species-dispersing agent.

In this activity you will examine the fruit of a seed plant. To determine the type of fruit, you need to know the following:

- How many carpels does the flower contain?
- How many flowers are required to produce the fruit?
- How many ovules does the ovary contain?

1. Obtain fruit samples illustrating the three main types of fruit (simple, aggregate, and multiple), as well as an apple, a razor blade, and a dissecting probe.
2. A fruit produced by one flower that has one carpel with one ovary is a **simple fruit**. Examine the sample of a simple fruit provided by your instructor. Does it have more than one seed?
3. Simple fruits can contain an ovary with one ovule or many ovules. Therefore, they can have one or many seeds. Name a type of simple fruit other than the one you've been examining.

4. A fruit produced by one flower with two or more carpels is an **aggregate fruit**. Examine the sample of an aggregate fruit provided by your instructor. Does it have more than one seed?

5. By definition, aggregate fruits must have more than one seed because they have multiple ovaries and therefore multiple ovules containing seeds. Name a type of aggregate fruit other than the one you've been examining.

6. A fruit produced by many flowers is a **multiple fruit**. Examine the sample of a multiple fruit provided by your instructor. Does it have more than one seed? Does it have more than one fruit?

7. By definition, a multiple fruit must have more than one fruit formed by the fusion of two or more flowers' ovaries. Name a type of multiple fruit other than the one you've been examining.

8. Apples are an exception to the rule. An apple is a simple fruit, but part of the stem forms the fruit. The ovary becomes surrounded by the stem growing around it. The stem portion of the apple remains soft and edible. The apple is classified as a **pome**. Study the development of the apple in Figure 20.9.
9. Follow your instructor's directions for dissecting the apple at your lab bench. Half of the class will cut an apple in a cross section; the other half will cut an apple in a longitudinal section.
10. On your dissected apple, identify the structures labeled in Figure 20.9.

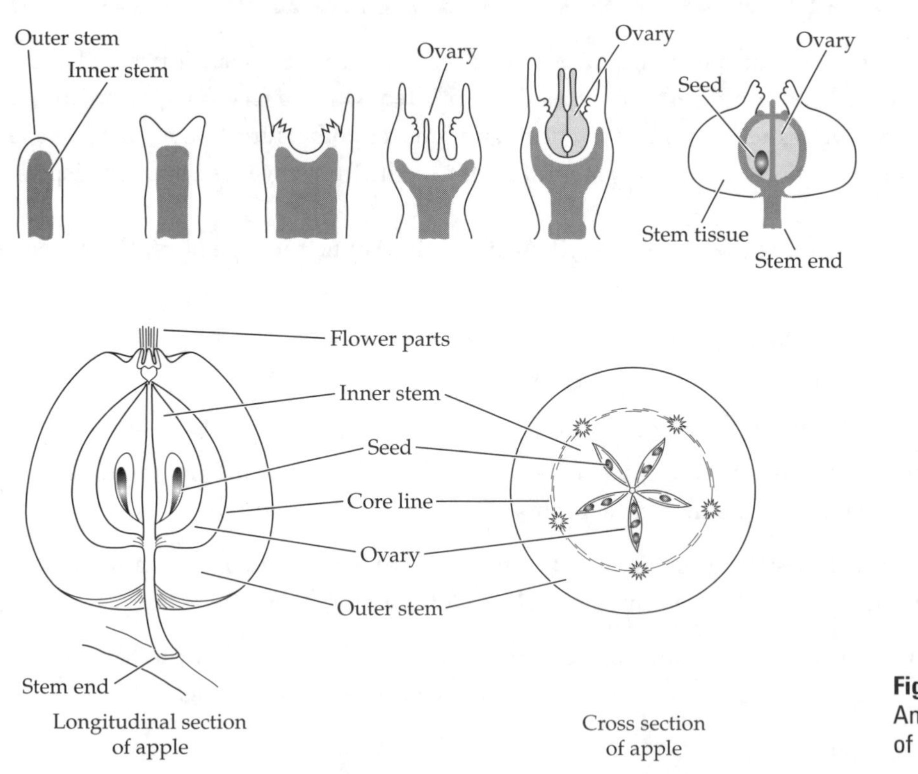

Figure 20.9
Anatomy and Development of an Apple

Concept Check

1. What are the three main types of fruits?

2. Relate the presence of multiple ovaries and ovules to the human female reproductive system.

3. How are apples an exception to the three traditional categories of fruit?

Key Terms

aggregate fruit (p. 20-18)
alternate leaf arrangement (p. 20-5)
alternation of generations (p. 20-12)
angiosperm (p. 20-1)
anther (p. 20-10)
autotroph (p. 20-1)
blade (p. 20-5)
carpel (p. 20-9)
cotyledon (p. 20-15)
dicot (p. 20-1)
diploid (2*n*) (p. 20-12)
double fertilization (p. 20-15)
embryo sac (p. 20-15)
endosperm (p. 20-15)
fibrous root system (p. 20-2)
filament (p. 20-10)
gamete (p. 20-12)
gametophyte (p. 20-12)
gymnosperm (p. 20-1)
haploid (*n*) (p. 20-12)
hermaphroditic (p. 20-10)
lateral bud (p. 20-3)
leaf (p. 20-5)
monocot (p. 20-1)
multiple fruit (p. 20-18)
node (p. 20-3)
opposite leaf arrangement (p. 20-5)
ovary (p. 20-10)
palmate (p. 20-7)
parallel (p. 20-7)
petal (p. 20-9)
petiole (p. 20-5)
pinnate (p. 20-7)
pollen tube (p. 20-12)
pome (p. 20-18)
primary growth (p. 20-4)
producers (p. 20-1)
root (p. 20-2)
secondary growth (p. 20-4)
seed coat (p. 20-15)
sepal (p. 20-9)
simple fruit (p. 20-15)
spore (p. 20-12)
sporophyte (p. 20-12)
stamen (p. 20-9)
stem (p. 20-3)
stigma (p. 20-10)
stipule (p. 20-5)
style (p. 20-10)
taproot system (p. 20-2)
terminal bud (p. 20-3)
venation (p. 20-7)
whorled leaf arrangement (p. 20-5)
xylem (p. 20-4)
zygote (p. 20-12)

Review Questions

1. Describe three functions of a root.

2. Why are leaves generally flat and broad?

3. How is secondary growth related to the age of woody plants?

4. Which type of growth—primary or secondary—is the terminal bud responsible for?

5. What evolutionary benefit does the fruit provide?

6. The stamen is composed of the ________________ and the ________________.

7. The carpel is composed of the ________________, the ________________, and the ________________.

8. Plants are self-feeders, or "autotrophs." Through the process of photosynthesis they use the energy in sunlight to make sugars, and are known as ________________.

9. What is a fruit?

10. What is the advantage of being a hermaphrodite?

11. What is double fertilization?

12. Monocots and dicots differ from embryo to adult. Describe some of the differences you examined in this lab.

13. Describe how a fruit can have more than one seed.

14. How does the life cycle of a plant differ from the life cycle of an animal?

LAB 21 Populations and Communities of Organisms

Objectives

- Determine the range of factors that affect organism populations in ecosystems.
- Determine the influence of carrying capacity on protist populations.
- Determine population estimates using the mark–recapture method.
- Analyze and construct a simplified terrestrial food web using owl pellets.
- Observe mutualistic symbiotic associations in termites.
- Apply satellite data to assess global net primary productivity.
- Compare bacterial populations through the use of nutrient agar plates.

Introduction

Living organisms do not exist in a vacuum. They interact with individuals of their own species (**population**) and with individuals from other species (**community**). These are the living, or **biotic**, components of an ecosystem. Organisms also interact with all the nonliving, or **abiotic**, components of the ecosystem—factors like temperature, water availability, nutrients, and energy. An **ecosystem** is therefore a community of organisms interacting with their physical environment. These interactions are dynamic and complex, changing in space and through time. Ecosystems are not closed systems operating independently of neighboring ecosystems. They are open systems, receiving inputs (organisms, nutrients, energy) from other ecosystems (both neighboring and distant) and providing inputs (organisms, nutrients, energy) to other ecosystems. This characteristic ensures that all ecosystems are interconnected either directly or indirectly and at different scales and times.

Ecology is one of the youngest branches of biology. Ecologists try to understand two basic characteristics of any ecosystem:

1. How energy flows through the ecosystem
2. How nutrients are cycled within the ecosystem

Almost all ecosystems receive their energy from the sun: Photosynthetic organisms known as **producers** use the sun's energy to make sugars. This energy can then be passed upward through **food chains** to many different **consumers** ranging from **primary consumers** like insects that feed on plants, to **secondary consumers** that feed on primary consumers, to **tertiary consumers** that in turn feed on secondary consumers. Each feeding level is known as a **trophic level**. Initial energy from the sun is passed upward in steps, with only about 10% of the energy being passed from one trophic level to the next—as illustrated in Figure 21.1, where an insect-eating bird is a secondary consumer and a bird-eating bird is a tertiary consumer. Ecologists study changes in ecosystems through time and are therefore interested in the amount of energy that producers make each year and can pass on to consumer animals. They use a measurement called **net primary productivity** (**NPP**): essentially, how much additional plant **biomass** (organisms per unit of area) is made each year or season in a given area. NPP is clearly different in summer and winter in temperate ecosystems, but it does not change very much in tropical ecosystems. **Nutrient** availability in temperate ecosystems will also change in different seasons, as factors like temperature and available biomass determine the activity of **decomposer** organisms, consumers that break down dead animal and plant tissues, making their chemical components available to the ecosystem again. A positive or negative change in the amount of plant biomass available each year will affect the entire food chain.

Figure 21.1
Idealized Energy Pyramid

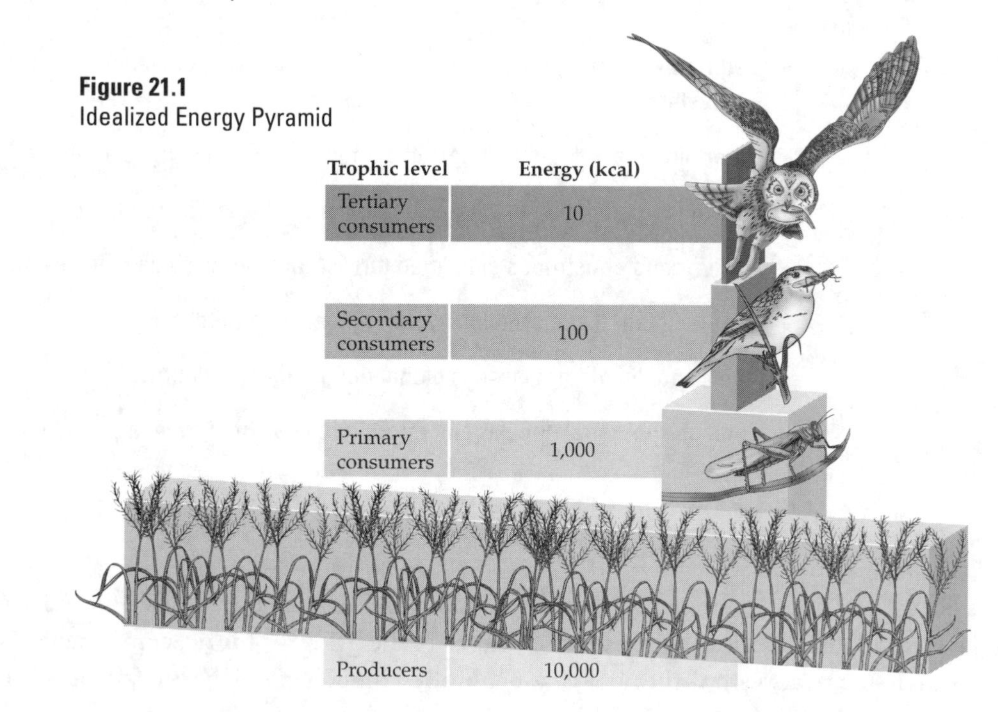

The populations of all producers and consumers in an ecosystem can expand and grow only if there is a plentiful supply of nutrients, food, and water. Plants require adequate sunlight, nutrients from the groundwater (including nitrates and phosphates), carbon dioxide in the atmosphere, and water to survive. Because they do not photosynthesize, animals need to eat to acquire their energy and nutrients and water. Low-level consumers that feed on plants (herbivores) need to be able to find the plant species that they can digest. High-level consumers that prey on animals (carnivores) need to be able to find prey animals that are big enough or numerous enough to satisfy the energy demands of their active lifestyles. All organisms, whether producers or consumers, lose metabolic heat as they use energy to maintain their own cellular processes. This loss of heat reduces the amount of energy available to an organism at the next level in the food chain (see Figure 21.1).

The population of any species of organism is therefore constrained by the levels of food, nutrients, and water made available by the ecosystem at any point in time. The maximum population size that can be supported indefinitely by the environment is known as the **carrying capacity** of an ecosystem for a given organism. In this lab you will explore the concepts of population growth and carrying capacity, and how the availability of nutrients affects populations. You will also learn how ecologists estimate population size, how certain decomposer organisms interact in symbioses with animals, and how global NPP is determined.

ACTIVITY 1 The Carrying Capacity of the Protist *Didinium*

Freshwater ponds, lakes, and rivers are teeming with microscopic representatives of the kingdom Protista. One group that is particularly well represented is the ciliates. This group includes bacterial-feeding protists of the genus *Paramecium* and its **predator**, the ciliate *Didinium*. Ciliates are fast-moving **protists** that glide over surfaces through the highly controlled undulations of cilia covering their outer surface like a field of wheat. Protists of the genus *Paramecium* are slipper-shaped; *Didinium* is shaped like a jet engine. *Didinium* can also move like a jet aircraft, able to rapidly pursue, capture, and ingest *Paramecium* **prey**.

The population size of any species depends on the sustainable availability of food. As an example, Figure 21.2 illustrates the carrying capacity of a laboratory population of paramecium (*Paramecium caudatum*): the population increases quickly at first, but then stabilizes at the maximum population size that can be supported by its environment. The **growth rate** of the population falls as the population size approaches carrying capacity because resources such as food become increasingly limited.

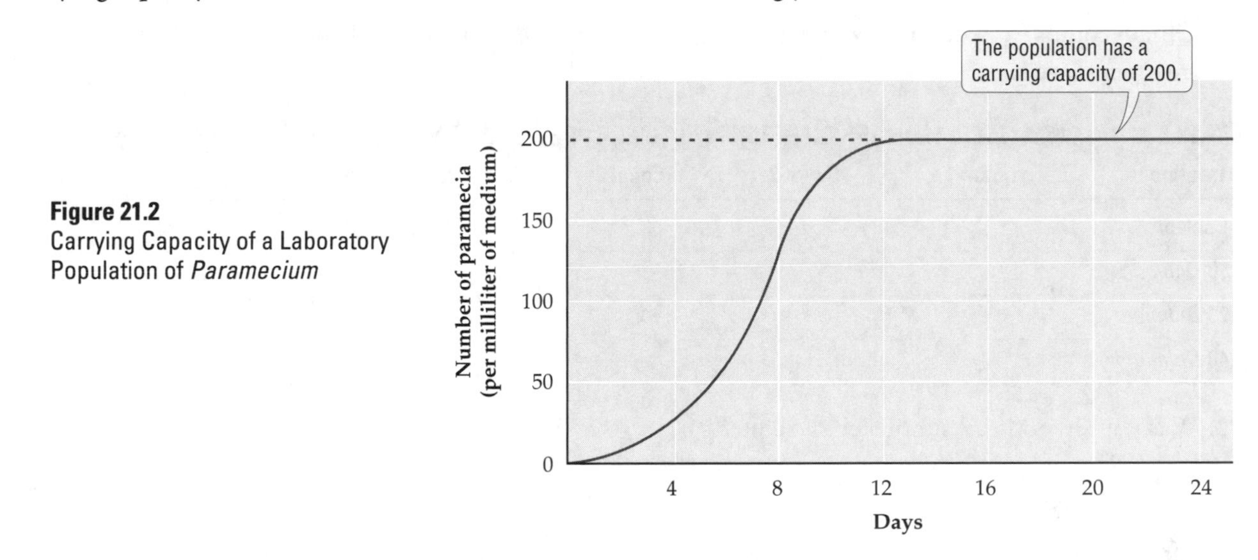

Figure 21.2
Carrying Capacity of a Laboratory Population of *Paramecium*

In this activity you will work with your classmates to determine how the population size of *Didinium* in a given environment affects the availability of its food source, *Paramecium caudatum*. Your instructor has cultured both organisms for 2 days at the following different concentrations of prey to predator:

- 200 individuals of *P. caudatum* only (control)
- 200 individuals of *P. caudatum* and 10 individuals of *Didinium*
- 200 individuals of *P. caudatum* and 20 individuals of *Didinium*
- 200 individuals of *P. caudatum* and 40 individuals of *Didinium*

1. Before you begin, formulate the hypothesis you're trying to test and write it here.

2. With the students in your lab group, obtain one of the four cultures listed above as directed by your instructor, along with two plastic pipettes, a test tube, three microscope slides, and three coverslips.

3. Using the pipette, withdraw 5 ml of water and protists from the bottom of the culture jar and put it in the test tube.

4. Return the culture jar for someone else to use.

5. Shake the test tube well while holding a finger or thumb over the opening.
6. Using another pipette, transfer one large drop from the test tube to a microscope slide.
7. Prepare a wet mount, carefully adding a coverslip by placing one side against the slide first, and then lowering the opposite side.
8. Using a compound microscope, count how many paramecia you can see at 100× total magnification. (Try searching in a zigzag manner, starting at one corner of the coverslip.) Record the number under "Count 1" for the appropriate culture in Table 21.1.
9. Make two more wet mounts from the same culture, and count how many paramecia you see at 100× total magnification in each one. Record the numbers under "Count 2" and "Count 3" for the appropriate culture in Table 21.1.
10. Calculate the average (mean) density for your culture and enter the value in Table 21.1.
11. Obtain counts from others in your lab class to complete the information in Table 21.1 for all four treatment levels.

TABLE 21.1 *PARAMECIUM* UNDER DIFFERENT LEVELS OF PREDATION BY *DIDINIUM*

Treatment	Count 1	Count 2	Count 3	Average (mean) density
Control				
10 *Didinium*				
20 *Didinium*				
40 *Didinium*				

12. Was your hypothesis supported or refuted?

13. What was the value of performing multiple counts?

14. In Figure 21.3, plot the average density of paramecia at each predator concentration. Label treatment type (control, 10 *Didinium*, 20 *Didinium*, and 40 *Didinium*) on the *x*-axis, and average density of *Paramecium* individuals on the *y*-axis.

15. Which treatment had the greatest predation rate?

Figure 21.3
Carrying Capacity of a Laboratory Population of *Didinium*

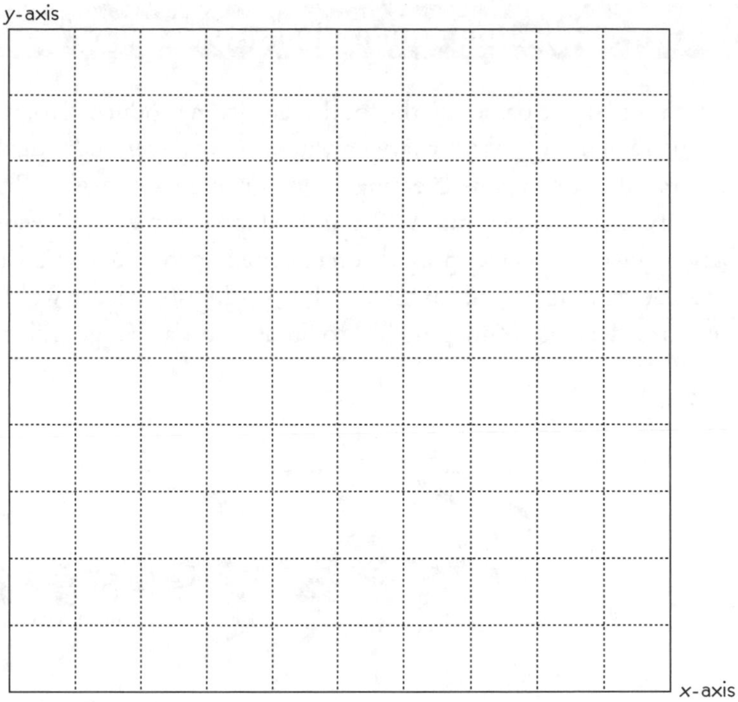

y-axis

x-axis

16. Which treatment had the greatest *Paramecium* survival rate?

17. Judging from the results of this experiment, which concentration of *Didinium* would be capable of surviving in a sustainable or ongoing way, given the available population of *Paramecium*?

Concept Check

1. What is the main factor that determines carrying capacity for a predator population?

2. Name two factors that determine carrying capacity for a prey population.

3. What is one predation avoidance strategy that *Paramecium* can use to reduce predation rates?

4. If prey number is vital to life for a predator, what can a predator do if the number of prey items in a given area is significantly reduced or they disappear altogether?

ACTIVITY 2 Estimating Population Size Using Mark–Recapture

Energy in ecosystems passes up the food chain, from producers, to primary consumers, to secondary consumers, finally ending with tertiary consumers. However, individual food chains have many cross-connections to other organisms, creating interconnected food webs (Figure 21.4). A **food web** connects all of the food chains in a community. To understand complex food webs, ecologists may simplify matters by focusing on just one species population in a specific trophic level. Making annual counts of an individual population over long periods of time enables ecologists to learn what factors might be causing population increases and decreases and, possibly, to predict how a change will cascade through an ecosystem.

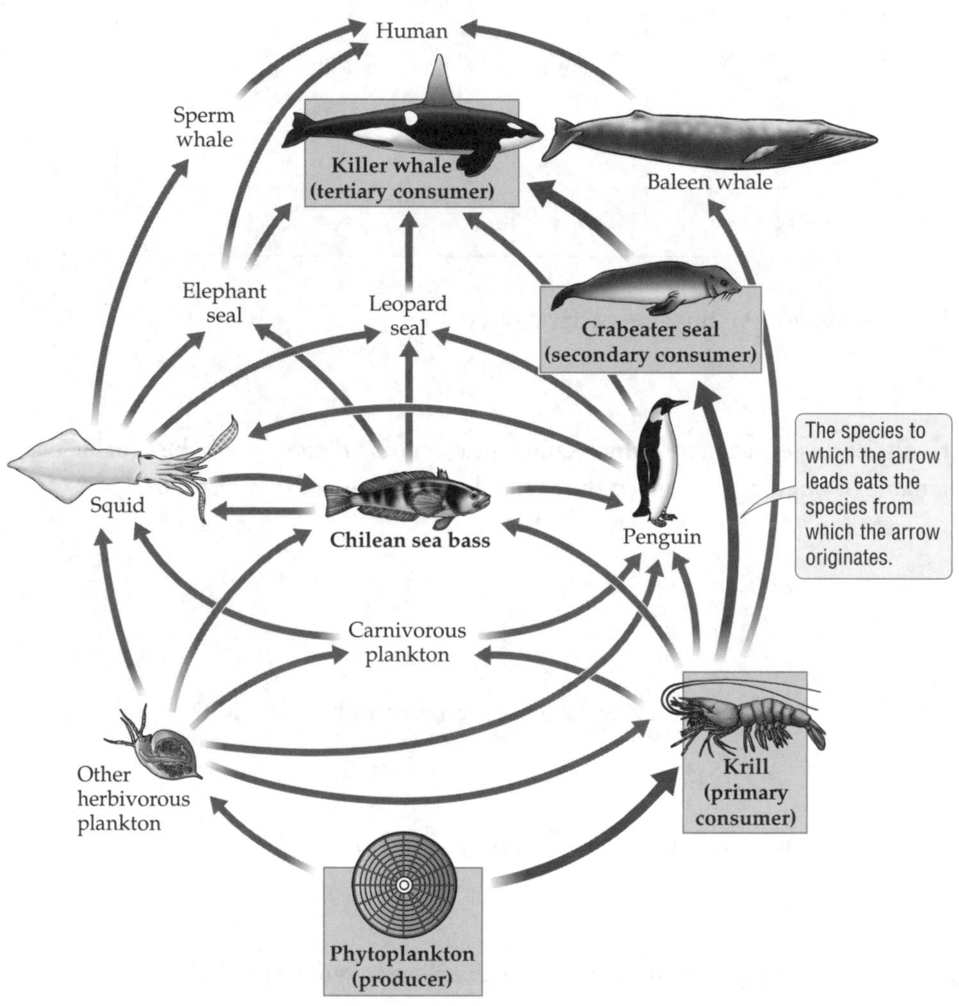

Figure 21.4
Simplified Food Web in the Antarctic Ocean

Counting the number of individuals of a species may be relatively easy when we're dealing with immobile organisms like plants or barnacles or rocks, but what about highly mobile animals or protists like *Paramecium* that can easily move from one area to another? One method that ecologists use is called **mark–recapture**. Using this method, ecologists capture and tag or mark a certain number of individual animals from a species' population. The marked animals are then returned to the wild. On a second visit, animals of the same species are captured and the number carrying markings from the first visit is noted. From this information the population size can be estimated by the following formula (the **Lincoln–Petersen method**):

$$N = \frac{MC}{R}$$

where

N = the estimate of total population size
M = the number of animals captured and marked initially
C = the number of animals captured the second time
R = the number of animals that were found to be marked the second time

In this activity, using beans as substitutes for animals that you might have captured in an ecosystem, you will estimate the population (number of beans) in a covered cardboard box or paper bag.

1. Obtain a container of beans and a permanent marker or China marker. Your instructor has told you that the container has 100 beans, but you are doubtful and wish to check! You will use the Lincoln–Petersen method (the formula above) to confirm your instructor's estimate.

2. Remove 20 beans from the container without looking into the container. Using the marking pen, mark each of the 20 beans with an "x." This step is your initial capture and marking, represented by M in the formula. Return the marked beans to the container. Shake the container to mix the beans, simulating movement.

3. Remove another 20 beans from the container without looking. These represent your second count, C. Count how many of these 20 beans are marked; these represent the marked and recaptured beans, R. Record the values of C and R below. Return the beans to the container. Shake the container to mix the beans.

   C =

   R =

4. Using the formula above and the values for C and R obtained in step 4, estimate the bean population size and enter that number in Table 21.2 as the first mark–recapture estimate. Show your calculations here.

TABLE 21.2 ESTIMATES OF BEAN POPULATION USING MARK–RECAPTURE

Mark–recapture number	Instructor's estimate	Mark–recapture estimate
1	100	
2	100	
3	100	

5. To make a second estimate of total bean population, remove another 20 beans from the container without looking (C). Note how many are marked (R). Record the C and R values below. Return the beans to the container. Shake the container to mix the beans.

   $C =$

   $R =$

6. Using the formula above and the values for C and R obtained in step 5, estimate the bean population size and enter that number in Table 21.2 as the second mark–recapture estimate. Show your calculations here.

7. To make a third estimate of total bean population, remove yet another 20 beans from the container without looking (C). Note how many are marked (R). Record the C and R values below. Return the beans to the container.

   $C =$

   $R =$

8. Using the formula above and the values for C and R obtained in step 7, estimate the bean population size and enter that number in Table 21.2 as the third mark–recapture estimate. Show your calculations here.

9. What is the average (mean) bean population from your three estimates?

10. Count all the beans in your container. What is the difference between your average estimated bean population and the actual bean population?

11. If you had removed only 10 beans each time, would your estimate have been more or less accurate? Why?

12. If you had removed 30 beans each time, would your estimate have been more or less accurate? Why?

Concept Check

1. For what sorts of animals would the mark–recapture method provide accurate population estimates?

2. For what sorts of animals would the mark–recapture method *not* provide accurate population estimates?

3. Would the length of time between counts make a difference in real animal populations? Why or why not?

4. Like everyone else, ecologists are restricted by the amount of time they have to conduct population estimates. Why, then, do you think mark–recapture is such an effective method of estimating population size?

ACTIVITY 3 Constructing a Food Web Using Owl Pellets

The barn owl (*Tyto alba*) is at the top of the food web in terrestrial ecosystems in the Pacific Northwest. The barn owl preys on a wide range of small animals, including rodents and birds. Rodents include voles, field mice, moles, pocket gophers, deer mice, and kangaroo rats. Barn owls swallow these prey animals whole, and any parts that cannot be digested they regurgitate as a pellet—on average, one per day. The pellets include fur, feathers, and bones. Analysis of the bones in a single pellet can allow ecologists to determine how many prey animals are eaten over the period of a year and to construct a simplified food web based on this information. For example, if analysis of 10 owl pellets reveals that they each contain 2 shrews, we can extrapolate, calculating that one barn owl could eat 730 shrews per year. If we maintain owl pellet analysis through time, we can detect a change in the food web—if, for example, the number of shrews eaten by a barn owl over one year falls to 365 or increases to 1,460. As ecologists know, the diet of the shrew is mainly insects, so any change in shrew numbers through time must provide an estimate of insect population numbers too. If the skeletons of voles were found in owl pellets, ecologists could estimate their populations in the same way. In creating a simplified food web based on the prey animals found in owl pellets, information from voles is important because these herbivores feed on plant matter. Any change in vole population estimates can provide clues to changes in NPP in the producer levels of the food web.

The undigested parts of the prey animals in regurgitated owl pellets were formed, during each animal's life, from significant amounts of energy from NPP. In terrestrial ecosystems, the bulk of NPP is directly consumed by bacteria and fungi, acting as decomposers of dead animal and plant material (Figure 21.5). The owl pellets, however, have prevented the bulk of the decomposer community from doing its work until the pellet has been regurgitated. The owl that regurgitated the pellet would have derived energy from the prey animals' soft tissues, and the decomposer community would have derived energy from the nondigested components of the pellet as the more resistant compounds were broken down by their enzymes. Energy is vital in ecosystems—energy sources are not wasted.

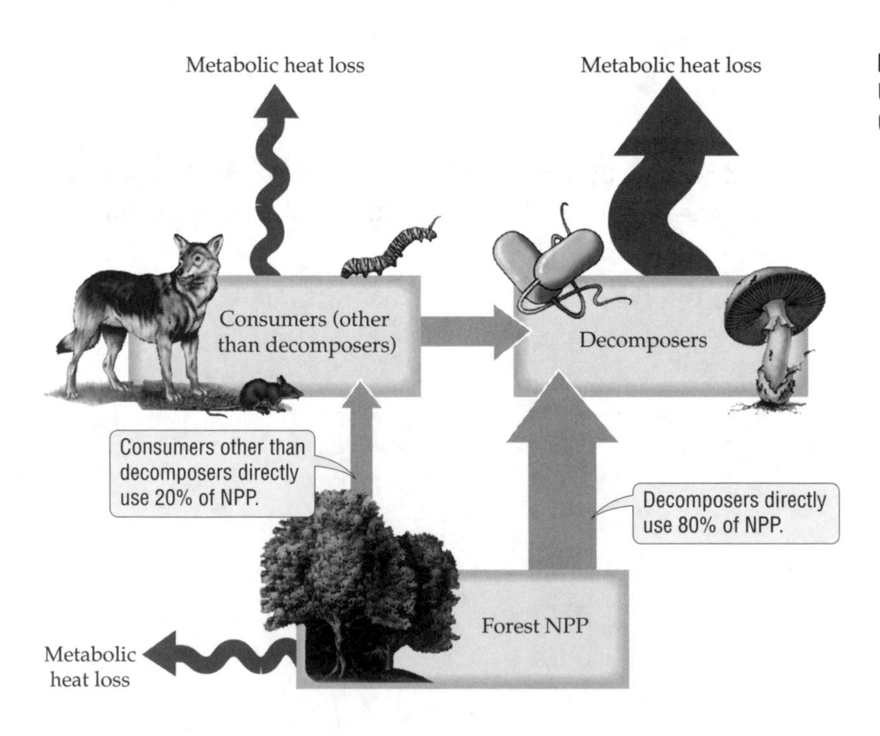

Figure 21.5
Use of NPP by Producers, Consumers, and Decomposers

In this activity you will analyze an owl pellet to determine which species the owl preyed upon and how many of each prey species were consumed. Your instructor placed owl pellets in beakers of water for 20 minutes and then drained the water from each pellet sample through cheesecloth into a large flask. The disassembled and drained pellet was then placed on paper towels on a tray to await your analysis.

1. With your lab partner(s) obtain an owl pellet, a key, some probes and forceps, and a clean sheet of white paper.
2. Using the probes and forceps, extract all bones from the pellet and lay them on the paper.
3. Using the keys provided by your instructor and with your instructor's guidance, separate out the skulls of the prey animals. If you find two skulls in your pellet, you have found two animals. If the skulls have no teeth but do have a beak, you must have two birds. If the skulls have teeth, they must be rodents.
4. Using the keys and pictorial guide provided by your instructor, complete Table 21.3 for your own data.

TABLE 21.3 PREY ANIMALS FOUND IN YOUR OWL PELLET

Prey species	Number of individuals

5. Using the data collected by the rest of your lab class, complete Table 21.4.

TABLE 21.4 CLASS DATA ON PREY ANIMALS

	Number of individuals										
Prey	Pellet 1	Pellet 2	Pellet 3	Pellet 4	Pellet 5	Pellet 6	Pellet 7	Pellet 8	Pellet 9	Pellet 10	Total
Vole											
Shrew											
Mole											
Bird											
Rat											
Total											

6. The illustration of a food web in Figure 21.4 should enable you to visualize the fact that energy from the sun is captured by plants and then transferred upward in a series of steps all the way to the top predator. The figure depicts an oceanic food web whose top predator is a killer whale. In the box below and with your instructor's guidance, construct a simplified food web based on the information gathered as a class from the owl pellets. Your food web has the barn owl as the top predator, with lower-level consumers including voles and field mice as herbivores (primary consumers), and shrews, moles, and rats as carnivores (secondary consumers). Start by writing "barn owl" at the top. Assuming you found an average of one shrew and two voles per pellet, and knowing that shrews feed on insects and voles feed on plant material, you could write "shrew" and "vole" slightly below "barn owl." Since NPP is due to plants in terrestrial ecosystems, you could write "plants" at the foot of the page. Since shrews feed on prey animals, including insects, you could write "insects" between "shrew" and "plants," representing another primary consumer group.

7. Remember that what you and your lab class have done is a simplification based on proxy evidence found in owl pellets. In reality, each trophic level has many different organisms, each individual consuming energy by feeding on organisms below it in the food web. The amount of energy available to all organisms at a given trophic level is clearly greater than the energy available to an individual organism in that trophic level. Recall also that the amount of energy available to a given trophic level decreases upward through the food web because energy is lost in each preceding trophic level as metabolic heat. In your simplified food web, connect the different trophic levels with lines to depict the flow of energy. Draw lines connecting plants to insects and voles. Draw lines connecting insects to shrews, and lines connecting both shrews and voles to the top predator, the barn owl. This is clearly a simplification, but it provides a general idea of how energy flows upward through this food web.

Food web based on owl pellets

8. According to analysis of your individual pellet, what was the main prey animal (the animal that left the most skeletal remains)?

9. According to the class data, what was the main prey animal?

10. How many of each main prey animal would you expect to be preyed on in a year?

Concept Check

1. What is the initial source of the energy held as NPP in producer communities?

2. Name one climatic event that can lead to reductions in plant biomass in terrestrial ecosystems. Explain why such an event would affect the entire food web.

3. Voles feed on plant material, and owls feed on numerous prey animals, including voles. Which of the two animals—vole or owl—has the most energy-rich diet?

4. What is one weakness of using owl pellet prey data to construct food webs?

ACTIVITY 4 Observing Mutualistic Symbiotic Associations in Termites

Ecosystems are dynamic systems with numerous interactions between living organisms and the abiotic, or non-living, environment. Many of the interactions between living organisms are forms of **symbiosis**. Symbiosis may be **mutualistic** (both organisms in the association benefit), it may be **commensal** (only one organism benefits but the other is not harmed), or it may be **parasitic** (one organism benefits but the other is harmed). Recall that two vital features of ecosystems are the flow of energy and the cycling of nutrients. In terrestrial ecosystems, nutrients like nitrates and phosphates are made available to producers after having been transformed by the decomposer community from organic matter (dead plant material, dead animals) back into the elemental form (nitrates and phosphates).

Bacteria and fungi carry out most decomposition, but other organisms are also capable of transforming organic matter into inorganic elements. Many organisms that are capable of digesting plant fibers and/or wood are found in mutualistic symbioses with larger animals. One frequently noted example is the multiflagellated protist *Trichonympha*, which exists in the guts of termites. Without these symbiotic partners and their ability to digest wood and to pass on the nutrients resulting from that process, the termite would die. The protist partner receives a steady energy supply (in the form of ingested wood) and a safe haven (the termite intestine) to live in. In this activity you will extract *Trichonympha* individuals from the guts of termites and examine them under the microscope.

1. Obtain a compound microscope, a microscope slide, Ringer's solution, a pipette, a termite, some needle probes, and a coverslip.
2. Using the pipette, transfer a drop of Ringer's solution to the slide.
3. Using either your fingers or forceps, place a single termite on the drop of Ringer's solution.
4. Using the needle probes, pull the termite apart by pulling on both ends.
5. Remove all the large parts of the termite, but leave the long tube that is the intestine. Check that you have done this by examining the slide under a compound microscope at low power.
6. Remove the slide from the microscope and pull apart the termite's intestine as carefully as possible. You are trying to distribute the multiflagellated *Trichonympha* individuals.
7. Carefully place a coverslip on top of the slide.
8. Return the slide to the microscope and observe under first low power and then high power. *Trichonympha* are "keyhole"-shaped protists with long flagella.
9. In the circle below, sketch a *Trichonympha* individual.

***Trichonympha* individual**

Concept Check

1. Describe the process of decomposition, and name two types of organisms that act as decomposers in ecosystems.

2. What is mutualism, and how do the two organisms in this lab activity contribute to this particular mutualistic association?

3. If an individual termite receives nutrients from *Trichonympha,* is this nutrient source lost to the rest of the food web? Explain your answer.

ACTIVITY 5 Using Satellites to Track Global NPP

The launch of the Soviet satellite *Sputnik* in late 1957 ushered in a period of frantic research and development in space travel. Humans have been transported to the moon and back, and satellites have taken photographs of distant planets. Ecologists have also benefited from satellites that have used instruments called reflectometers to determine chlorophyll concentrations at Earth's surface. One particular satellite—*SeaWiFS* (*Sea*-viewing *Wi*de *F*ield-of-view *S*ensor)—takes "pictures" of the ocean's surface to determine how much of the main photosynthetic pigment, **chlorophyll *a*,** is present at any given time. Scientists can use this information to determine where and when photosynthetic activity occurs in the oceans. If a certain part of the ocean has high photosynthetic activity, it will have high levels of chlorophyll *a*. In contrast, low photosynthetic activity will be reflected by low chlorophyll *a* levels. Figure 21.6 shows the NPP of marine ecosystems, where NPP is measured as grams of new biomass made each year in an area of one square meter.

NPP in the oceans changes in time and space. In temperate oceans, photosynthetic activity is low during cold winter months. As temperatures rise in the spring, photosynthetic activity increases. Just like the lawn in your backyard, however, producer populations require nutrients like nitrates and phosphates. Levels of NPP are highest in coastal areas, where these nutrients occur in higher concentrations, being brought into the oceans by rivers. Nutrients may also be brought to the surface of the ocean by currents called **upwellings**, as occurs along the west coast of South America. Producer organisms in the deep ocean far from land also need nutrients, including nitrates and phosphates. Nutrients in the deep ocean also come from decomposer communities on the ocean floor. However, many parts of the oceans are exceedingly deep (the deepest part of the ocean is over 10,000 meters deep), preventing these nutrients from reaching surface waters where producer organisms live. Variations in photosynthetic activity in time and space have significant impacts on oceanic food webs, with changes (positive or negative) in the producer level cascading upward through the food web.

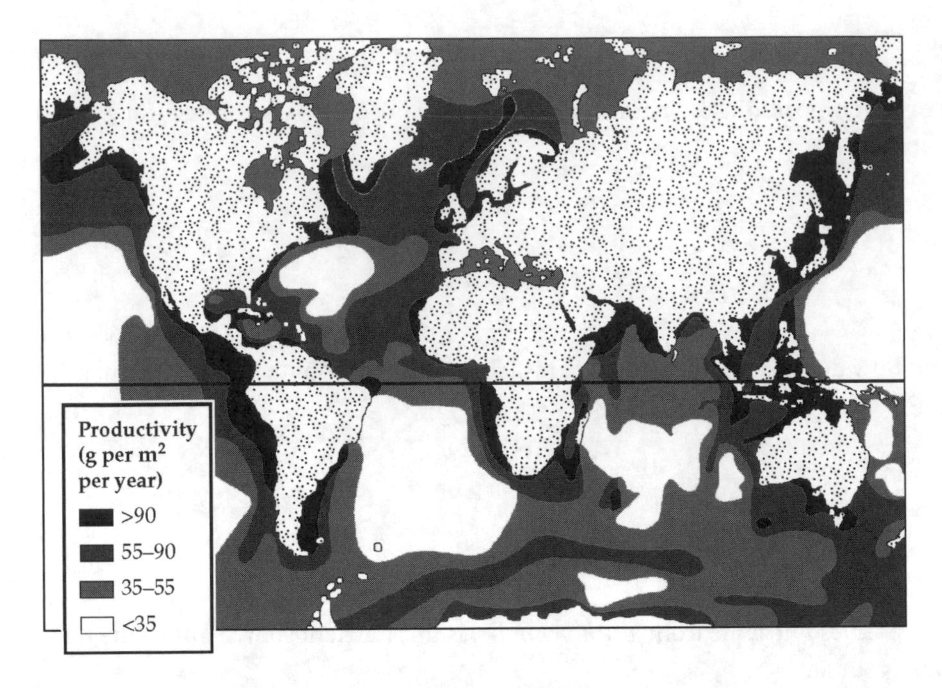

Figure 21.6
SeaWiFS Satellite Data for NPP in Marine Ecosystems

The producers found in oceans are not large like land plants; rather, they tend to be microscopic bacteria, diatoms, and dinoflagellates. Imagine the flow of energy from producers on the plains of the Serengeti in Africa. Primary consumers, including zebras, giraffes, and various antelope, feed on grasses and leaves of trees and use this energy to make tissues. They are, in turn, preyed on by top predators, including lions, cheetahs, and leopards. This ecosystem has only three trophic levels—for example, grass, zebra, and lion. Oceanic food webs have more trophic levels because the producer organisms are microscopic (bacteria, diatoms, and dinoflagellates), so the primary consumers are usually microscopic too. A sequence of animals in successive trophic levels through which energy flows might be this: diatom, krill, Chilean sea bass, penguin, and, as the top (or tertiary) consumer, the killer whale (see Figure 21.4).

In this activity you will use animations of the *SeaWiFS* satellite data on the Goddard Space Flight Center Web site to answer a series of questions about photosynthesis in the oceans.

1. Log on to the Web site: http://svs.gsfc.nasa.gov/search/Keyword/Chlorophyll.html. You should be at the "Scientific Visualization Studio" at the Goddard Space Flight Center.

2. Scroll down until you get to "Global Rotation of SeaWiFS Biosphere Decadal Average without Land." Click on this.

3. You will be taken to a new page with a paragraph at the top explaining how *SeaWiFS* has been collecting chlorophyll *a* data since 1997.

4. To view the animation, click on "512x288 (30 fps) MPEG-1 7 MB" under "Available formats" to the right of the first picture of Earth.

5. You can play and pause the animation using the button at the lower left of the new screen. Click on the play icon and watch the whole animation. High concentrations of chlorophyll *a* are depicted by green, yellow, orange, and red; low concentrations are depicted by shades of blue.

6. Replay the animation and stop it when you get to the southern Atlantic Ocean, with Africa on the right and South America on the left.

7. Why is chlorophyll *a* so high (red, orange, yellow) at the top of South America? (What river drains into the ocean here?)

8. Why is chlorophyll *a* so low in the middle of the southern Atlantic Ocean?

9. Continue playing the animation until you have the United States and Central America in the center.

10. Why is chlorophyll *a* so high (red, orange, yellow) at the northern end (the top end) of the Gulf of Mexico? (What river drains into the ocean here?)

11. Play the whole animation again.

12. Is there more photosynthesis (chlorophyll concentration) in the Northern Hemisphere or in the Southern Hemisphere? Why?

Concept Check

1. Why is there more photosynthesis in coastal ocean waters?

2. What influence do rivers have on photosynthesis?

3. Why is the deep, open ocean characterized by low photosynthetic rates?

4. Having seen the animation and the distribution of photosynthesis around the world's oceans, where do you expect the most diverse marine ecosystems to be?

ACTIVITY 6 Sampling Prokaryotic Populations

Prokaryotes were the first organisms to inhabit the Earth, about 3.5 billion years ago. Molecular analysis shows that they form two distinct domains: Bacteria and Archaea. Although small (usually less than 10 μm in length) and predominantly single-celled, prokaryotes are an extremely successful group found almost everywhere. Estimates suggest that Earth is home to some 5×10^{30} prokaryotic individuals, and millions more than the current 5,000 named prokaryotic species. Because of their small size, most of the time we do not see them; but if we provide a nutrient source and favorable incubation temperature, their rapid reproductive rates can be seen through the development of colonies of different shapes, colors, and sizes. Prokaryotes come in three general shapes: rods, spheres, and spiral (or corkscrew).

Prokaryotes are found on all exposed surfaces, including surfaces in your laboratory ranging from fume hoods to sinks and door handles. But they are especially abundant in the warm and moist human digestive system, especially the large intestine, where the important mutualist *Escherichia coli* lives. Each of us provides a home for as many as a 1,000 species of prokaryotes. The prokaryotic community inside the human body includes both normal residents and transient forms that we pick up by touching surfaces, including holding hands and especially kissing boyfriends and girlfriends!

For this activity your instructor swabbed two surfaces, the door handle to your lab and the center of one of the windows in the lab. Your instructor then transferred any microbial cells to petri dishes with nutrient agar and incubated them in a drawer in the lab for 1 week. You will analyze these door handle and window cultures to identify how many different colonies of prokaryotes are living in them.

ALTHOUGH THE PROKARYOTES YOU WILL OBSERVE IN THIS ACTIVITY ARE HARMLESS, THEY WILL BE IN HIGH CONCENTRATIONS IN THE PETRI DISHES. FOR THIS REASON, AVOID TOUCHING YOUR FACE AND MAKE SURE YOU WASH YOUR HANDS AFTER THE ACTIVITY.

1. Before you begin, formulate the hypothesis you're trying to test and write it here.

2. Obtain two petri dishes: one that has bacterial colonies from the door handle; the other, colonies from the window.

3. In each dish, inspect the colonies that have grown. (Some may be rather smelly!)

4. Count the number of total colonies in each dish. If the dish has many colonies, you might divide the dish into four parts by using a China marker or permanent marker and drawing the divisions on the bottom of the dish. Then you can count just one-fourth of the colonies and multiply by 4. In the first row of Table 21.5, record the total number of colonies for each dish.

NOTE: ANY COLONY THAT APPEARS "FUZZY" IS MOST LIKELY A FUNGAL COLONY AND SHOULD NOT BE COUNTED.

5. Now count the number of different-colored and different-shaped colonies in each dish. Record your counts in the second and third rows of Table 21.5.

TABLE 21.5 PROKARYOTIC COLONIES FROM TWO SURFACES

	Door handle	Window
Total number of colonies		
Number of different-colored colonies		
Number of different-shaped colonies		

6. Has your hypothesis been supported or refuted?

7. Which culture—the door handle culture or the window culture—has more colonies? Why?

8. Which would have more bacterial colonies—a swab of your palm or a swab of the back of your hand?

Concept Check

1. Why do you think the door handle had more colonies than the window glass?

2. Do you think you saw all colonies from the surfaces swabbed? Why or why not?

3. Having seen what was previously invisible to you, do you think there are more than 5,000 prokaryotic species on Earth?

4. Do you think the fact that we cannot see prokaryotes with the naked eye contributes to the ease with which bacterial diseases can be spread? Explain your answer.

Key Terms

abiotic (p. 21-1)
biomass (p. 21-2)
biotic (p. 21-1)
carrying capacity (p. 21-2)
chlorophyll *a* (p. 21-15)
commensal (p. 21-14)
community (p. 21-1)
consumer (p. 21-1)
decomposer (p. 21-2)
ecosystem (p. 21-1)
food chain (p. 21-1)
food web (p. 21-6)
growth rate (p. 21-3)
Lincoln–Petersen method (p. 21-7)
mark–recapture (p. 21-7)
mutualistic (p. 21-14)
net primary productivity (NPP) (p. 21-2)
nutrient (p. 21-2)
parasitic (p. 21-14)
population (p. 21-1)
predator (p. 21-3)
prey (p. 21-3)
primary consumer (p. 21-1)
producer (p. 21-1)
prokaryote (p. 21-18)
protist (p. 21-3)
secondary consumer (p. 21-1)
symbiosis (p. 21-14)
tertiary consumer (p. 21-1)
trophic level (p. 21-2)
upwelling (p. 21-15)

Review Questions

1. All individuals of the same species in an ecosystem represent a(n) ________________.
2. Temperature, nutrients, and energy are examples of ________________ components in an ecosystem.
3. What does "interconnected" mean in ecology?
4. The population of any animal is restricted by numerous factors, including available food sources and rate of predation. These two factors determine the ________________ of the population.
5. Ecologists can estimate the population size of mobile animals using ________________.
6. Name two prey animals of the barn owl.
7. What will happen to a termite if it loses its flagellated symbionts?
8. What factors prevent unlimited population growth?
9. How can scientists accurately measure the global role of producers?

10. What is one way to distinguish one bacterial colony from another?

11. A(n) ________________ is an example of an aquatic producer.

12. Net primary productivity represents the amount of energy that producers do not lose as metabolic heat. NPP is usually estimated as the amount of ________________ that the producers make available to an ecosystem at any point in time.

13. Think of how a vole uses the energy from plant material to build bone, soft tissue, and fur. This vole may then be eaten by a snake. With this scenario in mind, give two examples of how energy may be lost from one trophic level to another.

14. Producers in all ecosystems require nutrients to remain healthy. Why do nutrient levels in terrestrial ecosystems differ between seasons?

LAB **22**

The Impacts of Global Change on Ecosystems

Objectives

- Discuss the impact of humans on ecosystems.
- Demonstrate the impact of turbidity on the penetration of light in water bodies.
- Determine the influence of excessive nutrients on algal populations.
- Use simple water quality tests to check for pollutants in local water bodies and compare with tap and bottled water.
- Calculate carbon footprints to understand the human contribution to global change.
- Determine stomatal densities of plants to infer environmental influences.

Introduction

The impact of humans on the planet has never been more evident than it is now. The current human population is estimated at over 6.7 billion, having tripled since the beginning of the twentieth century. Of this total, almost half live within 150 km of the coast, and many near rivers that drain into the ocean (Figure 22.1). Our species is transforming the land and water we depend on, introducing pollutants to global nutrient cycles, and altering climate patterns by adding greenhouse gases to the atmosphere.

One hundred years ago, many people would have argued that the oceans were so vast they could soak up any **pollutants** (any contaminants that will lead to environmental degradation) that drained into them. The surface area of our oceans is approximately 361 million square kilometers (139,000 square miles). One can imagine how people who lived 100 years ago thought "dilution was the solution" to pollution! But our oceans have a finite area and volume, and our burgeoning population has added numerous pollutants. **Persistent organic pollutants,** or **POPs** (artificial chemicals that persist in nature), are now routinely found in animals in Antarctica, even though their industrial source may be thousands of miles away. These chemicals travel via atmospheric currents and are deposited in the oceans, where they are **biomagnified,** or found in higher concentrations at higher trophic levels in the food chain.

"Natural" materials that may seem harmless can also act as pollutants. In water bodies, suspended particles such as sand, mud, or wastes can block light and prevent photosynthesis when they are present in excessive quantities, as is often the case following the building of canals and levees that safeguard cities but increase water flow. Important nutrients in nature can also alter ecosystems significantly when they are added artificially. With over 3 billion people on our planet living near water bodies, human activities (addition of

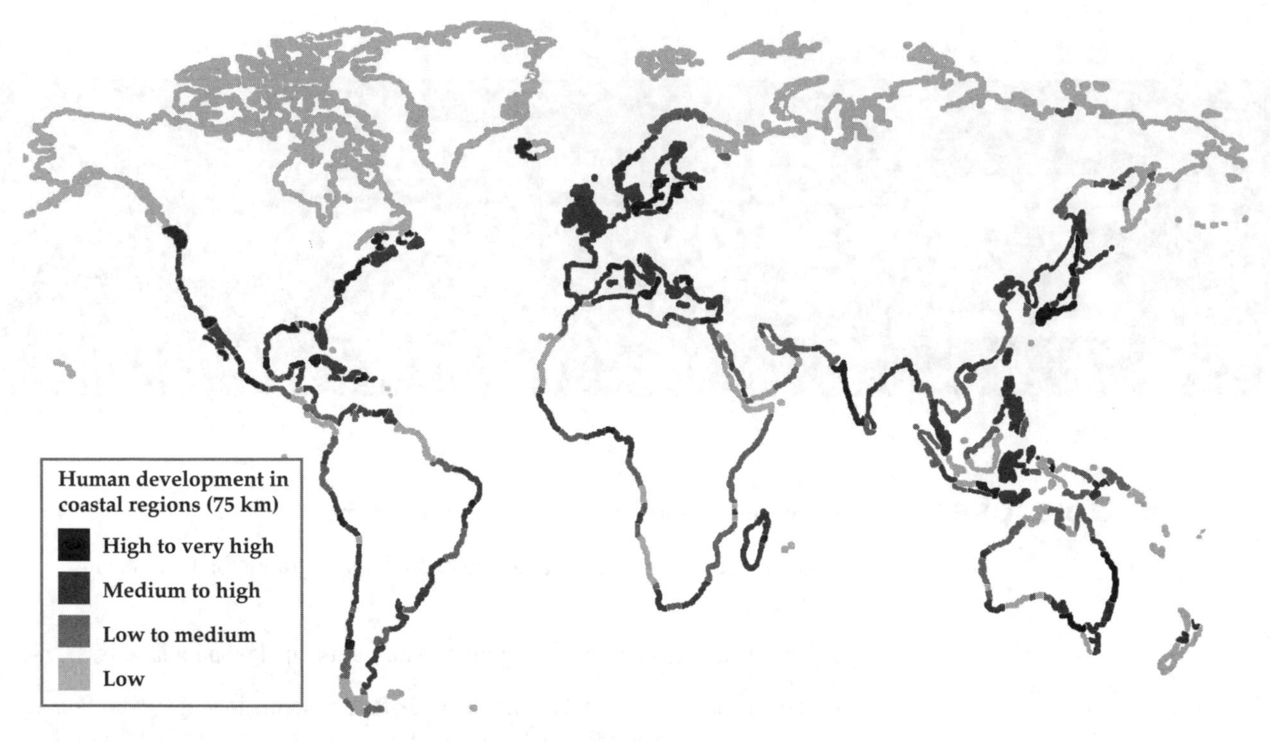

Figure 22.1
Human Development in Coastal Zones

fertilizers in agriculture and sewage waste) add significant amounts of **nitrates** and **phosphates** over and above those released naturally by ecosystems through the nitrogen cycle and the phosphorus cycle. One consequence of this type of pollution may be population explosions of microscopic algae, known as **algal blooms.** Although such blooms initially provide additional food for consumers, the sheer bulk of algal cells living and dying leads to reduced oxygen levels (**hypoxia**) on the ocean floor, as bacteria responsible for decomposition of the algal cells consume the available oxygen, creating "dead zones" like the one in the Gulf of Mexico.

One of the reasons life on Earth is possible is the "natural" **greenhouse effect** (Figure 22.2). **Greenhouse gases** like CO_2 and methane trap the heat radiated off the Earth's surface and prevent it from being lost to space.

The "enhanced" greenhouse effect is the buildup of greenhouse gases in the atmosphere, including CO_2 and methane, due to the activities of humans. Since the industrial revolution, CO_2 levels have risen 30 percent, and they are now estimated to be higher than at any time in the past 650,000 years. Figure 22.3 shows the rapid increase in atmospheric CO_2 levels, measured in parts per million (ppm), over the past 200 years. The dark circles depict results from direct measurements of CO_2 in the atmosphere; the light circles indicate CO_2 levels measured from air bubbles trapped in ice.

This human-induced, or "enhanced," greenhouse effect is often referred to as **global warming** and may result in shifts in the global temperature that could lead to significant changes in populations and the environments that support them. Consider the projection in Figure 22.4 illustrating how biomes in North America will be affected by a 4°C rise in global temperature. As a result of climate change, some species will migrate, some will adapt, and some will face extinction. The Intergovernmental Panel on Climate Change (IPCC) projects that global temperatures will increase between 1.4°C and 5.8°C by the year 2100. These changes are expected to vary considerably from region to region, with temperatures on land increasing faster than temperatures in the ocean. One only has to reflect on the 47 percent reduction in arctic sea ice cover during the summer months since 1980 to realize that temperatures are changing.

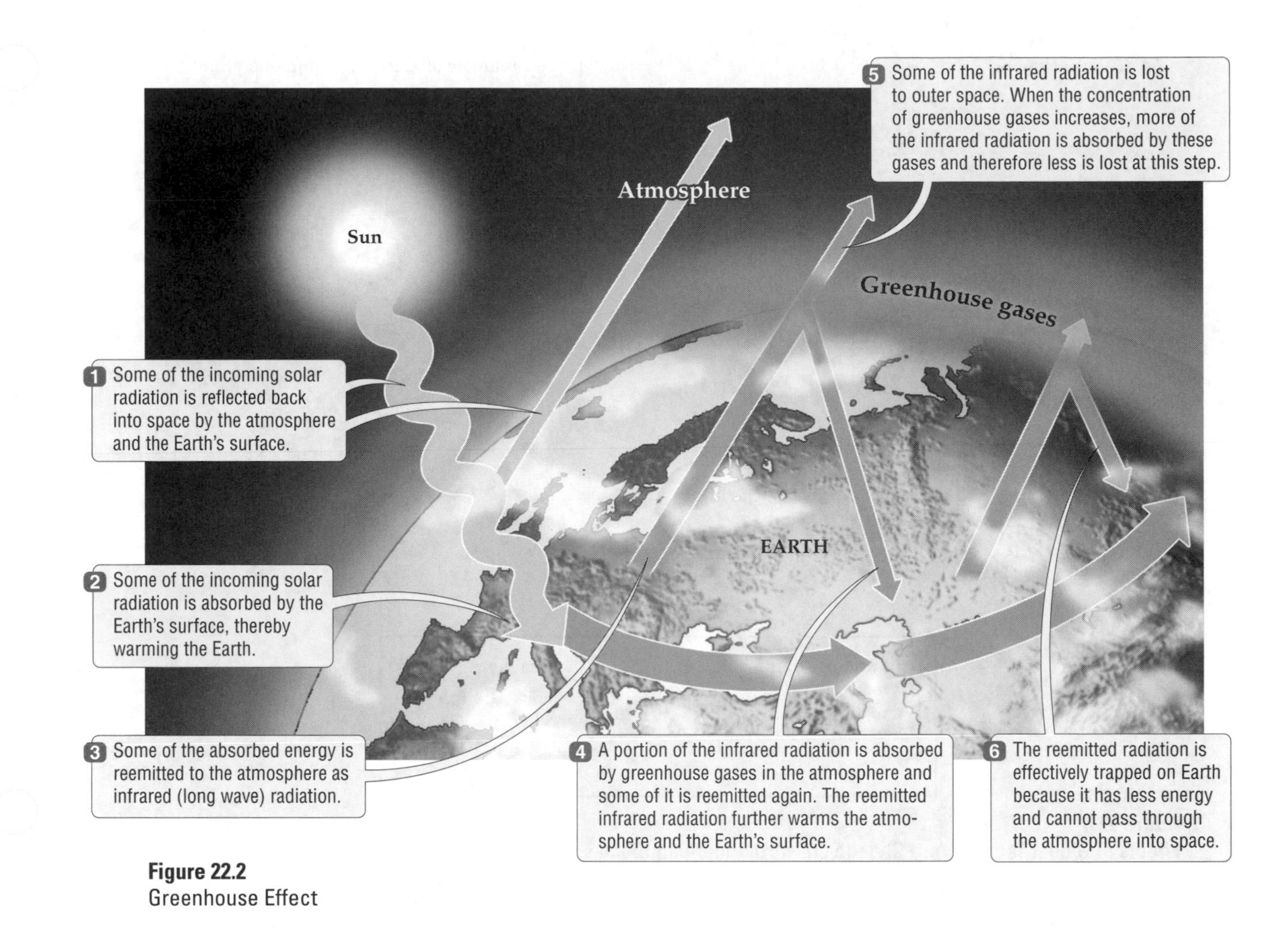

Figure 22.2
Greenhouse Effect

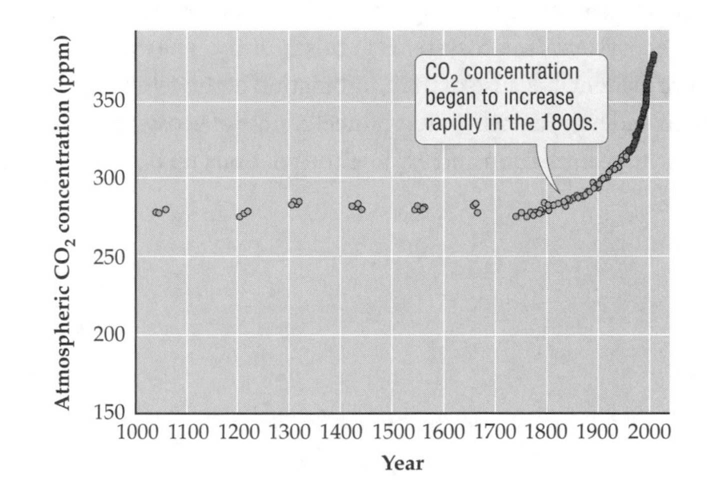

Figure 22.3
Rapid Rise of Atmospheric CO_2 Levels

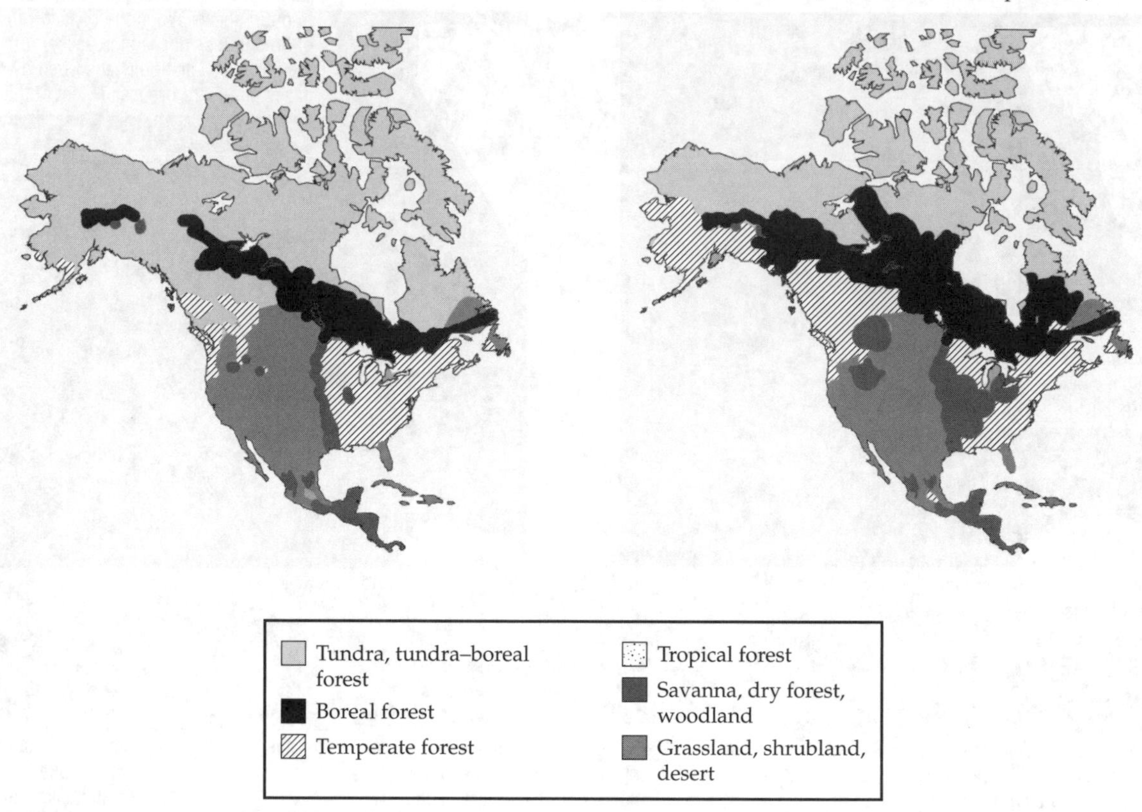

Figure 22.4
Effects of Global Temperature Change on North American Biomes

Recall that **ecosystems** are communities of organisms and their interactions with the physical environment. Ecosystems are interconnected systems with direct and indirect connections between all levels in a single ecosystem and between neighboring systems on many scales. The CO_2 emissions from advanced countries like the United States contribute significantly to atmospheric CO_2 levels around the planet, and thereby connect distant ecosystems to ecosystems in this country. The impact of human activities on ecosystems has never been more evident, because of the combination of the 6-billion-plus human population, the transforming activities discussed here, and the interconnectedness of ecosystems. In this lab you will examine the effects of land and water transformation and explore the relationship between human activity and global change.

ACTIVITY 1 Turbidity and Primary Producers

In the oceans, light can penetrate down to about 200 meters, as long as the water is not **turbid** (having particles in suspension). Suspended particles occur naturally for several reasons, including the washing of sediments and organic matter from rivers into the ocean. In **estuarine areas**, however—where fresh water borne by rivers and seawater meet—suspended particles (including mineral particles from soils, wastes, and microorganisms) can limit the penetration of light at certain times of the year to just a meter or so, acting in this sense as pollutants. In this case the impact on photosynthesis is caused not by an excess of nutrients, but by a reduction in light penetration. Humans have altered the natural channels of many rivers by constructing levees and adding concrete "dikes," accelerating the flow of those rivers and preventing the deposition of fine particles on floodplains. Animal wastes and associated microorganisms move into rivers from farmland and other human sources, including sewage treatment. These fine particles are then transported out to sea and increase the turbidity above natural levels.

The main factors that determine how long particles are suspended in water are their size and weight. Sediments that are washed down rivers into oceans may be coarse sands (just like your favorite beach) or fine silts and clays that make up the mud on those smelly mudflats you see on vacation. The finer sediments normally stay in suspension for some time. Organic particles like microorganisms and wastes can also remain suspended for lengthy periods, because of their light weight.

Scientists measure turbidity in several ways, but all methods rely on the penetration of light through a sample of water. You will use a spectrophotometer to measure the transmission of light at the main wavelength of light used by the plant pigment, chlorophyll *a*—namely, 430 nanometers (nm). A nanometer, the unit of measurement used for wavelengths of light, is approximately 1/25 millionth of an inch. A spectrophotometer can measure the absorbance or transmission of light at various wavelengths. (The colors of the rainbow represent different wavelengths of light—plants tend to use the blue and red colors to make sugars, and most plants reflect the green wavelengths, giving them their green color.) In this activity you will suspend different types of particles in water in a test tube to see how they affect the penetration of light. You will test samples of sand, mud, and multipurpose compost (to simulate wastes). Be sure to listen to your instructor's detailed instructions for using the spectrophotometer.

1. Before you begin, formulate the hypothesis you're trying to test and write it here.

2. Obtain a test tube rack, five test tubes, three spatulas, three beakers, three syringes, distilled water, and samples of sand, mud, and compost.

MAKE SURE YOU HANDLE THE TEST TUBES BY THEIR TIPS SO THAT OILS FROM YOUR FINGERS DO NOT ALTER THE READINGS.

Control

3. Fill one test tube three-fourths full with distilled water; this is your control. Fill a second test tube three-fourths full with distilled water; this is your "blanking," or calibration, test tube.

4. Set the wavelength knob on the spectrophotometer to 430 nm. All readings will be taken at 430 nm. (Your spectrophotometer will have been warmed up for 20 minutes.)

5. With the sample chamber empty and closed, use the "0" or "%T" control knob to set the instrument to 0% transmittance. This setting adjusts the internal electronics of the instrument.
6. Insert the blanking test tube (with distilled water) and, using the "100%T" knob on the front left of the instrument, adjust the transmittance until it reads 100%.
7. Insert the control test tube (with distilled water) and take transmittance readings at 10-second intervals for 60 seconds. Record these values in the first row of Table 22.1.

Sand

8. Pour 100 ml of distilled water into a beaker, add 2 grams of sand, and stir for 20 seconds.
9. Using a syringe, quickly transfer enough water from the beaker to fill the third test tube three-fourths full.
10. Place this test tube in the spectrophotometer and take readings at 10-second intervals for 60 seconds. Record these values in the second row of Table 22.1.

Mud

11. Pour 100 ml of distilled water into a beaker, add 2 grams of mud, and stir for 20 seconds.
12. Using a syringe, quickly transfer enough water from the beaker to fill the fourth test tube three-fourths full.
13. Place this test tube in the spectrophotometer and take readings at 10-second intervals for 60 seconds. Record these values in the third row of Table 22.1.

Compost

14. Pour 100 ml of distilled water into a beaker, add 2 grams of compost, and stir for 20 seconds.
15. Using a syringe, quickly transfer enough water from the beaker to fill the fifth test tube three-fourths full.
16. Place this test tube in the spectrophotometer and take readings at 10-second intervals for 60 seconds. Record these values in the last row of Table 22.1.

TABLE 22.1 TRANSMITTANCE READINGS AT VARIOUS LEVELS OF TURBIDITY AT 430 NM

Treatment	10 seconds	20 seconds	30 seconds	40 seconds	50 seconds	60 seconds
Control						
Sand						
Mud						
Compost						

17. Was your hypothesis supported or refuted?

18. Why does the test tube with mud have transmission readings lower than those of the test tubes with sand or compost?

19. Why does the test tube with sand have readings almost identical to those of the control test tube?

20. Why do you think the compost had higher readings than the mud?

Concept Check

1. What sorts of suspended particles lead to high turbidity?

2. What effect does increased turbidity have on fish populations in coastal waters?

3. Do you think turbidity levels are higher at certain times of year?

4. What human population factor has contributed significantly to increasing turbidity levels in coastal waters?

ACTIVITY 2 Eutrophication

In addition to sunlight and water, organisms that photosynthesize require nutrients, especially in the form of nitrates and phosphates. In pristine terrestrial ecosystems these nutrients are slowly released into the groundwater through decomposition; aquatic ecosystems receive them via decomposition occurring on the floor of a lake or ocean. Recall that the nitrogen cycle is an atmospheric cycle, while the phosphorus cycle is a sedimentary cycle. Living organisms require nitrogen to build cell components, including amino acids and DNA. Although nitrogen is plentiful in the atmosphere (78 percent), it needs to be transformed into a usable form (nitrates and ammonium) by bacteria-driven **nitrogen fixation**. In contrast, the phosphorus cycle has no atmospheric component; instead, the element is released slowly from sediments. Unlike the natural release of nitrogen and phosphorus, excess release of nutrients into water bodies by human activity happens much more quickly as a sudden pulse of nutrients. This excess release results from the increased use of nitrate- and phosphate-based fertilizers (for example, Miracle-Gro® All Purpose Plant Food has 24 percent nitrogen and 8 percent phosphorus), the location of waste treatment plants, and increasing amounts of animal manure and industrial wastes. Large seasonal influxes of these nutrients create algal blooms that can lead to the formation of "dead zones" like that found in the spring and summer in the Gulf of Mexico.

How does this happen? The increased nutrients enable **phytoplankton** populations (organisms like diatoms such as *Chaetoceros*, dinoflagellates such as *Ceratium*, and some bacteria) to reproduce rapidly (Figure 22.5). Small consumer organisms (especially **protists**) also have population explosions as their food supply increases. Almost all living organisms need to use oxygen to make cellular energy in the form of ATP. As the populations of these organisms increase, their consumption of oxygen does too, ultimately lowering the oxygen dissolved in water to levels that cannot support most life, creating **dead zones**. **Eutrophication** is the name given to this process, in which the nutrients added to water cause bacterial populations to increase and oxygen levels to plummet as a result.

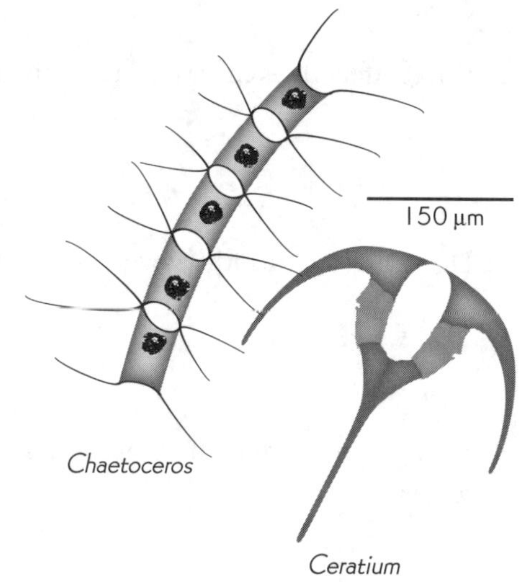

Figure 22.5
The Marine Protists *Chaetoceros* (a Diatom) and *Ceratium* (a Dinoflagellate)

In this activity you will simulate the effect of nitrates and phosphates on freshwater algal populations using a common plant fertilizer. You will have one control treatment (with no fertilizer) and three experimental treatments (with differing dosage levels of fertilizer: 0.5 grams/liter, 1 g/l, and 2 g/l. You will also make algal counts using a compound microscope.

Your instructor set up four treatment aquaria 3 days in advance. Each of the four small aquaria contains 4 liters of distilled water fertilizer in the concentrations specified above, and 10 ml of mixed algae (including *Chlamydomonas*, *Euglena*, and *Cyclotella*).

1. Before you begin, formulate the hypothesis you're trying to test and write it here.

2. Obtain four 500-ml flasks, four microscope slides, four coverslips, four plastic pipettes, spring water, and a permanent marker.

3. Label the four flasks "Control," "0.5 g/l," "1 g/l," and "2 g/l."

4. Transfer 200 ml of well-mixed water from each aquarium to the appropriate flasks.
5. Label the microscope slides "Control," "0.5 g/l," "1 g/l," and "2 g/l."
6. Mix the contents of each flask well by placing the palm of your hand tightly over opening and inverting the flask two or three times.
7. Using a plastic pipette, transfer a drop of water (and algae) from each of the four flasks onto the appropriate microscope slides and create a wet mount, carefully applying a coverslip by holding one edge of the coverslip against the slide and slowly lowering the opposite edge.
8. One by one, transfer the four slides to the microscope stage. For each slide, bring the algal cells into view using the low-power lens, and then change to the high-power objective lens (40×) so that you're observing the algal cells at 400× total magnification.
9. For each slide, count all the algal cells you see in one area, and record this number in Table 22.2, in the appropriate row under "Count 1." Then move the slide so that you see a new area under the coverslip, count the algal cells in that area, and record the number in the appropriate row under "Count 2." Do this twice more, and record the number of algal cells counted in the appropriate row under "Count 3" and "Count 4." Perform these four counts for each slide. For each slide, then, you will have counted the number of algal cells for four distinct areas at 400× total magnification.
10. For each treatment, calculate the average (mean) number of algal cells that you counted and record this number in Table 22.2.
11. Now consult with the rest of the class, and calculate the average number of algal cells counted for the class as a whole for each treatment. Record this number in the final column in Table 22.2.

TABLE 22.2 ALGAL COUNTS AT DIFFERENT NUTRIENT LEVELS

Treatment	Count 1	Count 2	Count 3	Count 4	Average (mean)	Class average (mean)
Control						
Fertilizer (0.5 g/l)						
Fertilizer (1 g/l)						
Fertilizer (2 g/l)						

12. Was your hypothesis supported or refuted?

13. Which treatment had the highest algal cell count, and why do you think this is so?

14. Why did you make four counts for each treatment?

Concept Check

1. Why are aquatic ecosystems so heavily affected by human activities?

2. What are the two most important nutrients required by producers?

3. What are the human-derived sources for these nutrients?

4. Why are many mobile organisms not able to escape from "dead zones" like the one found in the Gulf of Mexico?

ACTIVITY 3 Water Quality Sampling and Persistent Organic Pollutants

Cells in living organisms are made up of 70 to 90 percent water, and we live on a planet whose surface area is 70 percent water. Of all that water, however, only 2.5 percent is fresh water, and two-thirds of that is frozen in glaciers. Most of the fresh water we could use is in the form of groundwater or soil moisture, with less than 1 percent (of the 2.5 percent) found in lakes and rivers. The most important molecule to living organisms on land is in short supply, and the high concentration of human populations near these water bodies has led to significant transformations of water quality and flow through the addition of wastes and fertilizers and the building of canals and levees.

Concerns over the quality and safety of tap water have led in recent years to an explosion in the production of bottled water. The amount of bottled water sold worldwide currently exceeds 150 billion liters (1 gallon = 3.8 liters), and most of this water is sold in plastic bottles. Over 28 billion bottles of water were sold in the United States in 2005, creating an additional problem of where and how to dispose of them. Fortunately, our drinking water in the United States is strictly regulated by the Environmental Protection Agency (EPA) under the Safe Drinking Water Act. Standards have been set for about 90 contaminants, ranging from microbial organisms like *Giardia* (a protist) to elements like copper and lead. The EPA also tests for synthetic chemicals that arise from industrial activity, including chlorobenzene, dioxins, and polychlorinated biphenyls (PCBs). Many of these synthetic chemicals remain in nature for long periods of time and have been named persistent organic pollutants (POPs) for this reason. Although this activity focuses on water-borne pollutants, POPs have an atmospheric component that allows airborne particles to travel thousands of miles to distant ecosystems, where they become dissolved or suspended in water bodies. They may be ingested by low-level consumers and in this way passed up food chains.

In this activity you will work with your classmates to conduct water quality tests on three water samples. Your instructor has collected a water sample from a nearby water body (a river, lake, or pond) and will also provide a beaker of tap water and a beaker of bottled water. The simple water quality testing kit that you will use tests for the following:

- Iron
- Copper
- Nitrate
- Chlorine
- pH
- Hydrogen sulfide

Your instructor will assign you to a particular water sample (natural, tap, or bottled) to test, and you will compare your results with groups testing the other sample types.

1. Before you begin, formulate the hypothesis you're trying to test as a class and write it here.

2. In your groups, obtain a water sample from your instructor and a water testing kit.
3. Following the guide on the test kit and, using any extra glassware (test tubes, small beakers) you may need, conduct tests for the six components listed above, and record the results in Table 22.3. Record all but pH as parts per million (ppm).
4. Obtain the test results from another group in the class that performed tests on the other water samples, and record them in Table 22.3.

TABLE 22.3 SELECTED WATER QUALITY TESTS ON TAP WATER AND NATURAL WATER

	Iron (ppm)	Copper (ppm)	Nitrate (ppm)	Chlorine (ppm)	pH	Hydrogen sulfide (ppm)
Tap water						
Bottled water						
Natural water						

5. Was your hypothesis supported or refuted?

6. Why do you think the level of chlorine is higher in tap water?

7. If you found nitrate in natural water, what might its source be?

8. If you found hydrogen sulfide in natural water, what might its source be?

9. Do you drink bottled water? If so, why?

Concept Check

1. Why are most people in the United States happy to drink tap water?

2. Why are most people not happy to drink water from a local water body (pond or lake)?

3. How might we reduce the level of nitrates in local water bodies?

4. Do you think the negative disposal costs of bottled water or the cost of transporting the bottles from manufacturing to distribution sites offset any benefits of drinking it?

ACTIVITY 4 Calculating Carbon Footprints

Atmospheric CO_2 levels were approximately 280 parts per million (ppm) before the industrial revolution and are now 387 ppm. A greenhouse gas, CO_2 is emitted from a variety of sources, many of them associated with the burning of fossil fuels. For example, when you drive a car, the exhaust emissions include CO_2. When you heat or cool your home, the required energy often comes from the burning of fossil fuels in power stations. But have you ever stopped to think that your diet may play a role in CO_2 emissions? If you eat meat and shop for it at a supermarket chain, there is a strong chance that the source of your T-bone steak is a distant slaughterhouse. In winter, the lettuce that makes up the bulk of the salad you eat with your steak may be shipped from tropical locations. Your steak and lettuce then travel by truck to reach your supermarket delicatessen or produce counter, contributing to CO_2 emissions along the way. The release of CO_2 attributed to any one person is usually expressed in kilograms or tons and is known as a **carbon footprint**.

Lifestyles in advanced nations like the United States depend significantly on the direct or indirect burning of fossil fuels, adding to the global-warming problem. Many people in less developed nations, however, do not depend on large-scale burning of fossil fuels. Imagine a subsistence farmer in an underdeveloped part of the world who raises his/her family in a mud hut without any electricity. The family is fed from the small herd of animals and crops adjacent to its dwelling. The family does not own a car, and to travel to the next village it uses an ox-drawn cart. Because of the lack of electrical power, the family owns no electrical appliances—no TV, videogame console, or PC!

Consider another family in a more developed part of the world, but still far less developed than the advanced nations in western Europe or North America. The family does have electricity in its small home, but it does not own a car and has just bought its first TV. The father works in a nearby village, traveling there by bicycle. The mother walks to the local village market three times a week to buy fresh produce and locally slaughtered meat.

In this activity you will use an online calculator to determine carbon footprints. Most online carbon (or ecological) footprint calculators are created with advanced nations in mind, so you will not be able to establish the footprint of the subsistence farmer's family. You will, however, be able to determine the footprint of three hypothetical four-person families whose dependence on fossil fuels is quite different. Since your lifestyle may be different from those depicted in the three scenarios below, you will also calculate your own carbon footprint.

1. Before you begin, formulate a hypothesis about what kinds of behavior lead to reduced carbon footprints and write it here.

2. You will use the model provided by Earth Day Network at www.earthday.org/footprint-calculator. For each scenario that follows, work through the calculator at this Web site (using the "Enter Basic Information" option only) to determine the carbon footprint for the family described. The simulation provides an estimate of how many Earths would be required if everyone alive lived in the same manner as the family in each scenario.

Scenario 1

- A four-person family lives in a typical suburban home (running water and electricity) in the United States.
- The family owns two cars (an SUV and a minivan—average gas consumption 15 miles per gallon) that are used every day for taking kids to school and for commuting to work. The family travels an estimated 250 miles per week, occasionally riding with at least one passenger in the car.
- The family's home is 2,500 square feet.
- The family shops at a nearby supermarket chain, and all family members like meat. An estimated 80 percent of the family's food comes from more than 200 miles away.
- The family takes three holidays each year, traveling by air within the United States for 18 hours a year.
- The family has only just started recycling, and recycles half of its recyclable waste.

3. Run the simulation to determine the carbon footprint for the family in Scenario 1. In the circle below, draw the pie chart from the resulting "Your Ecological Footprint" page. How many Earths and tons of CO_2 would be required if everyone lived like this family?

Scenario 1 carbon footprint

4. Return to the Web site and click on "Edit Your Footprint." The family has decided to sell the SUV and take public transport to travel 150 miles per week. By how much has the family's carbon footprint decreased?

5. After your edit, how many Earths are required to support a global population exhibiting this family's carbon footprint?

Scenario 2

- A four-person family lives in an urban apartment building (with running water and electricity) in the United States.
- The family does not own a car, using public transportation for about 25 miles per week.
- The family's home is 1,500 square feet.
- The family shops at supermarkets, but every family member is vegetarian. An estimated 50 percent of the family's food comes from more than 200 miles away.
- The family takes one holiday a year, traveling by train.
- The family recycles 100% of its recyclable waste.

6. Run the simulation to determine the carbon footprint for the family in Scenario 2. In the circle below, draw the pie chart from the resulting "Your Ecological Footprint" page. How many Earths and tons of CO_2 would be required if everyone lived like this family?

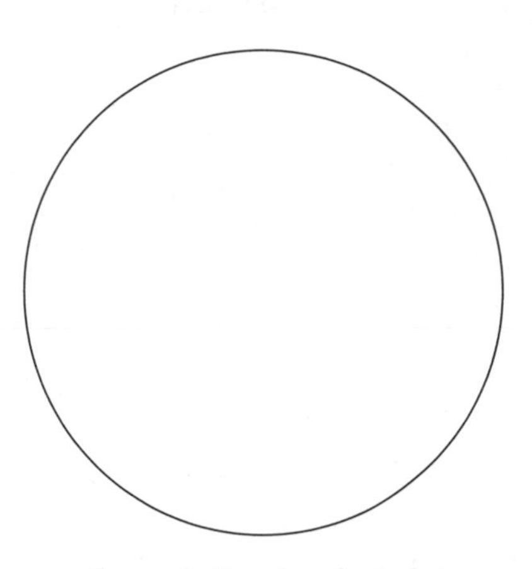

Scenario 2 carbon footprint

7. Return to the Web site and click on "Edit Your Footprint." The family has decided to move into the suburbs because crime has begun to rise in its neighborhood. The family has purchased a 2,000-square-foot home. How has its carbon footprint changed?

8. After your edit, how many Earths are required to support a global population exhibiting this family's carbon footprint?

Scenario 3

- A four-person family lives in a brand-new, "green"-designed home (with running water and electricity) in a suburb in the United States.
- The family owns one hybrid car (average gas consumption 45 miles per gallon) that is used every day for taking kids to school and for shopping (about 25 miles a week), but both parents use public transportation to commute to work, commuting about 30 miles per week.
- The family's home is 2,000 square feet.
- The family buys meat and fresh produce from a local farm, as well as some items from a supermarket. An estimated 20 percent of the family's food comes from 200 miles away.

- The family takes one holiday a year, traveling by air for 5 hours.
- The family recycles about 75% of its recyclable waste.

9. Run the simulation to determine the carbon footprint for the family in Scenario 3. In the circle below, draw the pie chart from the resulting "Your Ecological Footprint" page. How many Earths and tons of CO_2 would be required if everyone lived like this family?

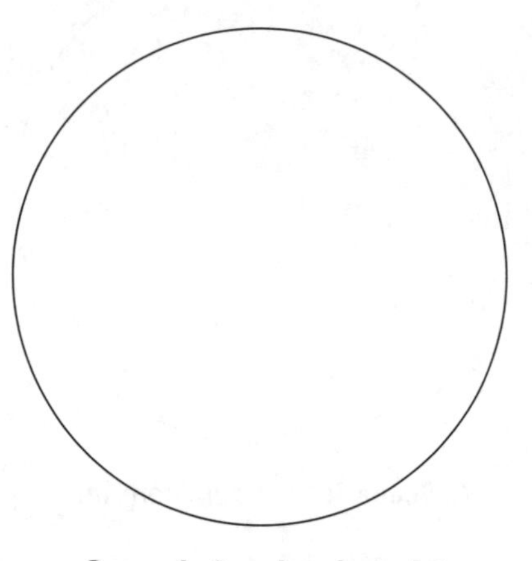

Scenario 3 carbon footprint

10. Return to the Web site and click on "Edit Your Footprint." The family has decided that it can no longer afford air travel and will use the train. How has its carbon footprint changed?

11. After your edit, how many Earths are required to support a global population exhibiting this family's carbon footprint?

Going Further

12. Was your hypothesis refuted or supported?

13. Return to the Web site and work through the simulation based on your own lifestyle.

14. In the circle below, draw the pie chart from the resulting "Your Ecological Footprint" page. How many Earths and tons of CO_2 would be required if everyone lived like you do?

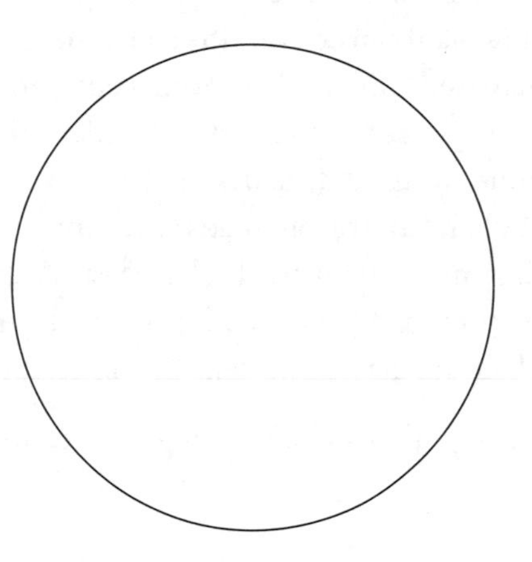

Your carbon footprint

Concept Check

1. What is the main source of greenhouse gases?

2. Name two ways that families can reduce their CO_2 emissions from transport with relative ease.

3. The simulation reveals the tonnage of CO_2 released by each family. Referring to the unedited "Your Ecological Footprint" page in each scenario, which category of natural-resource use accounts for the majority of CO_2 emissions?

4. Name two ways that you can personally reduce your carbon footprint.

ACTIVITY 5 Stomatal Densities and Pollution

Most land plants take up CO_2 from the atmosphere through small openings, usually found on the undersides of their leaves, called **stomata** (singular "stoma"). The carbon from this CO_2 is then used by the plants' chloroplasts to make sugars, from which the plant can convert the sun's energy into the cellular form of ATP. Studies show that plants subjected to increased CO_2 levels and other automobile emissions, like nitrous oxide and sulfurous oxide, have, among other responses, reduced stomatal densities on their leaves.

In this activity you will determine stomatal densities in plants (trees and/or shrubs) near busy roads (a busy urban intersection or a highway) and away from roads (the center of your college campus), and infer responses to these environmental pollutants. Your instructor has already selected leaves and placed them in trays marked "near busy road" and "away from road." Half of the class will determine stomatal densities from leaves picked near a busy road; the other half will determine densities from leaves picked away from a busy road.

1. Before you begin, formulate the hypothesis you're trying to test and write it here.

2. Obtain five leaves from either the tray marked "near busy road" or the tray marked "away from road" (as directed by your instructor), five microscope slides, clear nail varnish, clear packing tape, and scissors.
3. Place the leaves on the benchtop with the undersides facing upward.
4. Using the nail varnish, paint a square 2 cm by 2 cm on each leaf, to either the left or the right of the midrib (the main vein between the petiole and leaf tip), but not over the midrib. Let this square dry completely so that your finger does not stick when you press the square lightly—about 10 minutes (check for dryness by lightly touching it with your finger).
5. Firmly press the packing tape, sticky side down, onto each square. (You may press very firmly, but only if the varnish is dry.)
6. Peel back the tape on each of your leaves to reveal a leaf impression.
7. By pressing the tape (again, sticky side down) to a clean microscope slide, apply each leaf impression directly to a slide.
8. Using scissors, trim the tape to fit the slide.
9. Using Figure 22.6 as a reference, locate some stomata in your sample.

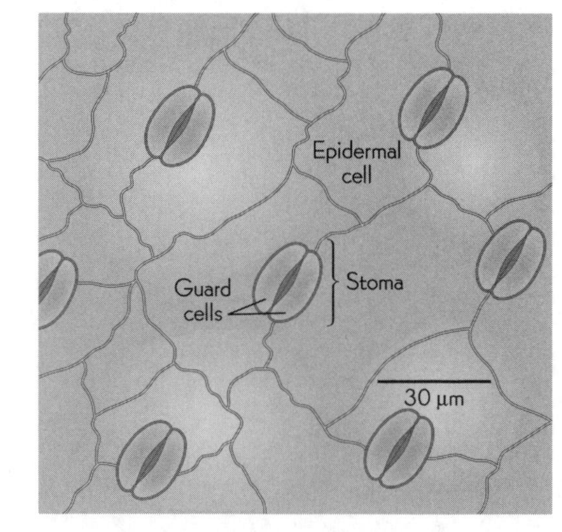

Figure 22.6
Stomata

10. Using a compound microscope at 400× total magnification, count all the stomata you see in the field of view for each leaf. (Check with your instructor if you have trouble locating stomata. At 400×, there are usually 10 to 40 stomata per slide.) Record the total number of stomata in the field of view for each leaf under "Number of stomata viewed" in Table 22.4.

11. At 400×, the diameter of your field of view is approximately 0.3 mm, giving a radius of 0.15 mm. The area of a circle is πr^2. So the total area of your field of view is 0.07 mm^2. However, you need to convert the total stoma densities to a standardized area of 1 mm^2, an area significantly larger than 0.07 mm^2. You will need to multiply your stoma counts by 14 to get the number of stomata per 1 mm^2. Consider the following example before you proceed:

    Example:

    John counts 31 stomata on the underside of his leaf at 400× magnification. At 400×, the radius of the field of view is 0.15 mm. John enters this value into the formula πr^2:

    $$\text{Area of field of view at } 400\times = 3.14 \times (0.15)^2$$
    $$= 3.14 \times 0.0225$$
    $$= 0.07 \text{ mm}^2$$

    To standardize his counts to 1 mm^2, John divides 1 by 0.07 to get 14.28 (but rounds it down to 14). He then multiplies 31 (the number of stomata) by 14:

    $$\text{Total stomata/mm}^2 = 31 \times 14 = 434$$

    a. For each of your five leaves, calculate the stomatal density per 0.07 mm^2, and record the value under "Number of stomata per 0.07 mm^2" in Table 22.4.
    b. For each leaf, multiply the value calculated in step (a) by 14, and record the value under "Number of stomata per 1 mm^2" in Table 22.4.

TABLE 22.4 STOMATAL DENSITIES OF PLANTS IN DIFFERENT LOCATIONS

	Number of stomata viewed	Number of stomata per 0.07 mm^2	Number of stomata per 1 mm^2
Near road			
Leaf 1			
Leaf 2			
Leaf 3			
Leaf 4			
Leaf 5			
Away from road			
Leaf 1			
Leaf 2			
Leaf 3			
Leaf 4			
Leaf 5			

12. What is the average (mean) stomatal density for all five leaves that you observed?Δ16

13. Obtain the stomatal densities from a group in your class that worked on the other location (either "near road" or "away from road"), and record these values in Table 22.4. What was the average density according to data from this group?

14. Was your hypothesis supported or refuted?

15. Do you think comparing one plant from each location is sufficient? Why or why not?

Concept Check

1. Why does the plant near the busy road have reduced stomatal densities?

2. What other factors may be affecting stomatal densities?

3. If atmospheric CO_2 levels continue to rise, do you think plants in more pristine locations will also potentially respond by reducing stomatal densities?

4. How do stomatal densities help us infer CO_2 levels?

Key Terms

algal bloom (p. 22-2)
biomagnification (p. 22-1)
carbon footprint (p. 22-12)
dead zone (p. 22-8)
ecosystem (p. 22-4)
estuarine area (p. 22-5)
eutrophication (p. 22-8)
global warming (p. 22-2)
greenhouse effect (p. 22-2)
greenhouse gas (p. 22-2)
hypoxia (p. 22-2)
nitrate (p. 22-2)
nitrogen fixation (p. 22-8)
persistent organic pollutant (POP) (p. 22-1)
phosphate (p. 22-2)
phytoplankton (p. 22-8)
pollutant (p. 22-1)
protist (p. 22-8)
stoma (p. 22-18)
turbidity (p. 22-5)

Review Questions

1. Name one human activity that affects ecosystems.

2. What are the main impacts of humans on ecosystems?

3. Why is the word "interconnectedness" so important in explaining the impact of humans on all ecosystems?

4. What is one of the impacts of increased CO_2 levels on land plants?

5. What are two of the main sources of turbidity in natural water bodies?

6. What are two sources of nutrients that may lead to eutrophication?

7. Name one body of water that is an example of a dead zone.

8. By what route do human-derived nutrients usually enter aquatic ecosystems?

9. Why is it vital that countries protect the quality of their fresh water supplies?

10. Where and when would you expect to find oceanic areas high in turbidity?

11. ___________________ is the name given to conditions of low dissolved oxygen.

12. What are two ways you could reduce your carbon footprint?

13. Why are the carbon footprints of people in advanced nations so much higher than those of people in underdeveloped nations?

LAB 23 Forensic Science

Objectives

- Conduct testing methods used to identify evidence found during a forensic investigation.
- Describe the difference between presumptive and confirmatory blood tests.
- Choose the correct method for evaluating evidence at a crime scene.
- Qualify fibers and hairs collected at a crime scene.
- Gather fingerprints from crime scene evidence and identify a potential suspect.
- Understand the complexity of crime scene investigation.

Introduction

With every major TV network broadcasting at least one, if not several, crime shows, citizens today have become amateur crime sleuths. This heightened public awareness of crime scene procedures puts added pressure on investigators to process crime scenes as meticulously as possible, not only because jurors are more educated on the interpretation of evidence, but also because the processing techniques themselves have become more refined and sensitive. Evidence collected and presented at a criminal case comes in two forms, **direct evidence** and **circumstantial evidence**. Direct evidence includes testimony from an eyewitness to the crime, the presence of a recording capturing the criminal act, or a confession to the crime. Any case that has no direct evidence is considered a circumstantial case. These cases are built on evidence collected at the crime scene, at places where the suspect or victim may have been before or after the crime, or at locations connected to the planning of the crime. Physical evidence is often the most important form of circumstantial evidence introduced during a trial and includes, but is not limited to, bodily fluids, hairs and fibers, and impression evidence such as fingerprints.

One of the greatest challenges of crime scene investigation is identifying all the items at the crime scene that are related to the crime. Stains made by bodily fluids may be invisible to the naked eye, but if such a stain is discovered and analyzed, it can yield many possible clues about the identity of who deposited it. **Presumptive tests** indicate whether it's possible that a particular substance is present. For instance, some tests indicate that a stain *may* be blood. **Confirmatory tests** reveal the true composition of a substance; these tests absolutely determine that a stain is a specific substance, such as human blood. A DNA match is the most powerful confirmatory test because it can definitively indicate that a specific individual contributed that biological material. *Note that the presence of DNA doesn't mean that an individual committed a crime; it merely indicates that that person left DNA at the site in question.* DNA evidence is therefore still classified as circumstantial, because it alone doesn't prove that the person implicated by the DNA committed the crime.

What is important in crime scene processing is the entire story; what does *all* the evidence tell the investigators about the crime? Physical evidence such as hairs, fibers, and fingerprints can lend credibility to hypotheses that an individual performed a specific act at the crime scene or touched a particular item. Edmond Locard, a French investigator from the early 1900s, developed the concept of transfer evidence, or **Locard's exchange principle**. This is the exchange of physical material when any two objects come into contact with each other. For instance, when an individual comes into contact with any surface—whether another human, a chair, or a glass—some physical evidence will be transferred from the individual to the other surface, and vice versa.

Hairs left at a crime scene can be shown to be consistent with an individual's own hair. Fibers collected at the crime scene can be shown to be consistent with a particular piece of clothing because of its chemical makeup, dimensions, and color. It is also possible to extract DNA from strands of hair, which are made from hardened cells. Finally, a variety of impressions can be recovered from a crime scene. Locard's exchange principle is most profoundly exemplified by fingerprints. These remnants of an individual are often invisible to the naked eye but can elucidate the presence and movements of an individual throughout a crime scene. Every human has unique fingerprints, and this uniqueness enables investigators to narrow down possible suspects.

As you perform the activities in this lab, imagine you are an investigator entering a crime scene where a male has been found assaulted and unconscious in his home. As you learn about the different types of evidence and tests, think of where the biological evidence might be located. You will perform different tests used by crime scene technicians to find blood evidence, you will see how hairs and fibers are collected and processed, and you will collect and analyze fingerprint evidence.

ACTIVITY 1 Presumptive Blood Tests

Many chemicals bind to different molecules in blood and then change color or release light to indicate the presence of blood. There are advantages and disadvantages associated with the use of these chemicals. The first substance you will use is leuco crystal violet (LCV), which is often used at crime scenes to enhance the appearance of blood on substrates of nonporous surfaces. LCV itself is clear, but when it comes into contact with blood or certain other biological substances, it turns a deep purple. Among the advantages of LCV are its ease of application using various spraying devices and the fact that it provides immediate visualization of the blood, which can be photographed to show fingerprints left within it. However, LCV may discolor a substrate, and such color changes may be disadvantageous to processing certain evidence.

The other substance you will use in this activity is luminol. LCV may be applied after luminol, but only if the suspected blood stain is **fixed** (prepared for processing) before luminol treatment. (In general, the sequence of presumptive tests is critical because administering some tests may preclude trying others.) Luminol can also be applied by spraying it onto an area of the crime scene suspected of having blood present. It reacts with the iron in the hemoglobin molecule to release blue light (Figure 23.1). This type of reaction is called **chemiluminescence** because photons are created by the chemical reaction.

The advantage of the luminol test is that it is extremely sensitive and can therefore be used to detect minute traces of blood, even when attempts have been made to wash away the evidence. However, it must be performed in complete darkness and requires respraying to illuminate the blood, which can cause significant loss of detail. Luminol also reacts with many other substances, such as bleach and metals other than iron. When a substance other than blood reacts positively, the results are considered a false positive, meaning that the reaction occurred because of cross-reactivity, not because of the presence of blood. A false negative occurs when blood is present but no color change or light production is seen.

In this activity you will conduct two presumptive blood tests to determine which stains are likely blood. Every student will receive an envelope containing a stain found at the crime scene. You will process this stain using the two presumptive tests. From the results of the presumptive tests, you will decide which samples merit the confirmatory test for human blood.

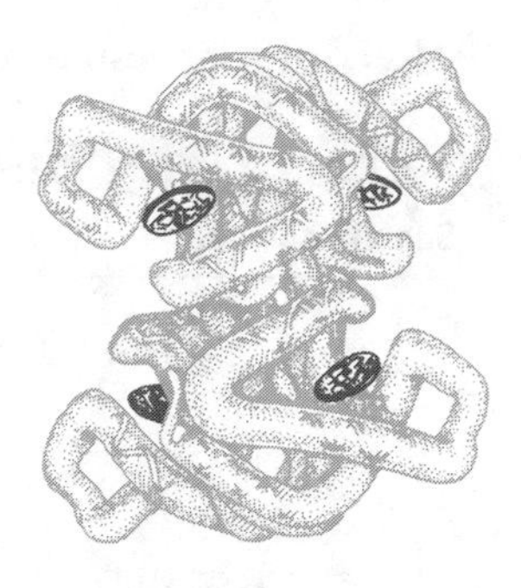

Figure 23.1
Hemoglobin

1. Obtain a crime scene envelope and a pair of gloves from your instructor.

WEAR THE GLOVES UNTIL THE EVIDENCE IS FINALLY DESTROYED, BOTH TO PROTECT YOURSELF FROM IT AND TO AVOID CONTAMINATING IT.

2. Once you are protected, take notes as a crime lab investigator would. What does the outside of the envelope say? What size and color is the envelope, and from whom did you receive it?

3. Record your sample number in Table 23.1.

4. Open the package and examine the substrate and the stain. What is the substrate? Describe the stain's shape, color, and size.

5. To fix and then process your stain using crystal violet (LCV), take your sample to a vacuum hood (a ventilated space). Fix your stain so that it can be processed using both luminol and leuco crystal violet (LCV). Spray the fixative provided by your instructor lightly over the stain. Does the stain change color?

6. To process the stain using luminol, take your sample and the luminol solution provided by your instructor to a darkened room, as this reaction requires complete darkness. Place your sample on a plastic tray along with a negative and positive control. Turn off the lights and spray the solution onto your sample and the control. Does the sample light up?

7. Record your results in Table 23.1. If your sample did illuminate, did you see any detail or pattern in the sample?

8. The LCV test can be performed back at your lab bench. Place your sample and the control on a plastic tray. Using the LCV, spray both your sample and the control lightly. Does your sample change color?

9. Record your results in Table 23.1. Does your sample warrant a confirmatory test for human blood?

TABLE 23.1 ACTIVITY 1: RESULTS FROM PRESUMPTIVE BLOOD DETECTION TESTS

Sample number	Luminol test	LCV test

Concept Check

1. What is a false positive? Is it better for a test to give a false positive or a false negative? Explain.

2. During crime scene investigations it is possible for an investigator to miss evidence. Describe at least three main reasons why.

3. How is a presumptive test useful to a crime scene investigator?

4. Why would these tests also show positive results for primate or other mammalian blood?

ACTIVITY 2 Confirmatory Blood Tests

Your blood contains substances that allow your body to recognize its own cells. These substances are surface molecules (often proteins) that jut out from your cells, waving like flags that say, "I am part of you." When a cell that is *not* part of you enters your body, this foreign cell waves a different flag, called an **antigen**. An antigen is any substance that stimulates the production of antibodies because it is detected as being a foreign intruder. **Antibodies** are specific proteins produced to recognize these antigens and mount an immune response. This kind of response occurs not only when bacteria and other microbes enter your body, but also if the wrong blood type is introduced into your body during a blood transfusion. The human population has different sugar-protein molecules projecting from their red blood cells. If red blood cells from the transfusion carry the wrong molecules, these molecules are considered antigens by the patient's immune system, and antibodies attack the foreign red blood cells, causing **agglutination**, or clumping of these cells. This is why you should know your **ABO blood type**. There are four different ABO blood types: A, B, AB, and O (Figure 23.2).

The letters "A" and "B" denote the presence of two distinct kinds of molecules on the surface of the red blood cell. So it follows that in type A blood all the red blood cells wave A molecules on their surface, the cells in type B blood wave B molecules on their surface, and type AB blood has cells that wave both A and B molecules on their surface. Type "O" denotes that neither A nor B molecules are present on the cells' surface. In addition, there is an antigen in the liquid (serum) of blood called the **Rh factor**; humans who have it are Rh positive (+), and those who don't are Rh negative (–). Blood typing is very useful for blood donations: Type O- blood doesn't cause an immune response in a person with any other blood type, so it is considered the universal donor. Type AB+ blood is the universal recipient: it will not attack any other blood type, because it won't detect any of the molecules as foreign. The blood groups and their ABO antigens and antibodies are outlined in Table 23.2.

The genetic mechanism of the ABO blood groups involves three alleles, any two of which may be present in one person. The three alleles are I^A, which produces antigen A; I^B, which produces antigen B; and *i*, which produces no antigens. I^A and I^B are both dominant over *i*, but they exhibit codominance toward each other; that is, neither one suppresses the other. The genotypes associated with each blood type are also given in Table 23.2.

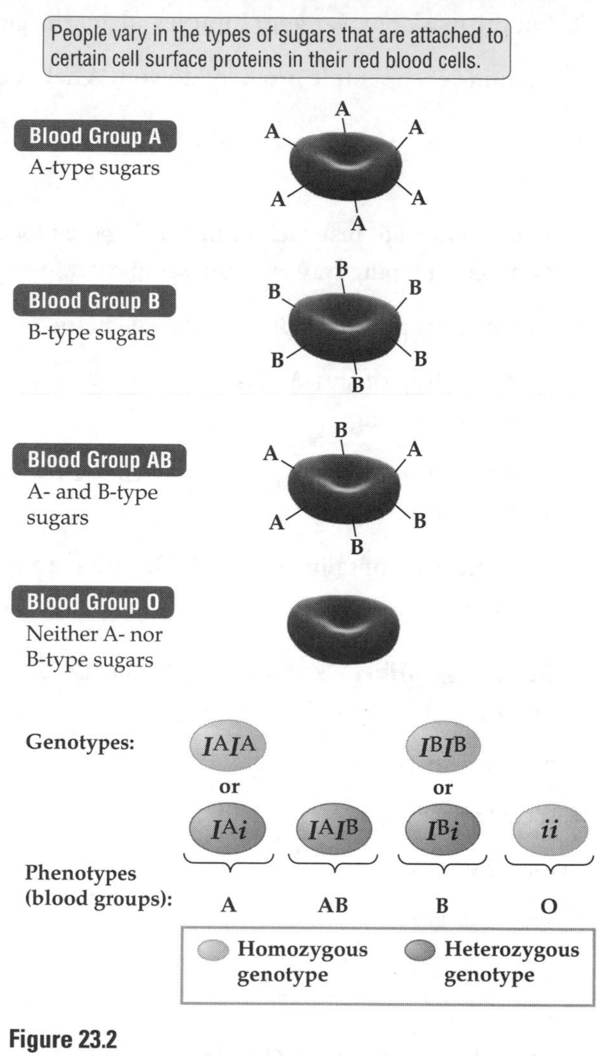

Figure 23.2
Blood Types

Prior to the development of DNA testing, forensic scientists used ABO typing on fluid recovered from a crime scene or victim to determine the donor of the fluid. Back at our investigation of the assault, the victim is in a coma and cannot provide the investigators with any leads. The investigators have discovered that the victim had problems with a coworker, a neighbor, a cousin, and an old girlfriend. In this activity you will receive a fluid sample from the crime scene and from each of these four suspects. You will determine which person could have contributed the fluid and should therefore be investigated further through DNA testing. In addition, you will learn how to calculate the statistical probability that an individual has a specific blood type.

TABLE 23.2 ABO BLOOD TYPES

Blood type	Antigens present	Antibodies present	Possible genotype
A	A	Anti-B	I^AI^A or I^Ai
B	B	Anti-A	I^BI^B or I^Bi
AB	A and B	None	I^AI^B
O	None	Anti-A and Anti-B	ii

1. Obtain a blood typing kit, the five fluid samples you will be testing, gloves, and a mask.
2. Put the mask on over your mouth and put the gloves on your hands.
3. Examine each sample provided to you. What does the fluid look like? Is it viscous?

4. You will need to test each sample using the blood typing kit. To prevent mixing up the samples, label each blood typing tray with the sample name.
5. Add one drop of a sample to each well in the first tray.
6. Add one drop of anti-A plasma to well 1. Do you see any clumping? In Table 23.3, note whether the sample is reacting.
7. Add one drop of anti-B plasma to well 2. Do you see any clumping? In Table 23.3, note whether the sample is reacting.
8. Add one drop of saline to well 3. Do you see any clumping? In Table 23.3, note whether the sample is reacting.

TABLE 23.3 ACTIVITY 2: BLOOD TYPING RESULTS

Sample name	Anti-A plasma	Anti-B plasma	Saline
Crime scene sample			
Suspect 1 sample			
Suspect 2 sample			
Suspect 3 sample			
Suspect 4 sample			

9. What blood type is the crime scene sample?

10. Would you send any of the suspect's fluid on for DNA testing? Why or why not?

11. From your instructor, obtain the frequency of your suspect's blood type in the human population. **Frequency**, in this context, is the number of times an event (in this case, blood type) occurs, divided by the total population that could contribute to the event. Write the frequency here.

12. In a criminal investigation, both the defense and the prosecution often like to emphasize the probability that people deposit their DNA, or in this case blood type, at crime scenes. This is where blood

type frequency comes into play, because not everyone in the world is likely to have been in contact with the crime scene. Instead of using the frequency in the entire human population, lawyers use a logical subset of the total population to obtain the **probability** within that subset. What is the probability that someone from a specific geographic area—such as the state you are in—will have your suspect's blood type? Find out your state's population (the number of individuals living there), and write it below. Multiply the frequency of the suspect's blood type by this population, and the result is the probability that an individual with the blood type in question is in your state, expressed as the number of individuals in your state matching that blood type.

13. Now calculate the corresponding probability for your city. Are these calculations accurate and precise? Why or why not?

Concept Check

1. Could a person with type A blood donate to any other blood type? Why or why not?

2. Would a person with type A+ blood who received a transfusion of type O^- blood have an immune response? Why or why not?

3. Where might a criminalist find blood evidence?

4. Given that blood typing doesn't identify specific individuals, how is it useful to a forensic scientist?

ACTIVITY 3 Hairs and Fibers

Hairs and fibers offer criminalists the ability to place people at the crime scene, and to describe what they were wearing and what they touched there. Hairs are clearly more powerful, because they may contain DNA if cells from the follicle (**follicular tag**) are intact. This follicular tag is where living cells are located, and DNA can usually be extracted. Even hair that doesn't have a follicle is useful, because it can provide information about a suspect to investigators. Both hair and fibers are very resilient and are often left at a crime scene, unbeknownst to the criminal.

Hair is composed of epithelial cells that become hardened (or keratinized), and growth occurs from cells within the hair **follicle** (Figure 23.3). The follicle is like a pocket where hair originates (located between layers of tissue called the dermis and hypodermis); the hair then protrudes through the epidermis

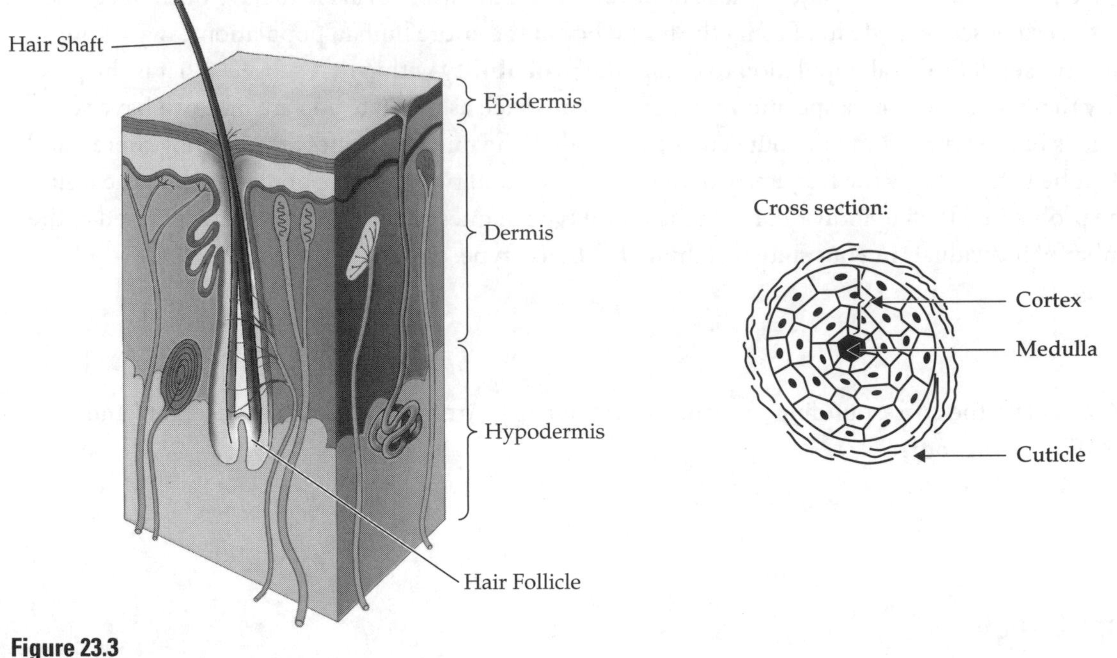

Figure 23.3
Hair Shaft and Cross Section

and finally penetrates the surface of the skin. The outermost structure of hair, the **cuticle**, is a clear, keratinized layer of dead cells that may look scaly under a microscope. In animals such as dogs and cats, this scaling is more obvious than in humans, with all of the scales pointing to the tip of the hair because the cells are layered on top of one another within the follicle as they are formed and pushed outward during growth. Beneath the clear layer of the cuticle, the **cortex** is where pigments are deposited and give color to hair. Pigment deposition varies across ethnic groups. Finally, the innermost structure of a hair is the **medulla**. The medulla may be present or absent, depending mainly on the source of the hair. If the medulla is present, this tube may have the same continuous color throughout the shaft, either because it is filled with air and appears dark, or because it has cells and appears lighter. Alternatively, the tube may not have a continuous color, in which case it is said to be fragmented, containing a combination of air and cells.

In addition to the physical features of hair, other evidence can be gleaned from hair found at a crime scene. A chemical analysis of the hair may be performed to determine whether the individual uses drugs and, if so, what type of drugs. Hairs come in different sizes, shapes, colors, and textures. These features may indicate whether a hair came from the head or another area of the body. Any chemical manipulation done to the hair can also help identify the contributor. When examining a hair, investigators must keep in mind that changes in color may be due to hair dye.

Fibers can be classified into **natural fibers** and **man-made fibers**. Most natural fibers from animal sources are hairs that have been extracted to form clothing or other products. These are visually similar to human hairs. Cotton, a plant product, is the most prevalent of all natural fibers. Under a microscope, cotton appears randomly twisted. Man-made fibers can be subclassified into **regenerated fibers** and **synthetic fibers**. Regenerated fibers are merely fibers made from cellulose, the main component of the cell walls of plants, which is extracted from a natural source and then reconfigured. Regenerated fibers include commonly known fibers such as rayon, acetate, and triacetate. Synthetic fibers are manufactured by linking together small repeating units called monomers, forming large macromolecules called polymers. The most common of these polymers include spandex, acrylic, nylon, and polyester. In man-made fibers, the diameter and color of the deposited fiber may correspond to one or more source fibers. There is no DNA in a fiber, so an investigator can say only that one fiber is a "match" to another known type of fiber—meaning they have common characteristics.

During the assault investigation, the crime scene technicians used a lint roller to collect any hairs or fibers on the victim's clothing. One suspect has emerged as the contributor of fluid found on the victim. Now the investigators have located this suspect's car, which was hidden at her mother's house. The car has uniquely colored fibers that make up the carpet. Investigators collect the fibers and also obtain a warrant to collect hairs from the suspect to compare to the hairs and fibers found on the victim's clothing. In this activity you will collect hair and fibers from your own clothing to understand the variety and complexity of evidence that may be collected at a crime scene.

1. Obtain a lint roller from your instructor.
2. Peel the first layer of tape off to ensure that you have a clean piece of tape.
3. Choose a piece of clothing to retrieve hairs and fibers from, and record this item in Table 23.4. Also provide a description of the clothing, including its color and size, but don't look at the label yet.

TABLE 23.4 ACTIVITY 3: CLOTHING ITEM DESCRIPTION

Clothing item	Description

4. Roll the lint roller along the entire piece of clothing in one direction and then in the opposite direction. Can you see any hairs and fibers on the roller?
5. Tear off the layer of tape that you rolled over your clothing. Place it sticky side up on the lab bench.
6. Choose an area in the middle of the tape and use a Sharpie to mark a 1-inch-by-1-inch square.
7. Under a fume hood, place a few drops of xylene in the marked area and let it incubate for 1 minute. This substance will dissolve the glue on the tape to allow you to transfer the evidence to a slide for examination.
8. After 1 minute, use the skinny edge of a microscope slide to scrape up all the deposited hairs and fibers. Do you see them being collected?
9. Transfer this material to another slide by wiping it onto the new slide as shown in Figure 23.4.

Figure 23.4
Slide Transfer

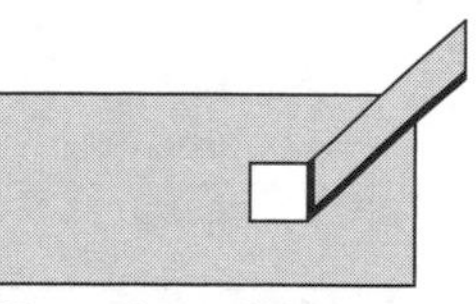

1. Use a clean slide to pick up hairs and fibers.

2. Wipe residue onto a second clean microscope slide for observation.

10. Place a drop of mounting medium onto the deposited fibers, and place a coverslip over your sample.

11. Examine your collection using a compound microscope, first scanning the sample to identify any hairs located on the slide. Refer to Figure 23.5*a* for guidance on the types of hairs you are likely to observe. Record the number of hairs of each type, and describe their color, length, and diameter, in Table 23.5. Do you see a medulla? If so, is it continuous or fragmented? Where do you think this hair is from? Is it human or from some other animal? If you believe it is human, where on the body do you think this hair originated? Why?

(a) Hairs

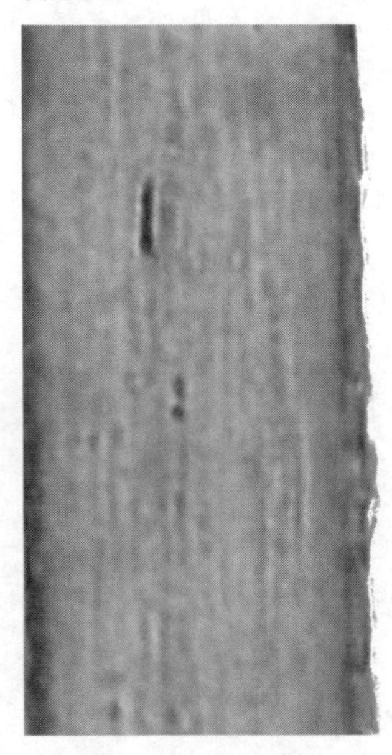

Human eyebrow hair

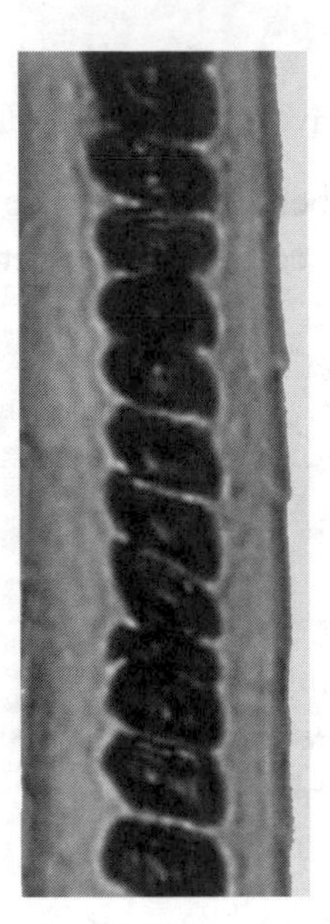

Dog hair

(b) Fibers

Cotton fiber

Acetate fiber

Polyurethane fiber

Figure 23.5
Microscope View of Hairs and Fibers

TABLE 23.5 ACTIVITY 3: MICROSCOPIC EXAMINATION OF HAIRS

Hairs found	Color	Length	Diameter	Medulla description	Possible source

12. Identify any fibers in your sample, classifying them as natural or man-made. Refer to Figure 23.5*b* for guidance on the types of fibers you are likely to observe. Do you see any irregularly twisted fibers? They should be classified as cotton. Record the number of fibers of each type, and describe their color, length, and diameter, in Table 23.6. Identify possible sources of these fibers. Do you think that all the fibers you found are from your own clothing?

TABLE 23.6 ACTIVITY 3: MICROSCOPIC EXAMINATION OF FIBERS

Fibers found	Color	Length	Diameter	Possible source

13. Look at your clothing tag and see if you correctly identified the types of fibers. How many fibers are actually listed? Do all the fibers you identified appear to be consistent with your garment?

14. Compare the number of hairs and fibers you identified with those the rest of the class found. Do you and your classmates have the same types or different types?

Concept Check

1. What is the visual difference between a hair and a fiber?

2. What is the visual difference between a natural fiber and a synthetic fiber?

3. How do regenerated and synthetic fibers differ?

ACTIVITY 4 Fingerprints

Fingerprints are formed by the buckling of the upper portion of the skin, the epidermis, over the lower layer, the dermis. As the epidermis and dermis form during fetal development, fingerprints form on the hands and feet. These structures likely evolved to improve sensory and gripping capabilities. The ridges, called **friction ridges** or **epidermal ridges**, and the valleys combine to provide valuable characteristics that are unique to each individual and helpful in forensic cases. But unlike DNA, fingerprints are even different in twins, because the fetuses have different experiences in the womb, leading to different formations of ridges.

Fingerprints are found in three different forms: (1) impressions in semisolids forming a three-dimensional remnant called a **plastic print**; (2) a completely visible print left as a result of an individual contacting a fluid and then touching an object, leaving what is called a **patent print**; or (3) a deposit resulting from contact by fingers, allowing the natural fluids excreted by the body to leave an outline of the ridges. This last type of fingerprint, called a **latent print**, is not clearly visible to the human eye, but an investigator can dust it with a powder or apply specific chemicals to provide enough contrast to make it visible. For the other types of fingerprints, photography is often useful in acquiring a digital image of the print. Overall, the most important factor for investigators is the type of surface where the fingerprint was left, because the surface type is what determines which technique must be used to accurately recover the print.

Fingerprints have different patterns generated by distinctive distances between ridges and varying lengths of ridges, allowing for unique identification of an individual from the print of even just one finger. The main formations identifiable in a fingerprint are **arches**, **loops**, and **whorls**; some of these patterns combine to form more complex patterns, such as the **tented arch**. Figure 23.6 shows an example of each of these patterns.

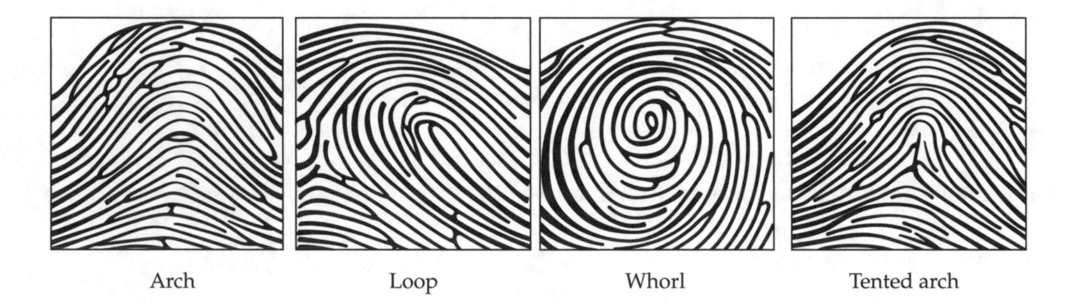

Figure 23.6
Fingerprint Patterns

The last pieces of evidence to emerge from our crime scene were fingerprints on a knife found in the alley behind the victim's house. The victim's blood was found on the knife. In this activity you will lift fingerprints from the evidence recovered from the crime scene and identify any unique patterns of the prints. You will then compare each print against a set of the suspect's fingerprints to determine whether the recovered print came from this suspect.

1. Try to identify your own fingerprint patterns by either visually inspecting your fingers—some individuals have very prominent ridges—or by using the ink provided by your instructor and transferring your prints to the boxes in Table 23.7. This will help familiarize you with the patterns you will be trying to identify. Do you have any whorls, loops, or arches?

TABLE 23.7 ACTIVITY 4: YOUR OWN FINGERPRINTS

Right hand (start with your thumb in the far-left box):

Left hand (start with your thumb in the far-right box):

2. Your crime scene items will be located on your lab bench on top of a lab mat to enable easy cleanup. Of the items on the mat, which one do you think is most likely to yield a good fingerprint? Why?

DON'T TOUCH THE ITEMS WITH YOUR BARE HANDS. DOING SO WOULD CLEARLY DEFEAT THE PURPOSE OF LIFTING FINGERPRINTS TO IDENTIFY PATTERNS IN A SUSPECT'S FINGERPRINT.

3. In Table 23.8, give each item a number and a name, and describe the item in detail. What shape is it? What color is it? How many items do you have?

TABLE 23.8 ACTIVITY 4: ITEMS FOR FINGERPRINT RECOVERY

Item number	Name	Description

4. Put gloves on your hands. Choose the item that you think will yield the best results. Remember: you have at your disposal only the basic fingerprint dusting technique.

5. After choosing the item, dip the fingerprinting brush into the dusting powder. Wipe the entire surface of the item using a light sweeping motion as you dust it.
6. Note whether there are any visible fingerprints. Cut the clear tape provided so that you have a piece twice as large as the area where you suspect a fingerprint is present. Apply the clear tape over the area, flanking each side with the extra tape. Once the tape is completely in contact with the area, remove the tape and place it in the space provided in Table 23.9 below.
7. Can you identify any fingerprints in the tape sample? If not, repeat step 6 using a different area of the item where you suspect fingerprints to be present.
8. After lifting all the fingerprints from the item, note in the "Description" column of Table 23.9 the patterns in the identifiable prints, the size of the prints, and whether they have any unique identifying marks.
9. Now compare these prints to the fingerprint cards provided by your instructor. Can you identify the individual who left prints on your item? If so, which person deposited fingerprints at the crime scene?

TABLE 23.9 ACTIVITY 4: FINGERPRINTS RECOVERED

Print	Description
1:	
2:	
3:	

Concept Check

1. How might a fingerprint be deposited at a crime scene?

2. What types of surfaces are most likely to provide liftable fingerprints? Why?

3. What are the three main types of patterns seen in fingerprints?

Key Terms

ABO blood type (p. 23-4)
agglutination (p. 23-4)
antibody (p. 23-4)
antigen (p. 23-4)
arch (p. 23-12)
chemiluminescence (p. 23-2)
circumstantial evidence (p. 23-1)
confirmatory test (p. 23-1)
cortex (p. 23-8)
cuticle (p. 23-8)
direct evidence (p. 23-1)
epidermal ridge (p. 23-11)
fingerprint (p. 23-11)
fixed (p. 23-2)
follicle (p. 23-7)
follicular tag (p. 23-7)
frequency (p. 23-6)
friction ridge (p. 23-11)
latent print (p. 23-12)
Locard's exchange principle (p. 23-2)
loop (p. 23-12)
man-made fiber (p. 23-8)
medulla (p. 23-8)
natural fiber (p. 23-8)
patent print (p. 23-12)
plastic print (p. 23-12)
presumptive test (p. 23-1)
probability (p. 23-7)
regenerated fiber (p. 23-8)
Rh factor (p. 23-5)
synthetic fiber (p. 23-8)
tented arch (p. 23-12)
whorl (p. 23-12)

Review Questions

1. Why do genetically identical humans have different fingerprints?

2. How are fingerprints formed?

3. A(n) ________________ is a print not visible to the human eye.

4. ________________ are protrusions from the epidermis found on the hands and feet.

5. What is the evolutionary advantage of human fingerprints?

6. What characteristics of hairs and fibers are useful when they are being examined as evidence?

7. What are the three parts of a hair?

8. How do monomers relate to fibers?

9. How might animal hairs be visually different from human hairs?

10. What are the main human blood types?

11. A(n) ________________ is a substance detected as foreign, eliciting an immune response that produces ________________.

12. Which blood type is the universal blood donor? Which blood type is the universal blood recipient?

13. A type of chemical reaction producing photons is called ________________.

14. What is the difference between a presumptive and a confirmatory test?

15. Examples of ________________ include an eyewitness, the presence of a video of the criminal act, or a confession to the crime.

16. Why are hairs potentially more explanatorily powerful than fibers?

17. If DNA from a suspect is found at the crime scene, does it prove that he or she engaged in the criminal activity? Why or why not?

LAB 24 Animal Behavior

Objectives

- Examine different types of animal behavior.
- Use tests to detect predictable or unpredictable behavior in organisms that is due to a specific stimulus.
- Identify how behaviors may be beneficial for organisms.
- Understand the limitations of the sensory systems of organisms.
- Consider the impact of humans on the behavior of other organisms and their potential evolution.
- Interpret the reaction of organisms when more than one stimulus is applied.

Introduction

A **behavior** is a coordinated response made by one organism in reaction to another organism or the physical environment. The organism's ability to respond is genetically based, but it can be influenced by nongenetic factors. Although even a single-celled organism may display a behavior, the easiest behaviors to recognize are those displayed by complex multicellular organisms such as humans. Responses by complex organisms are easily detected because of their integrated organ systems, which include intricate nervous and muscular systems to elicit quick, specific responses.

Functionally, a behavior involves an organism detecting an outside stimulus, interpreting the meaning of the stimulus, and then responding to the stimulus. Behavioral scientists identify these stimuli, observe the resulting behaviors, and catalog their functions. In addition, scientists may be able to trace behaviors through an evolutionary lineage, determining how these traits enable different groups to survive and reproduce, and therefore be better adapted to their environment.

The two main types of behaviors are fixed and learned. A **fixed behavior**, also called an innate behavior, is any genetically programmed, automatic response upon the first exposure to a stimulus. A common example of a fixed behavior is a kitten covering its excrement the first time it uses a litter box. No mother cat has to teach her kitten how this is done. This behavior is automatically performed after the kitten is done with a body function, but other fixed behaviors require a **releaser**, an outside stimulus that causes a fixed behavior to occur. For example, the red color on the wing of a male blackbird causes another male blackbird to defend his territory. This response is still a fixed behavior, because it is a genetically preprogrammed response; the blackbird father does not teach it to the blackbird son.

In contrast, a **learned behavior** is any response taught to the individual by another individual or acquired through experience. A dog can be taught tricks such as "sit" or "lie down" or "stay" via repeated verbal commands and/or signals. In a way, this is similar to how a student learns material for a test. By performing repetitious activities, the student learns and remembers information that can then be applied to questions on an exam. But students often study together, in which case their memory is influenced by the interaction with other individuals of the same species.

Social behavior comprises the behavioral interactions among members of a group, usually of the same species. Group members gain advantages not available to solitary animals, such as an improved ability to find resources efficiently and effectively, and a better defense against predators. Social groups may exhibit **altruistic behavior,** a type of social behavior in which individuals do things to help other members of their group survive or reproduce even while decreasing their own chances of doing so. In the end, such behavior is likely beneficial to altruistic individuals because they often assist relatives who carry some of the same genes they do, therefore increasing the likelihood that their genes will survive.

In this lab you will test a variety of stimuli on different types of organisms to determine whether a predictable behavior results.

ACTIVITY 1 Chemotaxis

Organisms may display specific behaviors related to their position in their environment. This type of orientation behavior often is a specific directional movement, or **taxis** (plural *taxes*). The result of taxis normally places the organism in a favorable area of its environment. Taxis can be based on many different factors, but one of the easiest to understand and test is **chemotaxis,** movement of an organism based on the detection of a chemical. In positive taxis, the organism moves toward the chemical because it likely indicates a food source; in negative taxis, the organism moves away from the chemical because it may sense danger or a toxic situation. Model organisms such as *Caenorhabditis elegans* (also known as *C. elegans*) (Figure 24.1) provide a quick and easy means to study these reactions.

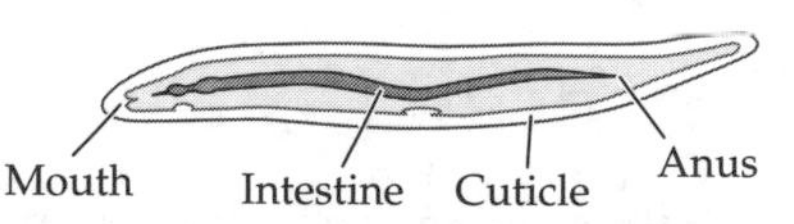

Figure 24.1
Caenorhabditis elegans

C. elegans is an invertebrate worm in the phylum Nematoda. These worms are used extensively in scientific research to study a variety of physiological reactions on a simpler scale. Since *C. elegans* doesn't have as complex a nervous system as humans have, researchers can experiment with it to narrow their focus on the causes and effects of behaviors, even locating the genes that are turned on or off to elicit a response. In this activity you will determine whether the behavior of *C. elegans* is affected by different compounds.

1. Obtain a sterile agar plate (a medium on which *C. elegans* can survive and easily move around), a dissecting microscope, a ruler, your test solutions, water, pipettes, and a marker.

2. In Table 24.1, record the solutions you will test in the "Test solution" column. Then formulate a question about how one of the chemicals may affect the movement of *C. elegans*. Develop a hypothesis and make predictions about the effect of this chemical on the behavior of *C. elegans*.

   Question: ____

   ____

   ____

Hypothesis: ____

____

____

Prediction(s): ____

____

____

3. Mark a dot in the middle of the bottom of the agar plate (the side containing the agar), and then mark a dot to its right and a dot to its left, each approximately 1 cm from the edge of the plate.

4. Flip the agar plate back over and transfer one drop of *C. elegans* from the culture bottle onto the agar above the center dot. Observe *C. elegans* under the microscope.

5. Place a drop of water on the dot marked to the left of the middle, and a drop of the test solution on the dot marked to the right.

6. Wait 30 minutes and observe the reaction of the worms. Are they moving? If so, in what direction? Do they stay in the same general area? Record your observations in Table 24.1. How many worms move toward the water, stay in the same original area, or move toward the drop of test solution?

TABLE 24.1 ACTIVITY 1: RESULTS FOR TAXIS OF WORMS

Test solution	Number of worms moving toward water	Number of worms staying in original area	Number of worms moving toward the solution

7. Repeat steps 3 through 6 with the remaining test solutions. Do you observe a difference in the behavior of the worms, depending on the type of solution?

8. After all the solutions have been tested, compare your results to those of the rest of the class. Do the results support your hypothesis?

9. Judging by your results, what would you test next to answer a similar question about this organism? Could you design this experiment with a different organism? If so, which organism, and would you have to change any aspects of the experimental design?

Concept Check

1. What is positive taxis?

2. Why would an organism exhibit negative taxis to an odor?

3. Which senses detect chemicals?

4. Why would chemotaxis be beneficial evolutionarily?

5. Why is it "good science" to compare your results to those obtained by your classmates?

ACTIVITY 2 Chemotaxis and Communication

Chemotaxis is important in helping individuals of some species find a mate, locate their group, or tell others in the group where food is located. For organisms such as termites, which have limited vision, chemical detection is essential in sorting out the location of food. After detecting food, termites can tell each other where the food is by using chemical communication.

Communication is a type of behavior that enables one individual to send information about the environment to others. Because of the complexity of organisms and their senses, communication can be visual, electromagnetic, auditory, physical, or chemical (through either smell or taste). One common chemical cue is a **pheromone**, a chemical signal produced by one individual to communicate its identity, its location, its physical condition, or a situation in its environment to others of the same species. Communication, just like other behaviors, has a variety of purposes, but its ultimate purpose is survival. In addition, communication is essential for certain species where group interaction can increase the evolutionary fitness of the average organism; these species have learned to live as a social network of individuals.

Interestingly, some everyday chemicals that humans use for the purpose of writing are very similar to chemicals that termites use to communicate with one another. Termites leave chemical trails to inform each other of what they have found, but an experimental design using the termites' own trails would be highly variable. In this activity you will mimic these communication trails, using the chemicals in writing implements to affect the behavior of termites.

1. Obtain a piece of Whatman paper, a ruler, a yellow highlighter, four different writing implements, and a termite.

2. In the top row of Table 24.2, record the writing implements you will test. Next, formulate a question about how the contents of the writing implements may affect the movement of the termites. Develop a hypothesis and make predictions about the effect of your items on the behavior of the termites.

   Question: ____

   ____

   ____

   Hypothesis: ____

   ____

   ____

   Prediction(s): ____

   ____

   ____

3. Place the Whatman paper on the lab bench and determine the middle point of the paper using the ruler. Mark this spot using a yellow highlighter. This is where you will place your termite.

4. Draw a 10-cm line away from the center point toward the edge of the paper with the first writing implement. In the opposite direction, draw another 10-cm line away from the center point with a second writing implement, creating a straight line. Next, create a "+" by turning the paper 90 degrees and using the last two writing implements in the same manner to draw two more lines perpendicular to the first two lines (Figure 24.2).

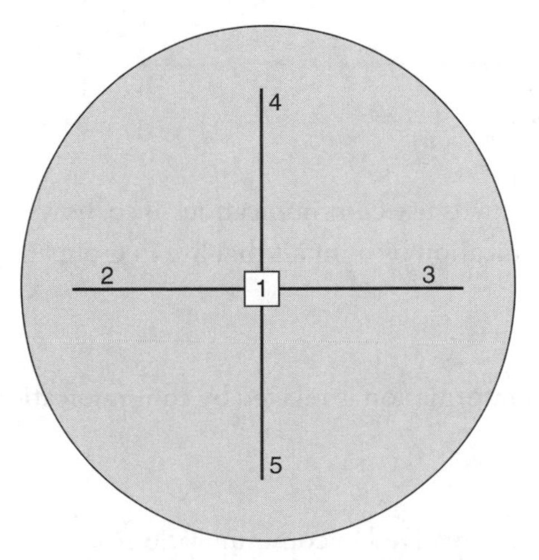

1. Mark created by highlighter
2. 10-cm line created by first writing implement
3. 10-cm line created by second writing implement
4. 10-cm line created by third writing implement
5. 10-cm line created by fourth writing implement

Figure 24.2
Whatman Paper Setup

5. Place one termite on the center dot made by the yellow highlighter.
6. Wait 2 minutes and note the reaction of the termite. Is it moving? Does it stay in the same general area? Record your result in Table 24.2.
7. Repeat the experiment with different individual termites of the same species or different organisms provided by your instructor, recording your results in Table 24.2. Do you observe a difference in the behavior of different species?

TABLE 24.2 ACTIVITY 2: RESULTS FOR CHEMOTAXIS OF ORGANISMS

Organism	Distance traveled along the line created by writing implement #1 (________)	Distance traveled along the line created by writing implement #2 (________)	Distance traveled along the line created by writing implement #3 (________)	Distance traveled along the line created by writing implement #4 (________)

8. After all the organisms have been tested, compare your results to those of the rest of the class. Do the results support your hypothesis?
9. Judging by your results, what would you test next to answer a similar question about this organism? Could you design this experiment with yet another organism? If so, which organism, and would you have to change any aspects of the experimental design?

Concept Check

1. Do solitary animals use communication? If so, how might communication by solitary animals differ from communication by animals that live in groups?

2. What type of information is relayed by communication?

3. What senses are involved in communication?

4. How can organisms use chemicals to communicate if they are physically separated?

ACTIVITY 3 Phototaxis

The onset of taxis is not always triggered by a chemical source. It may be caused by other stimuli, such as temperature, water, sound, pressure, or light. **Phototaxis** is the movement of an organism in a specific direction toward or away from light. Some organisms that exhibit phototaxis prefer to live in complete darkness, and therefore avoid lighted areas.

Planaria (Figure 24.3) are flatworms in the phylum Platyhelminthes. They have complex organ systems with a specialized sensory system including light-sensitive eyespots, which send information about the light in the environment to a simple brain. In this activity you will determine whether the planarian brain interprets light as a positive or negative stimulus.

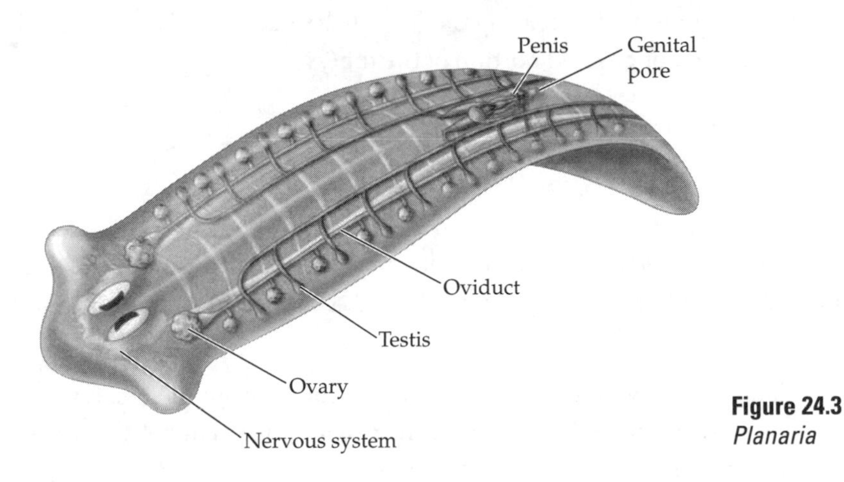

Figure 24.3
Planaria

1. Obtain two test tubes with screw caps, a sheet of aluminum foil, a ruler, a marker, spring water, and a plastic transfer pipette.

2. Formulate a question about how light may affect the movement of *Planaria*. Develop a hypothesis and make predictions about the effect of your items on the behavior of *Planaria*.

Question: ______________________________

Hypothesis: ______________________________

Prediction(s): ______________________________

3. Measure the length of each tube and mark the halfway point.
4. Fill both test tubes 75% full of spring water, and then, using the plastic pipette, transfer 10 *Planaria* from the culture tank into each test tube.
5. Fill the remainder of each tube with spring water, leaving only a small space for air at the top. Why do you have to use spring water for this experiment?

6. Place the tubes horizontally on the bench for 5 minutes to allow the *Planaria* to recover from being transferred. What are the *Planaria* doing?

7. Wrap half of each tube with aluminum foil, leaving the other half uncovered. Do you expect the *Planaria* to stay in the uncovered or covered portion of the tubes?

8. After 15 seconds, count how many *Planaria* are in the light half of each test tube and, therefore, how many are in the dark, and record these numbers in Table 24.3. Do this again at 30, 45, 60, and 75 seconds.
9. After recording all the data, average your results by adding the counts from the two test tubes and dividing by 2. Record these averages in Table 24.3. Were the results consistent between the test tubes? Are the *Planaria* exhibiting positive or negative taxis? Why?

10. Determine what percentage of your animals showed positive taxis at the end of the experiment and what percentage showed negative taxis.
    Positive taxis:
    Negative taxis:
11. Record the class average in Table 24.3. How do your results compare to the class average? Do the results support your hypothesis?

TABLE 24.3 ACTIVITY 3: RESULTS FOR PHOTOTAXIS OF *PLANARIA*

	15 s		30 s		45 s		60 s		75 s	
	Light	Dark	Light	Dark	Light	Dark	Light	Dark	Light	Dark
Test tube 1										
Test tube 2										
Average										
Class average										

12. Judging by your results, what would you test next to answer a similar question about this organism? Could you design this experiment with a different organism? If so, which organism, and would you have to change any aspects of the experimental design?

Concept Check

1. Why could light be a stimulus that an organism might avoid?

2. Why could an organism take longer to react to a stimulus?

3. Experiments using living organisms normally have varying results. What could explain the difference in results?

ACTIVITY 4 Combined Taxis

The combination of all the information gathered by organisms enables them to react appropriately to their environment. Many factors are involved in any individual's abilities. First, there is the underlying genetic basis of a behavior. An individual of one species may be more sensitive to a particular stimulus, and an external cue registered by one species may be undetectable by another. Sensitivity within a species is based on the number of sensory receptors possessed by an individual; the more receptors it has, the more sensitive it is to the stimulus. However, experience is also vital to survival. The past experience of the individual contributes to its overall response.

If an organism doesn't display a *specific* movement or orientation when exposed to a stimulus, the behavior is said to be **kinesis**, a nondirectional movement in response to a stimulus. For example, some organisms may not be affected by their orientation in relation to the ground, and therefore gravity has no effect on their behavior. Other organisms will always lie close to the bottom or top of their environment and are said to exhibit **geotaxis**, a specific movement toward or away from gravity. In this activity you will determine whether *Planaria* demonstrate geotaxis or phototaxis, and whether phototaxis and geotaxis reinforce each other or one form of taxis overpowers the other.

1. Obtain six test tubes with screw caps, a sheet of aluminum foil, a ruler, a marker, spring water, and a plastic transfer pipette.
2. Formulate a question about how light and gravity may affect the movement of *Planaria*. Develop a hypothesis and make predictions about the effect of your items on the behavior of *Planaria*. You might want to think first about how light and gravity may separately affect the movement of *Planaria*, and then about how these stimuli may affect *Planaria* in combination.

   Question: ____

   ____

   ____

   Hypothesis (light): ____

   ____

   ____

   Hypothesis (gravity): ____

   ____

   ____

   Hypothesis (both light and gravity): ____

   ____

   ____

   Prediction(s): ____

   ____

   ____

3. Measure the length of each tube and mark the halfway point.
4. Fill the test tubes 75% full of spring water and then, using the plastic pipette, transfer 10 *Planaria* from the culture tank into each test tube.
5. Fill the remainder of each tube with spring water, leaving only a small space for air at the top. Why do you have to use spring water for this experiment?
6. Place the tubes horizontally on the bench for 5 minutes to allow the *Planaria* to recover from being transferred. What are the *Planaria* doing?

7. You will be testing three different conditions to determine whether there is a connection between geotaxis and phototaxis. Test tubes 1 and 2 are for examining the effects of gravity alone. For this portion of the experiment, simply turn tubes 1 and 2 from a horizontal position to a vertical position.
8. Count the number of *Planaria* that are above the halfway point and the number that are below. In Table 24.4, record these numbers every 2 minutes for 10 minutes. Which movement is the positive taxis? Which movement is the negative taxis?
9. After recording all your data for test tubes 1 and 2, average the final results by adding together the "10 min" results from the two test tubes and dividing by 2. Record these averages in Table 24.4, in the column marked "Average at 10 min." Were the results consistent between the test tubes? Are the *Planaria* showing a positive or negative taxis? Why?
10. In test tubes 3 and 4 you will test one combination of phototaxis and geotaxis. In this second trial, place a piece of aluminum foil completely over the top half of the test tubes while they are still horizontal, and then when both tubes are ready, turn them vertical.
11. Count the number of *Planaria* that are above the halfway point (in the dark) and the number that are below (in the light). In Table 24.4, record these numbers every 2 minutes for 10 minutes. Are there consistent movements?
12. After recording all your data for test tubes 3 and 4, average the final results by adding together the "10 min" results from the two test tubes and dividing by 2. Record these averages in Table 24.4, in the column marked "Average at 10 min." Were the results consistent between the test tubes? Are these taxes in conflict with one another, or are they complementary?
13. In test tubes 5 and 6 you will test the opposite combination of phototaxis and geotaxis. In this third trial, place a piece of aluminum foil completely over the bottom half of the test tubes while they are still horizontal, and then when both tubes are ready, turn them vertical.
14. Count the number of *Planaria* that are above the halfway point (in the light) and the number that are below (in the dark). In Table 24.4, record these numbers every 2 minutes for 10 minutes. Are there consistent movements?
15. After recording all your data for test tubes 5 and 6, average the final results by adding together the "10 min" results from the two test tubes and dividing by 2. Record these averages in Table 24.4, in the column marked "Average at 10 min." Were the results consistent between the test tubes? Are these taxes in conflict with one another, or are they complementary?

TABLE 24.4 ACTIVITY 4: RESULTS FOR GEOTAXIS AND PHOTOTAXIS OF *PLANARIA*							
Test tube	**Taxis**	**2 min**	**4 min**	**6 min**	**8 min**	**10 min**	**Average at 10 min**
1	Above						Above (tubes 1 and 2):
	Below						
2	Above						Below (tubes 1 and 2):
	Below						
3	Above/dark						Above/dark (tubes 3 and 4):
	Below/light						
4	Above/dark						Below/light (tubes 3 and 4):
	Below/light						
5	Above/light						Above/light (tubes 5 and 6):
	Below/dark						
6	Above/light						Below/dark (tubes 5 and 6):
	Below/dark						

16. Judging by your results, do geotaxis and phototaxis reinforce each other, or does one response dominate the other? Why? Do the results support your hypothesis?

17. Again judging by your results, what would you test next to answer a similar question about this organism? Could you design this experiment with a different organism? If so, which organism, and would you have to change any of the experimental design?

Concept Check

1. How is geotaxis beneficial to an organism?

2. Why do some organisms react to a stimulus, whereas others do not?

3. What enables an individual to be more sensitive to a stimulus than other individuals of the same species are?

ACTIVITY 5 Environmental Impact

Humans have a great impact on the world around them. Their effect on other organisms often goes undetected, but it can have severe outcomes for these groups. Simple things like emptying out a soda bottle or fertilizing your garden can affect aquatic systems close by. Environmental effects on other organisms are sometimes difficult to measure directly, but indirect measures such as heart rate can still be informative. An increase in heart rate, for example, may be indicative of an increase in aggressive or fight-or-flight responses, whereas a lower heart rate may indicate paralysis. In this activity you will test how common chemicals affect the heart rate of an aquatic organism called *Daphnia* (Figure 24.4).

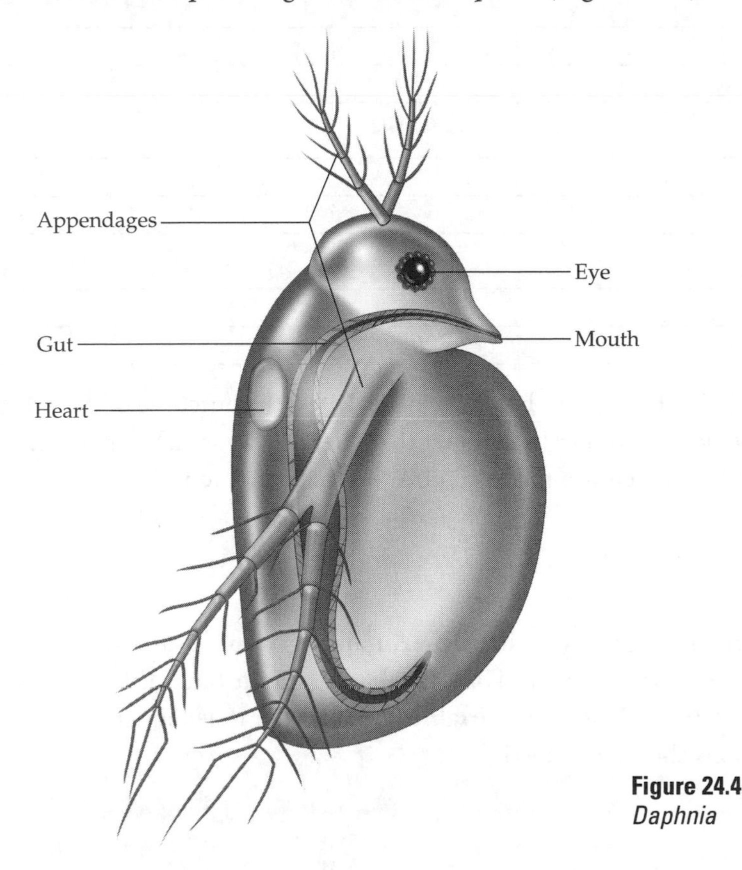

Figure 24.4
Daphnia

1. You have on your lab bench items often dumped either intentionally or unintentionally into bodies of water such as rivers, lakes, streams, oceans, and tidal basins. At the bench you will test the impact of one of these chemicals on the heart rate of *Daphnia*.

2. Observe the *Daphnia* and their normal behavior. The *Daphnia* are in spring water, which is their normal habitat, so they should behave as they would in nature. You will measure the average heart rate of *Daphnia* and then record any change in heart rate under different conditions. The heart is the circular organ located toward the eye and along the back of *Daphnia* (see Figure 24.4). Watch as the heart constricts: every constriction is considered a heartbeat. If the average heart rate is expected to be 200 beats per minute, what time span would be sufficient to collect heart rate data about *Daphnia*? Do you have to record for an entire minute?

3. From the following items, choose one to test its effect on *Daphnia*:
   - Caffeinated soft drink
   - Cigarette
   - Fertilizer

   (Each solution is provided at several different concentrations at the lower limit of lethality to the organisms.) Next, formulate a question about how one of the items provided may affect *Daphnia*. Develop a hypothesis and make predictions about the effect of your item.

   Question: ____
   ____
   ____

   Hypothesis: ____
   ____
   ____

   Prediction(s): ____
   ____
   ____

4. Obtain 12 culture tubes. Think about how to effectively test the outcome of the substance you chose on the behavior of *Daphnia*. To start, remember that scientists approach experimentation by having a control group in order to establish the baseline or normal conditions. What is your control group?

5. Determine the experimental group(s): The **experimental group** is typically a group or several groups in which the scientist has changed one variable per group. Since this factor is controlled by the scientist, it is called the ***independent* variable**. What is your independent variable? How many different concentrations should you test?

6. Obtain a means by which you can determine change. Examples include a microscope, a fancy mass spectrometer, or a ruler. Any change observed is the ***dependent* variable**. What will you use to determine the change in the dependent variable?

7. To ensure that the results recorded are valid, scientists often perform replicate treatments. If multiple identical treatments yield the same result, the experiment is then known to be more reliable, and its conclusions will be stronger. Table 24.5 has three rows for the control group and three rows for each experimental group, so you can run three replicates of each treatment. In the "Group" column of this table, record the conditions for all of your control and experimental groups.

8. Set up the experiment. Label the culture tubes with the appropriate names, according to what you wrote in Table 24.5.
9. Add *Daphnia* to each tube—but first think: how many *Daphnia* is it appropriate to test in each culture tube? Table 24.5 has three columns to record the heartbeat of three different animals, so you will need at least three *Daphnia* per tube. Be consistent in placing the same number into each culture tube. The most accurate way to add the correct number of *Daphnia* while ensuring that the spring water is at the same level in each tube is to first add the *Daphnia* and spring water to a graduated cylinder, measure 4 ml of water and *Daphnia* for each tube, and then add this measured amount to each labeled culture tube.
10. You will want to stagger your treatments so that you can record the organisms' heart rate after an equal exposure to each treatment. Therefore, add 10 ml of the appropriate solution for one treatment, and then wait 2 minutes and add the next, and so on, for each correctly marked tube.
11. After 5 minutes of exposure, remove three *Daphnia* and count how many heartbeats occur in 20 seconds for each individual. Record your results in Table 24.5. After all effects of the treatments have been recorded, average the heart rate across replicate treatments, and then convert this rate from heartbeats per 20 seconds to heartbeats per minute. Which treatment caused the greatest change in heart rate? Which caused the least? Were the replicates similar? Why or why not? What can account for differences between replicates?

TABLE 24.5 ACTIVITY 4: RESULTS OF *DAPHNIA* EXPERIMENT

Group	Number of heartbeats per 20 s for individual 1	Number of heartbeats per 20 s for individual 2	Number of heartbeats per 20 s for individual 3	Average number of heartbeats per 20 s	Average heart rate (beats/min)
Control group 1: ______________					
Control group 2: ______________					
Control group 3: ______________					
Experimental group 1a : ______________					
Experimental group 1b: ______________					
Experimental group 1c: ______________					

(continued on next page)

TABLE 24.5 Continued

Group	Number of heartbeats per 20 s for individual 1	Number of heartbeats per 20 s for individual 2	Number of heartbeats 2 per 20 s for individual 3	Average number of heartbeats per 20 s	Average heart rate (beats/min)
Experimental group 2a: ________					
Experimental group 2b: ________					
Experimental group 2c: ________					
Experimental group 3a: ________					
Experimental group 3b: ________					
Experimental group 3c: ________					

12. In Figure 24.5, label the *x*-axis with the independent variable (what you changed) and the *y*-axis with the dependent variable (what you measured). Remember to include the units of measurement where appropriate.

    a. How should you descriptively title your graph?

    b. Judging by your results, what scale should you use for each axis?

    c. Discuss with your group how to label the intervals of each axis.
    d. Plot your results on the graph. Does your group see any trend in your results?
    e. Do your data support your hypothesis? If you reject your hypothesis, how would you revise it?

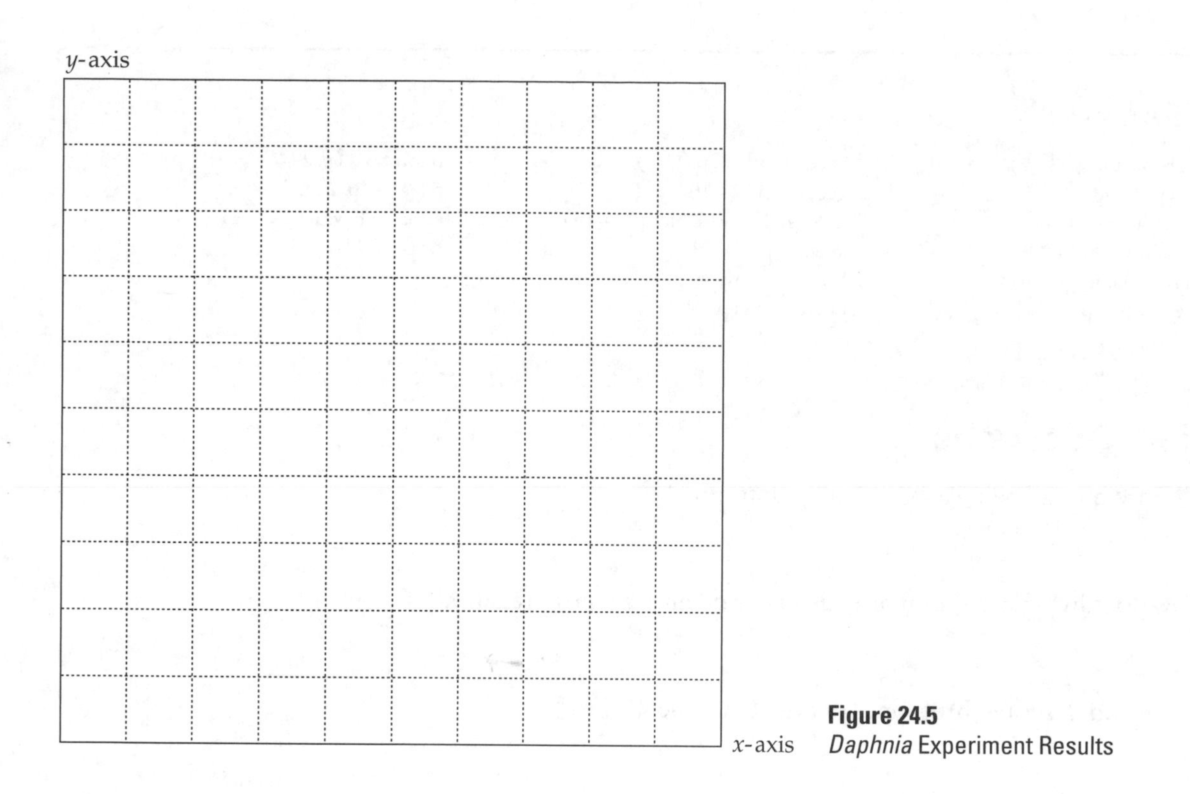

Figure 24.5
Daphnia Experiment Results

Concept Check

1. How did your control group help you to interpret the results of your experiment?

2. What information is important to include on a graph?

3. Did your replicates have the same results? If not, why not?

4. What is the value of performing replicate conditions?

5. Did displaying your data in the form of a graph assist your interpretation?

Key Terms

altruistic behavior (p. 24-2)
behavior (p. 24-1)
chemotaxis (p. 24-2)
communication (p. 24-4)
dependent variable (p. 24-14)
experimental group (p. 24-14)
fixed behavior (p. 24-1)
geotaxis (p. 24-10)
independent variable (p. 24-14)
kinesis (p. 24-10)
learned behavior (p. 24-2)
pheromone (p. 24-4)
phototaxis (p. 24-7)
releaser (p. 24-1)
social behavior (p. 24-2)
taxis (p. 24-2)

Review Questions

1. Why are human behaviors difficult to interpret?

2. How does living in a group provide better defense against predators?

3. Name and describe three types of orientation behaviors.

4. ________________ is a behavior that allows one individual to send information about the environment to others.

5. What is kinesis? How does it differ from taxis?

6. A(n) ________________ reaction occurs when an organism moves toward the stimulus.

7. A(n) ________________ reaction occurs when the organism moves away from the stimulus.

8. Describe a specific fixed behavior.

9. Describe a specific learned behavior.

10. How can a fixed behavior influence a learned behavior?

11. Give an example of how a fixed behavior may be an advantageous characteristic and therefore evolutionarily important.

12. Provide human examples of a fixed behavior and a learned behavior not described in the lab.

13. Define "social behavior." What types of organisms display social behavior? Why would organisms lead solitary lives?

14. ____________________ is a nondirectional movement in response to a stimulus.

15. How can humans influence the evolution of other organisms?

16. What combined senses do you use that influence your behavior?

17. What is the advantage of having more than one sense?

18. ____________________ is any stimulus that causes a fixed behavior to occur.

LAB **25**

Food Science

Objectives

- Explore the role of gluten in food preparation.
- Perform tests to determine differences and similarities of cooking methods.
- Understand how smell and taste combine to create the flavor of food.
- Examine the effects of salt on food preparation.
- Comprehend the importance of food safety during preparation and storage.

Introduction

Every organism needs it but employs different strategies to get it: *food*. **Food** is defined as the organic molecules used by organisms to produce fuel in the form of ATP, the molecule that powers cellular reactions. The vast array of organisms on Earth have several means of acquiring food. Some organisms, called **autotrophs**, make their own food from CO_2 in the air; other organisms, **heterotrophs**, must get food from outside of their bodies. Autotrophs come in two forms: they can assemble food using energy harnessed either from the sun (**photoautotrophs**) or from chemical reactions (**chemoautotrophs**). Correspondingly, heterotrophs must convert (digest) the organic molecules they obtain into a usable form using energy from the sun (**photoheterotrophs**) or from chemical reactions (**chemoheterotrophs**). Humans are considered chemoheterotrophs because we consume (eat) chemical compounds, such as carbohydrates, proteins, and fats, for nutrition and energy.

One extremely important objective for all heterotrophs when eating is *not* getting poisoned; another is taking in all the different molecules that the body requires. We accomplish these goals in part by having our senses of taste and smell steer us away from harmful food and lead us to the most nutritious food around us. In this lab you will explore compounds that lure us to certain foods, enhance the flavor of prepared foods, and prevent the contamination of foods by bacteria or allergens. Finally, you will experiment with different cooking methods that make food tasty and safe to eat.

BECAUSE SOME OF THESE ACTIVITIES INCLUDE TASTING, ANYONE WITH A KNOWN OR POSSIBLE FOOD ALLERGY SHOULD AVOID THE TASTING PORTION AND ONLY RECORD THE RESULTS OF THE CLASS.

ACTIVITY 1 Gluten

Many people have allergies of some sort. An **allergy** is an overreaction by the immune system to a substance that is not generally harmful but is detected as an invader. In response to an allergy-causing substance, the body sends cells, fluid to help the cells get there faster, and other molecules that can be helpful during an immune response. The area swells and becomes warm as the body continues its battle. A person can be born with an allergy or may become **sensitized**, meaning that although the first exposure to a substance doesn't cause a reaction, later exposures to the substance trigger an immune response, resulting in allergy symptoms. Sometimes the overreaction can be extremely severe, leading to anaphylactic shock, which results in constriction of the respiratory tubes, cramping, and dizziness, and requires immediate emergency assistance.

Food allergies are often difficult to diagnose, and they pose a challenge for the allergic individual when choosing what to eat. In addition, allergies are often confused with food **intolerance**, which is a problem digesting a compound within food. For instance, lactose intolerance is due to an inability to break down the sugar lactose during normal digestion because the affected individual doesn't produce an enzyme called lactase. Lactose instead passes into the colon, where it is digested by bacteria that produce gas as a by-product. Intolerance depends on the dose, so an individual will have a more severe response to a larger amount of the substance, whereas an allergy typically produces the same response regardless of the amount. Intolerance is not an immune response, so it is not an allergy.

One common substance that can cause an intolerance reaction, an allergy, and/or an autoimmune response (celiac disease) is gluten. **Gluten** is a complex compound formed from proteins found in the seeds of grains. The most common form of gluten is made of two wheat proteins—glutenin and gliadin—but it is also found in other grasslike grains. Gluten is elastic and gives "bounce" to the dough of flour made from these grains, which is in turn fashioned into food such as bread or pasta. Unfortunately for sensitive individuals, gluten is found not only in wheat-based products, but also in drugs, sauces, processed meats and foods, and even candies. During baking, different techniques allow gluten to form a more organized pattern and give the food strength. In this activity you will perform different techniques to determine the best methods to make bread.

1. Obtain all-purpose flour, yeast, water, salt, two mixing bowls, and a baking sheet.
2. Mix 355 ml of flour and 4.75 g of salt.
3. In a measuring cup, mix 177.5 ml of water and 4.75 g of yeast.
4. Pour the liquid into the flour and combine.
5. Once it is well combined, take a small piece of the dough (about 3 cm) and roll it into a ball. Measure the diameter of the ball and record it in Table 25.1.
6. Place the raw dough ball on the lab bench and roll it into a long cylinder. Grab it on both ends and pull. How long does the cylinder become before it breaks? Record the breaking length in Table 25.1.
7. After completing these measurements, place half of the dough directly onto a baking sheet. Put the baking sheet in the oven, and bake at 177°C–191°C for 30 minutes. When the loaf is done baking, note the outside color and texture, and then break open the bread and note the inside texture and color. Record your observations in Table 25.1.
8. Incubate the other half of the dough at 37°C for 30 minutes. After the incubation, take a small piece of the incubated dough (about 3 cm) and roll it into a ball. Measure the diameter of the ball and record it in Table 25.1.

9. Place the incubated dough ball on the lab bench and roll it into a long cylinder. Grab it on both ends and pull. How long does the cylinder become before it breaks? Record the breaking length in Table 25.1.
10. Place the incubated dough in the oven and bake it for the same amount of time as the first loaf. What do you think the difference between the two loaves will be? During the incubation, the gluten will organize itself into long, straight molecules, giving the resulting bread more elasticity. Also during this time, the yeast will produce gas. What outcome will this have? When the loaf is done baking, note the outside color and texture, and then break open the bread and note the inside texture and color. Record your observations in Table 25.1. How do these results compare to the first loaf?

TABLE 25.1 ACTIVITY 1: DOUGH AND LOAF OBSERVATIONS

	Diameter of dough ball (cm)	Breaking length of dough cylinder (cm)	Description of baked loaf
Raw dough			
Incubated dough			

11. For your next bowl, you will choose one variable to adjust and predict what the outcome will be. You can change time, temperature, or the quantity of an ingredient; you can add a new ingredient or take an ingredient out. For example, if more water is added to the mixture, larger holes will form in the loaf, and it will be flatter and chewier. What do you think the outcome would be if less water were added?
12. Develop your hypothesis and write it here.
13. Change your chosen variable, combine the ingredients as instructed in steps 2 through 4, and bake. When the experimental bread is done, compare your results to the original loaves you made. What are the differences? What are the similarities?

Concept Check

1. How can an individual develop an allergy?

2. Why does immediately heating a mixture produce a different result than letting the mixture incubate before heating it?

3. How does yeast make dough rise?

ACTIVITY 2 Cooking with Heat, Acid, and Microwaves

Cooking is a process of heating food to ready it for consumption. New technology and a better understanding of the cooking process have produced new techniques that ultimately have the same or similar results as the application of heat. Heat allows substances to dissolve and mix in a way that produces an overall flavor not achieved by mixing the same items together when they are cold. In addition, animal meat requires heating to kill off any potentially harmful microbes, and this heating changes the texture of the meat. **Temperature** is a measure of the heat in a system and is the means by which we determine whether food is done or ready to eat—that is, whether it has reached a high enough temperature to break down and kill off any microbes.

Heat speeds up the movement of molecules, leading to the breakdown of cellular structures and the denaturation of proteins. **Denaturation** is the unfolding of the three-dimensional shape of a protein. A microwave oven also increases molecular motion, but instead of using radiant heat as a stove or a conventional oven does, it uses electromagnetic radiation. The radiation uniformly excites polar molecules in the food. The most common polar molecule is water, which is found in almost all food. The excitation of these molecules causes them to move fast enough to degrade the surrounding cellular structures, and protein denaturation occurs. Denaturation can also be achieved by placing meat into citric acid, typically from a lemon or a lime, and marinating it for several hours. In this case, proteins denature because the low pH of citric acid breaks the bonds within them, causing them to unfold. These three methods—radiant heat, microwaves, and acid—support a variety of approaches to food preparation. In this activity you will compare these three techniques to understand their similarities and differences.

1. Obtain the following items from your instructor: fresh uncooked fish, three microwave-safe/oven-safe bowls, a knife, lemon juice, a razor blade, a dissecting microscope, and four petri dishes.

2. Cut the fish into seven pieces, each 2 cm square. Place two pieces into each bowl, saving the remaining piece for observation after the treatments. What is the general texture of the fish? What color is it?

3. Place one bowl in the conventional oven at 190°C for 30 minutes.
4. Place one bowl in the microwave on the "High" setting for 5 minutes.
5. Pour lemon juice over the fish in the remaining bowl, and place it in the refrigerator for 30 minutes.
6. After all the fish has been incubated for the allotted time, note the color and texture of the fish after each treatment, recording your observations in Table 25.2.
7. Place a slice of each treatment and the original remaining piece on a petri dish. Observe each piece of fish under the dissecting microscope and record the differences in Table 25.2.

TABLE 25.2 ACTIVITY 2: MICROSCOPIC OBSERVATIONS OF FISH

Fish	Color and texture after treatment	Appearance under the microscopic
Uncooked		
Oven-baked		
Microwaved		
Marinated in lemon juice		

Concept Check

1. How does marinating meat in citric acid "cook" it?

2. Why are some types of cookware microwave-safe, while other types aren't?

3. What is the purpose of cooking food?

4. Does marinating fish in lemon juice have the same antimicrobial effect as heating the fish?

ACTIVITY 3 Flavor versus Taste

This activity will demonstrate the genetic differences among you and your classmates by testing your ability to taste chemicals. "Taste" and "flavor" are terms that are often used interchangeably, but there are distinct differences between the two. **Taste** is one of the five familiar human senses, along with vision, hearing, smell, and touch. Just as there is variation in any population due to inherited traits affecting vision and hearing, there is also variation in taste. Humans can discern five basic tastes: salty, sour, sweet, bitter, and umami. Umami is a perception of "meatiness" associated with animal muscle.

The specialized sensory receptors for taste are located on the tongue in pockets called **taste buds** (Figure 25.1). These sensory cells are **chemoreceptors** because they can lock onto a variety of chemicals to produce one of the five basic tastes and send the appropriate message to the brain. The brain then interprets this message as a specific taste. But, you may be asking, what about the other things you "taste" when you eat? That's flavor.

Flavor is the combination of the information received from your taste buds and the information received from chemoreceptors in your nose. Just like your tongue, your nose contains receptors that lock onto specific chemicals, but the molecules are in the air instead of in food. As you chew your food, you release these chemicals into your nasal passages (Figure 25.2). Your nose has the ability to detect many more chemicals than your tongue can, and the chemicals tasted by the tongue combine with those smelled by the nose to be interpreted by the brain as flavor. In this activity you will compare how a substance smells and tastes and then interpret the flavor of that substance.

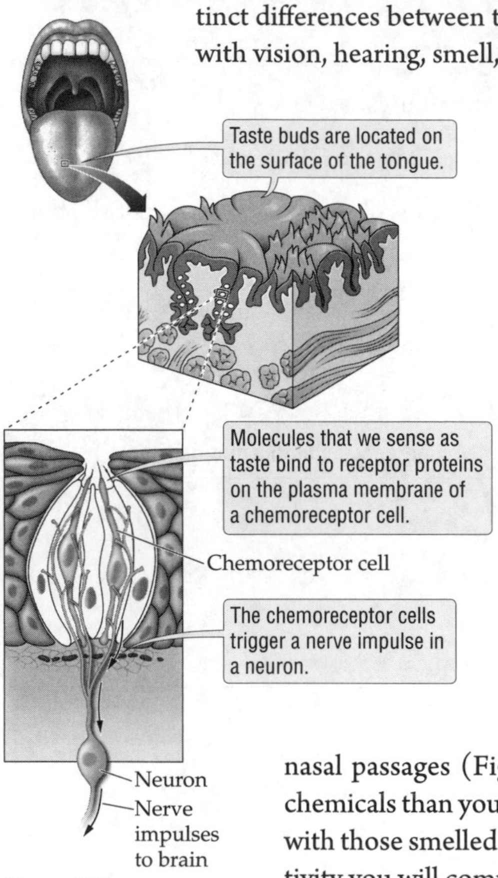

Taste buds are located on the surface of the tongue.
Molecules that we sense as taste bind to receptor proteins on the plasma membrane of a chemoreceptor cell.
Chemoreceptor cell
The chemoreceptor cells trigger a nerve impulse in a neuron.
Neuron
Nerve impulses to brain

Figure 25.1
Taste Buds

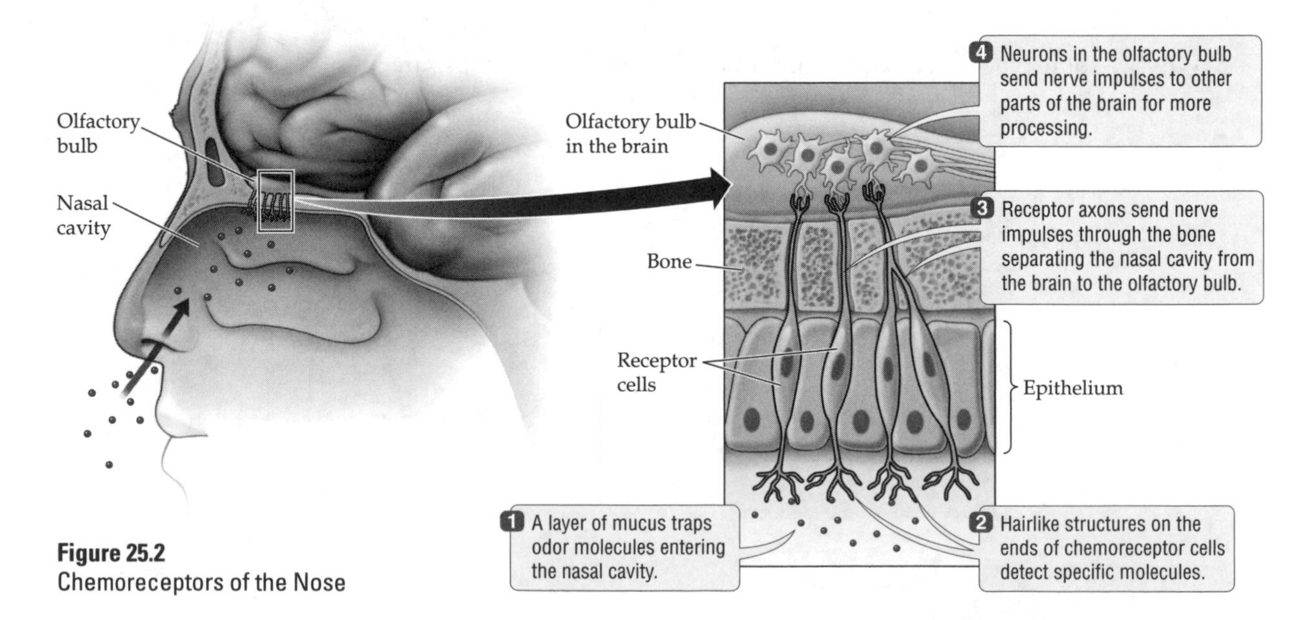

Figure 25.2
Chemoreceptors of the Nose

1. Choose one student who will administer the tests to the rest of the students (who will all be blindfolded). Blindfolding is necessary because seeing a substance influences our expectation of what it should taste and smell like, affecting our perception. The remainder of the instructions below are written to the student administering the test (the "administrator").
2. Blindfold everyone else who is participating in the test.
3. Obtain four substances to test and a clothespin for each participant. Make note of each substance's color, consistency, and possible identity, recording these observations in Table 25.3. Which substance do you think will be the most difficult to discern? Which one will be the least challenging?

TABLE 25.3 ACTIVITY 3: SUBSTANCES FOR TESTING

Substance	Description	Identity?
A		
B		
C		
D		

4. Determine which smells each student can detect. Students should not hear each other's answers, so keep them apart and have them whisper directly into your ear. Using a ruler, measure 13 cm from each participant's nose and open the container of substance A. Allow the participant to smell the substance for 30 seconds; then close the container. Ask the participant what he or she smells and record the answer under "Smell" in Table 25.4. Repeat this process with all other participants.
5. Repeat step 5 for each substance being tested. After participants have tasted one substance, have them rinse their mouth several times with water to help disperse and remove the old taste. Why is this important?
6. The second part of this activity examines the difference between taste and flavor. Keep the participants blindfolded and place the clothespins on their noses to prevent any air from entering their nasal passages. Hand each participant a substance (you don't need to do the substances in order, and it may be best to mix them up, so that the participant doesn't know the order by the smells they detected).
7. Have participants place the substance on their tongue and close their mouth. After 10 seconds, record what the participants taste by marking an "x" below each of the different tastes reported for that substance in Table 25.4.
8. After recording these tastes, have participants remove their clothespins and breathe out deeply through their noses. Do they immediately "taste" something different? This is flavor, the combination of smell and taste; record the flavor in the last column of Table 25.4. Repeat for each participant.

TABLE 25.4 ACTIVITY 3: RECORDING SMELL, TASTE, AND FLAVOR

			Taste					
Substance	**Student**	**Smell**	**Salty**	**Sour**	**Sweet**	**Bitter**	**Umami**	**Flavor**
A								
B								
C								
D								

9. Repeat steps 6 through 8 for each substance being tasted. Again, have the participants rinse their mouth several times with water between substances.

10. After all tasting has been performed, reveal the results to your participants. Compare the results from the entire class. How many participants got each substance initially correct by smell alone? How many participants got each substance correct by taste alone? What might have influenced the outcome of these results?

11. Which substance did the participants find the most difficult to discern by smell only? Was this the same substance you, the administrator, chose? Which substance was the least challenging? Was this the same substance you chose?

Concept Check

1. Why does a person taste only the five basic tastes when her nose is shut with a clothespin?

2. Why does one person have a more acute sense of smell than another person?

3. How are taste and smell different? How are they similar?

ACTIVITY 4 The Power of Salt

Sodium chloride, common table salt (Figure 25.3), is considered an essential seasoning by chefs around the world. Many people don't even take a bite of food before adding salt. But what function, if any, does salt have in our bodies? Both the sodium and chloride ions are constantly regulated by cells in different ways to help provide energy for reactions and to elicit a signal to send messages from one cell to the next. Different diseases, such as cystic fibrosis, affect this regulation, causing a complicated set of symptoms involving the respiratory, digestive, and reproductive systems. But it is not just genetic diseases that are important to contemplate while picking up a salt shaker. Having too much salt in your diet can lead to hypertension (high blood pressure), which can then lead to more serious circulatory system problems. The balance of ions is regulated by a host of mechanisms, but one of the main issues related to this balance is the composition of the solution surrounding your cells.

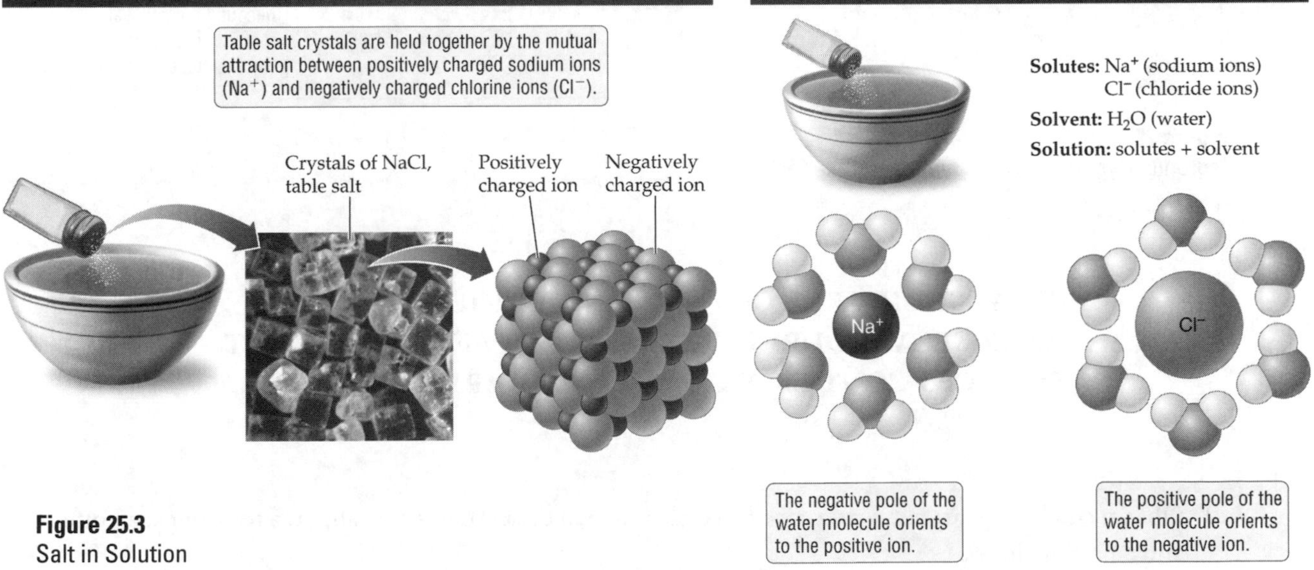

Figure 25.3
Salt in Solution

When cells are surrounded by solutions that have the same concentration of solutes as the cells' interior, the solution is said to be **isotonic** and the cells will neither gain nor lose water or solutes as a result. When the surrounding solution has a higher solute concentration than the cells, it is considered **hypertonic**; and when the surrounding solution has a lower solute concentration, it is **hypotonic** (Figure 25.4). In the hypertonic and hypotonic situations, there is a concentration imbalance between the solution surrounding the cells and the solution inside the cells, and it is possible for one solution to lose water or solutes to balance the difference. The movement of *water* across a selectively permeable membrane to balance the difference, as illustrated in Figure 25.4, is called **osmosis**; the movement of *solutes* to compensate for the difference it is called **diffusion**. These concepts are used in cooking preparation to either dry out foods or make them juicy. Brining (soaking in a salty solution) is a food preparation technique used to keep the tissue of meat and fish juicy. In this activity you will explore the effects salt has on the muscle of chicken.

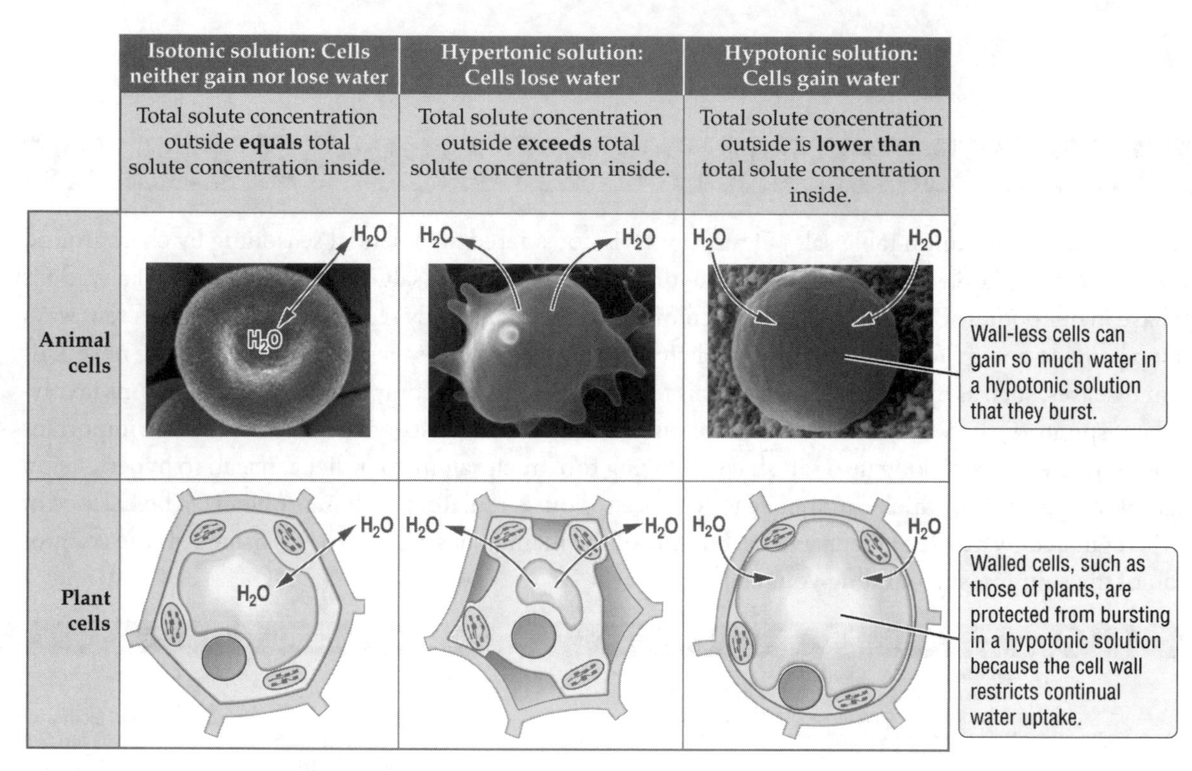

Figure 25.4
Isotonic, Hypertonic, and Hypotonic Solutions

BECAUSE CHICKEN CAN HAVE INFECTIOUS BACTERIA, WEAR GLOVES DURING THIS ACTIVITY AND BE SURE NOT TO CROSS-CONTAMINATE OTHER EQUIPMENT.

1. Obtain chicken meat, salt, water, five beakers, a ruler, a thermometer, a caliper, a toothpick, a knife, and a pair of gloves.
2. Put the gloves on and then cut the chicken into five 3-cm-square pieces. Note the texture, color, and general characteristics of the chicken, recording your observations at the top of Table 25.5.
3. Label the beakers from 1 to 5, corresponding to the treatments described in Table 25.5. You will use three controls during this activity: beaker 1 will not expose the chicken to anything, beaker 2 will

explore how the chicken reacts to exposure to water, and beaker 3 will expose the chicken to just salt. Which of these three is considered a negative control? Beakers 4 and 5 will contain salt solutions of different concentrations.

4. Measure the thickness and weight of each piece of meat. To measure the thickness, place the jaws of the caliper around the meat, and dial the caliper until you can let go of the piece of meat without its falling out of the caliper. Record the thickness and weight of each piece in the appropriate columns in Table 25.5, place each piece in front of its beaker.
5. Place one chicken piece in beaker 1. This piece will not be exposed to anything. Do you think this piece will lose water during its incubation?

6. Place a second chicken piece in beaker 2, and cover it with distilled water. What type of solution is the water surrounding the chicken—isotonic, hypertonic, or hypotonic?

7. Pour a coating of salt into beaker 3, and then place the next chicken piece in this salt-coated beaker. Fill the beaker with salt until the entire piece of chicken is covered. What do you think will happen to this piece of chicken?

8. Choose two salt concentrations to test on the last two chicken pieces, and note these concentrations at the top of the last two columns in Table 25.5. Make the dilutions, and then place one chicken piece in each of beakers 4 and 5, and pour the solutions over the chicken pieces. What types of solutions are these—isotonic, hypertonic, or hypotonic?

9. Put all beakers in a refrigerator for 1 hour.
10. While the chicken is refrigerating, heat an incubator (oven) to 121°C.
11. After the refrigeration is done, take the chicken pieces out of the beakers, keeping straight which beaker each piece belongs to, and dry the pieces with a paper towel. In Table 25.5, note any changes in the texture, color, or overall characteristics for each piece. Measure the thickness and weight of each piece as instructed in step 4, and record these measurements in Table 25.5.

12. Place the chicken pieces on an oven-safe plate, again keeping straight which beaker each piece belongs to, and put them in the oven. Heat the pieces until the internal temperature reaches 77°C.

13. Take the chicken pieces out of the oven. In Table 25.5, note the texture, color, and overall condition for each piece. Measure the thickness and weight of each piece again, as instructed in step 4, and record these measurements in Table 25.5.

TABLE 25.5 ACTIVITY 4: RESULTS OF CHICKEN TREATMENT

Step 2: General characteristics of uncooked/untreated chicken

	Beaker number and treatment				
	1: No exposure (control)	2: Exposure to distilled water (control)	3: Exposure to salt (control)	4: Salt solution of concentration ______ (experiment)	5: Salt solution of concentration ______ (experiment)
STEP 4: Measurements before any treatment					
Thickness of meat					
Weight of meat					
STEP 11: Measurements after beaker treatment and refrigeration					
Change in general characteristics					
Thickness of meat					
Weight of meat					
STEP 13: Measurements after cooking					
Change in general characteristics					
Thickness of meat					
Weight of meat					

14. Which beaker treatment caused the meat to lose the most weight during refrigeration (as measured after refrigeration but before cooking)? Which beaker treatment caused the meat to retain the most weight during refrigeration?

15. Which beaker treatment caused the meat to lose the most weight during cooking? Which beaker treatment caused the meat to retain the most weight during cooking?

16. Try to pierce the chicken with a toothpick. Do any of the pieces feel tough? Why do you think the meat would toughen? Why do you think these characteristics are relevant to how the chicken might taste?

Concept Check

1. Why may too much salt in your diet lead to hypertension?

2. What other factors may lead to hypertension?

3. How does brining affect meat?

ACTIVITY 5 Food Safety

Food safety is an issue not just for professional kitchens, but also for home cooks. Bacteria are everywhere and have a tendency to get into everything, so the factors that influence bacterial growth are important for everyone to appreciate. Other microbes include fungi, protists, and other eukaryotes. These microscopic organisms come in all shapes and sizes, but the ones most important to consider for food safety are those that can cause disease, called **pathogens**. Pathogens typically produce toxins that cause the resulting disease. Bacteria produce either an **exotoxin** (a secreted compound) or an **endotoxin** (a component of the bacterial cell).

Bacteria are also of special concern because of their reproductive abilities. All bacteria undergo a form of asexual reproduction in which they literally split into two new cells. This process, called **binary fission**, allows some bacteria to double their population in as little as 30 minutes. This rapid rate of reproduction is the major problem we face when trying to prevent bacterial growth during food preparation and storage. In addition, some bacteria are capable of producing a capsule called an **endospore** to protect them from harsh conditions; a temperature of 49°C is required to kill these protectively coated bacteria. In this activity you will test differently treated foods to determine the best method for storing food.

THIS ACTIVITY REQUIRES STERILE TECHNIQUES TO DETERMINE THE TRUE MICROBIAL CONTENT OF THE FOOD. IN ORDER TO PREVENT THE INTRODUCTION OF BACTERIA INTO THE EXPERIMENT FROM YOUR BODY OR FROM OBJECTS IN THE LAB, PUT GLOVES ON AND CLEAN YOUR LAB BENCH WITH THE DISINFECTANT PROVIDED.

1. Obtain food in a sealed container, sterile agar plates, sterile cotton swabs, and a marker.
2. Spray the outside of these objects and your gloves with the disinfectant, and dry them. Choose one student to record information for the group; in Table 25.6, record the food items you will test.
3. Open the containers, and note the color and consistency of the food and the conditions under which they were treated. Record how the food was treated, such as temperature and light exposure, in the "Treatment" column of Table 25.6.
4. Mark the bottom of each agar plate (the side containing the agar) with your initials, the date and time of inoculation, and the name of the food item you are testing.
5. Wearing gloves, open a sterile swab, open the food container, and wipe the swab tip across the top of the first food item. Don't jab the swab into the food, because we are mainly interested in understanding what is growing on the outside of the food. Then open the lid of an agar plate, and wipe the swab across the agar in a zigzag pattern without poking it.
6. Perform this same technique on all the foods you are testing. Do you think any of the foods are likely to have more growth than others? Why or why not?

7. After all the agar plates have been inoculated, place them at 37°C for the amount of time specified by your instructor.
8. After the allotted time, record your observations in the last two columns of Table 25.6. Are any of the plates clean of all growth? Why? Using the key provided by your instructor, determine whether your plates have only bacterial colonies or also contain fungi. Is it likely that any of the colonies didn't come from the food? Why? How could you test for this type of contamination?

TABLE 25.6 ACTIVITY 5: PLATE INOCULATION RESULTS

Food item description	Treatment	Number of colonies	Number of different types of colonies

Concept Check

1. How do bacteria reproduce?

2. What are disease-causing organisms called?

3. What can occur if you don't heat your food above 49°C?

4. What would heating or brining do to the metabolic "machinery" of microbes?

Key Terms

allergy (p. 25-2)
autotroph (p. 25-1)
binary fission (p. 25-13)
chemoautotroph (p. 25-1)
chemoheterotroph (p. 25-1)
chemoreceptor (p. 25-6)
cooking (p. 25-4)
denaturation (p. 25-4)
diffusion (p. 25-10)
endospore (p. 25-13)
endotoxin (p. 25-13)
exotoxin (p. 25-13)
flavor (p. 25-6)
food (p. 25-1)
gluten (p. 25-2)
heterotroph (p. 25-1)
hypertonic (p. 25-9)
hypotonic (p. 25-9)
intolerance (p. 25-2)
isotonic (p. 25-9)
osmosis (p. 25-10)
pathogen (p. 25-13)
photoautotroph (p. 25-1)
photoheterotroph (p. 25-1)
sensitized (p. 25-2)
taste (p. 25-6)
taste bud (p. 25-6)
temperature (p. 25-4)

Review Questions

1. How is intolerance different from an allergy?

2. What are the major outcomes of heating food during cooking?

3. How is marinating food in citric acid similar to cooking it on a stove or in a microwave?

4. ___________________ is an overreaction by the immune system to a substance that is not harmful but is detected as an invader.

5. Food ___________________ involves a problem with digesting a compound within food.

6. What is the most common polar molecule in food?

7. ___________________ is the unfolding of the three-dimensional shape of the protein.

8. What is the difference between flavor and taste?

9. What determines which chemicals you can taste or smell?

10. Name the five basic tastes. For each taste, name a food that contains it.

11. What is the difference between osmosis and diffusion?

12. What is hypertension?

13. Define “isotonic,” “hypertonic,” and “hypotonic.”

14. What is the benefit of drying food by surrounding it with salt?

15. ____________________ is a form of asexual reproduction producing two new cells as a result of splitting.

16. ____________________ are disease-causing microscopic organisms.

17. What is an exotoxin?

18. What is an endotoxin?